CAMBRIDGE LIBRARY COLLECTION

Books of enduring scholarly value

Mathematical Sciences

From its pre-historic roots in simple counting to the algorithms powering modern desktop computers, from the genius of Archimedes to the genius of Einstein, advances in mathematical understanding and numerical techniques have been directly responsible for creating the modern world as we know it. This series will provide a library of the most influential publications and writers on mathematics in its broadest sense. As such, it will show not only the deep roots from which modern science and technology have grown, but also the astonishing breadth of application of mathematical techniques in the humanities and social sciences, and in everyday life.

Scientific Papers

Lord Rayleigh served as President of Royal Society from 1905 to 1908, when he became Chancellor of the University of Cambridge. In 1904 he became the first professor at the Royal Institution to be awarded a Nobel Prize. He received the physics award while Ramsey, with whom Rayleigh had conducted the research and announced the discovery of argon, received the Nobel Prize for chemistry. In 1906 he published his electron fluid model of the atom, a modification of Thomson's 'plum pudding' proposal. This was superseded by a series of other (also invalid) models, until Bohr's atomic theory of 1913. In 1907 Rayleigh published a detailed observational study on how humans can perceive sound and distinguish the directions of pure and complex tones. His interest in optics also continued, with a 1907 analysis of the theoretical basis for unusual banding patterns arising when polarised light was shone on diffraction gratings. This volume includes his papers from 1902 to 1910.

Cambridge University Press has long been a pioneer in the reissuing of out-of-print titles from its own backlist, producing digital reprints of books that are still sought after by scholars and students but could not be reprinted economically using traditional technology. The Cambridge Library Collection extends this activity to a wider range of books which are still of importance to researchers and professionals, either for the source material they contain, or as landmarks in the history of their academic discipline.

Drawing from the world-renowned collections in the Cambridge University Library, and guided by the advice of experts in each subject area, Cambridge University Press is using state-of-the-art scanning machines in its own Printing House to capture the content of each book selected for inclusion. The files are processed to give a consistently clear, crisp image, and the books finished to the high quality standard for which the Press is recognised around the world. The latest print-on-demand technology ensures that the books will remain available indefinitely, and that orders for single or multiple copies can quickly be supplied.

The Cambridge Library Collection will bring back to life books of enduring scholarly value across a wide range of disciplines in the humanities and social sciences and in science and technology.

Scientific Papers

VOLUME 5: 1902–1910

BARON JOHN WILLIAM STRUTT RAYLEIGH

CAMBRIDGE
UNIVERSITY PRESS

CAMBRIDGE UNIVERSITY PRESS

Cambridge New York Melbourne Madrid Cape Town Singapore São Paolo Delhi

Published in the United States of America by Cambridge University Press, New York

www.cambridge.org
Information on this title: www.cambridge.org/9781108005463

This edition first published 1912
This digitally printed version 2009

ISBN 978-1-108-00546-3

SCIENTIFIC PAPERS

CAMBRIDGE UNIVERSITY PRESS

London: FETTER LANE, E.C.

C. F. CLAY, Manager

Edinburgh: 100, PRINCES STREET

Berlin: A. ASHER AND CO.

Leipzig: F. A. BROCKHAUS

New York: G. P. PUTNAM'S SONS

Bombay and Calcutta: MACMILLAN AND CO., Ltd.

SCIENTIFIC PAPERS

BY

JOHN WILLIAM STRUTT,

BARON RAYLEIGH,

O.M., D.Sc., F.R.S.,

CHANCELLOR OF THE UNIVERSITY OF CAMBRIDGE,
HONORARY PROFESSOR OF NATURAL PHILOSOPHY IN THE ROYAL INSTITUTION.

VOL. V.

1902—1910.

CAMBRIDGE:
AT THE UNIVERSITY PRESS
1912

CONTENTS.

ERRATA.

VOLUME I.

Page 144, line 6 from bottom. *For* D *read* D_1.

  ,,   442, line 9. *After* $\dfrac{\rho'-\rho}{\rho'}$ *insert* y.

  ,,   443, line 9. *For* (7) *read* (8).

  ,,   443, line 10. *For* η *read* ξ.

  ,,   446, line 10. *For* ϕ *read* ϕ'.

  ,,   448, line 5. *For* v *read* c.

  ,,   459, line 17. *For* 256, 257 *read* 456, 457.

  ,,   524. In the second term of equations (32) and following *for* ΔK^{-1} *read* $\Delta\mu^{-1}$.

  ,,   528, line 3 from bottom. *For* e^{int} *read* $e^{i(nt-kr_0)}$.

  ,,   538, line 11 from bottom. This passage is incorrect.

VOLUME II.

  ,,   197, line 19. *For* nature *read* value.

  ,,   240, line 22. *For* dp/dx *read* dp/dy.

  ,,   241, line 2. *For* du/dx *read* du/dy.

  ,,   244, line 4. *For* k/n *read* n/k.

  ,,   414, line 5. *For* favourable *read* favourably.

  ,,   551, first footnote. *For* 1866 *read* 1886.

VOLUME III.

  ,,   92, line 4. *For* Vol. I. *read* Vol. II.

  ,,   129, equation (12). *For* $e^{u(i-x)}dx$ *read* $e^{u(i-x)}du$.

  ,,   314, line 1. *For* (38) *read* (39).

  ,,   522, equation (31). *Insert as factor of last term* $1/R$.

  ,,   548, second footnote. *For* 1863 *read* 1868.

  ,,   569, second footnote. *For* alcohol *read* water.

  ,,   580, line 3. Prof. Orr remarks that a is a function of r.

VOLUME IV.

  ,,   277, equation (12). *For* dz *read* dx.

  ,,   299, first footnote. *For* 1887 *read* 1877.

  ,,   369, footnote. *For* 1890 *read* 1896.

  ,,   400, equation (14). A formula equivalent to this was given by Lorenz in 1890.

  ,,   418. In table opposite 6 *for* ·354 *read* ·324.

  ,,   556, line 8 from bottom. *For* reflected *read* rotated.

273.

INTERFERENCE OF SOUND.

[*Royal Institution Proceedings*, XVII. pp. 1—7, 1902;
Nature, LXVI. pp. 42—44, 1902.]

For the purposes of laboratory or lecture experiments it is convenient to use a pitch so high that the sounds are nearly or altogether inaudible. The wave-lengths (1 to 3 cm.) are then tolerably small, and it becomes possible to imitate many interesting optical phenomena. The ear as the percipient is replaced by the high pressure sensitive flame, introduced for this purpose by Tyndall, with the advantage that the effects are visible to a large audience.

As a source of sound a "bird-call" is usually convenient. A stream of air from a circular hole in a thin plate impinges centrically upon a similar hole in a parallel plate held at a little distance. Bird-calls are very easily made. The first plate, of 1 or 2 cm. in diameter, is cemented, or soldered, to the end of a short supply-tube. The second plate may conveniently be made triangular, the turned down corners being soldered to the first plate. For calls of medium pitch the holes may be made in tin plate. They may be as small as $\frac{1}{4}$ mm. in diameter, and the distance between them as little as 1 mm. In any case the edges of the holes should be sharp and clean. There is no difficulty in obtaining wave-lengths (complete) as low as 1 cm., and with care wave-lengths of ·6 cm. may be reached, corresponding to about 50,000 vibrations per second. In experimenting upon minimum wave-lengths, the distance between the call and the flame should not exceed 50 cm., and the flame should be adjusted to the verge of flaring*. As most bird-calls are very dependent upon the precise pressure of the wind, a manometer in immediate connection is practically a necessity. The pressure, originally somewhat in excess, may be controlled by a screw pinch-cock operating on a rubber connecting tube.

---

* *Theory of Sound*, 2nd ed. § 371.

In the experiments with conical horns or trumpets, it is important that no sound should issue except through these channels. The horns end in short lengths of brass tubing which fit tightly to a short length of tubing (A) soldered air-tight on the face of the front plate of the bird-call. So far there is no difficulty; but if the space between the plates be boxed in air-tight, the action of the call is interfered with. To meet this objection a tin-plate box is soldered air-tight to A, and is stuffed with cotton-wool kept in position by a *loosely* fitting lid at C. In this way very little sound can escape except through the tube A, and yet the call speaks much as usual. The manometer is connected at the side tube D. The wind is best supplied from a gas-holder.

With the steadily maintained sound of the bird-call there is no difficulty in measuring accurately the wave-lengths by the method of nodes and loops.

Fig. 1.

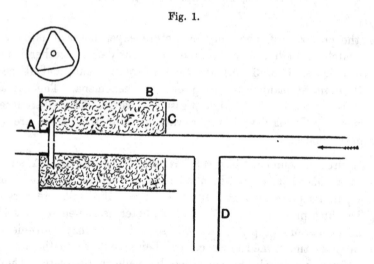

A glass plate behind the flame, and mounted so as to be capable of sliding backwards and forwards, serves as reflecting wall. At the plate, and at any distance from it measured by an *even* number of quarter wave-lengths there are nodes, where the flame does not respond. At intermediate distances, equal to *odd* multiples of the quarter wave-length, the effect upon the flame is a maximum. For the present purpose it is best to use nodes, so adjusting the sensitiveness of the flame that it only just recovers its height at the minimum. The movement of the screen required to pass over ten intervals from minimum to minimum may be measured, and gives at once the length of five complete progressive waves. For the bird-call used in the experiments of this lecture the wave-length is 2 cm. very nearly.

When the sound whose wave-length is required is not maintained, the application of the method is, of course, more difficult. Nevertheless, results of considerable accuracy may be arrived at. A steel bar, about 22 cm. long, was so mounted as to be struck longitudinally every two or three seconds by a small hammer. Although in every position the flame shows some uneasiness at the stroke of the hammer, the distinction of loops and nodes is perfectly evident, and the measurement of wave-length can be effected with an accuracy of about 1 per cent. In the actual experiment the wave-length was nearly 3 cm.

The formation of stationary waves with nodes and loops by perpendicular reflection illustrates interference to a certain extent, but for the full development of the phenomenon the interfering sounds should be travelling in the same, or nearly the same, direction. The next example illustrates the theory of Huyghens' zones. Between the bird-call and the flame is placed a glass screen perforated with a circular hole. The size of the hole, the distances, and the wave-length are so related to one another that the aperture just includes the first and second zones. The operation of the sounds passing these zones is antagonistic, and the flame shows no response until a part of the aperture is blocked off. The part blocked off may be either the central circle or the annular region defined as the second zone. In either case the flame flares, affording complete proof of interference of the parts of the sound transmitted by the aperture*.

From a practical point of view the passage of sound through apertures in walls is not of importance, but similar considerations apply to its issue from the mouths of horns, at least when the diameter of the mouth exceeds the half wave-length. The various parts of the sound are approximately in the same phase when they leave the aperture, but the effect upon an observer depends upon the phases of the sounds, not as they leave, but as they arrive. If one part has further to go than another, a phase discrepancy sets in. To a point in the axis of the horn, supposed to be directed horizontally, the distances to be travelled are the same, so that here the full effect is produced, but in oblique directions it is otherwise. When the obliquity is such that the nearest and furthest parts of the mouth differ in distance by rather more than one complete wave-length, the sound may disappear altogether through antagonism of equal and opposite effects. In practice the attainment of a complete silence would be interfered with by reflections, and in many cases by a composite character of sound, viz. by the simultaneous occurrence of more than one wave-length.

In the fog signals established on our coasts the sound of powerful sirens issues from conical horns of circular cross-section. The influence of obliquity is usually very marked. When the sound is observed from a sufficient distance at sea, a deviation of even 20° from the axial line entails a considerable

* [1901. See Vol. III. p. 31.]

loss, to be further increased as the deviation rises to 40° or 60°. The difficulty thence arising is met, in the practice of the Trinity House, by the use of two distinct sirens and horns, the axes of the latter being inclined to one another at 120°. In this way an arc of 180° or more can be efficiently guarded, but a more equable distribution of the sound from a single horn remains a desideratum.

Guided by the considerations already explained, I ventured to recommend to the Trinity House the construction of horns of novel design, in which an attempt should be made to spread the sound out horizontally over the sea, and to prevent so much of it from being lost in an upward direction. The solution of the problem is found in a departure from the usual circular section, and the substitution of an elliptical or elongated section, of which the short diameter, placed horizontally, does not exceed the half wave-length; while the long diameter, placed vertically, may amount to two wave-lengths or more. Obliquity in the *horizontal plane* does not now entail much difference of phase, but when the horizontal plane is departed from such differences enter rapidly.

Horns upon this principle were constructed under the supervision of Mr Matthews, and were tried in the course of the recent experiments off St Catherine's. The results were considered promising, but want of time and the numerous obstacles which beset large scale operations prevented an exhaustive examination.

On a laboratory scale there is no difficulty in illustrating the action of the elliptical horns. They may be made of thin sheet brass. In one case the total length is 20 cm., while the dimensions of the mouth are 5 cm. for the long diameter and $1\frac{1}{4}$ cm. for the shorter diameter. The horn is fitted at its narrow end to A (Fig. 1), and can rotate about the common horizontal axis. When this axis is pointed directly at the flame, flaring ensues; and the rotation of the horn has no visible effect. If now, while the long diameter of the section remains vertical, the axis be slewed round in the horizontal plane until the obliquity reaches 50° or 60°, there is no important falling off in the response of the flame. But if at obliquities exceeding 20° or 30° the horn is rotated through a right angle, so as to bring the long diameter horizontal, the flame recovers as if the horn had ceased sounding. The fact that there is really no falling off may be verified with the aid of a reflector, by which the sound proceeding at first in the direction of the axis may be sent towards the flame.

When the obliquity is 60° or 70° it is of great interest to observe how moderate a departure from the vertical adjustment of the longer diameter causes a cessation of effect. The influence of maladjustment is shown even more strikingly in the case of a larger horn. According to theory and observation a serious falling off commences when the tilt is such that the difference of distances from the flame of the two extremities of the long diameter reaches the half wave-length—in this case 1 cm. It is thus abundantly proved that the sound issuing from the properly adjusted elliptical cone is confined to a comparatively narrow belt round the horizontal plane and that in this plane it covers efficiently an arc of 150° or 160°.

Another experiment, very easily executed with the apparatus already described, illustrates what are known in Optics as Lloyd's bands. These bands are formed by the interference of the direct vibration with its very oblique reflection. If the bird-call is pointed toward the flame, flaring ensues. It is only necessary to hold a long board horizontally under the direct line to obtain a reflection. The effect depends upon the precise height at which the board is held. In some positions the direct and reflected vibrations co-operate at the flame and the flaring is more pronounced than when the board is away. In other positions the waves are antagonistic and the flame recovers as if no sound were reaching it at all. This experiment was made many years ago by Tyndall who instituted it in order to explain the very puzzling phenomenon of the "silent area." In listening to fog-signals from the sea it is not unfrequently found that the signal is lost at a distance of a mile or two and recovered at a greater distance in the same direction. During the recent experiments the Committee of the Elder Brethren of the Trinity House had several opportunities of making this observation. That the surface of the sea must act in the manner supposed by Tyndall cannot be doubted, but there are two difficulties in the way of accepting the simple explanation as complete. According to it the interference should always be the same, which is certainly not the case. Usually there is no silent area. Again, although according to the analogy of Lloyd's bands there might be a dark or silent place at a particular height above the water, say on the bridge of the *Irene*, the effect should be limited to the neighbourhood of the particular height. At a height above the water twice as great, or near the water level itself, the sound should be heard again. In the latter case there were some difficulties, arising from disturbing noises, in making a satisfactory trial; but as a matter of fact, neither by an observer up the mast nor by one near the water level, was a sound lost on the bridge ever recovered.

The interference bands of Fresnel's experiment may be imitated by a bifurcation of the sound issuing from A (Fig. 1). For this purpose a sort of T-tube is fitted, the free ends being provided with small elliptical cones, similar to that already described, whose axes are parallel and distant from another by about 40 cm. The whole is constructed with regard to symmetry, so that sounds of equal intensity and of the same phase issue from the two cones whose long diameters are vertical. If the distances of the burner from the mouths of the cones be precisely equal, the sounds arrive in the same phase and the flame flares vigorously. If, as by the hand held between, one of the sounds is cut off, the flaring is reduced, showing that with this adjustment the two sounds are more powerful than one. By an almost imperceptible slewing round of the apparatus on its base-board the adjustment above spoken of is upset and the flame is induced to recover its tall equilibrium condition. The sounds now reach the flame in opposition of phase and practically neutralise one another. That this is so is proved in a moment. If the hand be introduced between either orifice and the flame, flaring ensues, the sound not intercepted being free to produce its proper effect.

The analogy with Fresnel's bands would be most complete if we kept the sources of sound at rest and caused the burner to move transversely so as to occupy in succession places of maximum and minimum effect. It is more convenient with our apparatus and comes to the same thing, if we keep the burner fixed and move the sources transversely, sliding the base-board without rotation. In this way we may verify the formula, connecting the width of a band with the wave-length and the other geometrical data of the experiment.

The phase discrepancy necessary for interference may be introduced, without disturbing the equality of distances, by inserting in the path of one of the sounds a layer of gas having different acoustical properties from air. In the lecture carbonic acid was employed. This gas is about half as heavy again as air, so that the velocity of sound is less in the proportion of $1 : 1 \cdot 25$. If l be the thickness of the layer, the *retardation* is $\cdot 25\, l$; and if this be equal to the half wave-length, the interposition of the layer causes a transition from complete agreement to complete opposition of phase. Two cells of tin plate were employed, fitted with tubes above and below, and closed with films of collodion. The films most convenient for this purpose are those formed upon water by the evaporation of a few drops of a solution of celluloid in pear-oil. These cells were placed one in the path of each sound, and the distances of the cones adjusted to maximum flaring. The insertion of carbonic acid into *one* cell quieted the flame, which flared again when the second cell was charged so as to restore symmetry. Similar effects were produced. as the gas was allowed to run out at the lower tubes, so as to be replaced by air entering above*.

Many vibrating bodies give rise to sounds which are powerful in some directions but fail in others—a phenomenon that may be regarded as due to interference. The case of tuning forks (unmounted) is well known. In the lecture a small and thick wine-glass was vibrated, after the manner. of a bell, with the aid of a violin bow. When any one of the four vibrating segments was presented to the flame, flaring ensued; but the response failed

---

* In a still atmosphere the hot gases arising from lighted candles may be substituted for the layers of CO_2.

when the glass was so held at the same distance that its *axis* pointed to the flame. In this position the effects of adjacent segments neutralise one another and the aggregate is zero. Another example, which, strangely enough, does not appear to have been noticed, is afforded by the familiar open organ pipe. The vibrations issuing from the two ends are in the same phase as they start, so that if the two ends are equally distant from the percipient, the effects conspire. If, however, the pipe be pointed towards the percipient, there is a great falling off, inasmuch as the length of the pipe approximates to the half wave-length of the sound. The experiment may be made in the lecture-room with the sensitive flame and one of the highest pipes of an organ, but it succeeds better and is more striking when carried out in the open air with a pipe of lower pitch, simply listened to with the unaided ear of the observer. Within doors reflections complicate all experiments of this kind.

[1910. Some further discussion of interfering sources will be found in *Phil. Mag.* Sept. 1903, Art. 290 below.]

274.

SOME GENERAL THEOREMS CONCERNING FORCED VIBRATIONS AND RESONANCE.

[*Philosophical Magazine*, III. pp. 97—117, 1902.]

THE general equation for the small vibrations of a system whose configuration is defined by the generalized coordinates ψ_1, ψ_2,... may be written *

$$\frac{d}{dt}\frac{dT}{d\dot{\psi}} + \frac{dF}{d\dot{\psi}} + \frac{dV}{d\psi} = \Psi, \qquad \dots\dots\dots\dots(1)$$

where T, F, V, denoting respectively the kinetic energy, the dissipation function, and the potential energy, have the forms

$$\left.\begin{aligned}
T &= \tfrac{1}{2}a_{11}\dot{\psi}_1{}^2 + \tfrac{1}{2}a_{22}\dot{\psi}_2{}^2 + \dots + a_{12}\dot{\psi}_1\dot{\psi}_2 + \dots \\
F &= \tfrac{1}{2}b_{11}\dot{\psi}_1{}^2 + \tfrac{1}{2}b_{22}\dot{\psi}_2{}^2 + \dots + b_{12}\dot{\psi}_1\dot{\psi}_2 + \dots \\
V &= \tfrac{1}{2}c_{11}\psi_1{}^2 + \tfrac{1}{2}c_{22}\psi_2{}^2 + \dots + c_{12}\psi_1\psi_2 + \dots
\end{aligned}\right\}, \qquad \dots\dots\dots(2)$$

in which the coefficients a, b, c are constants.

If we substitute in (1) the values of T, F, and V, and write D for d/dt, we obtain a system of equations which may be put into the form

$$\left.\begin{aligned}
e_{11}\psi_1 + e_{12}\psi_2 + e_{13}\psi_3 + \dots &= \Psi_1 \\
e_{21}\psi_1 + e_{22}\psi_2 + e_{23}\psi_3 + \dots &= \Psi_2 \\
e_{31}\psi_1 + e_{32}\psi_2 + e_{33}\psi_3 + \dots &= \Psi_3 \\
\dots\dots\dots\dots\dots\dots\dots\dots\dots\dots\dots
\end{aligned}\right\}, \qquad \dots\dots\dots\dots(3)$$

where e_{rs} denotes the quadratic operator

$$e_{rs} = a_{rs}D^2 + b_{rs}D + c_{rs}. \qquad \dots\dots\dots\dots\dots(4)$$

And it is to be remarked that since

$$a_{rs} = a_{sr}, \qquad b_{rs} = b_{sr}, \qquad c_{rs} = c_{sr},$$

it follows that

$$e_{rs} = e_{sr}. \qquad \dots\dots\dots\dots\dots\dots(5)$$

---

* See *Theory of Sound*, Vol. I. §§ 82, 84, 104.

If we multiply the first of equations (3) by $\dot{\psi}_1$, the second by $\dot{\psi}_2$, &c., and then add, we obtain

$$\frac{d(T+V)}{dt} + 2F = \Psi_1\dot{\psi}_1 + \Psi_2\dot{\psi}_2 + \dots \quad \dots\dots\dots\dots(6)$$

In this the first term represents the rate at which energy is being stored in the system; $2F$ is the rate of dissipation; and the two together account for the work done upon the system in time dt by the external forces Ψ_1, Ψ_2, &c.

In considering forced vibrations of simple type we take

$$\Psi_1 = E_1 e^{ipt}, \qquad \Psi_2 = E_2 e^{ipt}, \qquad \&c., \quad \dots\dots\dots\dots(7)$$

and assume that ψ_1, ψ_2, &c. are also proportional to e^{ipt}. The coordinates are then determined by the system of algebraic equations resulting from the substitution in (4), (3) of ip for D. The most general motion possible under the assumed forces would require the inclusion of *free* vibrations, but (unless $F=0$) these die out as time progresses.

By the theory of determinants the solution of equations (3) may be expressed in the form

$$\left.\begin{array}{l} \nabla \cdot \psi_1 = \dfrac{d\nabla}{de_{11}}\Psi_1 + \dfrac{d\nabla}{de_{12}}\Psi_2 + \dots \\[2mm] \nabla \cdot \psi_2 = \dfrac{d\nabla}{de_{12}}\Psi_1 + \dfrac{d\nabla}{de_{22}}\Psi_2 + \dots \end{array}\right\}, \quad \dots\dots\dots\dots(8)$$

where ∇ denotes the determinant of the symbols e. If there be no dissipation, ∇, or as we may write it with fuller expressiveness $\nabla(ip)$, is an even function of ip vanishing when p corresponds to one of the natural frequencies of vibration. In such a case the coordinates ψ_1, &c. in general become infinite. When there is dissipation, $\nabla(ip)$ does not vanish for any (real) value of p. If we write

$$\nabla(ip) = \nabla_1(ip) + ip\,\nabla_2(ip), \quad \dots\dots\dots\dots(9)$$

in which ∇_1, ∇_2 are *even* functions of ip, ∇_2 depends entirely upon the dissipation, while if the dissipation be small, ∇_1 is approximately the same as if there were none.

As it will be convenient to have a briefer notation than that of (8), we will write

$$\left.\begin{array}{l} \psi_1 = A_{11}e^{ia_{11}}\Psi_1 + A_{12}e^{ia_{12}}\Psi_2 + \dots \\[2mm] \psi_2 = A_{21}e^{ia_{21}}\Psi_1 + A_{22}e^{ia_{22}}\Psi_2 + \dots \end{array}\right\}, \quad \dots\dots\dots\dots(10)$$

in which A, α are real and are subject to the relations

$$A_{rs} = A_{sr}, \qquad \alpha_{rs} = \alpha_{sr}. \quad \dots\dots\dots\dots(11)$$

In order to take account of the phases of the forces, we may suppose similarly that in (7)

$$E_1 = R_1 e^{i\theta_1}, \qquad E_2 = R_2 e^{i\theta_2}, \qquad \&c. \qquad \ldots\ldots\ldots\ldots(12)$$

Work Done.

If we suppose that but *one* force, say Ψ_1, acts upon the system, the values of the coordinates are given by the first terms of the right-hand members of (10). The work done by the force in time dt depends upon that part of $d\psi_1/dt$ which is in the same phase with it, corresponding to the part of ψ_1 which is in quadrature with the force. Thus, taking the real parts only of the symbolic quantities, so that

$$\Psi_1 = R_1 \cos(pt + \theta_1), \qquad \psi_1 = A_{11} R_1 \cos(pt + \theta_1 + \alpha_{11}), \qquad \ldots\ldots(13)$$

we have as the work done (on the average) in time t

$$- p A_{11} R_1^2 \int \cos(pt + \theta_1) . \sin(pt + \theta_1 + \alpha_{11}) \, dt,$$

or
$$- \tfrac{1}{2} p R_1^2 A_{11} \sin \alpha_{11} . t. \qquad \ldots\ldots\ldots\ldots\ldots(14)$$

As was to be expected, this is independent of θ_1.

Another expression for the same quantity may be obtained by considering how this work is dissipated. From (6) we see that

$$\int \Psi_1 \dot{\psi}_1 dt = 2 \int F dt = b_{11} \int \dot{\psi}_1^2 dt + b_{22} \int \dot{\psi}_2^2 dt + \ldots + 2 b_{12} \int \dot{\psi}_1 \dot{\psi}_2 dt + \ldots \ldots(15)$$

Taking again the real parts in (10), we have

$$\int \dot{\psi}_1^2 dt = \tfrac{1}{2} p^2 R_1^2 A_{11}^2 . t, \ldots\ldots\ldots\ldots\ldots\ldots\ldots\ldots\ldots\ldots(16)$$

$$\int \dot{\psi}_1 \dot{\psi}_2 dt = \tfrac{1}{2} p^2 R_1^2 A_{11} A_{12} \cos(\alpha_{11} - \alpha_{12}) . t, \qquad \ldots\ldots\ldots(17)$$

$$\int \dot{\psi}_2 \dot{\psi}_3 dt = \tfrac{1}{2} p^2 R_1^2 A_{12} A_{13} \cos(\alpha_{12} - \alpha_{13}) . t; \qquad \ldots\ldots\ldots(18)$$

so that by (15) the work dissipated in time t is

$$\tfrac{1}{2} p^2 R_1^2 t \{ b_{11} A_{11}^2 + b_{22} A_{21}^2 + \ldots + 2 b_{12} A_{11} A_{12} \cos(\alpha_{11} - \alpha_{12}) + \ldots \}. \qquad \ldots(19)$$

Equating the equivalent quantities in (14) and (19), we get

$$- p^{-1} A_{11} \sin \alpha_{11} = b_{11} A_{11}^2 + b_{22} A_{21}^2 + \ldots + 2 b_{12} A_{11} A_{12} \cos(\alpha_{11} - \alpha_{12}) + \ldots(20)$$

This assumes a specially simple form when F is a function of the squares only of $d\psi_1/dt$, &c.; so that b_{12}, &c. vanish.

In (14) we have calculated the work done by a force Ψ_1 acting alone upon the system. If other forces act, the expression for ψ_1 will deviate from (13); but in any case we may write

$$\Psi_1 = R_1 e^{i\theta_1}, \qquad \psi_1 = r_1 e^{i\phi_1}, \qquad \ldots\ldots\ldots\ldots\ldots(21)$$

and the work done in unit of time by the real part of Ψ_1 on the real part of ψ_1 will be

$$- \tfrac{1}{2} p R r \sin(\phi_1 - \theta_1), \qquad \ldots\ldots\ldots\ldots\ldots(22)$$

and depends upon the product of the moduli and the *difference* of phases.

If ψ_1 consist of two or more parts of the form (21), the work done is to be found by addition of the terms corresponding to the various parts.

One Degree of Freedom.

The theory of the vibrations of a system of one degree of freedom, resulting from the application of a given force, is simple and well known, but it will be convenient to make a few remarks and deductions.

The equation determining ψ in terms of Ψ is

$$(-ap^2 + c + ipb)\,\psi = \Psi\,;\quad\ldots\ldots\ldots\ldots\ldots\ldots\ldots(23)$$

so that in the notation of (10)

$$Ae^{ia} = \frac{1}{c - ap^2 + ipb}\,.\quad\ldots\ldots\ldots\ldots\ldots\ldots\ldots(24)$$

As in (14), the work done by the force in unit time is

$$\frac{\tfrac{1}{2}p^2 b\,\mathrm{Mod}^2\,\Psi}{(c - ap^2)^2 + p^2 b^2}\,,\quad\ldots\ldots\ldots\ldots\ldots\ldots\ldots(25)$$

and it reaches a maximum (b and p being given) when the tuning is such that $c - ap^2 = 0$, that is when the natural vibrations are isoperiodic with the forced vibrations. The maximum value itself is

$$\frac{\tfrac{1}{2}\mathrm{Mod}^2\,\Psi}{b}\,.\quad\ldots\ldots\ldots\ldots\ldots\ldots\ldots\ldots(26)$$

Let us now suppose that two forces act upon the system, one of which Ψ is given, while the second Ψ' is at disposal, and let us inquire how much work can be withdrawn by Ψ'. It will probably conduce to clearness if we think of an electric circuit possessing self-induction and resistance, and closed by a condenser, so as to constitute a vibrator. In this acts a given electromotive force Ψ of given frequency. At another part of the circuit another electromotive force can be introduced, and the question is what work can be obtained at that point. Of course any work so obtained, as well as that dissipated in the system, must be introduced by the operation of the given force Ψ.

It will suffice for the moment to take Ψ such that ψ *due to it* is unity, which will happen when $\Psi = A^{-1}e^{-ia}$. If Ψ' be $Re^{i\theta}$, the complete value of ψ is

$$\psi = 1 + ARe^{i(a+\theta)}.\quad\ldots\ldots\ldots\ldots\ldots\ldots(27)$$

The work done (in unit time) by Ψ' consists of two parts. That corresponding to the second term in (27) is the same as if Ψ' had acted alone and, as in (14), its value is

$$-\tfrac{1}{2}pR^2 A \sin a.$$

The work done by Ψ' upon the first part of ψ given in (27) is, as in (22),

$$-\tfrac{1}{2}pR \sin(-\theta).$$

The whole work done by Ψ' is found by adding these together; and the work *withdrawn* from the system by Ψ' is the *negative* of this, or

$$\tfrac{1}{2}pR^2A \sin \alpha - \tfrac{1}{2}pR \sin \theta. \quad\dots\dots\dots\dots\dots(28)$$

In this expression the first term is negative, and the whole is to be made a maximum by variation of R and θ. The maximum occurs when

$$\sin \theta = -1, \quad 2RA \sin \alpha = -1; \quad\dots\dots\dots(29)$$

and the maximum value itself is

$$-\frac{p}{8A \sin \alpha}. \quad\dots\dots\dots\dots\dots(30)$$

This corresponds to $\mathrm{Mod}\,\Psi = A^{-1}$; and as the work abstractable is proportional to $\mathrm{Mod}^2\,\Psi$, we have in general for the maximum

$$-\frac{pA^2 \,\mathrm{Mod}^2\,\Psi}{8A \sin \alpha}.$$

Now, as we see from the values of A and α in (24), or otherwise by (20),

$$-p^{-1}A \sin \alpha = bA^2;$$

and thus the maximum work that can be abstracted is

$$\frac{\mathrm{Mod}^2\,\Psi}{8b}. \quad\dots\dots\dots\dots\dots\dots\dots(31)$$

It may at first occasion surprise that the work obtainable should be independent of a and c, upon which the behaviour of the system as a resonator depends. But the truth is that by suitable choice of Ψ' we have in effect *tuned* the system, and so reduced it to the condition of evanescent a and c—in the electrical illustration to a merely resisting circuit. Had we assumed the evanescence of a and c from the beginning, we could of course have arrived more simply at the expression (31).

In the case of maximum withdrawal of energy the complete symbolical value of ψ in (27) becomes

$$\psi = 1 + \frac{ie^{i\alpha}}{2 \sin \alpha} = 1 - \tfrac{1}{2} + \tfrac{1}{2}i \cot \alpha, \quad\dots\dots\dots(32)$$

the part of the complete value which is in the same phase as before being halved.

It is not difficult to recognise that the result as to the maximum work abstractable admits of further generalization. So far we have considered the case of a single degree of freedom, *e.g.*, a single electric circuit. Other degrees of freedom, *e.g.*, neighbouring electric circuits, do not affect the result, provided that the forces in them all vanish and that the only dissipation is that already considered. If in equations (3) all the quantities b except b_{11} vanish, as well as the forces Ψ_2, Ψ_3, &c., the second, third, and following equations determine *real* ratios between all the other coordinates and ψ_1, and virtually reduce the system to a single degree of freedom. The reaction of the other parts of the system will influence the force Ψ_1' required in order to abstract most work, but not the maximum value itself.

Several Degrees of Freedom.

Hitherto we have supposed that the force by means of which work is abstracted is of the same type as that which is supposed to be given and by means of which work is introduced into the system; but in the investigations which follow it will be our object to trace the effect of voluntary operation upon one coordinate ψ_1, while the system is subject to given forces Ψ_2, Ψ_3, &c., operating upon the remaining coordinates. To explain what is meant the more clearly, let us consider the simple case of two electric circuits influencing one another by induction. Each circuit may be supposed to be closed by a condenser, so as to constitute, when considered by itself, a simple vibrator. If a given periodic electromotive force (Ψ_2) act in the second circuit, the current in the first circuit would depend upon the various elements of the compound system. Let it be proposed to inquire what work can be withdrawn at the first circuit by electromotive forces (Ψ_1) there applied. For simplicity it will be supposed that the first circuit has no resistance ($b_{11} = 0$).

When Ψ_2 acts alone, the ψ_1 due to it is given by

$$\psi_1 = A_{12}e^{i\alpha_{12}}\Psi_2. \quad\dots\dots(33)$$

If we ascribe to Ψ_2 the value $A_{12}^{-1}e^{-i\alpha_{12}}$, the ψ_1 due to it will be unity, and this for the present purpose is the simplest supposition to make. When a force Ψ_1 is introduced, the complete value of ψ_1 will deviate from unity, but Ψ_2 is supposed to retain throughout the above prescribed value. If $\Psi_1 = R_1e^{i\theta_1}$, as in (12), the complete expression of ψ_1 is

$$\psi_1 = 1 + A_{11}R_1e^{i(\theta_1+\alpha_{11})}. \quad\dots\dots(34)$$

The work done by Ψ_1 upon this is composed of two parts. If the real components of the expressions be retained, the first is

$$-p\int R_1\cos(pt+\theta_1).\sin pt\,dt = \tfrac{1}{2}pR_1\sin\theta_1.t;$$

and the second is, as in (14),

$$-\tfrac{1}{2}pR_1^2A_{11}\sin\alpha_{11}.t,$$

a quantity necessarily positive.

Thus altogether the work done by Ψ_1 in unit time is

$$\tfrac{1}{2}pR_1\sin\theta_1 - \tfrac{1}{2}pR_1^2A_{11}\sin\alpha_{11}. \quad\dots\dots(35)$$

The work that may be *abstracted* from the circuit is the negative of this, and it is to be made a maximum by variation of R_1 and θ_1. We must take

$$\sin \theta_1 = -1, \qquad 2R_1 A_{11} \sin \alpha_{11} = -1; \quad \dots\dots\dots\dots\dots(36)$$

and the maximum work abstractable will be

$$\tfrac{1}{4} p R_1, \text{ or } -\frac{p}{8 A_{11} \sin \alpha_{11}}. \quad \dots\dots\dots\dots\dots(37)$$

The symbolic expression for ψ_1 becomes at the same time from (22)

$$\psi_1 = 1 - \tfrac{1}{2} + \tfrac{1}{2} i \cot \alpha_{11}, \quad \dots\dots\dots\dots\dots\dots(38)$$

so that the part of ψ_1 in the same phase as when $\Psi_1 = 0$ is half as great as before.

So far as the results embodied in (37), (38), and (39) are concerned, it is a matter of indifference whether the prescribed $\psi_1 = 1$ when $\Psi_1 = 0$ is due to Ψ_2 only, or to Ψ_2 acting in conjunction with forces Ψ_3, &c., corresponding to further degrees of freedom. But for the present we will suppose that there are only two degrees of freedom.

The maximum work that can be abstracted by Ψ_1, when Ψ_2 is given, may be expressed as

$$-\frac{p \operatorname{Mod}^2 \psi_1}{8 A_{11} \sin \alpha_{11}}, \quad \dots\dots\dots\dots\dots\dots(39)$$

where ψ_1 is due to Ψ_2 acting alone. By (33)

$$\operatorname{Mod}^2 \psi_1 = A_{12}{}^2 \operatorname{Mod}^2 \Psi_2,$$

and by (20) with $b_{11} = 0$, $b_{12} = 0$,

$$-p^{-1} A_{11} \sin \alpha_{11} = b_{22} A_{21}{}^2,$$

so that the maximum work obtainable is simply

$$\frac{\operatorname{Mod}^2 \Psi_2}{8 b_{22}}. \quad \dots\dots\dots\dots\dots\dots(40)$$

That it is independent not only of a_{11}, c_{11}, a_{22}, c_{22}, but also of the coefficients of mutual influence a_{12}, c_{12}, is very remarkable. To revert to the electrical example, the work abstractable in the first circuit (devoid of resistance) when a given electromotive force acts in the second, is independent of the value of the coefficient of mutual induction. If indeed this coefficient be very small, the supposition of zero resistance becomes more and more unpractical on account of the large currents which must then be supposed to flow. But the theoretical result remains true, when b_{11} is diminished without limit. In view of its independence of so many circumstances that might at first be supposed material, it may now not be surprising to note that (40) coincides with (31), that is that the work obtainable in the second circuit is the same as might have been obtained in the first where the given force itself acts.

The existence of further degrees of freedom than those corresponding to the given force Ψ_2 and the disposable force Ψ_1 makes no difference to (39). And so long as Ψ_2, Ψ_1 are the only forces in operation, we have still

$$\text{Max. work} = -\frac{pA_{12}^2\,\text{Mod}^2\,\Psi_2}{8A_{11}\sin\alpha_{11}}. \qquad (41)$$

If further all the coefficients b vanish, except b_{22}, (40) remains unaffected. If, however, we suppose that b_{33}, b_{44}, &c. are finite, while b_{11}, b_{12}, b_{13}, b_{23}, &c. still vanish, (20) gives

$$-p^{-1}A_{11}\sin\alpha_{11} = b_{22}A_{21}^2 + b_{33}A_{31}^2 + \dots, \qquad (42)$$

and the expression for the maximum work becomes

$$\frac{\tfrac{1}{8}A_{12}^2\,\text{Mod}^2\,\Psi_2}{b_{22}A_{12}^2 + b_{33}A_{13}^2 + \dots}. \qquad (43)$$

Since b_{33}, &c. are positive, the value of (43) is less than when b_{33}, &c. vanish.

The expression (43) is necessarily more complicated than (40); but a simple result may again be stated if we suppose that given forces act *successively* of the second, third, and following types, provided they be of such magnitudes that they would severally (the non-corresponding resistances vanishing) allow the same work to be abstracted by Ψ_1, that is provided

$$\frac{\text{Mod}^2\,\Psi_2}{b_{22}} = \frac{\text{Mod}^2\,\Psi_3}{b_{33}} = \dots\dots = \frac{\text{Mod}^2\,\Psi}{b}. \qquad (44)$$

On this supposition the *sum* of the energies abstractable in the various cases has the value

$$\frac{\text{Mod}^2\,\Psi}{8b}, \qquad (45)$$

of the same form as before.

In the electrical application we have to consider any number of mutually influencing circuits, of which the first is devoid of resistance. The electromotive forces acting successively in the other circuits are to be inversely as the square roots of the resistances of those circuits, *i.e.* such as would do the same amount of work on each circuit supposed to be isolated and reduced (*e.g.* by suitable adjustment of the associated condenser) to a mere resistance. The sum of all the works abstractable in the first circuit is then the same as if there were no other circuits than the first and second; or, again, as if the second circuit were isolated and it were allowed to draw work from it.

Action of Resonators.

We now abandon the idea of drawing work from the system by means of Ψ_1, and on the contrary impose the condition that Ψ_1 shall do no work, positive or negative. The effect of Ψ_1 is then equivalent to a change in the inertia a_{11}, or spring c_{11}, associated with this coordinate and the operation may be regarded as a tuning, or mistuning, of the system. If, as before, ψ_1, due to the given force Ψ_2, be unity, and $\Psi_1 = R_1 e^{i\theta_1}$, the complete value of ψ_1 is that given in (34), and (35) represents the work done altogether by Ψ_1 in unit time. Equating this to zero, we get as the relation between R_1 and θ_1,

$$A_{11} R_1 \sin \alpha_{11} = \sin \theta_1, \dots\dots\dots\dots\dots\dots(46)$$

and the part of ψ_1 due to Ψ_1 is

$$\frac{\sin \theta_1 e^{i(\theta_1 + \alpha_{11})}}{\sin \alpha_{11}}. \quad \dots\dots\dots\dots\dots\dots(47)$$

The modulus of this is a maximum when $\sin \theta_1 = \pm 1$, and the value of the maximum is $\operatorname{cosec} \alpha_{11}$. In this case (47) becomes

$$-1 + i \cot \alpha_{11}, \dots\dots\dots\dots\dots\dots(48)$$

and the complete value of ψ_1 is

$$\psi_1 = i \cot \alpha_{11}, \dots\dots\dots\dots\dots\dots(49)$$

in quadrature with the former value, viz., 1.

We may regard the state of things now defined as being in a sense the greatest possible disturbance of the original state of things. If the system be quite out of resonance, forces and displacements are nearly in the same phase, and α_{11} is small. The altered ψ_1 is then a large multiple of the original value.

The work done by Ψ_1 on the complete value of ψ_1 is zero by supposition; but the work done upon the part of ψ_1 due to *itself* is by (14) in unit time

$$-\tfrac{1}{2} p R_1^2 A_{11} \sin \alpha_{11} = -\frac{\tfrac{1}{2} p}{A_{11} \sin \alpha_{11}}.$$

This corresponds to the original $\psi_1 = 1$, or $\Psi_2 = A_{12}^{-1} e^{-i\alpha_{12}}$. If the prescribed value of Ψ_2 be now left open, we have as the work in question

$$-\frac{\tfrac{1}{2} p A_{12}^2 \operatorname{Mod}^2 \Psi_2}{A_{11} \sin \alpha_{11}}; \quad \dots\dots\dots\dots\dots(50)$$

and this by (20) is the same as

$$\frac{\tfrac{1}{2} A_{12}^2 \operatorname{Mod}^2 \Psi_2}{b_{22} A_{12}^2 + b_{33} A_{13}^2 + \dots}, \quad \dots\dots\dots\dots\dots(51)$$

b_{11}, b_{12}, &c. being supposed to be zero. This expression differs from (43) only as regards the numerical factor. If b_{33}, &c. also vanish, (51) becomes

$$\frac{\operatorname{Mod}^2 \Psi_2}{2 b_{22}} \quad \text{simply.}$$

If in (51) we introduce the suppositions of (44), we get as in (45) for the sum of all the values

$$\frac{\operatorname{Mod}^2 \Psi}{2b}. \quad \dots\dots\dots\dots\dots\dots(52)$$

In an interesting paper entitled "An Electromagnetic Illustration of the Theory of Selective Absorption of Light in a Gas*," Prof. Lamb has developed a general law for the maximum energy emitted by a resonator situated in a uniform medium when submitted to incident plane waves. "The rate at which energy is carried outwards by the scattered waves is, in terms of the energy-flux in the primary waves,

$$\frac{2n+1}{2\pi}\lambda^2, \quad\ldots\ldots\ldots\ldots\ldots\ldots\ldots\ldots\ldots(53)$$

where λ is the wave-length, and n is the order of the spherical harmonic component of the incident waves which is effective." Prof. Lamb remarks that the law expressed by (53) "is of a very general character, and is independent of the special nature of the conditions to be satisfied at the surface of the sphere. It presents itself in the elastic solid theory, and again in the much simpler acoustical problem where there is synchronism between plane waves of sound and a vibrating sphere on which they impinge."

The generality claimed by Lamb for (53) seemed to me to indicate a still more general theorem in the background; and it was upon this suggestion that the investigations of the preceding pages were developed. An initial difficulty, however, stood in the way. The occurrence of n, a quantity special to the spherical problem, seemed to constitute a limitation; and the further question suggested itself as to why the efficiency of the resonator should rise with increasing n. For example, why in the acoustical problem should a resonator formed by a rigid sphere, moored to a fixed point by elastic attachments ($n=1$), be three times as effective as the simple resonator ($n=0$), for which the theory is given by my book on Sound, § 319?

The answer may be found in a slightly different presentation of the matter. In the above example the rigid sphere is supposed to be *symmetrically* moored to a fixed point, and the vibration actually assumed is in a direction parallel to that of propagation of the incident waves. Three degrees of freedom are really involved here, while the more typical case will be that in which the motion is limited to *one* direction. The efficiency of the resonator will then be proportional to the square of the cosine (μ) of the angle between the direction of vibration and that of the incident waves; and the *mean* efficiency will bear to the *maximum* efficiency ($\mu=1$) a ratio equal to that of

$$\int_0^1 \mu^2 \, d\mu : \int_0^1 d\mu,$$

that is of $\frac{1}{3}$. Thus, if the vibration in the case of $n=1$ be limited to *one* direction, the *mean* efficiency of the resonator is the *same* as when $n=0$; and a similar conclusion will hold good in all cases. In this way the factor

* *Cambr. Trans.* Vol. XVIII. p. 348 (1899).

$2n + 1$ is eliminated, and the statement assumes a form more nearly capable of generalization to all vibrating systems.

Now that a general theorem (52) has been demonstrated, it will be of interest to trace its application to some case of a uniform medium, for which purpose we may take the simple acoustical resonator. But this deduction is not quite a simple matter, partly on account of the extension to infinity, and also, I think, for want of a more general theory of waves in a uniform medium than any hitherto formulated. If the object be merely to obtain a result, it is far more easily attained by a special investigation from the formulæ of the Theory of Sound, on the lines indicated by Prof. Lamb. It may perhaps be well to sketch the outline of such an investigation.

The time factor e^{ikat} being suppressed, the velocity-potential ϕ of the primary waves is (§ 334) e^{ikx}, or $e^{ikr\mu}$, and the harmonic component of the nth order has the expression

$$\phi_n = (2n + 1) P_n(\mu) . P_n\left(\frac{d}{d . ikr}\right)\left(\frac{\sin kr}{kr}\right), \quad \ldots\ldots(53\,\text{bis})$$

while (§ 329) the corresponding expression for the divergent secondary waves is

$$\psi_n = (-1)^n ka_n P_n(\mu) . P_n\left(\frac{d}{d . ikr}\right)\left(\frac{\cos kr - i \sin kr}{kr}\right). \quad \ldots(54)$$

Accordingly

$$\phi_n + \psi_n = \{2n + 1 - (-1)^n ika_n\} P_n\left(\frac{d}{d . ikr}\right)\frac{\sin kr}{kr}$$

$$+ (-1)^n ka_n P_n\left(\frac{d}{d . ikr}\right)\frac{\cos kr}{kr}. \quad \ldots\ldots\ldots\ldots\ldots\ldots\ldots(55)$$

Now the only condition imposed upon the appliances introduced at the surface of the sphere is that they shall do no work. The velocity is $d\phi/dr + d\psi/dr$, and the pressure, proportional to $d\phi/dt + d\psi/dt$, is in quadrature with (55). All therefore that is required is that (55) and its derivative with respect to r be in the same phase, or that the ratio of these symbolic expressions be *real*. Since P_n is a wholly odd or wholly even function, this requirement is satisfied if

$$\frac{2n + 1 - (-1)^n ika_n}{(-1)^n ka_n} \quad \text{be real.}$$

If A_n, which may be complex, be written $Ae^{i\alpha}$, we get

$$- kA = -(-1)^n (2n + 1) \sin \alpha. \quad \ldots\ldots\ldots\ldots\ldots(56)$$

Thus A is a maximum when

$$\sin \alpha = -(-1)^n, \ldots\ldots\ldots\ldots\ldots\ldots\ldots(57)$$

and the maximum value is

$$A = (2n + 1)/k. \quad \ldots\ldots\ldots\ldots\ldots\ldots\ldots(58)$$

By (57), (58) $a_n = -(-1)^n i (2n+1)/k,$ ....(59)

so that in (55) $2n + 1 - (-1)^n ika_n = 0,$ ....(60)

but $\phi_n + \psi_n$ does not itself vanish.

If the incident plane waves are regarded as due to a source at a great distance R, we have in correspondence with (53)

$$\phi = -\frac{Re^{-ikR}}{R}, \quad\text{...........................(61)}$$

with which we may compare

$$\psi_n = \frac{a_n e^{-ikr}}{r} \cdot P_n(\mu). \quad\text{..................(62)}$$

The work emitted from the primary source being represented by $R^2 \int_{-1}^{+1} d\mu$, that emitted, or rather diverted, by the resonator will be

$$\text{Mod}^2 a_n \int_{-1}^{+1} P_n^2(\mu) \, d\mu.$$

Now $\int_{-1}^{+1} P_n^2(\mu) \, d\mu = \dfrac{2}{2n+1},$ and $\int_{-1}^{+1} d\mu = 2.$

Also $$\text{Mod}^2 a_n = \frac{(2n+1)^2}{k^2}, \quad\text{......................(63)}$$

so that the ratio of works is

$$\frac{2n+1}{k^2 R^2} \cdot \quad\text{...........................(64)}$$

This agrees with the result given in *Theory of Sound*, § 319 for a symmetrical resonator ($n = 0$).

In order to express (64) in terms of the energy-flux (per unit area) of the primary waves at the place of the resonator, we have only to multiply (64) by the area ($4\pi R^2$) of the sphere of radius R. If we restore $2\pi/\lambda$ for k, we get as the equivalent of (64)

$$(2n+1)\lambda^2/\pi. \quad\text{............................(65*)}$$

If we limit the resonator to one definite harmonic vibration of order n and suppose that the primary waves may be incident indifferently in all directions, the *mean* of the values of (65) is λ^2/π simply, as follows from known properties of the spherical functions.

---

* It will be observed that (65) is the double of the value (53) above quoted.

Dec. 17.—I have since learned that Prof. Lamb's calculations for the acoustical problem have already been published. See *Math. Soc. Proc.* Vol. xxxii. p. 11, 1900, where equation (44) is identical with (65) above. Reference may also be made to Lamb, *Math. Soc. Proc.* Vol. xxxii. p. 120, 1900.

Before we can apply the general theorem (52) to an independent investigation of these results, it is necessary to consider the connexion between the formulæ for plane and spherical waves; and for this purpose it is desirable to use a method which, if not itself quite general, is of a character susceptible of generalization. If ϕ denote the velocity-potential due to a "force" ΦdV acting at the element of volume dV and proportional to the periodic introduction and abstraction of fluid at that place, we may write

$$\phi = B\Phi dV \frac{e^{-ikr}}{r}, \quad \dots\dots\dots\dots\dots(66)$$

where $k = 2\pi/\lambda$ and B is some multiplier, which may be complex. The time factor e^{ikat} is suppressed. In order to obtain plane waves we may suppose that Φ acts uniformly over the whole slice between x and $x + dx$. The effect may be calculated as in a well-known optical investigation. If $\rho^2 = r^2 - x^2$, the element of volume is $2\pi\rho\, d\rho\, dx$, or $2\pi r\, dr\, dx$; and for the plane waves

$$\phi = B\Phi\, dx \int_x^\infty 2\pi e^{-ikr}\, dr = B\Phi\, dx\, \frac{2\pi e^{-ikx}}{ik}. \quad \dots\dots\dots(67)$$

Here Φ acts at $x = 0$, and

$$\phi_0 = \frac{2\pi B}{ik}\,\Phi\, dx, \qquad \dot\phi_0 = 2\pi a B\Phi\, dx. \quad \dots\dots\dots\dots(68)$$

Since ϕ_0 must be in the same phase as Φ, it follows that B is real.

We have now to consider the work done in generating the plane waves per unit of time and per double unit area of wave-front. For this we have

$$\frac{dx}{t}\int_0^t \Phi\dot\phi_0\, dt = \pi a B\, (dx)^2 \operatorname{Mod}^2 \phi; \quad \dots\dots\dots\dots(69)$$

or since by (67) $$\operatorname{Mod}\phi = \frac{2\pi B\, dx}{k}\operatorname{Mod}\Phi, \quad \dots\dots\dots\dots(70)$$

we get for the work propagated in *one* direction per unit of area of wave-front

$$\frac{k^2 a}{8\pi B}\operatorname{Mod}^2\phi. \quad \dots\dots\dots\dots\dots(71)$$

Reverting now to (66), we see that for divergent waves

$$\operatorname{Mod}^2\phi = \frac{B^2}{r^2}\operatorname{Mod}^2(\Phi dV),$$

or $$4\pi r^2 \operatorname{Mod}^2\phi = 4\pi B^2 \operatorname{Mod}^2(\Phi dV).$$

Accordingly by (71), since at a sufficient distance the distinction between plane and divergent waves disappears, the work emitted in unit time by a point-source ΦdV is

$$\tfrac{1}{2}k^2 a B \operatorname{Mod}^2(\Phi dV). \quad \dots\dots\dots\dots(72)$$

It may be observed that in order to preserve a better correspondence between "force" and "coordinate" a somewhat different interpretation is here put upon Φ from that adopted in *Theory of Sound*, § 277. If we compare our present (71) with (10), § 245, we find that

$$B = \frac{1}{4\pi\rho}; \quad \dots\dots\dots\dots\dots\dots\dots\dots\dots(73)$$

so that according to the present interpretation of Φ, (66) gives

$$\phi = \frac{1}{4\pi\rho} \Phi dV \frac{e^{-ikr}}{r}, \quad \dots\dots\dots\dots\dots\dots(74)$$

whereas in the notation of § 277

$$\phi = \frac{1}{4\pi a^2} \Phi dV \frac{e^{-ikr}}{r}.$$

We are now to some extent prepared for the application of (52), but the difficulty remains that (52) deals in the first instance with a finite system subject to dissipative forces; whereas the uniform medium is infinite, and need not be supposed subject to any forces truly dissipative. There is, however, no objection to the introduction of a small dissipative force of the character supposed in the general theorem, that is, proportional everywhere to $d\phi/dt$. Under this influence plane waves are attenuated as they advance; the law of attenuation being represented by the introduction into (67) of the factor e^{-ax}, where α is a small quantity, real and positive.

The connexion between α and b may be investigated by considering the action of the dissipative force $-b\phi_0$ operative over a slice δx at $x = 0$ in causing the attenuation. By (67) the effect at x of this force is represented by

$$\delta\phi = -\frac{2\pi B e^{-ikx}}{ik} b\,\phi_0\,\delta x,$$

so that

$$\delta\phi/\phi = -2\pi ab B\,\delta x.$$

By supposition this must be the same as $-\alpha\,\delta x$; and accordingly

$$\alpha = 2\pi ab B. \quad \dots\dots\dots\dots\dots\dots\dots\dots(75)$$

If we use this result to eliminate B from (72), we get as the work emitted from a point-source

$$\frac{k^2 \alpha}{4\pi b} \operatorname{Mod}^2(\Phi dV). \quad \dots\dots\dots\dots\dots\dots(76)$$

The formula (52) expresses the sum of all the works emitted by the resonator when submitted successively to all the various forces Ψ, subject themselves to conditions (44). These conditions are satisfied in the present case if we identify each Ψ with the force ΦdV acting over the various *equal* elements dV into which infinite space may be divided, the value of Φ being everywhere the same. Each point-source is regarded as the origin of plane waves which fall upon the resonator. The efficiency of the sources which lie in a given direction still depends upon the distance, the waves as they reach the resonator being attenuated by the resistance and also in the usual manner according to the law of inverse squares.

Let us compare the efficiency of the element ΦdV at distance r with the efficiency of an equal element at distance unity, the value of α being so small that no perceptible attenuation due to it occurs in distance unity. The element of volume

$$dV = d\sigma . d(\tfrac{1}{3}r^3), \quad \dots\dots\dots\dots\dots\dots\dots\dots\dots(77)$$

in which for the present $d\sigma$ [the element of angular area] is kept unchanged. The efficiency of the element at distance r varies as

$$(\Phi dV)^2 . e^{-2\alpha r} . r^{-2};$$

and $\qquad \Sigma(\Phi dV)^2 e^{-2\alpha r} r^{-2} = \Phi^2 d\sigma dV \int_0^\infty e^{-2\alpha r} dr = \dfrac{d\sigma}{2\alpha dV}(\Phi dV)^2.$

Hence for the sum of all the elements lying within $d\sigma$ we have

$$\dfrac{d\sigma}{2\alpha dV} \times \text{efficiency of } (\Phi dV) \text{ at distance unity.}$$

This has now to be again integrated with respect to $d\sigma$. The result may be expressed by the statement that the sum of all the works emitted by the resonator is

$$\dfrac{4\pi}{2\alpha dV} \times \text{mean work emitted by resonator corresponding to the various}$$

positions of the point-source on the sphere $r = 1$.

By the theorem (52) this sum of all the works is also expressed by

$$\dfrac{(\Phi dV)^2}{2b dV},$$

or in accordance with (76) is equal to

$$\dfrac{2\pi}{k^2 \alpha dV} \times \text{work emitted by } (\Phi dV) \text{ itself.}$$

We see therefore that the mean work emitted by the resonator for positions of the point-sources distributed uniformly over the sphere $r = 1$ is equal to the work emitted by each of the point-sources themselves divided by k^2. If the point-sources are supposed to lie at a distance r in place of unity, the divisor becomes $k^2 r^2$ in place of k^2.

Although the above deduction may stand in need of some supplementing before it could be regarded as rigorous at all points, it suffices at any rate to show that the general theorem (52) really does include the more special cases which suggested it. In some applications, *e.g.* to an elastic solid, we should have at first to suppose the forces introduced at any element of volume dV to act in various directions, but no great complication thence arises, and the general result finally takes the same form.

Energy stored in Resonators.

In preceding investigations we have been concerned with energy emitted from a resonator. We now turn to the consideration of some general theorems relating to the energy *stored*, as it were, in the resonator when the applied forces have frequencies in the neighbourhood of the natural frequency of the resonator. And we will treat first the simple case of one degree of freedom.

As in (2) we have

$$T = \tfrac{1}{2} a \dot{\psi}^2, \qquad F = \tfrac{1}{2} b \dot{\psi}^2, \qquad V = \tfrac{1}{2} c \psi^2, \quad \dots\dots\dots(78)$$

giving as the equation of vibration

$$a \ddot{\psi} + b \dot{\psi} + c \psi = \Psi = E e^{ipt}. \quad \dots\dots\dots\dots(79)$$

The time factor being suppressed, the solution of (79) is

$$\psi = \frac{E}{c - a p^2 + i p b}, \quad \dots\dots\dots\dots(80)$$

whence

$$\mathrm{Mod}^2 \, \psi = \frac{\mathrm{Mod}^2 \, \Psi}{a^2 (n^2 - p^2)^2 + p^2 b^2}, \quad \dots\dots\dots(81)$$

n, equal to $\sqrt{(c/a)}$, being the value of p corresponding to maximum resonance. If, as we suppose, b is very small, the important values of $\mathrm{Mod}^2 \, \psi$ are concentrated in the neighbourhood of $p = n$, and we may substitute n for p in the term $p^2 b^2$. Also $n^2 - p^2$ may be identified with $2n (n - p)$. Accordingly (81) becomes

$$\mathrm{Mod}^2 \, \psi = \frac{1}{n^2 b^2} \, \frac{\mathrm{Mod}^2 \, \Psi}{1 + \dfrac{4 c^2}{n^4 b^2} (p - n)^2}. \quad \dots\dots\dots(82)$$

We now suppose that $\mathrm{Mod}^2 \, \Psi$ is constant, while p varies over the small range in the neighbourhood of n for which alone $\mathrm{Mod}^2 \, \psi$ is sensible, and inquire as to the sum of the values of $\mathrm{Mod}^2 \, \psi$. Since

$$\int_{-\infty}^{+\infty} \frac{dx}{1 + a^2 x^2} = \frac{\pi}{a}, \quad \dots\dots\dots\dots(83)$$

we find

$$\int \mathrm{Mod}^2 \, \psi \, dp = \frac{\pi \, \mathrm{Mod}^2 \, \Psi}{2 b c},$$

or again

$$\tfrac{1}{2} c \int \mathrm{Mod}^2 \, \psi \, dp = \frac{\pi}{4 b} \, \mathrm{Mod}^2 \, \Psi. \quad \dots\dots\dots(84)$$

On the left $\tfrac{1}{2} c \, \mathrm{Mod}^2 \, \psi$ represents the potential energy of the system at the phase of maximum displacement, which is the same as the nearly constant total energy, so that (84) gives the integral of this total energy as p passes through the value which calls out the maximum and (by supposition) very great resonance.

The most remarkable feature of (84) is perhaps that the integral is independent of a and c. Large values of these quantities will increase the energy of the system at the point where $p = n$; but on the other hand this maximum falls off more rapidly as p departs from the special value.

We pass next to the more difficult considerations which arise when the force Ψ_2 is of one kind, while the coordinate ψ_1 on which the resonance principally depends is of another. In the first instance we shall suppose that there are no other than these two degrees of freedom.

If in equation (3) we assume Ψ_1, ψ_3, ψ_4, &c. to vanish, we get

$$\frac{\psi_1}{\Psi_2} = \frac{e_{12}}{e_{12}^2 - e_{11}e_{22}}, \dots\dots\dots\dots\dots(85)$$

where e_{11}, e_{12}, e_{22} have the values given in (4) with ip substituted for D. We suppose further that $b_{12} = 0$, $b_{11} = 0$, so that the dissipation depends entirely on b_{22}. With these simplifications the numerator of (85) becomes

$$e_{12} = c_{12} - p^2 a_{12}, \dots\dots\dots\dots\dots(86)$$

and for the denominator (taken negatively)

$$e_{11}e_{22} - e_{12}^2 = (c_{11} - p^2 a_{11})(c_{22} - p^2 a_{22})$$
$$- (c_{12} - p^2 a_{12})^2 + ip b_{22}(c_{11} - p^2 a_{11}). \dots\dots\dots\dots(87)$$

If n be one of the values of p corresponding to maximum resonance, the real part of (87) vanishes when $p = n$; so that

$$(c_{11} - n^2 a_{11})(c_{22} - n^2 a_{22}) - (c_{12} - n^2 a_{12})^2 = 0, \dots\dots\dots(88)$$

or written as a quadratic in n^2,

$$c_{11}c_{22} - c_{12}^2 - n^2(a_{11}c_{22} + a_{22}c_{11} + 2a_{12}c_{12}) + n^4(a_{11}a_{22} - a_{12}^2) = 0. \dots(89)$$

By subtraction of (89), (87) may be written

$$e_{11}e_{22} - e_{12}^2 = -(p^2 - n^2)(a_{11}c_{22} + a_{22}c_{11} + 2a_{12}c_{12})$$
$$+ (p^4 - n^4)(a_{11}a_{22} - a_{12}^2) + ip b_{22}(c_{11} - p^2 a_{11}). \dots\dots\dots(90)$$

If b_{22} were zero, ψ_1 would become infinite for $p = n$. If we assume that b_{22}, while not actually zero, is still relatively very small, the values of p in the neighbourhood of n retain a preponderating importance; and we may equate p to n with exception of the factor $(p - n)$. Thus (86), (90) become

$$e_{12} = c_{12} - n^2 a_{12}, \dots\dots\dots\dots\dots(91)$$

$$e_{11}e_{22} - e_{12}^2 = -2n\,(p-n)\,\{a_{11}c_{22} + a_{22}c_{11} - 2a_{12}c_{12} - 2n^2\,(a_{11}a_{22} - a_{12}^2)\}$$
$$+ in\,b_{22}\,(c_{11} - n^2a_{11})$$
$$= -\frac{2\,(p-n)}{n}\,\{c_{11}c_{22} - c_{12}^2 - n^4\,(a_{11}a_{22} - a_{12}^2)\} + in\,b_{22}\,(c_{11} - n^2a_{11}),\dots(92)$$

use being made of (89).

From (91), with use of (88),

$$\text{Mod}^2\,e_{12} = (c_{11} - n^2a_{11})\,(c_{22} - n^2a_{22}); \quad \dots\dots\dots\dots\dots(93)$$

and from (92)

$$\text{Mod}^2\,(e_{11}e_{22} - e_{12}^2) = \frac{4\,(p-n)^2}{n^2}\,\{c_{11}c_{22} - c_{12}^2 - n^4\,(a_{11}a_{22} - a_{12}^2)\}^2$$
$$+ n^2b_{22}^2\,(c_{11} - n^2a_{11})^2. \quad \dots\dots\dots\dots\dots(94)$$

If we now, as for (84), carry out the integration with respect to p, $\text{Mod}\,\Psi_2$ being constant, we find from (85)

$$\frac{\int \text{Mod}^2\,\psi_1\,dp}{\text{Mod}^2\,\Psi_2} = \frac{\pi\,(c_{22} - n^2a_{22})}{2b_{22}\,\{c_{11}c_{22} - c_{12}^2 - n^4\,(a_{11}a_{22} - a_{12}^2)\}}. \quad \dots\dots\dots(95)$$

So far we have assumed merely that the compound system is in high resonance when $p = n$; but more than this is required in order to arrive at a simple result. We must further assume that the coefficients of interconnexion a_{12}, c_{12} are small (b_{12} has been already made zero), so that the resonating coordinate may vibrate with a considerable degree of independence. We are also to suppose that n corresponds to these comparatively independent vibrations, so that $c_{11} - n^2a_{11} = 0$ approximately, while $c_{22} - n^2a_{22}$ is relatively large. These simplifications reduce the bracket in the denominator of (95) to

$$c_{11}c_{22} - n^4a_{11}a_{22}, \quad \text{or to} \quad c_{11}(c_{22} - n^2a_{22});$$

whence we obtain finally

$$\frac{\tfrac{1}{2}c_{11} \int \text{Mod}^2\,\psi_1\,dp}{\text{Mod}^2\,\Psi_2} = \frac{\pi}{4b_{22}}. \quad \dots\dots\dots\dots\dots(96)$$

In this expression, which is of the same form as (84), the numerator on the left may be considered to represent the integrated energy of the *resonator*. It must not be overlooked that the suppositions involved are to some extent antagonistic. For example, the coefficient of b_{22} in (90) is treated as constant when p varies, although $(c_{11} - n^2a_{11})$ is small. The theorem should be regarded as one applicable in the limit when b_{22} is exceedingly small.

If there be more than two degrees of freedom, the result is unaffected, provided that the forces Ψ_3, Ψ_4, &c. of the new types vanish and that the only dissipation is that represented by b_{22}. By the 3rd, 4th, &c. of (3) the new coordinates may be eliminated. In this process b_{22} is undisturbed, and everything remains as if there were only two coordinates as above.

The idea of the integration with respect to p is borrowed from a paper by Prof. Planck (*Ann. d. Phys.* I. p. 99, 1900), in which is considered the behaviour of an infinitely small electromagnetic resonator under incident plane waves. The proof of the general theorem covering Prof. Planck's case would require a process similar to that by which (51) was established. Subject to the condition

$$\frac{\text{Mod}^2\,\Psi_2}{b_{22}} = \frac{\text{Mod}^2\,\Psi_3}{b_{33}} = \ldots\ldots = \frac{\text{Mod}^2\,\Psi}{b}, \quad \ldots\ldots\ldots(97)$$

we might expect to find, as in (52),

$$\Sigma \tfrac{1}{2} c_{11} \int \text{Mod}^2\,\psi_1\, dp = \frac{\pi\, \text{Mod}^2\,\Psi}{4b}. \quad \ldots\ldots\ldots\ldots(98)$$

275.

ON THE LAW OF THE PRESSURE OF GASES BETWEEN 75 AND 150 MILLIMETRES OF MERCURY.

[*Phil. Trans.* CXCVIII. A, pp. 417—430, 1902.]

IN a recently published paper* I have examined, with the aid of a new manometer, the behaviour of gases at very low pressures, rising to 1·5 millims. of mercury, with the result that Boyle's law was verified to a high degree of precision. There is, however, a great gap between the highest pressure there dealt with and that of the atmosphere—a gap which it appeared desirable in some way to bridge over. The sloping manometer, described in the paper referred to, does not lend itself well to the use of much greater pressures, at least if we desire to secure the higher proportional accuracy that should accompany the rise of pressure. The present communication gives the results of observations, by another method, of the law of pressure in gases between 75 millims. and 150 millims. of mercury. It will be seen that for air and hydrogen Boyle's law is verified to the utmost. In the case of oxygen, the agreement is rather less satisfactory, and the accordance of separate observations is less close. But even here the departure from Boyle's law amounts only to one part in 4000, and may perhaps be referred to some reaction between the gas and the mercury. In the case of argon too the deviation, though very small, seems to lie beyond the limits of experimental errors. Whether it is due to a real minute departure from Boyle's law, or to some complication arising out of the conditions of experiment, must remain an open question.

In the case of pressures not greatly below atmosphere, the determination with the usual column of mercury read by a cathetometer (after Regnault) is sufficiently accurate. But when the pressure falls to say one-tenth of an atmosphere, the difficulties of this method begin to increase. The guiding idea in the present investigation has been the avoidance of such difficulties

* *Phil. Trans.* Vol. 196, A, p. 205, Feb. 1901. [Vol. IV. p. 511.]

by the use of manometric gauges combined in a special manner. The object is to test.whether when the volume of a gas is halved its pressure is doubled, and its attainment requires two gauges indicating pressures which are in the ratio of 2 : 1. To this end we may employ a pair of independent gauges as nearly as possible similar to one another, the similarity being tested by combination in parallel, to borrow an electrical term. When connected below with one reservoir of air and above with another reservoir, or with a vacuum, the two gauges should reach their settings simultaneously, or at least so nearly that a suitable correction may be readily applied. For brevity we may for the present assume precise similarity. If now the two gauges be combined *in series*, so that the low-pressure chamber of the first communicates with the high-pressure chamber of the second, the combination constitutes a gauge suitable for measuring a doubled pressure.

The Manometers.

The construction of the gauges is modelled upon that used extensively in my researches upon the density of gases, so far as the principle is concerned, although of course the details are very different. In fig. 1 A and B represent [about ½] size the lower and upper chambers. As regards the glass-work, these communicate by a short neck at D as well as by the curved tube ACB. Through the neck is carried the glass measuring-rod FDE, terminating downwards at both ends in carefully prepared points E, F. The rod is held, at D only, with cement, which also completely blocks up the passage, so that when mercury stands in the curved tube the upper and lower chambers are isolated from one another. The use of the gauge is fairly obvious. Suppose for example that it is desired to adjust the pressure of gas in a vessel communicating with G to the standard of the gauge. Mercury standing in C, H is connected to the pump until a vacuum is established in the upper chamber. From a hose and reservoir attached below, mercury is supplied through I until the point F and its image in the mercury surface nearly coincide. If E coincides with its image, the pressure is that defined; otherwise adjustment must be made until the points E, F both coincide with their images, or as we shall say until both mercury surfaces are *set*. The pressure then corresponds to the column of mercury whose height is the length of the measuring-rod between the points E, F. The verticality of EF is tested with a plumb-line.

The measuring-rods appear somewhat slender; but it is to be remembered that the instruments are used under conditions that are almost constant. So far as the comparison of one gas with another is concerned, the qualification "almost" may indeed be omitted. The coincidence of the points and their images is observed with the aid of four magnifiers of 20 millims. focus, fixed in the necessary positions.

General Arrangement of Apparatus.

In fig. 2 is represented the connection of the manometers with one another and with the gas reservoirs. The left-hand manometer can be connected above through F with the pump or with the gas supply. The lower chamber A communicates with the upper chamber D of the right-hand manometer and with an intermediate reservoir E, to which, as to the manometers, mercury can be supplied from below. The lower chamber C of the

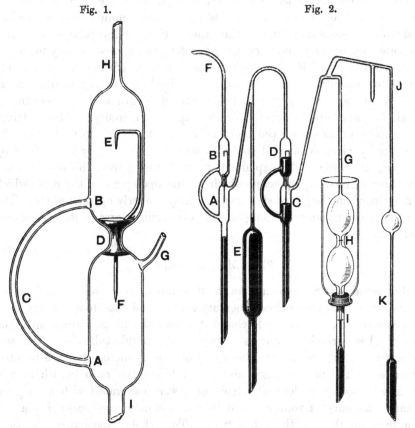

Fig. 1.   Fig. 2.

right-hand manometer is connected with the principal gas reservoir. This consists of two bulbs, each of about 129 cub. centims. capacity, connected together by a neck of very narrow bore. Three marks are provided, one G above the upper bulb, a second H on the neck, and a third I below the lower bulb, so adjusted that the included volumes are nearly equal. The use of the side-tube JK will be explained presently.

When, as shown, the mercury stands at the lower mark, the double volume is in action and the pressure is such as will balance the mercury in one (the right-hand) manometer.   A vacuum is established in the upper chamber D from which a way is open through AB to the pump.   When the mercury is raised to the middle mark H, the volume is halved, and the pressure to be dealt with is doubled.   Gas sufficient to exert the single pressure (75 millims.) must be supplied to the intermediate chamber, which is now isolated from the pump by the mercury standing up in AB.   Both manometers can now be set, and the doubling of the pressure verified.

The communication through F with the pump is free from obstruction, but on a side-tube a three-way tap is provided communicating on the one hand with the gas supply and on the other with a vertical tube, more than a barometer-height long and terminating below under mercury, by means of which a wash-out of the generating vessels can be effected when it is not desired to evacuate them.   The five tubes leading downwards from A, E, C, I, K are all over a barometer-height in length and are terminated by suitable hoses and reservoirs for the supply of mercury.   When settings are actually in progress, the mercury in the hoses is isolated from the reservoirs by pinch-cocks and the adjustment of the supply is effected by squeezing the hoses.   As explained in my former paper, the final adjustment must be made by squeezers which operate upon parts of the hoses which lie flat upon the large wooden mercury tray underlying the whole.   The adjustment being somewhat complicated, a convenient arrangement is almost a necessity.

The Side Apparatus.

By the aid of these manometers the determination of pressure is far more accurate than with the ordinary mercury column and cathetometer, but since the pressures are defined beforehand, the adjustment is thrown upon the volume.   The variable volume is introduced in the side-tube JK.   This was graduated to $\frac{1}{2}$ cub. centim., in the first instance by mercury from a burette. Subsequently the narrow parts above and below the bulb (which as will presently be seen are alone of importance) were calibrated with a weighed column of mercury of volume equal to $\frac{1}{2}$ cub. centim. and occupying about 80 millims. of the length of the tube.   The whole capacity of the tube between the lowest and highest marks was $20\frac{1}{2}$ cub. centims.   The object of this addition is to meet a difficulty which inevitably presents itself in apparatus of this sort.   The volume occupied by the gas cannot be limited to the capacities susceptible of being accurately gauged.   Between the upper mark G and the mercury surface in C when set, a volume is necessarily included which cannot be gauged with the same accuracy as the volumes between G and H and between H and I.   The simplest view of the side apparatus is that it is designed to measure this volume.   In the notation subsequently used V_3 is the volume included when the mercury stands at C, at G, and at the top mark J. Let us suppose that with a certain quantity of gas imprisoned it is necessary in order to set the manometer CD, the upper chamber being vacuous, to add to V_3 a further volume V_5, amounting to the greater part of the capacity of the side-tube, so that the whole volume is $V_3 + V_5$. When the second manometer is brought into use, the volume must be halved, for which purpose the mercury is raised through the bulb until it stands somewhere in the upper tube. The whole volume is now $V_3 + V_4$. And since

$$V_3 + V_5 = 2(V_3 + V_4),$$

we see that

$$V_3 = V_5 - 2V_4,$$

which may be regarded as determining V_3, V_4 and V_5 being known. A somewhat close accommodation is required between V_3, about 19 cub. centims. in my apparatus, and the whole contents of the side-tube.

General Sketch of Theory.

As the complete calculation is rather complicated on account of the numerous temperature corrections, it may be convenient to give a sketch of the theory upon the assumption that the temperature is constant, not only throughout the whole apparatus at one time, but also at the four different times concerned. We shall see that it is not necessary to assume Boyle's law, even for the subsidiary operations in the side-tube.

$V_1 =$ volume of two large bulbs together between I and G (about 258 cub. centims.),

$V_2 =$ volume of upper bulb between G and H,

$V_3 =$ volume between C, G and highest mark J on side-tube,

$V_4 =$ measured volume on upper part of J from highest mark downwards,

$V_5 =$ measured volume, including bulb, of side apparatus from highest mark downwards,

$P_1 =$ small pressure (height of mercury in right-hand manometer),

$P_2 =$ large pressure (sum of heights of mercury in two manometers).

In the first pair of operations when the large bulbs are in use, the pressure P_1 corresponds to the volume $(V_1 + V_3 + V_5)$ and the pressure P_2 corresponds to $(V_2 + V_3 + V_4)$, *the quantity of gas being the same*. Hence the equation

$$P_1(V_1 + V_3 + V_5) = BP_2(V_2 + V_3 + V_4), \quad \dots\dots\dots\dots(1)$$

B being a numerical quantity which would be unity according to Boyle's law. In the second pair of operations with a *different quantity of gas* but with the *same pressures*, the mercury stands at G throughout, and we have

$$P_1(V_3 + V_5') = BP_2(V_3 + V_4'); \quad \dots\dots\dots\dots\dots(2)$$

whence by subtraction

$$P_1(V_1 + V_5 - V_5') = BP_2(V_2 + V_4 - V_4'). \quad \dots\dots\dots\dots(3)$$

From this equation V_3 has been eliminated and B is expressed by means of P_1/P_2, and the actually gauged volumes

$$V_1, \quad V_2, \quad V_5 - V_5', \quad V_4 - V_4'.$$

It is important to remark that only the *differences*

$$(V_5 - V_5'), \quad (V_4 - V_4')$$

are involved. The first is measured on the lower part of the side-apparatus and the second on the upper part, while the capacity of the intervening bulb does *not* appear.

If the principal volumes V_1 and V_2 are nearly in the right proportion, there is nothing to prevent both $V_5 - V_5'$ and $V_4 - V_4'$ from being very small. When the temperature changes are taken into account, V_3, V_4, V_5 are not fully eliminated, but they appear with coefficients which are very small if the temperature conditions are good.

Thermometers.

As so often happens, much of the practical difficulty of the experiment turned upon temperature. The principal bulbs were drowned in a water-bath which could be effectively stirred, and so far there was no particular impediment to accuracy. But the other volumes could not so well be drowned, and it needed considerable precaution to ensure that the associated thermometers would give the temperatures concerned with sufficient accuracy. As regards the side-tube, a thermometer associated with its bulb and wrapped well round with cotton-wool was adequate. A third thermometer was devoted to the space occupied by the manometers and the tube leading from C to J. It was here that the difficulty was greatest on account of the proximity of the observer. Three large panes of glass with enclosed air spaces were introduced as screens, and although the temperature necessarily rose during the observations, it is believed that the rise was adequately represented in the thermometer readings. A single small gas flame, not allowed to shine directly upon the apparatus, supplied the necessary illumination, being suitably reflected from four small pieces of looking-glass fixed to a wall behind the glass points of the manometers.

As regards the success of the arrangement for its purpose, it is to be remembered that by far the larger part of any error that might arise is *eliminated in the final result*, since it is only a question of a *comparison* of observations with and without the large bulbs. Any systematic error made in the first case as regards the temperature of the undrowned capacities will be repeated in the second, and so lose its importance. A similar remark applies to any deficiency in the comparison of the three thermometers with one another.

Comparison of Large Bulbs.

This comparison needs to be carried out with something like the full precision aimed at in the final result, although it is to be noted that an error enters to only the half of its proportional amount, since we have to do not with the ratio of the capacities of the two bulbs, but with the ratio of the capacity of the upper bulb to the capacity of the two bulbs together. Thus if the volume of the upper bulb be unity and that of the lower $(1 + \alpha)$, the ratio with which we are concerned is $2 + \alpha : 1$, differing from $2 : 1$ by the proportional error $\frac{1}{2}\alpha$.

To adjust the capacities to approximate equality and to determine the outstanding difference, the double bulb was mounted vertically, in connection above with a Töpler pump and below with a stop-cock, such as is used with a mercury burette. The "marks" were provided by small collars of metal embracing tubing of 3 millims. bore and securely cemented, to the lower edge of which the mercury could be set as in reading barometers. A measuring flask, with a prolonged neck consisting of uniform tubing of 6 millims. diameter, was prepared having nearly the same capacity as the bulbs. The mercury required at a known temperature to fill the upper bulb between the marks was run from the tap into this flask. Air specks being removed, the flask was placed in a water-bath and the temperature varied until the mercury stood at a fixed mark upon the neck of the flask. Subsequently the mercury required at the same temperature to fill the lower bulb between the middle and the lower marks was measured in the same way. On a mean of two trials it was found that the flask needed to be $2°\cdot4$ C. warmer in the second case than in the first, showing that the capacity of the lower bulb was a little the smaller. Taking the relative expansions of mercury and glass for one degree to be $\cdot00016$, we get as the proportional difference $\cdot00038$. Thus in the notation already employed,

$$V_1 : V_2 = 2 - \cdot00038 = 1\cdot99962. \quad \dots\dots\dots\dots\dots(4)$$

It appeared that so far as the measurements were concerned this ratio should be correct to at least $\frac{1}{20000}$; but disturbances due to pressure introduce uncertainty of about the same order.

Comparison of Gauges.

A simple method of comparing the gauges is to combine them in parallel so that the pressures in the lower chambers are the same, and also the pressures in the upper chambers, and then to find what slope must be given to the longer measuring-rod in order that its effective length may be equal to that of the shorter rod maintained vertical. The mercury can then be set to coincidence with all four points, and the equality of the gauges so arranged actually tested. It is afterwards an easy matter to calculate back so as to find the proportional difference of heights when both measuring-rods are vertical. Preliminary experiments of this kind upon the gauges, mounted on separate levelling stands and connected by india-rubber tubing, had shown that the difference was about $\frac{1}{800}$ part.

It would be possible, having found by the combination in parallel an adjustment to equality, to maintain the same sloped position during the subsequent use when the gauges must be combined in series. But in this case it would hardly be advisable to trust to wood-work in the mounting. At any rate in my experiments the gauges were erected with measuring-rods vertical, an arrangement which has at least the advantage that a displacement is of less importance as well as more easily detected. At the close of the observations upon the various gases it became necessary to compare the gauges with full precision.

For this purpose, they were connected (without india-rubber) in parallel, the upper chambers of both being in communication with the pump, and the lower chambers of both in communication with the gas reservoirs GI. Had the lengths of the measuring-rods been absolutely equal, this equality would be very simply proved by the possibility of so adjusting the pressure of the gas and the supply of mercury to the two manometers that all four mercury surfaces could be *set* simultaneously. It was very evident that no such simultaneous setting was possible, and the problem remained to evaluate the small outstanding difference. To pass from one manometer to the other, either the volume or the temperature [of the gas] had to be varied.

In principle it would perhaps be simplest to keep the volume constant and determine what difference of temperature (about half a degree) would be required to make the transition. But the temperature of the undrowned parts (now increased in volume) could not be ascertained with great precision, so that I preferred to vary the volume and to trust to alternations backwards and forwards for securing that the mean temperature in the two cases to be compared should not be different. Thus in one set, including seven observations following continuously, four alternate observations were settings with one manometer and three were settings with the other. According to the thermometers, the mean temperature in the first case was for the drowned volume 11°·38 and for the (much smaller) undrowned volume 12°·76. In the second case the corresponding temperatures were 11°·39 and 12°·80, so that the differences could be neglected. The volume changes were effected in the side-tube JK, and the mean difference in the two cases was ·411 cub. centim. It will be understood that in order to define the volume *both* manometers were always set *below*. The whole volume was reckoned at 294 cub. centims., of which about 258 cub. centims. represents the capacity of the bulbs GI drowned in the water-bath. According to these data the proportional difference in the lengths of the measuring-rods, equal to the proportional difference of the above determined volumes, is ·00140. Two other similar sets of observations gave ·00136, ·00137; so that the mean adopted value is ·00138. The measuring-rod of the manometer on the *right*, fig. 2, is the longer.

As in the case of the volumes, any error in the above comparison is halved in the actual application. If H_2 be the length of the rod in the right-hand manometer, H_1 the length in the left, we are concerned only with the ratio $H_1 + H_2 : H_2$. And from the value above determined we get

$$\frac{H_1 + H_2}{H_2} = 1\cdot99862. \quad \dots\dots\dots\dots\dots\dots\dots\dots(5)$$

The Observations.

In commencing a set of observations the first step is to clear away any residue of gas by making a high vacuum throughout the apparatus, the mercury being lowered below the manometers and bulbs. The mercury having been allowed to rise into the pump head of the Töpler, the gas to be experimented on is next admitted to a pressure of about 75 millims. This occupies the manometers, the bulbs, and part of the capacity of the intermediate chamber E. The passage through the right-hand manometer is then closed by bringing up the mercury to the neighbourhood of C, and by rise of mercury from I to H the pressure is doubled in the upper bulb. The next step is to cut off the communication between A and B, and to renew the vacuum in B. If the right amount of gas has been imprisoned, it is now possible to make a setting, the mercury standing at A, C, H, and in the side apparatus somewhere in the upper tube below J. If, as is almost certain to be the case in view of the narrowness of the margin, a suitable setting cannot be made, it becomes necessary to alter the amount of gas. This can usually be effected, without disturbing the vacuum, by lowering the mercury at C and allowing gas to pass in pistons in the curved tube CD either from the intermediate chamber to the bulbs, or preferably in the reverse direction.

When the right amount of gas has been obtained, the observations are straightforward. On such occasion six readings were usually taken, extending over about an hour, during which time the temperature always rose, and the means were combined into what was considered to be one observation.

A complete set included four observations with the large bulbs at 150 millims. pressure and four at 75 millims. To pass to the latter the mercury must be lowered from H to I and in the left-hand manometer, and the pump worked until a vacuum is established in D. It was considered advisable to break up one of the sets of four; for example, after two observations at 150 millims. to take four at 75 millims., and afterwards the remaining pair at 150 millims. In this way a check could be obtained upon the quantity of gas, of which some might accidentally escape, and there were also advantages in respect of temperature changes. These eight observations with the large bulbs were combined with four in which the side apparatus was alone in use, the mercury standing all the while at G. Of these, two related to the 75 millims. pressure and two to the 150 millims. Finally, the means were taken of all the corresponding observations.

The following table shows in the notation employed the correspondence of volumes and temperatures:

I.	V_1	θ_1	V_3	τ_1	V_5	t_1
II.	V_2	θ_2	V_3	τ_2	V_4	τ_2
III.	—	—	V_3	τ_3	V_5'	t_3
IV.	—	—	V_3	τ_4	V_4'	τ_4

In the first observation V_1 is the volume of the two large bulbs and θ_1 the temperature of the water-bath, reckoned from some convenient neighbouring temperature as a standard. V_3 is the ungauged volume already discussed whose temperature τ_1 is given by the upper thermometer. V_5 is the (larger) volume in the side apparatus whose temperature t_1 is that of the lower thermometer. In the second observation V_2 is the volume of the upper bulb and θ_2 its temperature. V_4 is the volume in the side apparatus whose temperature, as well as that of V_3, is taken to be τ_2, the mean reading of the upper thermometer. III. and IV. represent the corresponding observations in which the large bulbs are not filled. The reading of the water-bath thermometer is in every case denoted by θ, that of the upper thermometer by τ, and that of the lower thermometer by t. The temperature of the columns of mercury in the manometers is also represented by τ.

As an example of the actual quantities, the observations on air between October 28 and November 5 may be cited. The values of V_1 and V_3 are approximate. As appears from the formulæ, V_3 occurs with a small coefficient, as does also V_1, except in the ratio $V_1 : V_2$ otherwise provided for. We have

$$V_1 = 258 \cdot 4, \qquad V_3 = 19 \cdot 05;$$
$$V_4 = \quad \cdot 810, \qquad V_5 = 20 \cdot 493;$$
$$V_4 - V_4' = \quad \cdot 0841, \quad V_5 - V_5' = \quad \cdot 0266;$$
$$\theta_1 = -\cdot 077, \qquad \theta_2 = -\cdot 059; \qquad t_1 = \quad \cdot 257, \qquad t_3 = \cdot 141;$$
$$\tau_1 = \quad \cdot 092, \qquad \tau_2 = \quad \cdot 186, \qquad \tau_3 = -\cdot 033, \qquad \tau_4 = \cdot 100.$$

The volumes are in cubic centimetres and the temperatures are in Centigrade degrees reckoned from 14°.

An effort was made, and usually with success, to keep all the temperature differences small, and especially the difference between θ_1 and θ_2. It is desirable also so to adjust the quantities of gas in the two cases that $V_4 - V_4'$, $V_5 - V_5'$ shall be small.

The Reductions.

The simple theory has already been stated, but the actual reductions are rather troublesome on account of the numerous temperature corrections. These, however, are but small.

We have first to deal with the expansion of mercury in the manometers. If, as in (5), the actual heights of the mercury (at the same temperature) be H_1, H_2, we have for the corresponding *pressures* $H/(1 + m\tau)$, where $m = \cdot00017$. Thus in the notation already employed

$$P_1 = \frac{H_2}{1 + m\tau_1}, \quad \text{or} \quad \frac{H_2}{1 + m\tau_3},$$

and

$$P_2 = \frac{H_1 + H_2}{1 + m\tau_2}, \quad \text{or} \quad \frac{H_1 + H_2}{1 + m\tau_4}.$$

The quantity of gas at a given pressure occupying a known volume is to be found by dividing the volume by the absolute temperature. Hence each volume is to be divided by $1 + \beta\theta, 1 + \beta\tau, 1 + \beta t$, as the case may be, where β is the reciprocal of the absolute temperature taken as a standard. Thus in the above example for air (p. 36),

$$\beta = \frac{1}{273 + 14} = \frac{1}{287}.$$

Our equations, expressing that the quantities of gas are the same at the single and at the double pressure, accordingly take the form

$$\frac{H_2}{1 + m\tau_1}\left\{\frac{V_1}{1 + \beta\theta_1} + \frac{V_3}{1 + \beta\tau_1} + \frac{V_5}{1 + \beta t_1}\right\} = \frac{B(H_1 + H_2)}{1 + m\tau_2}\left\{\frac{V_2}{1 + \beta\theta_2} + \frac{V_3 + V_4}{1 + \beta\tau_2}\right\},$$

$$\frac{H_2}{1 + m\tau_3}\left\{\frac{V_3}{1 + \beta\tau_3} + \frac{V_5'}{1 + \beta t_3}\right\} = \frac{B(H_1 + H_2)}{1 + m\tau_4}\frac{V_3 + V_4'}{1 + \beta\tau_4},$$

where B is the numerical quantity to be determined—according to Boyle's law identical with unity.

By subtraction we deduce

$$\frac{1}{(1 + m\tau_1)(1 + \beta\theta_1)} - \frac{BV_2(H_1 + H_2)}{V_1 H_2(1 + m\tau_2)(1 + \beta\theta_2)}$$

$$= \frac{V_3}{V_1} \left\{ \frac{B(H_1 + H_2)}{H_2(1 + m\tau_2)(1 + \beta\tau_2)} - \frac{B(H_1 + H_2)}{H_2(1 + m\tau_4)(1 + \beta\tau_4)} \right.$$

$$\left. - \frac{1}{(1 + m\tau_1)(1 + \beta\tau_1)} + \frac{1}{(1 + m\tau_3)(1 + \beta\tau_3)} \right\}$$

$$+ \frac{B(H_1 + H_2)V_4}{H_2 V_1} \left\{ \frac{1}{(1 + m\tau_2)(1 + \beta\tau_2)} - \frac{1}{(1 + m\tau_4)(1 + \beta\tau_4)} \right\}$$

$$- \frac{V_5}{V_1} \left\{ \frac{1}{(1 + m\tau_1)(1 + \beta t_1)} - \frac{1}{(1 + m\tau_3)(1 + \beta t_3)} \right\}$$

$$+ \frac{B(H_1 + H_2)(V_4 - V_4')}{H_2 V_1} \frac{1}{(1 + m\tau_4)(1 + \beta\tau_4)}$$

$$- \frac{V_5 - V_5'}{V_1} \frac{1}{(1 + m\tau_3)(1 + \beta t_3)} \quad \dots\dots\dots\dots\dots\dots(6)$$

The first three terms on the right, viz., those in V_3, V_4, V_5, vanish if $\tau_1 = \tau_3$, $\tau_2 = \tau_4$, $t_1 = t_3$. In the small terms we expand in powers of the small temperatures (τ, t), and further identify $B(H_1 + H_2)/H_2$ with 2. The five terms on the right then assume the form

$$\frac{V_3}{V_1} \left\{ (m + \beta)(\tau_1 - \tau_3 - 2\tau_2 + 2\tau_4) + \beta^2 (2\tau_2^2 - 2\tau_4^2 - \tau_1^2 + \tau_3^2) \right\}$$

$$- \frac{2V_4}{V_1} \left\{ (m + \beta)(\tau_2 - \tau_4) - \beta^2 (\tau_2^2 - \tau_4^2) \right\}$$

$$- \frac{V_5}{V_1} \left\{ m(\tau_3 - \tau_1) + \beta(t_3 - t_1) + \beta^2 (t_1^2 - t_3^2) \right\}$$

$$+ \frac{2(V_4 - V_4')}{V_1} \left\{ 1 - (m + \beta)\tau_4 + \beta^2\tau_4^2 \right\}$$

$$- \frac{V_5 - V_5'}{V_1} \left\{ 1 - m\tau_3 - \beta t_3 + \beta^2 t_3^2 \right\},$$

in which $m\beta$ and m^2 are neglected, while β^2 is detained. In point of fact, the terms of the second degree were seldom sensible.

Taking the data above given for the observations on air October 28— November 5, we find

Term in V_3		$= - \cdot000012$
" V_4		$= - \cdot000002$
" V_5		$= + \cdot000034$
" $(V_4 - V_4')$		$= + \cdot000652$
" $(V_5 - V_5')$		$= - \cdot000103$
		$+ \cdot000569$

For the first term on the left of (6), we find

$$\frac{1}{(1 + m\tau_1)(1 + \beta\theta_1)} = 1.000256*;$$

so that $$B = \frac{V_1 H_2 (1 + m\tau_2)(1 + \beta\theta_2)}{V_2 (H_1 + H_2)} \times .999687,$$

or when the numerical values are introduced from (4), (5),

$$B = 1.00002.$$

The deviation from Boyle's law is quite imperceptible.

It may be noted that a value of B exceeding unity indicates an excessive compressibility, such as is manifested by carbonic acid under a pressure of a few atmospheres.

The Results.

Little now remains but to record the actual results. All the gases were, it is needless to say, thoroughly dried.

Air.

Date.	B.
April 15–29, 1901	.99986
May 22–28, 1901	1.00003
October 28–November 5, 1901 ...	1.0002
Mean	.99997

Hydrogen.

Date.	B.
July 6–13, 1901	.99999
July 16–23, 1901	.99996
Mean	.99997

The hydrogen was first absorbed in palladium, from which it was driven off by heat as required.

Oxygen.

Date.	B.
June 7–17, 1901	1.00022
July 21–July 1, 1901	1.00044
September 18–30, 1901	1.00005
October 10–18, 1901	1.00027
Mean	1.00024

* [1901. A misprint of sign is here corrected.]

The two first fillings of oxygen were with gas prepared by heating permanganate of potash contained in a glass tube and sealed to the remainder of the apparatus. The desiccation was, as usual, by phosphoric anhydride. In the third and fourth fillings the gas was from chlorate of potash and had been stored over water.

Nitrous Oxide.

Date.	B.
July 31–August 5, 1901	1·00059
August 8–24, 1901	1·00074
Mean	1·00066

Argon.

Date.	B.
December 28–January 1, 1902 ...	1·00024
January 2–9, 1902	1·00019
Mean	1·00021

The argon was from a stock which had been carefully purified some years ago and has since stood over mercury. In this case the two sets of observations recorded related to the same sample of gas imprisoned in the apparatus. In all other cases the gas was renewed for a new set of observations.

With regard to the accuracy of the results it was considered that systematic errors should not exceed $\frac{1}{10000}$. In the comparison of one gas with another most of the systematic errors are eliminated, and the mean of two or three sets should be accurate according to the standard above stated. That nitrous oxide should show itself more compressible than according to Boyle's law is not surprising, but there appear to be deviations also in the cases of oxygen and argon. Whether these deviations are to be regarded as real departures from Boyle's law, or are to be attributed to some complication relating to the glass or the mercury cannot be decided. At any rate they are very minute. It will be noted that the oxygen numbers are not so concordant as they ought to be. I am not in a position to suggest an explanation, and the discrepancies were hardly large enough to afford a handle for further investigation.

If we are content with a standard of $\frac{1}{5000}$, we may say that air, hydrogen, oxygen, and argon obey Boyle's law at the pressures concerned and at the ordinary temperatures (10° to 15°)*.

Throughout the investigation I have been efficiently assisted by Mr Gordon, to whom I desire to record my obligations.

* [1910. For carbonic oxide B=1·00005 ; see *Phil. Trans.* A, 204, p. 351.]

276.

ON THE PRESSURE OF VIBRATIONS.

[*Philosophical Magazine*, III. pp. 338—346, 1902.]

THE importance of the consequences deduced by Boltzmann and W. Wien from the doctrine of the pressure of radiation has naturally drawn increased attention to this subject. That æthereal vibrations must exercise a pressure upon a perfectly conducting, and therefore perfectly reflecting, boundary was Maxwell's deduction from his general equations of the electromagnetic field; and the existence of the pressure of light has lately been confirmed experimentally by Lebedew. It seemed to me that it would be of interest to inquire whether other kinds of vibration exercise a pressure, and if possible to frame a general theory of the action.

We are at once confronted with a difference between the conditions to be dealt with in the case of æthereal vibrations and, for example, the vibrations of air. When a plate of polished silver advances against waves of light, the waves indeed are reflected, but the medium itself must be supposed capable of penetrating the plate; whereas in the corresponding case of aerial vibrations the air as well as the vibrations are compressed by the advancing wall. In other cases, however, a closer parallelism may be established. Thus the transverse vibrations of a stretched string, or wire, may be supposed to be limited by a small ring constrained to remain upon the equilibrium line of the string, but capable of sliding freely upon it. In this arrangement the string passes but the vibrations are compressed, when the ring moves inwards.

We will commence with the very simple problem of a pendulum in which a mass C is suspended by a string. B is a ring [which embraces the string] constrained to the vertical line AD and capable of moving along it; $BC = l$, and θ denotes the angle between BC and AD at any time t. If B is held at rest, BC is an ordinary pendulum, and it is supposed to be executing small vibrations; so that $\theta = \Theta \cos nt$, where $n^2 = g/l$. The tension of the string is approximately W, the weight of the bob; and the force tending to push B upwards is at time t $W(1 - \cos\theta)$. Now this expression is closely related to the potential energy of the pendulum, for which

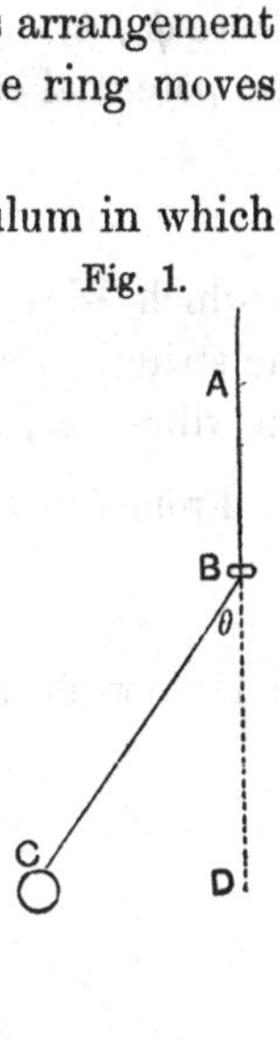

$$V = Wl(1 - \cos\theta).$$

Fig. 1.

The mean upward force upon B is accordingly equal to the mean value of $V \div l$; or since the mean value of V is half the constant total energy E of the system, we conclude that the mean force (L), driving B upwards, is measured by $\frac{1}{2}E/l$.

From the equation

$$L = \tfrac{1}{2}E/l \dots\dots\dots\dots\dots\dots\dots\dots\dots(1)$$

it is easy to deduce the effect of a *slow* motion upwards of the ring. The work obtained at B must be at the expense of the energy of the system, so that

$$dE = -Ldl = -\tfrac{1}{2}Edl/l.$$

By integration

$$E = E_1 l^{-\frac{1}{2}}, \dots\dots\dots\dots\dots\dots\dots\dots(2)$$

where E_1 denotes the energy corresponding to $l = 1$. From (2) we see that by withdrawing the ring B until l is infinitely great, the whole of the energy of vibration may be abstracted in the form of work done by B, and this by a uniform motion in which no regard is paid to the momentary phase of the vibration.

The argument is nearly the same for the case of a stretched string vibrating transversely in one plane. The string itself may be supposed to be unlimited, while the vibrations are confined by two rings of which one may be fixed and one movable.

If the origin of x be at one end of a string of length l, the transverse displacement [and velocity] may be expressed by

$$y = \phi_1 \frac{\sin \pi x}{l} + \phi_2 \sin \frac{2\pi x}{l} + \dots, \dots\dots\dots\dots\dots(3)$$

$$\dot{y} = \dot{\phi}_1 \frac{\sin \pi x}{l} + \dot{\phi}_2 \sin \frac{2\pi x}{l} + \dots, \dots\dots\dots\dots\dots(4)$$

where $\phi_1, \phi_2, \dots$ are coefficients depending upon the time. For the kinetic and potential energies we have respectively (*Theory of Sound*, § 128)

$$T = \tfrac{1}{4}\rho l \sum_{s=1}^{s=\infty} \dot{\phi}_s^2, \qquad V = \tfrac{1}{4}Wl \sum_{s=1}^{s=\infty} \frac{s^2 \pi^2}{l^2} \phi_s^2, \dots\dots\dots\dots(5)$$

in which W represents the constant tension and ρ the longitudinal density of the string. For each kind of ϕ the sums of T and V remain constant during the vibration; and the same is of course true of the totals given in (5).

From (3)

$$\frac{dy}{dx} = \frac{\pi}{l}\left(\phi_1 \cos \frac{\pi x}{l} + 2\phi_2 \cos \frac{2\pi x}{l} + \dots\right),$$

so that when $x = l$

$$\frac{dy}{dx} = \frac{\pi}{l}\left(-\phi_1 + 2\phi_2 - 3\phi_3 + \dots\right).$$

Accordingly the force tending to drive out the ring at $x = l$ is at time t

$$\tfrac{1}{2} W . \frac{\pi^2}{l^2} (-\phi_1 + 2\phi_2 - 3\phi_3 + \ldots)^2,$$

or in the mean taken over a long interval,

$$\tfrac{1}{2} W . \text{Mean} \, \Sigma \, \frac{s^2 \pi^2}{l^2} \, \phi_s^2.$$

Comparing with (5), we see that the mean force L has the value $2l \times$ mean V; or since mean $V =$ mean $T = \tfrac{1}{2} E$, E denoting the constant total energy,

$$L = E/l. \quad \ldots\ldots\ldots\ldots\ldots\ldots\ldots\ldots\ldots(6)$$

The force driving out the ring is thus numerically equal to the *longitudinal density of the energy*.

This result may readily be extended to cases where the vibrations are not limited to one plane; and indeed the case in which the plane of the string uniformly revolves is especially simple in that T and V are then constant with respect to time.

If the ring is allowed to move out slowly, we have

$$dE = - L \, dl = - E \, dl/l,$$

or on integration

$$E = E_1 l^{-1}, \quad \ldots\ldots\ldots\ldots\ldots\ldots\ldots\ldots(7)$$

analogous to (5), though different from it in the power of l involved. If l increase without limit, the whole energy of the vibrations may be abstracted in the form of work done on the ring.

We will now pass on to consider the case of air in a cylinder, vibrating in one dimension and supposed to obey Boyle's law according to which $p = a^2 \rho$. By the general hydrodynamical equation (*Theory of Sound*, § 253 a),

$$\varpi = \int \frac{dp}{\rho} = -\frac{d\phi}{dt} - \tfrac{1}{2} U^2, \quad \ldots\ldots\ldots\ldots\ldots\ldots(8)$$

where ϕ denotes the velocity-potential and U the resultant velocity at any point; so that in the present case, if we integrate over a long interval of time,

$$a^2 \int \log p \, dt + \tfrac{1}{2} \int U^2 \, dt \quad \ldots\ldots\ldots\ldots\ldots(9)$$

retains a constant value over the length of the cylinder. If p_0 denote the pressure that would prevail throughout, had there been no vibrations, $p - p_0$ is small and we may replace (9) by

$$a^2 \int \left\{ \frac{p - p_0}{p_0} - \tfrac{1}{2} \frac{(p - p_0)^2}{p_0^2} \right\} dt + \tfrac{1}{2} \int U^2 \, dt. \quad \ldots\ldots\ldots(10)$$

The expression (10) has accordingly the same value at the piston as for the mean of the whole column of length l. Now for the mean of the whole column

$$\int (p - p_0)\, dx = 0 ;$$

and thus if p_1 denote the value of p at the piston where $x = l$,

$$a^2 \int \left\{ \frac{p_1 - p_0}{p_0} - \tfrac{1}{2} \frac{(p_1 - p_0)^2}{p_0^2} \right\} dt = - \frac{a^2}{2l} \iint \frac{(p - p_0)^2}{p_0^2}\, dx\, dt + \frac{1}{2l} \iint U^2\, dx\, dt.$$

$$\ldots\ldots(11)$$

It is not difficult to prove that the right-hand member of (11) vanishes. Thus, expressing the motion in terms of ϕ, suppose that

$$\phi = \cos \frac{s\pi x}{l} \cos \frac{s\pi a t}{l}. \qquad\ldots\ldots\ldots\ldots(12)$$

Then

$$p - p_0 = \rho_0\, d\phi/dt, \quad U = d\phi/dx ;$$

and since $p_0 = a^2 \rho_0$, we get

$$\frac{1}{2l} \iint \left\{ \left(\frac{d\phi}{dx}\right)^2 - \frac{1}{a^2}\left(\frac{d\phi}{dt}\right)^2 \right\} dx\, dt,$$

and this vanishes by (12). Accordingly

$$\int (p_1 - p_0)\, dt = \int \frac{(p_1 - p_0)^2}{2p_0}\, dt. \qquad\ldots\ldots\ldots\ldots(13)$$

Again by (12)

$$\int \left(\frac{d\phi}{dt}\right)^2_l dt = \frac{2}{l} \iint \left(\frac{d\phi}{dt}\right)^2 dx\, dt,$$

so that

$$\int (p_1 - p_0)\, dt = \frac{1}{p_0 l} \iint (p_1 - p_0)^2\, dx\, dt = \frac{\rho_0}{l} \iint U^2\, dx\, dt.$$

Now $\rho_0 \iint U^2\, dx\, dt$ represents twice the mean total kinetic energy of the vibrations or, what is the same, the constant total energy E. Thus if L denote the mean additional force due to the vibrations and tending to push the piston out,

$$L = E l^{-1}. \qquad\ldots\ldots\ldots\ldots\ldots\ldots(14)$$

As in the case of the string, the total force is measured by the longitudinal density of the total energy; or, if we prefer so to express it, the additional *pressure* is measured by the volume-density of the energy.

In the last problem, as well as in that of the string, the vibrations are in one dimension. In the case of air there is no difficulty in the extension to two or three dimensions. Thus, if aerial vibrations be distributed equally in all directions, the pressure due to them coincides with *one-third* of the volume-density of the energy. In the case of the string, where the vibrations are transverse, we cannot find an analogue in three dimensions; but a membrane with a flexible and extensive boundary capable of slipping along the surface, provides for two dimensions. If the vibrations be equally distributed in the plane, the force outwards per unit length of contour will be measured by one-half of the superficial density of the total energy.

A more general treatment of the question may be effected by means of Lagrange's theory. If l be one of the coordinates fixing the configuration of a system, the corresponding equation is

$$\frac{d}{dt}\left(\frac{dT}{dl'}\right) - \frac{dT}{dl} + \frac{dV}{dl} = L, \quad \dots\dots\dots\dots\dots\dots(15)$$

where T and V denote as usual the expressions for the kinetic and potential energies. On integration over a time t_1

$$\int \frac{L\,dt}{t_1} = \frac{1}{t_1}\left[\frac{dT}{dl'}\right] + \frac{1}{t_1}\int\left(\frac{dV}{dl} - \frac{dT}{dl}\right) dt.$$

If dT/dl' remain finite throughout, and if the range of integration be sufficiently extended, the integrated term disappears, and we get

$$\int \frac{L\,dt}{t_1} = \frac{1}{t_1}\int\left(\frac{dV}{dl} - \frac{dT}{dl}\right) dt. \quad \dots\dots\dots\dots\dots\dots(16)$$

On the right hand of (16) the differentiations are partial, the coordinates other than l and all the velocities being supposed constant.

We will apply our equation (16) in the first place to the simple pendulum of fig. 1, l denoting the length of the vibrating portion of the string BC. If x, y be the horizontal and vertical coordinates of C,

$$x = l \sin \theta, \quad y = l - l \cos \theta;$$

and accordingly if the mass of C be taken to be unity,

$$T = \tfrac{1}{2} l'^2 (2 - 2 \cos \theta) + l'\theta' . l \sin \theta + \tfrac{1}{2}\theta'^2 l^2, \quad \dots\dots\dots(17)$$

l', θ' denoting dl/dt, $d\theta/dt$. Also

$$V = gl (1 - \cos \theta). \quad \dots\dots\dots\dots\dots\dots(18)$$

From (17), (18)

$$\frac{dV}{dl} = g (1 - \cos \theta), \quad \frac{dT}{dl} = l'\theta' \sin \theta + \theta'^2 l. \quad \dots\dots\dots\dots(19)$$

These expressions are general; but for our present purpose it will suffice if we suppose that l' is zero, that is that the ring is held at rest. Accordingly

$$\frac{dV}{dl} = \frac{V}{l}, \quad \frac{dT}{dl} = \frac{2T}{l},$$

and (16) gives

$$\int \frac{L\,dt}{t_1} = \frac{1}{t_1}\int \frac{V - 2T}{l}\, dt. \quad \dots\dots\dots\dots\dots(20)$$

On the right hand of (20) we find the mean values of V and of T. But these mean values are equal. In fact

$$\int V\,dt = \int T\,dt = \tfrac{1}{2} E t_1, \quad \dots\dots\dots\dots\dots\dots(21)$$

if E denote the total energy. Hence, if L now denote the mean value,

$$L = -\tfrac{1}{2}E/l, \quad \dots\dots\dots\dots\dots\dots(22)$$

the negative sign denoting that the mean force necessary to hold the ring at rest must be applied in the direction which tends to diminish l, i.e. downwards. In former equations (1), (6), (14), L had the reverse sign.

We will now consider more generally the case of one dimension, using a method that will apply equally whether for example the vibrating body be a stretched string, or a rod vibrating flexurally. All that we postulate is homogeneity of constitution, so that what can be said about any part of the length can be said equally about any other part. In applying Lagrange's method the coordinates are l the length of the vibrating portion, and ϕ_1, ϕ_2, &c. defining, as in (3), the displacement from equilibrium during the vibrations. As functions of l, we suppose that

$$V \propto l^m, \quad T \propto l^n. \quad \dots\dots\dots\dots\dots(23)$$

Thus, if L be the force corresponding to l, we get by (16)

$$\int \frac{L\,dt}{t_1} = \frac{1}{t_1}\int \left(\frac{mV}{l} - \frac{nT}{l}\right)dt,$$

in which

$$\int V dt = \int T dt = \tfrac{1}{2}E \cdot t_1,$$

E representing as before the constant total energy. Accordingly, L now representing the mean value,

$$L = \frac{(m-n)E}{2l}. \quad \dots\dots\dots\dots\dots(24)$$

In the case of a medium, like a stretched string, propagating waves of all lengths with the same velocity, $m = -1$, $n = 1$, and $L = -E/l$, as was found before.

In the application to a rod vibrating flexurally, $m = -3$, $n = 1$, so that

$$L = -2E/l. \quad \dots\dots\dots\dots\dots(25)$$

If $m = n$, L vanishes. This occurs in the case of the line of disconnected pendulums considered by Reynolds in illustration of the theory of the group velocity*, and the circumstance suggests that L represents the tendency of a group of waves to spread. This conjecture is easily verified. If in conformity with (13) we suppose that

$$V = V_0\, l^m\, \phi_1{}^2, \quad T = T_0\, l^n\, \dot{\phi}_1{}^2,$$

and also that

$$\phi_1 = \sin\frac{2\pi t}{\tau}, \quad \dot{\phi}_1 = \frac{2\pi}{\tau}\cos\frac{2\pi t}{\tau},$$

---

* See *Proc. Math. Soc.* IX. p. 21 (1877); this collection, I. p. 322. Also *Theory of Sound*, Vol. I. Appendix.

τ being the period of the vibration represented by the coordinate ϕ_1, we obtain, remembering that the sum of T and V must remain constant,

$$V_0 l^m = T_0 l^n . 4\pi/\tau^2.$$

This gives the relation between τ and l. Now v, the wave-velocity, is proportional to l/τ; so that

$$v \propto l^{1 - \frac{1}{2}n + \frac{1}{2}m}. \quad \dots\dots\dots\dots\dots\dots\dots\dots(26)$$

Thus, if u denote the group-velocity, we have by the general theory

$$u/v = \tfrac{1}{2}n - \tfrac{1}{2}m ; \quad \dots\dots\dots\dots\dots\dots\dots\dots(27)$$

and in terms of u and v by (24)

$$L = - \frac{uE}{vl} . \quad \dots\dots\dots\dots\dots\dots\dots\dots(28)$$

Boltzmann's theory is founded upon the application of Carnot's cycle to the radiation inclosed within movable reflecting walls. If the pressure (p) of a body be regarded as a function of the volume v*, and the absolute temperature θ, the general equation deduced from the second law of thermodynamics is

$$\frac{dp}{d \log \theta} = M, \quad \dots\dots\dots\dots\dots\dots\dots\dots(29)$$

where $M dv$ represents the heat that must be communicated while the volume alters by dv and $d\theta = 0$. In the application of (29) to radiation we have evidently

$$M = U + p, \quad \dots\dots\dots\dots\dots\dots\dots\dots(30)$$

where U denotes the density of the energy—a function of θ only. Hence†

$$\frac{dp}{d \log \theta} = U + p. \quad \dots\dots\dots\dots\dots\dots\dots\dots(31)$$

If further, as for radiation and for aerial vibrations,

$$p = \tfrac{1}{3}U, \quad \dots\dots\dots\dots\dots\dots\dots\dots(32)$$

it follows at once that

$$d \log U = 4 d \log \theta,$$

whence

$$U \propto \theta^4, \quad \dots\dots\dots\dots\dots\dots\dots\dots(33)$$

the well-known law of Stefan. It may be observed that the existence of a pressure is demanded by (31), independently of (32).

If we generalize (32) by taking

$$p = \frac{1}{n} U, \quad \dots\dots\dots\dots\dots\dots\dots\dots(34)$$

where n is some numerical quantity, we obtain as the generalization of (33)

$$U \propto \theta^{n+1}. \quad \dots\dots\dots\dots\dots\dots\dots\dots(35)$$

---

* Now with an altered meaning.

† Compare Lorentz, *Amsterdam Proceedings*, Ap. 1901.

It is an interesting question whether any analogue of the second law of thermodynamics can be found in the general theory of the pressure of vibrations, whether for example the energy of the vibrations of a stretched string is partially unavailable in the absence of appliances for distinguishing *phases*. It might appear at first sight that the conclusion already given, as to the possibility of recovering the whole energy by mere retreat of the inclosing ring, was a proof to the contrary. This argument, however, will not appear conclusive, if we remember that a like proposition is true for the energy of a gas confined adiabatically under a piston. The residual energy of the molecules may be made as small as we please, but the completion of the cycle by pushing the piston back will restore the molecular energy unless we can first abolish the infinitesimal residue remaining after expansion, and this can only be done with the aid of a body at the absolute zero of temperature. It would appear that we may find an analogue for temperature, so far as the vibrations of *one* system are concerned; but, so far as I can see, the analogy breaks down when we attempt a general theory.

[1910. See further *Phil. Mag.* Sept. 1905 "On the Momentum and Pressure of Gaseous Vibrations, and on the Connection with the Virial Theorem."]

277.

ON THE QUESTION OF HYDROGEN IN THE ATMOSPHERE.

[*Philosophical Magazine*, III. pp. 416—422, *April* 1902.]

IT will be remembered that M. Armand Gautier, as the result of very elaborate investigations, was led to the conclusion that air, even from the Atlantic, contains by volume nearly two parts in 10,000 of free hydrogen. The presence of so much hydrogen, nearly two-thirds of the carbonic acid which plays such an important part, is of interest in connexion with theories pointing to the escape of light constituents from the planetary atmospheres. Besides the free hydrogen, M. Gautier found in the air of woods and towns considerable quantities of hydrocarbons yielding CO_2 when led over hot copper oxide.

Spectroscopic Evidence.

In the *Philosophical Magazine* for Jan. 1901[*], I described some observations upon the spectrum of sparks taken in dried air at atmospheric pressure, which seemed "to leave a minimum of room for the hydrogen found by M. Gautier." Subsequently (April 1901), these experiments were repeated with confirmatory results. The spectra, taken from platinum points, of pure country air and the same to which $\frac{2}{10,000}$ of hydrogen had been added were certainly and easily distinguished by the visibility of the C-line. An improvement was afterwards effected by the substitution of aluminium points for platinum. A strong preliminary heating reduced the C-line with a stream of pure dried air to the least yet seen, only just continuously visible, and contrasting strongly with the result of substituting the air to which the two parts in 10,000 of hydrogen had been added.

To air from outside one thousandth part of hydrogen was introduced and allowed time to mix thoroughly. Excess of chlorine was then added, and

[*] [This Collection, Vol. IV. p. 496.]

after a while the whole was exposed to strong sunshine, after which the superfluous chlorine was removed by alkali. Tested in the spectroscope, this sample showed only about the same signs of hydrogen as the pure air, indicating that the added hydrogen had effectively been removed—a result which somewhat surprised me.

As there now appeared to be a margin for further discrimination, three samples were prepared, the first pure air, the second air to which was added $\frac{1}{10,000}$ of hydrogen, and the third air with addition of $\frac{2}{10,000}$ of hydrogen. In the spectroscope the three were just certainly distinguishable, showing C in the right order. The chlorine-treated mixture showed about the same as the pure air. On repetition with fresh samples these results were confirmed.

In my former note I mentioned that nitrous oxide and oxygen showed the C-line as much as, if not more than, air. I cannot say whether this result is inevitable, but the gases were prepared with ordinary care. In the more recent repetition N_2O showed the C-line about the same as the air to which $\frac{1}{10,000}$ H_2 had been added. Oxygen, prepared from permanganate, showed C *much* more than does pure air; and there was not much change when oxygen from mixed chlorates of potash and soda was substituted.

The impurity in the oxygen, if it be an impurity*, does not appear to be easily removed. The visibility of C was not perceptibly diminished by passage of either kind of oxygen over hot copper oxide. On the other hand the air containing $\frac{2}{10,000}$ of added hydrogen was reduced by the same treatment to equality with pure air. Possibly the impurity is a hydrocarbon not readily burnt.

Neither by treatment with chlorine could oxygen from either source be freed from the property of exhibiting the C-line.

The spectroscopic evidence here set forth is certainly far from suggesting that air, previously to any addition, already contains two parts in 10,000 of free hydrogen. The passage from two to three parts in 10,000 might possibly produce the observed change of visibility which followed the introduction of one ten-thousandth of hydrogen; and the behaviour with chlorine and hot copper oxide is not absolutely inconsistent with the initial hydrogen. But the reconciliation seems to involve coincidences of little *à priori* probability.

---

* It is possible that traces of hydrogen, derived from the electrodes or from the glass, show more in oxygen than in air.

Determinations by Combustion.

In M. Gautier's experiments large volumes of dried air were passed through tubes containing copper oxide heated in a specially constructed furnace, the water formed being collected in suitable phosphoric tubes and accurately weighed. In unsystematic experiments the source of the water so collected might be doubtful, but it is explained that the apparatus was tested with pure dry *oxygen*, and that under these conditions the phosphoric tube showed no increase of weight exceeding ·1 mg. The work was evidently very careful and thorough; and the impression left upon the mind of the reader is that the case is completely made out. Indeed, had I been acquainted with the details, as set forth in *Ann. d. Chimie*, t. XXII., Jan. 1901, at an earlier stage, I should probably have attempted no experiments of my own. It so happens, however, that I had already begun some work, which has since been further extended, and which has yielded results that I find rather embarrassing, and am even tempted to suppress. For the conclusion to which these determinations would lead me is that the hydrogen in country air is but a small fraction, perhaps not more than one eighth part, of that given by M. Gautier. Although I am well aware that my experience in these matters is much inferior to his, and that I may be in error, I think it proper that some record should be made of the experiments, which were carefully conducted with the assistance of Mr Gordon and many times repeated.

The quantity of air upon which I operated was almost uniformly 10 litres much less than was used by M. Gautier. A glass aspirating bottle, originally filled with water, was discharged upon the lawn, so that the water was replaced by fresh *country* air. During an experiment the air was driven forward, at the rate of about 1¼ litres per hour, by water entering below. After traversing a bubbler charged with alkali, it was desiccated first by passing over a surface of sulphuric acid, and subsequently by phosphoric anhydride. Next followed the hot copper oxide, contained in a hard glass tube and heated by an ordinary combustion-furnace. Next in order followed the U-tube charged with phosphoric anhydride whose increase of weight was to indicate the absorption of water, formed in or derived from the furnace-tube. The U-tube was protected upon the further (down-stream) side by other phosphoric tubes. It was provided with glass taps and was connected on either side by short pieces of thick rubber of which but little was exposed to the passing air. The counterpoise in the balance was a similar closed phosphoric tube of very nearly the same volume, and allowance was made for the pressure and temperature of the air included in the working-tube at the moment when the taps were closed.

Two parts in 10,000 of free hydrogen, *i.e.* 2 c.c., yield on combustion the same volume of water-vapour, and of this the weight would be 1·5 mg. to be collected in the phosphoric tube. Any water, due to hydrocarbons originally present in the air and oxidized in the furnace-tube, would be additional to the above 1·5 mg.

The earlier experiments, executed at the end of 1900 and beginning of 1901, gave results which I found it difficult to interpret. The gain of weight from the passage of 10 litres of fresh air was about ·4 mg., that is, far too little; and, what was even more surprising, this gain was not diminished when the air after passage was collected and used over and over again. The gain appeared to have nothing to do with hydrogen originally present in the air, being maintained, for example, when a single litre of air was passed round and round eight or nine times. Neither did the substitution of oxygen for air make any important difference. Subsequently it was found that the gain was scarcely diminished when the furnace remained cold during the passage of the air.

Warned by M. Gautier, I was prepared for a possible gain of weight due to retention of oxygen; but this gain ought to be additional, and should not mask the difference between air containing and not containing free hydrogen. Faulty manipulation might be expected to entail an excessive rather than a defective gain; and the only cause to which I could attribute the non-appearance of the hydrogen was a failure of the copper oxide to do its work. The sample which I had employed was of the kind sold as granulated. M. Gautier himself found a considerable length of copper oxide necessary to complete the action. The question is, of course, not one of length merely, but rather of the *time* during which the travelling gas remains in close proximity to the oxidizing agent. Taking into account the slower rate of passage in litres per hour, it would seem that my arrangement had the advantage in this respect.

An attempt was made to improve the phosphoric anhydride by a preliminary heating for many hours to 260° in a current of dry air, somewhat as recommended by M. Gautier, but the results were not appreciably altered. When one remembers the experiments of Baker, from which it appears that, if all is thoroughly dry, heated phosphorus does not combine with oxygen, it is difficult to feel confidence in this process.

It was certain that the sample of phosphoric anhydride hitherto employed in this work was inferior. When a tube of it which had been used for some time (in other work) and had become gummy at the ends, was strongly heated over a spirit flame, occasional flashes could be seen in the dark. Treatment in the cold with *ozonized* air seemed to effect an improvement; but experiments in this direction were not pursued to a definite conclusion in consequence of the discovery that when another sample of phosphoric anhydride was substituted for that hitherto in use the anomaly disappeared. Thus in four trials where 10 litres of air were passed without a furnace, the gains were:

Nov. 26, 1901	− ·00016
„ 29, „	− ·00003
„ 30, „	+ ·00014
Dec. 2, „	+ ·00006
Mean............	·00000

thus on the whole no gain of weight. The errors would appear somewhat to exceed ·1 mg., but it may be noted that in this and following tables the real error is liable to appear exaggerated. If in consequence of an error of weighing, or of allowance for weight of included air, or of a varied condition of the outer surface of the tube, a recorded gain is too high, the next is likely to appear too low.

In the operations which followed, the furnace-tube was charged with copper oxide prepared *in situ* by oxidation of small pieces of thin copper foil with which the tube was packed. Examination once or twice after a breakage showed that the oxidation was not complete, but no measurements were taken until there was no appreciable further absorption of oxygen. After the copper oxide has been exposed to the air of the room, many hours' heating to redness in a current of dry air are required to remove the adherent moisture.

The results, referring in each case to 10 litres of fresh *country* air, are as follows, the weights being in gms.:

Dec. 6, 1901	+ ·00023
„ 9, „	+ ·00042
„ 11, „	+ ·00010
„ 13, „	+ ·00014
„ 14, „	+ ·00025
„ 15, „	− ·00001
„ 17, „	+ ·00031
„ 18, „	+ ·00028
„ 20, „	+ ·00025
„ 24, „	+ ·00016
Mean............	+ ·00021

It will be seen that the mean water collected is only about one seventh of that corresponding to the complete combustion of the hydrogen, according to M. Gautier's estimate of the amount.

As has already been suggested, a defective gain of weight can hardly be explained by faulty manipulation. The important question is as to the efficiency of the copper oxide. Did my furnace-tube allow the main part of the free hydrogen to pass unburnt? The question is one that can hardly be answered directly, but I may say that variations of temperature (within moderate limits) did not influence the result.

What it is possible to examine satisfactorily is the effect of small additions of hydrogen to the air as collected. In my later experiments the added hydrogen was only 1 c.c., that is, $\frac{1}{10,000}$ by volume, or half the quantity originally present according to M. Gautier. The hydrogen was first diluted in a gas pipette with about 100 c.c. of air and allowed time to diffuse. The 10 litre aspirating-bottle being initially full of water, the diluted hydrogen was introduced at the top, and was followed by 10 litres of air from the open, after which the mixture stood over night, precautions which had been found sufficient to ensure a complete mixture in the spectroscopic work. The results were:

Dec. 19, 1901	+ ·00091
„ 23, „	+ ·00084
„ 27, „	+ ·00103
Mean............	+ ·00093

The additional gain is thus ·00072, very nearly the full amount (·00075) corresponding to the 1 c.c. of added hydrogen. We may say then that the copper oxide was competent to account for a small *addition* of hydrogen to air.

Following a suggestion from the spectroscopic experiments, I have examined the effect of treatment with *chlorine*. To 10 litres of air 10 c.c. (in one case 5 c.c.) of hydrogen were added, followed by excess of chlorine and exposure to sunshine. The excess of chlorine having been removed by alkali, the air was desiccated and passed through the furnace as usual. The resulting gains of weight in the phosphoric tube were:

Dec. 31, 1901	+ ·00011
Jan. 2, 1902	+ ·00050
„ 6, „	+ ·00044
„ 8, „	+ ·00030
Mean............	+ ·00034

According to these figures, the treatment with chlorine after hydrogen left the gains somewhat larger than in the case of pure air; but the sunshine, especially on Jan. 2, was feeble, and the difference of ·13 mg. may be the consequence of incomplete insolation. In any case the added hydrogen was *very nearly* removed by the chlorine.

In two experiments 10 litres of pure air were subjected to the chlorine treatment, and after removal of excess were mixed with 1 c.c. of hydrogen. The gains of weight recorded in the phosphoric tubes were:

Jan. 1, 1902	+ ·00082
„ 7, „	+ ·00069
Mean............	+ ·00075

The difference between this and ·00093 would suggest that perhaps some original hydrogen had been removed by chlorine, but the amount is very small.

I shall best fulfil my intention if I refrain from attempting to sum up the whole of the evidence. So far as my own work is concerned, the natural inference from it would be that the free hydrogen in country air does not exceed $\frac{1}{30,000}$ of the volume. If I may make a suggestion, it would be in favour of working with some such quantity as 10 litres, collected in glass bottles, and of comparisons between fresh air and air already passed once or twice through the furnace-tube.

278.

DOES CHEMICAL TRANSFORMATION INFLUENCE WEIGHT?

[*Nature*, LXVI. pp. 58, 59, 1902.]

In *Nature* (Vol. LXIV. p. 181, 1901)* I directed attention to experiments by Heydweiller (*Drude Ann.* Vol. V. p. 394) from which he inferred that some chemical transformations, such as the solution of copper sulphate in water, were attended by real, though minute, changes of weight, and I pointed out certain difficulties involved in the acceptance of this statement. In connexion with another subject, it has lately occurred to me that such changes of weight would really be in opposition to the laws of thermodynamics, and I propose now briefly to sketch the argument from which this opposition appears.

It is known† that by suitable arrangements the dissolution of salt may be effected reversibly at a given temperature. During the process, a certain amount of work is gained and a certain amount of heat at the given temperature has to be supplied. In the reverse process, of course, an equal amount of work has to be performed and an equal amount of heat is recovered. The temperature being given, these operations are not affected (it is assumed) by the height above the earth's surface at which they may be supposed to take place.

Conceive now that the temperature is uniform throughout and that the materials are initially at a low level and in one state (A). Let them be raised to a high level and there be transformed into the other state (B). Subsequently let them be brought down to the low level and transformed back into state A. The reverse transformations above and below compensate one another thermodynamically, and *if the weights are the same in the two states*, so do the operations of raising and lowering. But if the weights in states A and B are different, the cycle of operations may be so executed that *work is gained*. Such a difference of weight is therefore excluded, unless,

---

* [Vol. IV. p. 549.]

† "On the Dissipation of Energy," *Nature*, XI. p. 454, 1875; *Scientific Papers*, Vol. I. p. 238.

indeed, hitherto unsuspected thermal effects accompany a rising or falling against or with gravity. It is scarcely necessary to say that we are not here concerned with the differences of temperature and pressure which may actually be met with at different levels over the earth's surface.

There are many chemical transformations which cannot easily be supposed to take place reversibly. But this, though it might complicate the statement, does not affect the essence of the argument; and the conclusion appears to be general.

If the reasoning here put forward be accepted, it increases the difficulty of admitting the reality of such changes of weight as have been suspected, and it justifies a severe criticism of experimental arrangements. In my former letter I pointed out a possible source of error.

It is to be hoped that the matter may soon be cleared up, for it is scarcely creditable to science that doubt should hang over such a fundamental question. But for my own part I would wish to say that I fully recognize how much easier it is to criticize than to experiment.

279.

IS ROTATORY POLARIZATION INFLUENCED BY THE EARTH'S MOTION?

[Philosophical Magazine, IV. pp. 215—220, 1902.]

THE question whether the rotation of the plane of polarization of light propagated along the axis of a quartz crystal is affected by the direction of this axis relatively to that of the earth's orbital motion, is of considerable theoretical importance. According to an investigation of Lorentz, an effect of the first order might be looked for. Such an effect would be rendered apparent by comparing the rotations when the direction of propagation of the light is parallel to that of the earth's motion and in the reverse direction, and it might amount to $\frac{1}{10,000}$ of the whole rotation*. According to Larmor's theory† there should be no effect of the first order.

The question was examined experimentally many years ago by Mascart‡, who came to the conclusion that the reversal of the ray left the rotation unchanged to $\frac{1}{20,000}$ part. In most of the experiments, however, the accuracy was insufficient to lend support to the above conclusion.

Dr Larmor (*l. c.* p. 220) having expressed the opinion that it might be desirable to re-examine the question, I have made some observations which carry the test as far as can readily be done. It appears that the rotation is certainly not altered by $\frac{1}{100,000}$ part, and probably not by the half of this, when the direction of propagation of the light is altered from that of the earth's motion to the opposite direction.

I should scarcely have been able to carry the test to so satisfactory a point, had it not been for the kindness of Prof. MacGregor, who allowed me the use of certain valuable quartz crystals belonging to the Edinburgh collection of apparatus. These crystals, five in number, are all right-handed, and measure about 50 mm. each in the direction of the optical axis, to which the polished faces are approximately perpendicular. They were prepared for

---

* This fraction representing approximately the ratio of the velocity of the earth in its orbit to the velocity of light.

† *Æther and Matter*, Cambridge, 1900.

‡ *Annales de l'École Normale*, Vol. I. p. 157 (1872).

Prof. Tait, and were employed by him for his "rotatory polarization spectro-scope of great dispersion *." For the most part they are nearly free from blemish, and well adapted to the purpose in view.

In principle the experiment is very simple, scarcely differing from ordinary polarimetry, as, for example, in determining the rotation due to sugar and other active bodies. But the apparatus needs to be specially mounted upon a long stiff board, itself supported upon a point, so that the absolute direction of the light may be reversed without danger of even the slightest relative displacement of the parts. The board swings round in the horizontal plane; and if its length is directed from east to west, or from west to east, observa-tions taken at noon (in June) correspond pretty accurately to propagation of the light with or against the earth's motion in its orbit. Similar com-parisons at 6 o'clock are nearly independent of the earth's motion.

In another respect the experiment is peculiar on account of the enormous amount of the rotation to be dealt with. For sodium light the rotation is $22°$ per millimetre of quartz, so that the whole rotation is $5500°$, or more than 15 complete revolutions. In the preliminary experiments, with one of the crystals only, sodium light was employed; but the observations were unsatisfactory, even although the light was resolved into a spectrum. If the flame was well supplied with salt, the extinction of the D-line by suitable adjustment of the nicol still left the *neighbouring* region of the spectrum so bright as to prejudice the observation by lessening the sensitiveness of the eye. This effect, which is quite distinct from what is ordinarily called the broadening of the D-lines and can be made still more pronounced by stimulating the flame with oxygen, does not appear to present itself in any other method of observation, and is of interest in connexion with the theory of luminous emission. A very moderate rotation of the nicol revives the D-lines sufficiently to extinguish the neighbouring spectrum, just as the first glimpse of the limb of the sun after a total eclipse extinguishes the corona †.

When all five quartzes were brought into use it was hopeless to expect good results from a soda-flame. From the fact that the rotation is as λ^{-2} we see that there must be $11°$ difference of rotation for the two D-lines, so that a satisfactory extinction is out of the question. For the observations about to be recorded a so-called vacuum-tube, charged with *helium*, was employed, the yellow line (situated close to the D-lines) being chosen. It was actuated by a Ruhmkorff coil and four Grove cells, situated at some distance away.

* *Nature*, Vol. xxii. 1880; Tait's *Scientific Papers*, Vol. i. p. 423.

† *July* 6.—A doubt having suggested itself as to whether this effect might not be due to an actual whitening of the Bunsen flame, such as sometimes occurs rather unexpectedly, the experiment was repeated with a flame of pure *hydrogen*. The region of the spectrum in the neighbourhood of D was even brighter than before. An attempt to produce an analogous effect with *lithium* was a failure, apparently in consequence of insufficient brightness of the flame.

The various parts, all mounted upon the pivoted board, will now be specified in order. First came the helium tube with capillary vertical, then at a distance of 25 cm. a collimating spectacle-lens, followed by the polarizing nicol. The field of view presented by this nicol was contracted to a circular aperture 7 mm. in diameter, and was further divided into two parts by a "sugar-cell." This cell was the same as that formerly used in a cognate research on the rotation of the plane of polarization in bisulphide of carbon under magnetic force*. "The polarimeter employed is on the principle of Laurent, but according to a suggestion of Poynting (*Phil. Mag.* July 1880) the half-wave plate of quartz is replaced by a cell containing syrup, so arranged that the two halves of the field of view are subjected to small rotations differing by about 2°. The difference of thickness necessary is best obtained by introducing into the cell a piece of thick glass, the upper edge of which divides the field into two parts. The upper half of the field is then rotated by a thickness of syrup equal to the entire width of the cell (say ½ inch), but in the lower half of the field part of the thickness of syrup is replaced by glass, and the rotation is correspondingly less. With a pretty strong syrup a difference of 2° may be obtained with a glass $\frac{3}{16}$ inch [inch = 2·54 cm.] thick. For the best results the operating boundary should be a true plane perpendicular to the face. The pieces used by me, however, were not worked, being simply cut with a diamond from thick plate glass; and there was usually no difficulty in finding a part of the edge sufficiently flat for the purpose, *i.e.* capable of exhibiting a field of view sharply divided into two parts....By this use of sugar, half-shade polarimeters may be made of large dimensions at short notice and at very little cost. The syrup should be filtered (hot) through paper, and the cell must be closed to prevent evaporation."

The light next traversed the quartz crystals, each mounted upon a small stand admitting of adjustment in azimuth and level so as to bring the optical axis into parallelism with the line of vision. The analysing nicol, mounted near the end of the board, was distant 102 cm. from the polarizer. After passing the nicol the light traversed in succession a direct-vision prism of sufficient aperture and a small opera-glass focussed upon the sugar-cell. The aperture limiting the field had been so chosen that, as seen through the spectroscope, the yellow image under observation was sufficiently separated from the neighbouring red and green images corresponding to other spectral lines of helium. The position of the analysing nicol was read with a vernier to tenths of a degree—an accuracy which just sufficed, and the setting could be made by causing the two halves of the field of view afforded by the sugar-cell to appear *equally* dark.

A good deal of time was spent in preliminary experiment before the best

---

* *Phil. Trans.* CLXXVI. p. 343 (1885); *Scientific Papers*, Vol. II. p. 363.

procedure was hit upon. It is necessary that the optic axes of the crystals be adjusted pretty accurately to the line of vision, and this in several cases involved considerable obliquity of the terminal faces. In these adjustments the sugar-cell and its diaphragm are best dispensed with, the crystals being turned until the rotation required to darken the field is a minimum and the darkness itself satisfactory. When the first crystal has been adjusted, a second is introduced and adjusted in its turn, and so on. In some cases a further shift of the crystal parallel to itself was required in order to remove an imperfection from the part of the field to be utilized. In the end a fairly satisfactory darkness was attained, but decidedly inferior to that obtainable when the quartzes were removed. Part of the residual light may have been due to want of adjustment; but more seemed to originate in imperfections in the quartzes themselves.

In my former experiments upon bisulphide of carbon advantage was found from a device for rocking the plane of polarization through a small constant angle*. During the observations now under discussion this effect was obtained by the introduction of a second sugar-cell, not divided into two parts or seen in focus, just in front of the analysing nicol. The cell was mounted so that it could slide horizontally in and out up to fixed stops. The thickness of the cell being sufficient, the strength of the syrup was adjusted to the desired point. Thus when the nicol was correctly set, the upper half of the field was *just distinctly* the brighter when the cell was *in*, and the lower half with *equal distinctness* the brighter when the cell was *out*, the object to be aimed at in the setting of the nicol being the *equality* of these small *differences*. For the results now to be given the setting of the nicol was by myself and the reading of the vernier was by Mr Gordon. A second observer is a distinct advantage.

As a specimen, chosen at random, I will give in full all the readings made in the neighbourhood of noon on June 19. Five readings were taken in each position and then the board was reversed. The headings "East" and "West" indicate the end at which the observer was sitting; "East" therefore meaning that the course of the light was from West to East.

The mean of the three "Easts" is 45·75, and of the two "Wests" is 45·71; so that

$$E - W = + ·04°.$$

All these numbers are in decimals of a degree. The progressive alteration in the readings corresponds to the rise of temperature. It would appear that, as was natural, the quartzes lagged somewhat behind the thermometer.

* *Loc. cit.*; *Scientific Papers*, Vol. II. p. 366.

TABLE I.

Time 11h 30m Temp. 17°·4 East	Time 11h 50m Temp. 17°·7 West	Time 12h 5m Temp. 17°·9 East	Time 12h 15m Temp. 17°·9 West	Time 12h 25m Temp. 17°·9 East
45·7	45·4	45·6	45·9	46·0
45·5	45·9	45·8	45·7	46·1
45·5	45·4	45·5	45·9	46·1
45·6	45·7	45·6	45·7	46·0
45·6	45·7	45·7	45·8	46·0
45·58	45·62	45·64	45·80	46·04

TABLE II.—Noon.

Date	E.—W.
June 17.........	+ ·03
„ 18.........	− ·05
„ 19.........	+ ·04
Mean......	+ ·007

Three sets of observations were taken at noon, and the results are recorded in Table II. In two other sets taken about 6h the differences $E - W$ were even less. The comparison of the two hours serves to check possible errors, *e.g.* of a magnetic character, such as might be caused by the magnetism of the Ruhmkorff coil, if insufficiently distant.

It seems certain that at neither hour does the difference $E - W$ actually amount to $\frac{1}{20}$ of a degree, *i.e.* to $\frac{1}{100,000}$ of the whole rotation. In all probability the influence of the reversal is much less, if indeed it exists at all.

P.S. Since the above observations were made, I see from the *Amsterdam Proceedings* (May 28, 1902) that Lorentz maintains his opinion against the criticism of Larmor. Lorentz's theoretical result contains an unknown quantity which might be adjusted so as to make the influence of the earth's motion evanescent; but for this special adjustment there appears to be no theoretical reason. I hope that the above experimental demonstration of the absence of effect, to a high order of accuracy, will be found all the more interesting.

280.

DOES MOTION THROUGH THE ÆTHER CAUSE DOUBLE REFRACTION?

[*Philosophical Magazine*, IV. pp. 678—683, 1902*.]

THE well-known negative result of the Michelson-Morley experiment in which interference takes place between two rays, one travelling to and fro in the direction of the earth's motion, and the other to and fro in a perpendicular direction, is most naturally interpreted as proving that the æther in the laboratory shares the earth's motion. But other phenomena, especially stellar aberration, favour the opposite theory of a stationary æther. The difficulty thus arising has been met by the at first sight startling hypothesis of FitzGerald and Lorentz that solid bodies, such as the stone platform of Michelson's apparatus, alter their relative dimensions, when rotated, in such a way as to compensate the optical change that might naturally be looked for. Larmor (*Æther and Matter*, Cambridge, 1900) has shown that a good case may be made out for this view.

It occurred to me that such a deformation of matter when moving through the æther might be accompanied by a sensible double refraction; and as the beginning of double refraction can be tested with extraordinary delicacy, I thought that even a small chance of arriving at a positive result justified a careful experiment. Whether the result were positive or negative, it might at least afford further guidance for speculation upon this important and delicate subject.

So far as liquids are concerned, the experiment is of no great difficulty, and the conclusion may be stated that there is no double refraction of the order to be expected, that is comparable with 10^{-8} of the single refraction†. But the question arises whether experiments upon liquids really settle the matter. Probably no complete answer can be given, unless in the light of

---

* Read before Section A of the British Association at Belfast.

† $10^{-8} = (10^{-4})^2$, where 10^{-4} is the ratio of the velocity of the earth in its orbit to the velocity of light.

some particular theory of these relations. But it may be remarked that the liquid condition is no obstacle to the development of double refraction under electric stress, as is shown in Dr Kerr's experiments.

The apparatus was mounted upon the same revolving board as was employed for somewhat analogous experiments upon the rotation in quartz (*Phil. Mag.* Vol. iv. p. 215, 1902)*. Light, at first from the electric arc but later and preferably from lime heated by an oxyhydrogen jet, after passing a spectacle-lens so held as to form an image of the source upon the analysing nicol, was polarized by the first nicol in a plane inclined to the horizontal at 45°. The liquid, held in a horizontal tube closed at the ends by plates of thin glass, was placed, of course, between the nicols. When at 12 o'clock the board stands north and south, the earth's motion is transverse and the situation is such as to exhibit any double refraction which may ensue. It might be supposed, for instance, that luminous vibrations parallel to the earth's motion, *i.e.* east and west, are propagated a little differently from those whose direction is transverse to the earth's motion, *i.e.* vertical. But if the board be turned through a right angle so as to point east and west, both directions of vibration for light passing the tube are transverse to the earth's motion, and therefore no double refraction could manifest itself. The question is whether turning the board from the north and south position to the east and west position makes any difference. In no case is any effect to be expected from a rotation through 180°, and such effect as a rotation through 90° may entail must be of the *second* order in the ratio which expresses the velocity of the earth relatively to that of light.

It should not be overlooked that according to the theory of a stationary æther, we have to do not only with the motion of the earth in its orbit, but also with that of the sun in space. The latter is supposed to be much the smaller, and to be directed towards the constellation Hercules. In the month of April, when successful experiments were first made, the two motions would approximately conspire.

If the suggested double refraction, due to the earth's motion, were large enough, it would suffice to set the analysing nicol to extinction in one position of the board, and to observe the revival of light consequent on a rotation of the latter through 90°. But a more delicate method is possible and necessary. Between the polarizing nicol and the liquid column we introduce a strip of glass whose length is horizontal and transverse to the board. This strip, being supported (at two points) near the middle of its length, and being somewhat loaded at its ends, is in a condition of strain, and causes a revival of light except in the neighbourhood of a horizontal band along the "neutral axis." Above and below this band the strained

* [*Supra* p. 58.]

condition of the glass produces just such a double refraction as might be caused by the motion of the liquid through the æther, so that the existence of the latter would be evidenced by a displacement of the dark band upwards or downwards. In order the better to observe a displacement, two horizontal wires are disposed close to the bent glass so as just to inclose the band, and a small opera-glass focussed upon these is introduced beyond the analysing nicol. The slightest motion of the band is rendered evident by changes in the feeble illumination just inside the wires.

The board is mounted upon a point so as to revolve with the utmost freedom. The point is carried on the table and faces upwards. The bearing is a small depression in an iron strap, rigidly attached to the board, and raised sufficiently to give stability. The gas-leading tubes are connected in such a manner as to give rise to no forces which could appreciably vary as the board turns.

Observations were made upon bisulphide of carbon in a tube 76 cms. long, and upon water in a tube $73\frac{1}{2}$ cms. long. In neither case could the slightest shift of the band be seen on rotation of the board from the north-south position to the east-west position, whether at noon or at 6 P.M. The time required to pass from one observation to the other did not exceed 15 seconds, and the alternate observations were repeated until it was quite certain that nothing could be detected.

Of course the significance of this result depends entirely upon the delicacy of the apparatus, and it is worth little without an estimate of the smallest double refraction that would have been detected. It may even be objected that the investigation stands self-condemned. In consequence of the earth's magnetism there must be a rotation of the plane of polarization when the light traverses the bisulphide of carbon in the north and south position; and this effect, it may be argued, ought to manifest itself upon rotation of the board.

To take the objection first, it is easy to calculate the rotation of the plane of polarization. For one C.G.S. unit of magnetic potential the rotation in CS_2 at 18° is ·042 minute of angle[*]. In the present case the length is 76 cms. and the earth's horizontal force is ·18; so that the whole rotation to be expected[†] is

$$76 \times ·18 \times ·042 = ·58'.$$

So small a rotation of the plane, which would show itself, if at all, by a *fading* and not by a *displacement* of the band, is below the limit of observation.

The delicacy of the apparatus for its purpose may, indeed, be inferred indirectly from the rotation of the nicol found necessary to engender a

---

[*] *Phil. Trans.* CLXXVI. p. 343 (1885); *Scientific Papers*, Vol. II. p. 377.

[†] The difference between astronomical and magnetic north is here neglected.

marked revival of light at the darkest part of the band. If θ be this angle, the revived light is $\sin^2 \theta$, expressed as a fraction of the maximum obtainable with parallel nicols. In the actual observation the nicols remain accurately crossed, and the question is as to the effect of a double refraction causing *e.g.* a retardation of vertical vibrations relatively to horizontal ones. If this retardation amounted to $\frac{1}{2}\lambda$, λ being the wave-length, the effect would be the same as of a rotation of the nicol through 90°. In general, a retardation of phase ϵ, in place of π, gives a revival of light measured by $\sin^2 (\frac{1}{2}\epsilon)$. If the revivals of light in the two cases be the same, we may equate θ to $\frac{1}{2}\epsilon$. Hence if we find that rotation θ produces a sensible effect in lessening the darkness at the darkest place, we may infer that there is delicacy sufficient to detect a relative retardation of 2θ due to double refraction. This comparison would apply if the test for double refraction were made by simple observation of the revival of light. As actually carried out by location of the band, the test must be many times more delicate.

It was found that a marked fading of the band attended a rotation of the nicol through 4′. According to this ϵ would be $\frac{1}{450}$; or since a retardation of $\frac{1}{2}\lambda$ corresponds to $\epsilon = \pi$, a retardation amounting to $\frac{1}{1400} \times \frac{1}{2}\lambda$ should be perceptible many times over, regard being paid to the superior delicacy of the method in which a band is displaced relatively to fixed marks.

Another and perhaps more satisfactory method of determining the sensitiveness was by introducing a thin upright strip of glass which could be compressed in the direction of its length by small loads. These loads were applied symmetrically in such a manner as to cause no flexure. The double refraction due to the loads is of exactly the character to be tested for, and accordingly this method affords a very direct check. If the load be given, the effect is independent of the length of the strip and of its thickness along the line of vision, but is inversely as the width. The strip actually employed had a width of 15 mm.; and the application (or removal) of a total of 50 gms. caused a marked shifting of the band, while 25 gms. was just perceptible with certainty.

To interpret this we may employ some results of Wertheim (Mascart's *Traité d'Optique*, t. II. p. 232), who found that it requires a load of 10 kilograms per millimetre of width to give a relative retardation of $\frac{1}{2}\lambda$, so that with the actual strip the load would need to be 150 kilograms. The retardation just perceptible is accordingly $\frac{1}{2}\lambda \div 6000$. This may be considered to agree well with what was expected from the effect of rotating the nicol.

We have now only to compare the relative retardation which would be detected with the whole retardation incurred in traversing the 76 cm. of bisulphide of carbon. In this length there are contained 1,200,000 wave-lengths of yellow light, or 2,400,000 half wave-lengths. The retardation due to the refraction may be reckoned at ·6 of this, or 1,440,000 half wave-lengths. Thus the double refraction that might be detected, estimated as a fraction of the whole refraction, is $1·2 \times 10^{-10}$. The effect to be expected is of the order 10^{-8}, so that there is nearly 100 times to spare. The above relates to the bisulphide of carbon. With the water the delicacy of the test was somewhat less.

When it is attempted to replace the liquid by solid matter, the difficulties of experiment are greatly increased. The best results that I have been able to obtain were with built up thicknesses of plate-glass. A sufficient thickness in one piece is liable to exhibit too much double refraction from the effect of internal strains. A number of triangular pieces of plate-glass, no larger than necessary, and about 6 mm. thick, were put together in a trough to a total thickness of about 110 mm. The interstices between the faces being filled up with bisulphide of carbon, the internal reflexions were sufficiently reduced. One difficulty is to get quit of motes and threads which adhere to the glass and become extraordinarily conspicuous. Advantage was thought to be derived from shaking up the bisulphide of carbon with strong sulphuric acid. At the best the residual motes and specks in the glass interfere very seriously with the observation, and the loss of light due to imperfect transparency operates in the same direction. The least load upon the upright strip that could be detected with certainty was now 100 grms., so that as compared with the observations upon liquid there was a loss of delicacy of four times. In addition to this, the effect to be expected is reduced in the proportion of 7 : 1, that being the ratio of lengths traversed by the light. Thus in all we lose 28 times as compared with the liquid. In the latter case we calculated a margin of 100 times, so that here there would remain a margin of about 3 times.

A subsequent attempt was made to increase the total thickness of the combined glasses to about 220 mm., but no real advantage was gained. The loss of light and increase of disturbance from motes and residual double refraction prejudiced the delicacy in about the same proportion as the length of path was increased.

But although the results of the observations upon solids are very much less satisfactory than in the case of liquids, enough remains to justify us in concluding that even here there is no double refraction (of the order to be expected) due to motion through the æther.

281.

ON THE DISTILLATION OF BINARY MIXTURES.

[*Philosophical Magazine*, IV. pp. 521—537, 1902.]

AT various times during the past twenty years I have turned my attention to the theory of distillation, and have made experiments upon a question, as to which information seemed to be almost entirely lacking, viz., the relation between the strengths of liquid and vapour which are in equilibrium with one another when a binary mixture is subjected to distillation. In order to be intelligible I must set forth a little in detail some matters which are now fairly well known and understood, although they were not so at the time when my notes were written.

Distillation of a Pure Liquid.

The temperature of the saturated vapour just over the liquid depends upon the pressure. If the end of the condenser-tube, *e.g.*, of the Liebig type, be open, the pressure is of necessity nearly atmospheric. Suppose that in this tube a piston, moving freely, separates pure vapour from pure air. Then the whole wall of the condenser on the vapour side is almost at boiling-point. If we imagine the piston removed, the air and vapour may mix, and it is now the total pressure which is atmospheric. Wherever the temperature is below boiling there must be admixture of air sufficient to bring up the pressure.

Two or more Liquids which press independently.

This is the case of liquids like water and bisulphide of carbon whose vapour-pressures are simply added So long as the number of ingredients remains unchanged, the composition of the vapour rising from the boiling mixture is a function of the temperature (or total pressure) only. Hence in simple distillation the composition of the distillate remains constant until perhaps one constituent of the liquid (not necessarily the most volatile) is exhausted. At this point the distillate, as well as the boiling-temperature, changes discontinuously and the altered values are preserved until a second constituent is exhausted, and so on. None of the separate distillates thus obtained would be altered by repetition of the process at the same pressure.

Liquids which form true Mixtures.

The above is as far as possible from what happens in the case of miscible liquids, e.g., water and common alcohol. Here the composition of the vapour, as well as the boiling-point under given pressure, depends upon the composition of the liquid, and all three will in general change continuously as the distillation proceeds. But, so long as the total pressure is fixed, to a given composition of the liquid corresponds a definite composition of the vapour; and it is the function of experiment to determine the relation between the two. The results of such experiments may be exhibited graphically upon a square diagram (e.g. figs. 3 and 4, pp. 76 and 79) in the form of a curve stretching between opposite corners of the square, the abscissa of any point upon the curve representing the composition of the liquid and the ordinate representing the composition of the vapour in equilibrium with it. For the pure substances at the ends of the scale, represented by opposite corners of the square, the compositions of liquid and vapour are necessarily the same.

The character of the separation capable of being effected by distillation depends in great measure upon whether or not the curve meets the diagonal at any intermediate point, as well as at the extremities. If there be no such intersection, the curve lies entirely in the upper (or in the lower) triangular half of the square, so that for all mixtures the distillate is richer (or poorer) than the liquid. As the distillation of a limited quantity of mixed liquid proceeds, the composition of the residue moves always in one direction and must finally approach one or other condition of purity.

If on the other hand the curve crosses the diagonal, the point of intersection represents a state of things in which the liquid and vapour have the same composition, so that distillation ceases to produce any effect. This happens for example with a solution of hydrochloric acid at a strength of 20 per cent. (fig. 3) and with aqueous alcohol at a strength of 96 per cent. By no process of distillation can originally weak alcohol be strengthened beyond the point named, and if (Le Bel) we start with still stronger alcohol (prepared by chemical desiccation) the effect of distillation is reversed. The vapour being now weaker (in alcohol) than the liquid, the residue in the retort strengthens until it reaches purity.

In the case of substances which have no tendency to mix, e.g., water and bisulphide of carbon, the composition of the vapour is, as we have seen, always the same. The representative curve, reducing to a straight line parallel to the axis of abscissæ, or rather to the broken line AEFD (fig. 1), necessarily crosses the diagonal. The point of intersection (H) represents a condition of things in which the compositions of the liquid and vapour are the same. As distillation proceeds, the residue retains its composition, and both ingredients are exhausted together.

If we commence with a liquid containing CS_2 in excess of the above proportion, the excess gradually increases until nothing but CS_2 remains

Fig. 1.

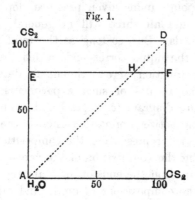

behind. In the same way, if the water be originally in excess, the excess accentuates itself until the (finite) residue is pure water. The critical condition is thus in a sense unstable, and can only be realized by adjustment beforehand.

The conclusions drawn above may be generalized. Whatever may be the ingredients of a binary mixture, in the upper triangular half of the square the vapour is stronger (we will say) than the liquid, in the lower half weaker. Hence, as the liquid distils away, progress from a point in the upper half is towards diminishing abscissæ, and in the lower half towards increasing abscissæ. When, as in fig. 1, the curve in its course from A to D crosses AD from left to right, the condition represented by the point of intersection H is unstable. When, as in the case of hydrochloric acid (fig. 3), the crossing takes place from right to left, $i.e.$ from the lower half to the upper half of the square, the progress from points in the neighbourhood is always $towards$ the point of intersection, so that the state represented thereby is $stable$. We may sum up by saying that if, as the liquid strengthens, the vapour having been weaker than the liquid becomes the stronger, the point of transition, representing constant distillation, is stable; but if the vapour having been at first the stronger becomes the weaker, then the point of transition is unstable.

The question presents itself, whether as the liquid strengthens (in a particular ingredient) the vapour necessarily strengthens with it. Does the curve on our diagram slope everywhere upwards on its course from A to D? Although a formal proof may be lacking, it would seem probable that this must be so when the ingredients mix in all proportions. A limiting case is when two ingredients do not mix at all, *e.g.*, water and bisulphide of carbon, or when the mixture divides itself into two parts of constant composition as when ether and water are associated in certain proportions. In these cases the composition of the vapour is constant for the whole or for a part of the range (Konowalow), and the representative curve is without slope.

Konowalow's Theorem.

An important connexion has been formulated by Konowalow* between the vapour-pressure, regarded as a function of the composition of the liquid with which it is in equilibrium, and the existence of a point of constant distillation. "The pressure of the vapour from a fluid consisting of two different substances is in general a function of the composition of the mixture.... Let such a mixture, confined in a closed space, be maintained at a constant temperature. We may conceive this space bounded by fixed walls and by a movable piston. The conditions of stable equilibrium are then (1) that the external pressure operative upon the piston should be equal to the pressure of the saturated vapour at the given temperature; (2) that by increase, or diminution, of the vapour space the pressure should become respectively not greater, or not less, than the external pressure. In expansion the vapour-pressure can thus either remain constant, or become smaller. On the basis of this law we can establish a relation between the composition of the liquid and that of the vapour."

Before proceeding further I must remark that the principle, as stated, appears to need elucidation. Why should the equilibrium of the piston under a constant load be stable? There must of course be some position of stable equilibrium for a given load and temperature; but this might, for all that appears, correspond to complete evaporation of the liquid or to complete condensation of the vapour.

The following argument, however, suffices to show that Konowalow's principle is a necessary consequence of the second law of Thermodynamics. Suppose that the cylinder in which are contained the given liquid and vapour communicates by a lateral channel (fig. 2) with a large reservoir filled with liquid of similar composition, and that all are maintained at the prescribed temperature. As a first operation close the tap between the vessels, and then let the piston rise a little. The motion is supposed to be so slow that equilibrium prevails throughout. The result of the expansion may be that the compositions of the liquid and of the vapour undergo a change. Now open the tap, and allow diffusion to take place, if necessary, until equilibrium is again established. On account of the large quantity of liquid in the reservoir the pressure is sensibly restored to its original value

* Wied. *Ann.* xiv. p. 48 (1881).

and remains undisturbed as the piston is slowly pushed back to its first position. During this cycle of operations work cannot be gained; and thus is excluded the possibility of a rise of pressure during the expansion. It follows that a fall of pressure cannot accompany compression.

Upon the basis of this principle Konowalow proceeds as follows: Suppose that at a particular composition-ratio the pressure of vapour increases as the liquid becomes richer in a specified component. In this case the expansion of the mass cannot enrich the liquid; for if this result occurred the pressure would rise, which we have proved it cannot do. During the expansion fresh vapour is formed; and if the composition of the vapour were poorer than that of the liquid, the latter would inevitably be enriched by the operation. We conclude that at the point in question the vapour cannot be poorer than the liquid. In like manner if the vapour-pressure falls with increasing richness

Fig. 2.

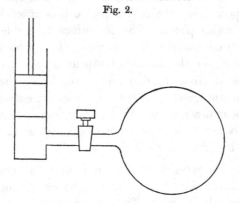

of liquid, compression of a given mass cannot enrich the liquid, and this requires that the vapour be not richer than the corresponding liquid. If we suppose the vapour-pressure to be plotted as a function of richness of corresponding liquid, we may express these results by saying that rising parts of the pressure-curve can have no representation in the lower triangle of our former diagrams (where vapour is poorer than liquid), and that falling parts cannot be represented in the upper triangle.

It is now evident that the passage from a rising to a falling part of the pressure-curve can only occur when the vapour is neither richer nor poorer than the liquid, and we arrive at Konowalow's important theorem that *any mixture, which corresponds to a maximum or minimum of vapour-pressure, has (at the temperature in question) the same composition as its vapour.*

The particular case in which one ingredient is wholly involatile is worth a moment's notice. The vapour over a solution of salt in water can never have the same composition as the liquid; and from this we may conclude that the vapour-pressure has no maximum or minimum, or rather that there is no transition anywhere between rising and falling.

The converse of Konowalow's theorem is also not without importance. Consider two mixtures of slightly differing composition, one of which is richer than its vapour and the other poorer. Expansion of the first entails an enrichment of the liquid, and during the operation the pressure cannot rise. Expansion of the second impoverishes the liquid, and again the pressure cannot rise. The curve exhibiting pressure as a function of composition (of liquid), if it slopes at all at the two points, must slope in opposite directions. Hence by approaching nearer and nearer to the point where the compositions of vapour and liquid are the same, we see that the vapour-pressure must there be stationary in value.

An example of the use of the converse theorem is afforded by the consideration of mixtures of water and common alcohol. The question of the existence of a mixture having the same composition as its vapour is not easily settled directly, but the recent observations of Noyes and Warfel[*] show conclusively that the mixture containing 96 per cent. of alcohol by weight has a minimum boiling-point, and accordingly distils without change. It may be noted that the curve given by Konowalow himself would point to the contrary conclusion.

In the practical conduct of distillation it is the pressure that is constant rather than the temperature. Inasmuch, however, as pressure always rises with temperature, a maximum or minimum pressure when temperature is given necessarily corresponds with a minimum or maximum temperature when pressure is given. In the case of a solution of hydrochloric acid, for example, the thermometer marks a maximum temperature at the point where the solution distils without change.

Calculation of Residue.

Before proceeding to the experimental part of this paper it may be well to explain further the significance of the curves exhibiting the relative compositions of liquid and vapour. If w represent the whole quantity (weight) of liquid, say alcohol and water, remaining in the retort at any time, y the quantity of one ingredient (alcohol), the abscissa ξ of the curve is y/w. As the distillation proceeds for a short time w becomes $w + dw$, and y becomes $y + dy$[†]; and the composition of the vapour, that is the ordinate η of the diagram, is dy/dw. Thus

$$\xi = y/w, \qquad \eta = dy/dw,$$

[*] Am. Chem. Soc. xxiii. p. 463 (1901).

[†] dw and dy being negative.

while the functional relation between ξ and η is given by the curve, and may be analytically expressed by $\eta = f(\xi)$. Thus

$$\frac{d(w\xi)}{dw} = f(\xi),$$

whence

$$\log \frac{w}{w_0} = \int_{\xi_0}^{\xi} \frac{d\xi}{f(\xi) - \xi},$$

w_0, ξ_0 being corresponding values of w and ξ.

When ξ is small the curve is often approximately straight. If we set $f(\xi) = \kappa\xi$ we find

$$\xi/\xi_0 = (w/w_0)^{\kappa-1}.$$

For example, in the case of alcohol and water, we have for very weak mixtures $\eta = 12\xi$ approximately, so that $\kappa = 12$. As the distillation proceeds, w diminishes and ξ soon becomes exceedingly small. The halving of w implies a diminution of ξ in the ratio of $2^{11} : 1$. The residue in the retort thus approximates rapidly to pure water.

On the other hand, in the case of acetic acid and water κ is about $\frac{3}{4}$. When weak acetic acid is distilled the residue *strengthens*, but the earlier stages of the process are covered by the formula given, which now assumes the form

$$\xi/\xi_0 = (w_0/w)^{\frac{1}{4}}.$$

In order to double the strength of the liquid remaining in the retort, 15/16 of it would have to be distilled away, or again, in order to increase the strength in the ratio of $3 : 2$, the distillation must proceed until the liquid is reduced in the ratio of $16 : 81$, or nearly of $1 : 5$. An experiment of this sort upon acetic acid is recorded below.

Observations.

The experimental results about to be given were obtained by simple distillation of mixtures of known composition. In order to avoid too rapid a change of composition, somewhat large quantities were charged into a retort and were kept in vigorous ebullition. By special jacketing arrangements security was taken that the upper part of the retort should be maintained at a distinctly higher temperature than the liquid, so that there could be no premature condensation which would vitiate the result. *All* the vapour rising from the liquid must be condensed in the specially provided Liebig condenser and be collected as distillate. Subject to this condition, and in view of the rapid stirring effected by the rising vapour, it would seem safe to assume that the distillate really represents the vapour which is in equilibrium with the liquid at the time in question. The compositions of the liquid and vapour are of course continually changing as the distillation proceeds.

The distillates (including the first drop) were collected in 50 c.c. measuring flasks. It will save circumlocution to speak of a particular case, and I will take that of alcohol and water, for which the analyses were made by specific gravity. The successive collections of 50 c.c. show an increasing specific gravity corresponding to a diminishing strength. The specific gravity of each gives the total weight, and the strength, deduced from tables, allows us to calculate the alcohol and water in each collection. The total alcohol and water originally present in the retort being known in the same way, we are able to deduce by subtraction the quantities remaining in the retort at each stage, and thus to compare the strengths of corresponding liquid and vapour. In the reduction any particular distillate is considered to correspond with the mean condition of the liquid before and after its separation therefrom.

If the process above sketched could be absolutely relied on, it would be possible, starting with a strong spirit in the retort, to obtain from one distillation data relating to a great variety of strengths. But this method is not to be recommended, as the errors would tend to accumulate. The first 50 c.c., condensed under somewhat abnormal conditions, was not used directly, but only to allow for the change going on in the retort. The 2nd, 3rd, and 4th collections were usually calculated so as to show the strengths of these distillates in comparison with that of the liquid, but they were regarded rather as checks upon one another than as independent results relating to an altered state of affairs.

Alcohol and Water.

Observations upon mixtures of water and ethyl alcohol, sufficient to give a nearly complete curve, were made in 1891 and again in 1898 with good general agreement. The specific gravities were found in the balance with a bottle of 20 c.c. capacity, and the calculations of strength were by Mendeleef's tables with appropriate temperature correction. The results of the second series are given in the accompanying table and are exhibited as a curve, A, in fig. 3. The strengths are throughout reckoned by weight.

The observation of May 4 thus signifies that to a liquid containing by weight 1·97 per cent. of alcohol there corresponds a vapour containing by weight 17·5 per cent. of alcohol. From the results of May 24 we see that when the liquid reaches 92 per cent. the vapour is but little the stronger, and the difference practically disappears at 95 per cent. Indeed according to May 25 the vapour is a little the weaker at this point. The difference, however, is not to be trusted, since the difficulties of manipulation, depending partly upon the attraction of strong alcohol for aqueous vapour, are much

Date 1898	Strength of Liquid	Strength of Vapour
May 4.........	·01970	·1750
„ 3.........	·03982	·3159
„ 5.........	·0601	·3979
„ 9.........	·0988	·5145
„ 10.......	·2586	·6803
„ 13.........	·4562	·7412
„ 16.........	·6606	·7976
„ 17.........	·7739	·8414
„ 20.........	·8221	·8622
„ 23.........	·8594	·8849
„ 24.........	·9241	·9284
„ 25.........	·9555	·9545

Fig. 3.

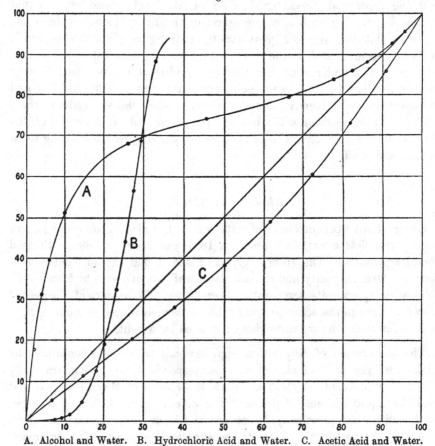

A. Alcohol and Water.   B. Hydrochloric Acid and Water.   C. Acetic Acid and Water.

increased at this stage. It was these difficulties and the uncertainty as to what exactly happened with spirit stronger than 95 per cent. that retarded the publication of the work. I had intended to make further experiments upon this point, but the matter was postponed from time to time. The observations of Noyes and Warfel (*l. c.*) seem now to remove all doubt. The existence of a minimum boiling-temperature for a strength of 96 per cent. shows that the curve there crosses the diagonal. Between this point and 100 per cent. the vapour is the weaker, and the curve lies in the lower triangular half of the square. But the deviation from the diagonal in this region is probably extremely small.

The following from Noyes and Warfel's table may be useful:

Strength	Boiling-point	Strength	Boiling-point
100	78·300	85	78·645
99	78·243	80	79·050
98	78·205	75	79·505
97	78·181	65	80·438
96	78·174	55	81·77
95	78·177	48	82·43
94	78·195	35	83·87
93	78·227	26	85·41
92	78·259	20	87·32
91	78·270	10	91·80
90	78·323	0	100·00

Hydrochloric Acid and Water.

One of the ingredients of the mixture being gaseous under ordinary conditions, the observations are limited to that portion of the curve for which the strength of the liquid does not exceed 35 per cent., unless freezing appliances are called into play. No attempt was made in the present experiments to pass the above limit, the object being merely to determine with moderate accuracy that part of the curve with which we are usually concerned in the laboratory. It was known from the experiments of Roscoe and others that the curve would cross the diagonal at the strength of about 20 per cent.

The general plan of the work was the same as in the case of alcohol and water, but the strengths were usually determined chemically. In the case of the stronger acids it was not possible to condense the vapour at atmospheric temperature; and I contented myself with a calculation in which the strength of the vapour was inferred from observations of the quantity and strength of the liquid in the retort before and after the operation. Results obtained in this way are doubtless of minor accuracy.

It may be worth while to reproduce in tabular form the data relating to the weakest acid.

Distillation of Hydrochloric Acid.—Sept. 13, 1898.

No.	Volume in c.c.	Specific gravity	Total weight	Percentage of HCl	Weight of HCl	Weight of H_2O	HCl remaining in retort	H_2O remaining in retort	Total remaining	Percentage of HCl remaining
0	1800	1·031	1855·8	6·0	111·4	1744·4	...	...	...	6·0
1	250	1·0013	250·32	0·26	0·65	249·7	110·7	1494·7	1605·4	6·9
2	250	1·0023	250·58	0·45	1·12	249·4	109·6	1245·3	1354·9	8·1
3	250	1·0038	250·95	0·76	1·90	249·0	107·7	996·3	1103·9	9·8
4	250	1·0076	251·90	1·53	3·86	248·0	104·0	748·3	852·3	12·2

The first column contains the numbers of the successive distillates from 1 to 4, the entry 0 referring to the mixture with which the retort was originally charged. The volume of this mixture was 1800 c.c. of specific gravity 1·031 and of 6·0 per cent. strength. Of the total weight 1855·8 gms., 111·4 gms. is hydrochloric acid and 1744·4 gms. water. In like manner the volume of the first distillate is 250 c.c., the specific gravity 1·0013, the total weight 250·32 gms., of which 0·65 gms. is HCl and 249·7 gms. H_2O. The residue in the retort after the first 250 c.c. has been distilled over is accordingly composed of 111·4 − 0·65 or 110·7 gms. HCl and 1744·4 − 249·7 or 1494·7 gms. H_2O, making 1605·4 gms. in all. At this stage the percentage strength of the liquid remaining in the retort is 6·9. The strengths of the liquid in the retort after the 1st, 2nd, ... 4th distillates have been removed are found in this way to be 6·9, 8·1, 9·8, and 12·2 per cent. The first distillate, whose strength is 0·26 per cent., thus corresponds with a liquid whose strength varied from 6·0 to 6·9 per cent., on an average 6·45 per cent. We thus obtain the following corresponding strengths:

Percentage Strengths.

Liquid	Vapour
6·45	0·26
7·50	0·45
8·95	0·76
11·00	1·53

It is hardly worth while to record all the separate results. In addition o the above the following will suffice for the construction of the curve.

In the three last the strengths of the distillates were not directly observed, but were calculated from the condition of the liquid before and after as already mentioned.

Percentage Strengths.

Date (1898)	Liquid	Vapour
Sept. 21.........	14·5	4·46
„ 28.........	18·0	12·8
Oct. 10.........	20·3	18·9
„ 12.........	23·3	32·4
„ 14.........	25·2	44·2
„ 15.........	27·4	56·6
„ 18.........	29·0	68·8
„ 19.........	32·8	88·3

The results are plotted in Curve *B*, fig. 3.

Ammonia and Water.

In this case the analysis was by specific gravity and the results were somewhat rough, the intention being merely to obtain an approximation to the form of the curve. On this account they are plotted upon a smaller

Fig. 4.

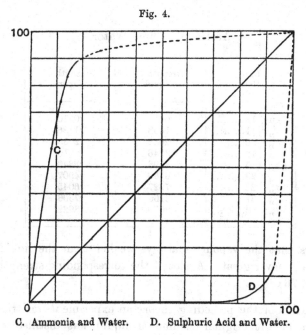

C. Ammonia and Water.   D. Sulphuric Acid and Water.

scale *C*, fig. 4, the dotted portion of the curve being conjecturally added to indicate the progress towards the corner of the square.

Sulphuric Acid and Water.

The distillates were here determined chemically. From acid in the retort of less strength than 60 per cent. the distillate failed to redden litmus. From 75 per cent. acid the distillate contained about one-thousandth part of H_2SO_4. From 81 per cent. the distillate contained 1·6 per cent.; from 90 per cent. the distillate contained 7·1 per cent.; and from 93 per cent. liquid the distillate contained 12·8 per cent. of acid. The curve is given in D, fig. 4, the dotted portion for strengths of liquid greater than 93 per cent. being conjectural.

Acetic Acid and Water.

This case was examined as likely to exemplify a very different behaviour from any of the others, since it was known that these substances are not easily separated by distillation. The retort was charged with 1000 c.c. of mixture and two distillates were collected of 150 c.c. each. The analyses were conducted chemically and the results calculated as already explained. Thus the first distillate was considered to correspond with the mean strength in the retort before and after its separation. The following were the results obtained:

Acetic Acid and Water.

Date 1902	Strength of Liquid	Strength of Vapour
Aug. 18.........	·0677	·0510
„ 11.........	·1458	·1136
„ 6.........	·2682	·2035
„ 18.........	·3746	·2810
„ 5.........	·4998	·3849
„ 23.........	·6156	·4907
„ 22.........	·7227	·6045
„ 21.........	·8166	·7306
„ 20.........	·9070	·8622

It appears that the vapour is always weaker than the liquid, but that the difference is never great. A plot of the corresponding strengths is given in C (fig. 3, p. 76).

In illustration of the preceding theory an experiment was tried in which three-quarters of the original volume of liquid was distilled over. The original liquid consisted of 1000 c.c. of acid of density 1·010, and of strength (as determined chemically) ·0757, representing as usual the proportion of the weight of acetic acid to the whole weight. The residue measuring 250 c.c. was of density 1·016 and of strength ·1100. From these data we find

$$\log (\xi/\xi_0) = \cdot 1624, \qquad \log (w_0/w) = \cdot 5995,$$

whence

$$1 - \kappa = \cdot 27, \qquad \kappa = \cdot 73.$$

The number denoted by κ represents the ratio of strengths of vapour and liquid when weak mixtures are distilled.

A new Apparatus with uniform Régime.

In the theory and experiments so far considered the distillation has always been supposed to be simple, that is, the vapour rising from the boiling liquid is supposed to be removed and to be condensed as a whole, so that the distillate has the same composition as the vapour leaving the boiling liquid. In practice, as is well known, this condition is often and advantageously violated. A preliminary partial condensation of the vapour in the still-head frees it from some of the less volatile ingredient; and, when the residue is condensed and collected, the more volatile ingredient is obtained in a nearer approach to purity. Prof. S. Young has shown that the principle is more effectively carried out if the still-head be maintained at a suitable temperature.

Even with a preliminary partial condensation in the still-head, the "fractionation" of a mixture is usually regarded as a very tedious operation. The stock of mixture in the retort is constantly changing its composition as the distillation and partial condensation proceed, and no uniform *régime* can be established. Although theoretical simplicity and practical convenience are not always conjoined, a uniform *régime* seems very desirable, and it excludes the usual arrangement in which the whole supply of mixture is charged into the retort. The return into the retort of the liquid first condensed from the original vapour is also objectionable.

The problem of distillation may be stated to be the separation from a binary mixture of the whole of the two components in, as nearly as may be, a state of purity. There is no theoretical reason why this should not be effected at one operation; but for this purpose the mixture must be fed in continuously and not at the place of highest or lowest temperature. A description of the procedure followed in some illustrative experiments will make the nature of the process plain.

The mixtures actually employed were of water and common alcohol. The choice was perhaps not a happy one, as in consequence of the peculiar properties of strong alcohol it was unlikely that a distillate could be obtained stronger than about 90 per cent. As regards apparatus, the retort and still-head are replaced by a long length (12 metres) of copper tubing, 15 mm. in diameter. This is divided into two parts, arranged in spirals, like the worms of common condensers, and mounted in separate iron pails. The lower and longer spiral was surrounded with water which was kept boiling. The water surrounding the upper spiral was maintained at a suitable temperature, usually 77° C. The copper tubes forming the two spirals were connected by a straight length of glass, or brass, tubing of somewhat greater bore, and provided with a lateral junction through which the material could be supplied. The connecting piece and the spirals were so arranged that the entire length was on a slight and nearly uniform gradient, rising from near the bottom of the lower pail to the top of the upper pail. On leaving the latter the tube turned downwards and was connected with an ordinary Liebig's condenser capable of condensing the whole of the vapour which entered it. At the lower end of the system of tubing the watery constituent is collected. In strictness the receiver should be connected air-tight and be maintained at 100°. In distilling the stronger mixtures (60 or 75 per cent. alcohol) this precaution was found advisable or necessary; but in the case of the weaker ones the water could be allowed to discharge itself through a short length of pipe whose end was either exposed to the atmosphere or slightly sealed by the liquid in the receiver.

The feed of the mixture was arranged as a visible and rather rapid succession of drops, and was maintained at a uniform rate. In the case of the stronger mixtures the evaporating power of the lower coil was hardly sufficient, and was assisted by applying heat to the feed, so that a good proportion was evaporated before reaching the main tube. The weaker mixtures on the other hand could be fed in without any preliminary heating. The uniform *régime* should be maintained long enough to ensure that the liquids collected at the two ends shall be fairly representative and not complicated by anything special that may happen before the uniform *régime* is established.

During the operation every part of the tube (not too near the ends) is occupied by a double stream—an ascending stream of vapour and a descending stream of liquid. Between these streams an exchange of material is constantly taking place, the liquid, as it descends, becoming more aqueous and the vapour, as it rises, becoming more alcoholic. In view of the slowness of the feed and the length of the tube, we may regard the liquid and the vapour as being everywhere in approximate equilibrium. At the lower end, since the pressure is atmospheric and the temperature scarcely below 100°, there can be little alcohol; for similar reasons at the upper end there cannot be much water, although the exclusion is here less complete on account of the peculiar character of the curve representing the relation of compositions at this extreme.

Experiments were tried with four different strengths of mixture—20, 40, 60, and 75 per cent. of alcohol. In all cases the water was collected nearly pure, never containing more than ½ per cent. of alcohol. The alcoholic part condensed from the upper end varied but little. The weakest (from the 20 per cent. mixture) was of 89 per cent., and the strongest was of 90·3 per cent. All strengths are given by *weight*, and were calculated by Mendeleef's tables from the observed specific gravities with suitable temperature correction. The watery constituent which, not having been evaporated, sometimes looked a little dirty, was usually redistilled so as to obviate any risk of its purity being over-estimated. In some cases it could not be distinguished from pure water.

The apparatus illustrates very well the principles of ideal distillation, and its performance may be regarded as satisfactory. When once the conditions, as to preliminary heating (if necessary) and as to rate of feed, have been found for a particular mixture, the continued working is almost self-acting, or at any rate could be made so without much difficulty; and it is probable that separations, otherwise very troublesome, could be easily effected by use of it.

282.

NOTE ON THE THEORY OF THE FORTNIGHTLY TIDE.

[*Philosophical Magazine*, v. pp. 136—141, 1903.]

AN adequate calculation *à priori* of the tide of fortnightly period—that which depends upon the moon's motion in declination—would be of great interest as affording the means, by comparison with observation, of determining the extent to which the solid earth yields to the tide-generating force. On the assumption that the fortnightly tide over an absolutely rigid earth would be sensibly equal to its "equilibrium value," Prof. G. Darwin[*] has estimated that the actual rigidity must be at least as great as that of steel, in accordance with the earlier surmises of Lord Kelvin.

But is an "equilibrium theory" adequate? The known properties of a system vibrating about a configuration of thoroughly stable equilibrium would certainly suggest an affirmative answer, when it is considered that a fortnight is a long period in comparison with those of the more obvious free oscillations. It is to be remembered, however, that a tidally undisturbed sea is not in equilibrium, and that in virtue of the rotation of the earth the tides are really oscillations about a condition of *steady motion*. In Laplace's theory the rotation of the earth is taken fully into account, but the sea must be supposed to cover the entire globe, or at any rate to be bounded only by coasts running all round the globe along parallels of latitude. The resulting differential equation was not solved by Laplace, who contented himself with remarking that in virtue of friction the solution for the case of fortnightly and (still more) semi-annual tides could not differ much from the "equilibrium values."

The sufficiency of Laplace's argument has been questioned, and apparently with reason, by Darwin[†], who accordingly resumed Laplace's differential equation in which frictional forces are neglected. Taking the case of an ocean of uniform depth completely covering the globe and following the indications of Lord Kelvin[‡], he arrives at a complete evaluation of Laplace's

---

* Thomson and Tait's *Natural Philosophy*, 2nd ed. Vol. I. pt. II. p. 400 (1883).
† *Proceedings of the Royal Society*, Vol. XLI. p. 337 (1886).
‡ *Phil. Mag.* Vol. L. p. 280 (1875).

"Oscillation of the First Species." A summary of Darwin's work has been given by Lamb* from which the following extracts are taken. The equilibrium value of the fortnightly tide being

$$\overline{\zeta} = H' \left(\tfrac{1}{3} - \mu^2\right),$$

the actual tide for a depth of 7260 feet is found to be

$$\zeta/H' = \cdot1515 - 1\cdot0000\,\mu^2 + 1\cdot5153\,\mu^4 - 1\cdot2120\,\mu^6$$
$$- \cdot2076\,\mu^{10} + \cdot0516\,\mu^{12} - \cdot0097\,\mu^{14} + \cdot0018\,\mu^{16} - \cdot0002\,\mu^{18},$$

whence at the poles $(\mu = \pm 1)$

$$\zeta = -\tfrac{2}{3}H' \times \cdot154,$$

and at the equator $(\mu = 0)$

$$\zeta = \tfrac{1}{3}H' \times \cdot455.$$

Again, for a depth of 29040 feet, we get

$$\zeta/H' = \cdot2359 - 1\cdot000\,\mu^2 + \cdot5898\,\mu^4$$
$$- \cdot1623\,\mu^6 + \cdot0258\,\mu^8 - \cdot0026\,\mu^{10} + \cdot0002\,\mu^{12},$$

making at the poles $\qquad \zeta = -\tfrac{2}{3}H' \times \cdot470,$

and at the equator $\qquad \zeta = \tfrac{1}{3}H' \times \cdot708.$

It appears that with such oceans as we have to deal with the tide thus calculated is less than half its equilibrium amount.

The large discrepancy here exhibited leads Darwin to doubt whether "it will ever be possible to evaluate the effective rigidity of the earth's mass by means of tidal observations."

From the point of view of general mechanical theory, the question at once arises as to what is the meaning of this considerable deviation of a long-period oscillation from its equilibrium value. A satisfactory answer has been provided by Lamb†; and I propose to consider the question further from this point of view in order to estimate if possible how far an equilibrium theory may apply to the fortnightly tides of the actual ocean.

The tidal oscillations are included in the general equations of small vibrations, provided that we retain in the latter the so-called gyrostatic terms. By a suitable choice of coordinates, as in the usual theory of normal coordinates, these equations may be reduced to the form

$$\left.\begin{aligned}
a_1\ddot{q}_1 + c_1 q_1 \quad\quad + \beta_{12}\dot{q}_2 + \beta_{13}\dot{q}_3 + \ldots &= Q_1, \\
a_2\ddot{q}_2 + c_2 q_2 + \beta_{21}\dot{q}_1 \quad\quad + \beta_{23}\dot{q}_3 + \ldots &= Q_2, \\
a_3\ddot{q}_3 + c_3 q_3 + \beta_{31}\dot{q}_1 + \beta_{32}\dot{q}_2 \quad\quad + \ldots &= Q_3,
\end{aligned}\right\} \quad \ldots(1)$$

in which $\qquad\qquad \beta_{rs} = -\beta_{sr}. \qquad \ldots(2)$

---

* *Hydrodynamics*, § 210, Cambridge, 1895.
† *Hydrodynamics*, §§ 196, 198, 207.

From these we may fall back upon the case of small oscillations about stable equilibrium by omitting the terms in β; but in general tidal theory these terms are to be retained. If the oscillations are free, the quantities Q, representing impressed forces, are to be omitted.

If the coefficients β are *small*, an approximate theory of the free vibrations may be developed on the lines of *Theory of Sound*, § 102, where there are supposed to be small dissipative (but no rotatory) terms. For example, the frequencies are unaltered if we neglect the *squares* of the β's. Further, the next approximation shows that the frequency of the slowest vibration is *diminished* by the operation of the β's; or more generally that the effect of the β's is to cause the values of the various frequencies to *repel* one another*.

To investigate forced vibrations of given period we are to assume that all the variables are proportional to $e^{i\sigma t}$, where σ is real. If the period is very long, σ is correspondingly small, and the terms in $\ddot{q}$ and $\dot{q}$ diminish generally in importance relatively to the terms in q. In the limit the latter terms alone survive, and we get

$$q_1 = Q_1/c_1, \qquad q_2 = Q_2/c_2, \quad \&c., \quad \dots\dots\dots\dots(3)$$

which are the "equilibrium values." But, as Prof. Lamb has shown, exceptions may arise when one or more of the c's vanish. This state of things implies the possibility of steady motions of disturbance in the absence of impressed forces. For example, if $c_2 = 0$, we have as a solution $\dot{q}_2 = $ constant, with

$$q_1 = -\beta_{12}\dot{q}_2/c_1, \qquad q_3 = -\beta_{32}\dot{q}_2/c_3, \quad \&c.$$

In illustration Prof. Lamb considers the case of two degrees of freedom, for which the general equations are

$$a_1\ddot{q}_1 + c_1 q_1 + \beta\dot{q}_2 = Q_1, \qquad a_2\ddot{q}_2 + c_2 q_2 - \beta\dot{q}_1 = Q_2; \quad \dots\dots\dots(4)$$

supposing that $c_2 = 0$ and also that $Q_2 = 0$, while Q_1 remaining finite is proportional to $e^{i\sigma t}$, as usual. We find

$$q_1 = \frac{a_2 Q_1}{a_2(c_1 - a_1\sigma^2) + \beta^2}, \qquad \dot{q}_2 = \frac{\beta Q_1}{a_2(c_1 - a_1\sigma^2) + \beta^2}; \quad \dots\dots\dots(5)$$

so that in the case of a disturbance of very long period, when σ approaches zero,

$$q_1 = \frac{Q_1}{c_1 + \beta^2/a_2}, \qquad \dot{q}_2 = \frac{\beta Q_1}{a_2 c_1 + \beta^2}. \quad \dots\dots\dots\dots(6)$$

Since a_2 is positive, q_1 is *less* than its equilibrium value; and it is accompanied by a motion of type q_2, although there is no extraneous force of the latter type.

It is clear then that in cases where a steady motion of disturbance is possible the outcome of an extraneous force of long period may differ greatly

* See Art. 283 below.

from what the equilibrium theory would suggest. It may, however, be remarked that the particular problem above investigated is rather special in character. In illustration of this let us suppose that there are *three* degrees of freedom, and that c_2, c_3, Q_2, Q_3 are evanescent. The equations then become

$$(c_1 - \sigma^2 a_1) q_1 + i\sigma\beta_{12} q_2 + i\sigma\beta_{13} q_3 = Q_1,$$
$$-\sigma a_2 q_2 + i\beta_{21} q_1 + i\beta_{23} q_3 = 0,$$
$$-\sigma a_3 q_3 + i\beta_{31} q_1 + i\beta_{32} q_2 = 0;$$

whence, regard being paid to (2),

$$q_1 \left\{ (c_1 - \sigma^2 a_1) + \frac{\sigma^2 (a_3 \beta_{12}{}^2 + a_2 \beta_{13}{}^2)}{\sigma^2 a_2 a_3 - \beta_{23}{}^2} \right\} = Q_1. \quad \dots\dots\dots(7)$$

When $\sigma = 0$, the value of q_1 reduces to Q_1/c_1, unless $\beta_{23} = 0$, so that in general the equilibrium value applies. But this is only so far as regards q_1. The corresponding values of q_2, q_3 are

$$q_2 = -q_1 \beta_{31}/\beta_{32}, \qquad q_3 = -q_1 \beta_{21}/\beta_{23}; \quad \dots\dots\dots\dots(8)$$

and thus the equilibrium solution, considered as a whole, is finitely departed from. And a consideration of the general equations (1) shows that it is only in very special cases that there can be any other outcome when the possibility of steady motion of disturbance is admitted.

It thus becomes of great importance in tidal theory to ascertain what steady motions are possible, and this question also has been treated by Lamb (§ 207). It may be convenient to repeat his statement. In terms of the usual polar coordinates Laplace's equations are

$$\frac{du}{dt} - 2nv \cos \theta = -\frac{g}{a} \frac{d}{d\theta} (\zeta - \bar{\zeta}), \quad \dots\dots\dots\dots\dots(9)$$

$$\frac{dv}{dt} + 2nu \cos \theta = -\frac{g}{a \sin \theta} \frac{d}{d\omega} (\zeta - \bar{\zeta}), \quad \dots\dots\dots(10)$$

$$\frac{d\zeta}{dt} = -\frac{1}{a \sin \theta} \left\{ \frac{d (hu \sin \theta)}{d\theta} + \frac{d (hv)}{d\omega} \right\}, \quad \dots\dots\dots(11)$$

where u, v are the velocities along and perpendicular to the meridian, ζ is the elevation at any point, $\bar{\zeta}$ the equilibrium value of ζ, a denotes the earth's radius, n the angular velocity of rotation, and h the depth of the ocean at any point. To determine the free steady motions, we are to put $\bar{\zeta} = 0$ as well as du/dt, dv/dt, $d\zeta/dt$. Thus

$$u = -\frac{g}{2na \sin \theta \cos \theta} \frac{d\zeta}{d\omega}, \qquad v = \frac{g}{2na \cos \theta} \frac{d\zeta}{d\theta}; \quad \dots\dots(12)$$

and

$$\frac{d (h \sec \theta)}{d\theta} \frac{d\zeta}{d\omega} - \frac{d (h \sec \theta)}{d\omega} \frac{d\zeta}{d\theta} = 0. \quad \dots\dots\dots(13)$$

If $h \sec \theta$ be constant, (13) is satisfied identically. In any other case a restriction is imposed upon ζ. If h be constant or a function of the latitude only, ζ must be independent of ω; in other words the elevation must be symmetrical about the polar axis. In correspondence therewith u must be zero and v constant along each parallel of latitude.

In the application to an ocean completely covering the earth, such as is considered in Darwin's solution, the above conditions are easily satisfied, and the free steady motions, thus shown to be possible, explain the large deviation of the calculated fortnightly tide from the equilibrium value. What does not appear to have been sufficiently recognized is the extent to which this state of things must be disturbed by the limitations of the actual ocean. Since v must be constant along every parallel of latitude, it follows that a single barrier extending from pole to pole would suffice to render impossible all steady motion; and when this condition is secured a tide of sufficiently long period cannot deviate from its equilibrium value. Now the actual state of things corresponds more nearly to the latter than to the former ideal. From the north pole to Cape Horn the barriers exist; and thus it is only in the region south of Cape Horn that the circulating steady motion can establish itself. It would seem that this restricted and not wholly un-obstructed area would fail to disturb greatly the state of things that would prevail, were every parallel of latitude barred. If this conclusion be admitted, the theoretical fortnightly tide will not differ materially from its equilibrium value, and Darwin's former calculation as to the earth's rigidity will regain its significance.

Some caution is required in estimating the weight of the argument above adduced. Though there were no free disturbance possible of *infinitely* long period, it would come to the same, or to a worse, thing if free periods existed comparable with that of the forces, which is itself by hypothesis a long period. On this account a blocking of every parallel of latitude by small detached islands would not suffice, although meeting the theoretical requirement of the limiting case.

It would serve as a check and be otherwise interesting if it were possible to calculate the fortnightly tide for an ocean of uniform depth *bounded by two meridians*. The solution must differ widely from that appropriate to an unlimited ocean; but, although the conditions are apparently simple, it does not seem to be attainable by Laplace's methods. A similar solution for the semi-diurnal tide would be interesting for other reasons.

In any case I think that observations and reductions of the fortnightly tide should be pursued. Observation is competent to determine not merely the general magnitude of the tide but the law as dependent upon latitude and longitude. Should the observed law conform to that of the equilibrium theory, it would go a long way to verify *à posteriori* the applicability of this theory to the circumstances of the case.

283.

ON THE FREE VIBRATIONS OF SYSTEMS AFFECTED WITH SMALL ROTATORY TERMS.

[*Philosophical Magazine*, v. pp. 293—297, 1903.]

By a suitable choice of coordinates the expressions for the kinetic and potential energies of the system may be reduced to the forms

$$T = \tfrac{1}{2} a_1 \dot{\phi}_1^2 + \tfrac{1}{2} a_2 \dot{\phi}_2^2 + \dots , \qquad \dots\dots\dots\dots\dots\dots(1)$$

$$V = \tfrac{1}{2} c_1 \phi_1^2 + \tfrac{1}{2} c_2 \phi_2^2 + \dots . \qquad \dots\dots\dots\dots\dots\dots(2)$$

If there be no dissipative forces, the equations of free vibration are

$$\left.\begin{array}{l} a_1 \ddot{\phi}_1 + c_1 \phi_1 + \beta_{12}\dot{\phi}_2 + \beta_{13}\dot{\phi}_3 + \dots = 0 \\[2mm] a_2 \ddot{\phi}_2 + c_2 \phi_2 + \beta_{21}\dot{\phi}_1 + \beta_{23}\dot{\phi}_3 + \dots = 0 \\[2mm] \dots\dots\dots\dots\dots\dots\dots\dots\dots\dots\dots \end{array}\right\} , \qquad \dots\dots\dots\dots(3)$$

where $\beta_{rs} = -\beta_{sr}$; and under the restriction contemplated all the quantities β are *small*.

If in equations (3) we suppose that the whole motion is proportional to $e^{i\sigma t}$,

$$\left.\begin{array}{l} (c_1 - \sigma^2 a_1)\,\phi_1 + i\sigma\beta_{12}\phi_2 + i\sigma\beta_{13}\phi_3 + \dots = 0 \\[2mm] (c_2 - \sigma^2 a_2)\,\phi_2 + i\sigma\beta_{21}\phi_1 + i\sigma\beta_{23}\phi_3 + \dots = 0 \\[2mm] \dots\dots\dots\dots\dots\dots\dots\dots\dots\dots\dots \end{array}\right\} ; \qquad \dots\dots\dots\dots(4)$$

and it is known that whatever may be the magnitudes of the β's, the values of the σ's are real. The *frequencies* are equal to $\sigma/2\pi$.

If there were no rotatory terms, the above system of equations would be satisfied by supposing one coordinate ϕ_r to vary suitably, while the remaining coordinates vanish. In the actual case there will be *in general* a corresponding solution in which the value of any other coordinate ϕ_s will be small relatively to ϕ_r.

Hence if we omit the terms of the *second* order in β, the rth equation becomes

$$(c_r - \sigma_r^2 a_r)\,\phi_r = 0, \qquad \dots\dots\dots\dots\dots\dots(5)$$

from which we see that σ_r is approximately the same as if there were no rotatory terms.

From the sth equation we obtain

$$(c_s - \sigma_r^2 a_s)\, \phi_s + i\sigma_r \beta_{sr} \phi_r = 0,$$

terms of the second order being omitted; whence

$$\phi_s : \phi_r = -\frac{i\sigma_r \beta_{sr}}{c_s - \sigma_r^2 a_s} = \frac{i\sigma_r \beta_{sr}}{a_s(\sigma_r^2 - \sigma_s^2)}, \quad \ldots\ldots\ldots\ldots(6)$$

where on the right the values of σ_r, σ_s from the first approximation (5) may be used. This equation determines the altered type of vibration; and we see that the coordinates ϕ_s are in the same phase, but that this phase differs by a quarter period from the phase of ϕ_r.

We have seen that when the rotatory terms are small, the value of σ_r may be calculated approximately without allowance for the change of type; but by means of (6) we may obtain a still closer approximation, in which the squares of the β's are retained. The rth equation (4) gives

$$a_r \sigma_r^2 = c_r + \Sigma \frac{\sigma_r^2 \beta_{rs}^2}{a_s(\sigma_r^2 - \sigma_s^2)}. \quad \ldots\ldots\ldots\ldots\ldots(7)$$

Since the squares of the σ's are positive, as well as a_r, a_s, c_r, we recognize that the effect of β_{rs} is to increase σ_r^2 if σ_r^2 be already greater than σ_s^2, and to diminish it if it be already the smaller. Under the influence of the β's the σ's may be considered to *repel* one another. If the smallest value of σ_r be finite, it will be lowered by the action of the rotatory terms*.

The vigour of the repulsion increases as the difference between σ_r and σ_s diminishes. If σ_r and σ_s are equal, the formulæ (6), (7) break down, unless indeed $\beta_{rs} = 0$. It is clear that the original assumption that ϕ_s is small relatively to ϕ_r fails in this case, and the reason is not far to seek. When two normal modes have exactly the same frequency, they may be combined in any proportions without alteration of frequency, and the combination is as much entitled to be considered normal as its constituents. But the smallest alteration in the system will in general render the normal modes determinate; and there is no reason why the modes thus determined should not differ finitely from those originally chosen.

A simple example is afforded by a circular membrane vibrating so that one diameter is nodal. When all is symmetrical, any diameter may be chosen to be nodal; but if a small excentric load be attached, the nodal diameter must either itself pass through the load or be perpendicular to the diameter that

---

* This conclusion was given in *Phil. Mag.* v. p. 138 (1903) [p. 86 above], but without some reservations presently to be discussed. Similar reservations are called for in *Theory of Sound*, §§ 90, 102.

does so (*Theory of Sound*, § 208). Under the influence of the load the two originally coincident frequencies separate.

In considering the modifications required when equal frequencies occur, it may suffice to limit ourselves to the case where *two* normal modes only have originally the same frequency, and we will suppose that these are the first and second. Accordingly, the coincidence being supposed to be exact,

$$c_1/a_1 = c_2/a_2 = \sigma_0^2. \quad \dots\dots\dots\dots\dots\dots(8)$$

The relation between ϕ_1 and ϕ_2 and the altered frequencies are to be obtained from the first two equations of (3), in which the terms in ϕ_3, ϕ_4, &c. are at first neglected as being of the second order of small quantities. Thus

$$(c_1 - \sigma^2 a_1)\,\phi_1 + i\sigma\beta_{12}\phi_2 = 0, \quad (c_2 - \sigma^2 a_2)\,\phi_2 - i\sigma\beta_{12}\phi_1 = 0, \quad \dots\dots(9)$$

in which the two admissible values of σ^2 are given by

$$(c_1 - a_1\sigma^2)(c_2 - a_2\sigma^2) - \sigma^2\beta_{12}^2 = 0. \quad \dots\dots\dots\dots(10)$$

If one of the factors of the first term, *e.g.* the second, be finite, β_{12}^2 may be neglected and a value of σ^2 is found by equating the first factor to zero; but in the present case both factors are small together. On writing σ_0 for σ in the small term, (10) becomes

$$(\sigma^2 - \sigma_0^2)^2 = \sigma_0^2\beta_{12}^2/a_1 a_2, \quad \dots\dots\dots\dots\dots(11)$$

so that

$$\sigma^2 - \sigma_0^2 = \pm\,\sigma_0\beta_{12}/\sqrt{(a_1 a_2)}, \quad \dots\dots\dots\dots(12)$$

or

$$\sigma = \sigma_0 \pm \tfrac{1}{2}\beta_{12}/\sqrt{(a_1 a_2)}. \quad \dots\dots\dots\dots(13)$$

The disturbance of the frequency from its original value is now of the *first* order in β_{12}, and one frequency is raised and the other depressed by the same amount.

As regards the ratios in which ϕ_1, ϕ_2 enter into the new normal modes, we have from (9)

$$\frac{\phi_1}{\phi_2} = \frac{a_2(\sigma_0^2 - \sigma^2)}{i\sigma_0\beta_{12}} = \pm\,i\,\sqrt{(a_2/a_1)}. \quad \dots\dots\dots\dots(14)$$

From (14) we see that in the new normal vibrations the two original coordinates are combined so as to be in quadrature with one another, and in such proportion that the energies of the constituent motions are equal.

The value of any other coordinate ϕ_s accompanying ϕ_1 and ϕ_2 in vibration σ is obtained from the sth equation (4). Thus, squares of β's being neglected,

$$(c_s - \sigma^2 a_s)\,\phi_s + i\sigma\beta_{s1}\phi_1 + i\sigma\beta_{s2}\phi_2 = 0, \quad \dots\dots\dots\dots(15)$$

in which, if we please, we may substitute for ϕ_2 in terms of ϕ_1 from (14).

For the second approximation to σ we get from (15) and the two first equations (4)

$$\left\{c_1 - \sigma^2 a_1 - \Sigma \frac{\sigma^2 \beta_{1s}^2}{c_s - \sigma^2 a_s}\right\} \phi_1 + \left\{i\sigma\beta_{12} + \Sigma \frac{\sigma^2 \beta_{1s}\beta_{s2}}{c_s - \sigma^2 a_s}\right\} \phi_2 = 0,$$

$$\left\{c_2 - \sigma^2 a_2 - \Sigma \frac{\sigma^2 \beta_{2s}^2}{c_s - \sigma^2 a_s}\right\} \phi_2 + \left\{i\sigma\beta_{21} + \Sigma \frac{\sigma^2 \beta_{2s}\beta_{s1}}{c_s - \sigma^2 a_s}\right\} \phi_1 = 0,$$

in which the summation extends to all values of s other than 1 and 2. In the coefficients of the second terms it is to be observed that $\beta_{12} = -\beta_{21}$, and that $\beta_{1s}\beta_{s2} = \beta_{2s}\beta_{s1}$; so that the determinant of the equations becomes

$$\left\{c_1 - \sigma^2 a_1 - \Sigma \frac{\sigma^2 \beta_{1s}^2}{c_s - \sigma^2 a_s}\right\} \left\{c_2 - \sigma^2 a_2 - \Sigma \frac{\sigma^2 \beta_{2s}^2}{c_s - \sigma^2 a_s}\right\} - \sigma^2 \beta_{12}^2 = 0, \quad \dots(16)$$

terms of the fourth order in β being omitted. In (16) $c_1 - \sigma^2 a_1$, $c_2 - \sigma^2 a_2$ are each of the order β. Correct to the third order we obtain with the use of (12)

$$(\sigma^2 - \sigma_0^2)^2 - \sigma_0^2 \frac{\beta_{12}^2}{a_1 a_2} \mp \frac{\sigma_0 \beta_{12}^3}{(a_1 a_2)^{\frac{3}{2}}} \pm \frac{\sigma_0^3 \beta_{12}}{(a_1 a_2)^{\frac{3}{2}}} \Sigma \frac{a_1 \beta_{2s}^2 + a_2 \beta_{1s}^2}{c_s - \sigma_0^2 a_s} = 0, \quad \dots(17)$$

whence

$$\sigma^2 - \sigma_0^2 = \pm \sigma_0 \frac{\beta_{12}}{\sqrt{(a_1 a_2)}} + \frac{\frac{1}{2}\beta_{12}^2}{a_1 a_2} - \frac{\frac{1}{2}\sigma_0^2}{a_1 a_2} \Sigma \frac{a_1 \beta_{2s}^2 + a_2 \beta_{1s}^2}{c_s - \sigma_0^2 a_s}. \quad \dots\dots(18)$$

In (18) β_{12} is supposed to be of not higher order of small quantities than β_{1s}, β_{2s}. For example, we are not at liberty to put $\beta_{12} = 0$.

In the above we have considered the modification introduced by the β's into a vibration which when undisturbed is one of two with equal frequencies. If the type of vibration under consideration be one of those whose frequency is not repeated, the original formulæ (6), (7) undergo no essential modification.

In the following paper some of the principles of the present are applied to a hydrodynamical example.

284.

ON THE VIBRATIONS OF A RECTANGULAR SHEET OF ROTATING LIQUID.

[*Philosophical Magazine*, v. pp. 297—301, 1903.]

THE problem of the free vibrations of a rotating sheet of gravitating liquid of small uniform depth has been solved in the case where the boundary is circular*. When the boundary is rectangular, the difficulty of a complete solution is much greater; but I have thought that it would be of interest to obtain a partial solution, applicable when the angular velocity of rotation is *small*.

If ζ be the elevation, u, v the component velocities of the relative motion at any point, the equations of free vibration, when these quantities are proportional to $e^{i\sigma t}$, are

$$i\sigma u - 2nv = -g\, d\zeta/dx, \qquad i\sigma v + 2nu = -g\, d\zeta/dy, \quad \dots\dots\dots(1)$$

and

$$\frac{d^2\zeta}{dx^2} + \frac{d^2\zeta}{dy^2} + \frac{\sigma^2 - 4n^2}{gh}\zeta = 0, \quad \dots\dots\dots\dots\dots(2)$$

in which n denotes the angular velocity of rotation, h the depth of the water (as rotating), and g the acceleration of gravity. The boundary walls will be supposed to be situated at $x = \pm \frac{1}{2}\pi$, $y = \pm y_1$.

When n is evanescent, one of the principal vibrations is represented by

$$u = \cos x, \qquad v = 0; \quad \dots\dots\dots\dots\dots\dots(3)$$

and ζ is proportional to $\sin x$, so that

$$\sigma^2 = gh. \quad \dots\dots\dots\dots\dots\dots\dots(4)$$

This determines the frequency when $n = 0$. And since by symmetry a positive and a negative n must influence the frequency alike, we conclude that (4) still holds so long as n^2 is neglected. Thus to our order of approximation the

---

* Kelvin, *Phil. Mag.* Aug. 1880; Lamb, *Hydrodynamics*, §§ 200, 202, 203.

frequency is uninfluenced by the rotation, and the problem is reduced to finding the effect of the rotation upon that mode of vibration to which (3) is assumed to be a first approximation. The equation for ζ is at the same time reduced to

$$\frac{d^2\zeta}{dx^2} + \frac{d^2\zeta}{dy^2} + \zeta = 0. \quad \dots\dots\dots\dots\dots\dots\dots(5)$$

Since v is itself of the order n, the first of equations (1) shows that u, as well as ζ, satisfies (5).

Taking u and v as given in (3) and the corresponding ζ as the first approximation, we add terms u', v', ζ', proportional to n, whose forms are to be determined from the equations

$$i\sigma u' = - g \, d\zeta'/dx, \quad \dots\dots\dots\dots\dots\dots\dots\dots\dots(6)$$

$$i\sigma v' = - g \, d\zeta'/dy - 2n \cos x, \quad \dots\dots\dots\dots\dots(7)$$

$$(d^2/dx^2 + d^2/dy^2 + 1)\,(\zeta', u', v') = 0. \quad \dots\dots\dots\dots(8)$$

They represent in fact a motion which would be possible in the absence of rotation under forces parallel to v and proportional to $\cos x$. This consideration shows that u' is an odd function of both x and y, and v' an even function. If we assume

$$u' = A_2 \sin 2x + A_4 \sin 4x + \dots, \quad \dots\dots\dots\dots\dots(9)$$

the boundary condition to be satisfied at $x = \pm \frac{1}{2}\pi$ is provided for, whatever functions of y A_2, A_4, &c. may be. If we eliminate ζ' from (6), (7), we find

$$\frac{dv'}{dx} = \frac{du'}{dy} + \frac{2n}{i\sigma} \sin x = \frac{dA_2}{dy} \sin 2x + \frac{dA_4}{dy} \sin 4x + \dots + \frac{2n}{i\sigma} \sin x \, ;$$

or, on integration,

$$v' = \frac{2ni}{\sigma} \cos x - \frac{dA_0}{dy} - \tfrac{1}{2} \cos 2x \frac{dA_2}{dy} - \tfrac{1}{4} \cos 4x \frac{dA_4}{dy} - \dots, \quad \dots\dots(10)$$

dA_0/dy being the constant of integration. In (10) the A's are to be so chosen that $v' = 0$ when $y = \pm y_1$ for all values of x between $-\frac{1}{2}\pi$ and $+\frac{1}{2}\pi$.

From (8) we see that A_2, A_4, &c. are to be taken so as to satisfy

$$\frac{d^2A_2}{dy^2} - 3A_2 = 0, \qquad \frac{d^2A_4}{dy^2} - 15A_4 = 0, \quad \&c.,$$

or, since the A's are odd functions of y,

$$A_2 = B_2 \sinh(\sqrt{3} \cdot y), \qquad A_4 = B_4 \sinh(\sqrt{15} \cdot y), \quad \&c.$$

Also
$$A_0 = B_0 \sin y.$$

In these equations the B's are absolute constants.

The boundary conditions at $y = \pm y_1$ now take the form

$$0 = \frac{2ni}{\sigma} \cos x - B_0 \cos y_1$$

$$- \tfrac{1}{2}\sqrt{3} \, . \, B_2 \cosh (\sqrt{3} \, . \, y_1) \, . \, \cos 2x$$

$$- \tfrac{1}{4}\sqrt{15} \, . \, B_4 \cosh (\sqrt{15} \, . \, y_1) \, . \, \cos 4x - \dots, \quad \dots\dots\dots(11)$$

which can be satisfied if $\cos x$ be expressed between the limits of x in the series

$$\cos x = C_0 + C_2 \cos 2x + C_4 \cos 4x + \dots. \quad \dots\dots\dots(12)$$

By Fourier's theorem we find that (12) holds between $x = -\tfrac{1}{2}\pi$ and $x = +\tfrac{1}{2}\pi$, if

$$C_0 = \frac{2}{\pi}, \quad C_2 = \frac{4}{3\pi}, \quad C_4 = -\frac{4}{15\pi}, \quad C_{2m} = -(-1)^m \frac{4}{(4m^2 - 1)\pi} \quad \dots(13)$$

The B's are thus determined by (11), and we get

$$A_0 = \frac{2ni}{\sigma} \frac{2}{\pi} \frac{\sin y}{\cos y_1}, \qquad A_2 = \frac{2ni}{\sigma} \frac{2}{\sqrt{3}} \frac{4}{3\pi} \frac{\sinh (\sqrt{3} \, . \, y)}{\cosh (\sqrt{3} \, . \, y_1)},$$

$$A_{2m} = \frac{2ni}{\sigma} \frac{2m}{\sqrt{(4m^2 - 1)}} \frac{4(-1)^{m+1}}{(4m^2 - 1)\pi} \frac{\sinh (y \sqrt{4m^2 - 1})}{\cosh (y_1 \sqrt{4m^2 - 1})}.$$

Hence, finally, for the complete values of u and v to this order of approximation

$$u = \cos x + \frac{2ni}{\sigma} \left\{ \frac{8 \sin 2x}{3 \sqrt{3} \, . \, \pi} \frac{\sinh (\sqrt{3} \, . \, y)}{\cosh (\sqrt{3} \, . \, y_1)} + \dots \right\}, \quad \dots\dots\dots(14)$$

$$v = \frac{2ni}{\sigma} \left\{ \cos x - \frac{2}{\pi} \frac{\cos y}{\cos y_1} - \frac{4}{3\pi} \cos 2x \frac{\cosh (\sqrt{3} \, . \, y)}{\cosh (\sqrt{3} \, . \, y_1)} + \dots \right\} \quad \dots\dots(15)$$

The limiting values of x have been supposed, for the sake of brevity, to be $\pm \tfrac{1}{2}\pi$. If we denote them by $\pm x_1$, we are to replace x, y, y_1 in (14), (15) by $\tfrac{1}{2}\pi x/x_1, \tfrac{1}{2}\pi y/x_1, \tfrac{1}{2}\pi y_1/x_1$. At the same time (4) becomes

$$\sigma^2 = \frac{\pi^2 gh}{4x_1{}^2}. \quad \dots\dots\dots\dots\dots\dots\dots(16)$$

As was to be expected, the small terms in (14), (15) are in quadrature with the principal term. The success of the approximation requires that the frequency of revolution be small in comparison with that of vibration.

If y_1 be such that $\cos (\tfrac{1}{2}\pi y_1/x_1)$ vanishes, or even becomes very small, the solution expressed in (14), (15) fails. This happens, for example, when the boundary is square, so that $y_1 = x_1$. The inference is that the assumed solution (3) does not, or rather does not continue to, represent the facts of the case as a first approximation.

From the principles explained in the previous paper, or independently, it is evident that in the case of the square (3) must be replaced by

$$u = \cos x, \qquad v = \pm i \cos y, \qquad \dots\dots\dots\dots\dots(17)$$

corresponding to which

$$\zeta = \frac{\sigma_0}{g}(- i \sin x \pm \sin y). \qquad \dots\dots\dots\dots\dots(18)$$

These values satisfy all the conditions when there is no rotation, and $\sigma_0 = \sqrt{(gh)}$, as in (4). For the second approximation we retain these terms, adding to them u', v', ζ', which are to be treated as small. So far, the procedure is the same as in the formation of (6), (7); but now we must be prepared for an alteration of σ from its initial value σ_0 by a quantity of the first order. Hence, with neglect of n^2,

$$i(\sigma - \sigma_0) \cos x + i\sigma_0 u' \mp 2ni \cos y = - g \, d\zeta'/dx, \qquad \dots\dots\dots(19)$$

$$\mp (\sigma - \sigma_0) \cos y + i\sigma_0 v' + 2n \cos x = - g \, d\zeta'/dy. \qquad \dots\dots\dots(20)$$

These equations are the same as would apply in the absence of rotation if we suppose impressed forces to act parallel to u and v proportional to

$$i(\sigma - \sigma_0) \cos x \mp 2ni \cos y, \qquad \dots\dots\dots\dots\dots(21)$$

$$\mp (\sigma - \sigma_0) \cos y + 2n \cos x, \qquad \dots\dots\dots\dots\dots(22)$$

respectively.

The complete solution of (19), (20) to the first order of n would lead to rather long expressions. The point of greatest interest is the alteration of frequency, and this can perhaps be most easily treated by a simple mechanical consideration. The forces given in (21), (22) must be such as not wholly to disturb the initial motion (17) with which they synchronize. Accordingly (21) must be free from a component capable of stimulating a vibration similar to $u = \cos x$, and in like manner (22) must be incapable of stimulating a motion similar to $v = \cos y$. The necessary conditions are

$$\iint \cos x \, \{(\sigma - \sigma_0) \cos x \mp 2n \cos y\} \, dx \, dy = 0,$$

$$\iint \cos y \, \{(\sigma - \sigma_0) \cos y \mp 2n \cos x\} \, dx \, dy = 0,$$

the integration being taken over the whole area. On account of the symmetry the two conditions coincide; and it is sufficient to integrate for x and y between the limits 0 and $\frac{1}{2}\pi$. Thus

$$(\sigma - \sigma_0) \cdot \tfrac{1}{2}\pi \cdot \tfrac{1}{2}\pi = \pm 2n \cdot 1 \cdot 1,$$

so that

$$\sigma - \sigma_0 = \pm \frac{16n}{\pi^2}. \qquad \dots\dots\dots\dots\dots\dots(23)$$

Since n and σ are of the same dimensions, this result holds good, whatever may be the side of the square.

It may be of interest, and serve as a confirmation of the above procedure, to mention that when applied to the principal vibration in a rotating *circular* trough it gives

$$\sigma - \sigma_0 = \pm \frac{2n}{z_1^2 - 1}, \quad \ldots\ldots\ldots\ldots\ldots\ldots\ldots(24)$$

where z_1 is the first root of $J_1'(z) = 0$, equal to $1\cdot841$, so that

$$\sigma - \sigma_0 = \pm \frac{2n}{2\cdot38}. \quad \ldots\ldots\ldots\ldots\ldots\ldots\ldots(25)$$

An accordant result may be deduced from the analysis given by Lamb, § 203, by putting $s = 1$, and taking account of the properties of the function J_1. The corresponding value of ζ is given by

$$\zeta = e^{i\sigma t} J_1(kr) \{\cos\theta + i\sin\theta\}. \quad \ldots\ldots\ldots\ldots\ldots(26)$$

[1911. This subject is pursued in "Notes respecting Tidal Oscillations upon a Rotating Globe," *Proceedings Royal Society*, A, Vol. LXXXII. p. 448, 1909.]

285.

ON THE SPECTRUM OF AN IRREGULAR DISTURBANCE.

[*Philosophical Magazine*, v. pp. 238—243, 1903.]

IN my paper "On the Character of the Complete Radiation at a given Temperature"*, I have traced the consequences of supposing white light to consist of a random aggregation of impulses of certain specified types, and have shown how to calculate the distribution of energy in the resulting spectrum. The argument applies, of course, to all vibrations capable of propagation along a line, and it is convenient to fix the ideas upon the transverse vibrations of a stretched string. Suppose that this is initially at rest in its equilibrium position and that velocities represented by $\phi(x)$ are communicated to the various parts. The whole energy is proportional to $\int_{-\infty}^{+\infty} \{\phi(x)\}^2 dx$; and it is desired to know how this energy is distributed among the various components into which the disturbance may be analysed. By Fourier's theorem,

$$\pi \phi(x) = \int_0^\infty f_1(k) \cos kx \, dk + \int_0^\infty f_2(k) \sin kx \, dk, \quad \dots\dots\dots(1)$$

where

$$f_1(k) = \int_{-\infty}^{+\infty} \cos kv \, \phi(v) \, dv, \quad f_2(k) = \int_{-\infty}^{+\infty} \sin kv \, \phi(v) \, dv. \quad \dots\dots(2)$$

It was shown that the desired information is contained in the formula

$$\int_{-\infty}^{+\infty} \{\phi(x)\}^2 dx = \frac{1}{\pi} \int_0^\infty [\{f_1(k)\}^2 + \{f_2(k)\}^2] \, dk. \quad \dots\dots\dots(3)$$

As an example, we may take an impulse localized in the neighbourhood of a point, and represented by

$$\phi(x) = e^{-c^2 x^2}. \quad \dots\dots\dots\dots\dots\dots\dots\dots(4)$$

Equation (1) becomes

$$e^{-c^2 x^2} = \frac{1}{c\sqrt{\pi}} \int_0^\infty e^{-k^2/4c^2} \cos kx \, dk, \quad \dots\dots\dots\dots(5)$$

---

* *Phil. Mag.* XXVII. p. 460 (1889); *Scientific Papers*, III. p. 268.

while for the distribution of energy in the spectrum by (3)

$$\int_{-\infty}^{+\infty} e^{-2c^2x^2}\,dx = \frac{1}{c^3}\int_0^{\infty} e^{-k^2/2c^2}\,dk. \quad\ldots\ldots\ldots\ldots\ldots\ldots(6)$$

"If an infinite number of impulses, similar (but not necessarily equal) to (4) and of arbitrary sign, be distributed at random over the whole range from $-\infty$ to $+\infty$, the intensity of the resultant for an absolutely definite value of k would be indeterminate. Only the *probabilities* of various resultants could be assigned. And if the value of k were changed, by however little, the resultant would again be indeterminate. Within the smallest assignable range of k there is room for an infinite number of independent combinations. We are thus concerned only with an average, and the intensity of each component may be taken to be proportional to the total number of impulses (if equal) without regard to their phase-relations. In the aggregate vibration, the law according to which the energy is distributed is still for all practical purposes that expressed by (6)."

The factor $e^{-c^2x^2}$ in the impulse was introduced in order to obviate discontinuity. The larger c is supposed to·be, the more highly localized is the impulse. If we suppose c to become infinite, the impulse is infinitely narrow, and the disturbances at neighbouring points, however close, become independent of one another. It would seem therefore from (6) that in the spectrum of an absolutely irregular disturbance (where the ordinates of the representative curve are independent at all points) the energy between k and $k + dk$ is proportional to dk simply, or that the energy curve is a straight line *when k is taken as abscissa.* If we take the wave-length λ (to which k is reciprocal) as abscissa, the ordinate of the energy curve would be as λ^{-2}.

The simple manner in which dk occurs in Fourier's theorem has always led me to favour the choice of k, rather than of λ, as independent variable. This may be a matter of convenience or of individual preference; but something more important is involved in the alternative of whether the energy of absolutely arbitrary disturbance is proportional to dk or to $d\lambda$. In Prof. Schuster's very important application of optical methods to the problems of meteorology, which seems to promise a revolution in that and kindred sciences, the latter is the conclusion arrived at. "Absolute irregularity would show itself by an energy-curve which is independent of the wave-length; *i.e.*. a straight line when the energy and wave-length or period are taken as rectangular coordinates..."*. It is possible that the discrepancy may depend upon some ambiguity; but in any case I have thought that it would not be amiss to reconsider the question, using a different and more elementary method.

For this purpose we will regard the string as fixed at the two points

---

* "The Periodogram of Magnetic Declination, &c.," *Camb. Phil. Trans.* xviii. p. 108 (1899).

$x = 0$ and $x = l$. The possible vibrations are then confined to the well-known "harmonics," and k is limited to an infinite series of detached values forming an arithmetical progression. The general value of the displacement y at time t is

$$y = \Sigma \sin \frac{s\pi x}{l} \left(A_s \cos \frac{s\pi a t}{l} + B_s \sin \frac{s\pi a t}{l} \right), \dots \dots \dots (7)$$

in which a is the velocity of propagation and s is one of the series $1, 2, 3 \dots$. From (7) the constant total energy $(T + V)$ is readily calculated. Thus (*Theory of Sound*, § 128) if M denote the whole mass, τ_s the period of component s,

$$T + V = \pi^2 M . \Sigma \frac{A_s{}^2 + B_s{}^2}{\tau_s{}^2}, \dots \dots \dots \dots (8)$$

an equation which gives the distribution of energy among the various modes.

The initial values of y and $\dot{y}$ are

$$y_0 = \Sigma A_s \sin \frac{s\pi x}{l}, \quad \dot{y}_0 = \frac{\pi a}{l} \Sigma s B_s \sin \frac{s\pi x}{l};$$

whence

$$A_s = \frac{2}{l} \int_0^l y_0 \sin \frac{s\pi x}{l} \, dx; \quad B_s = \frac{2}{\pi a s} \int_0^l \dot{y}_0 \sin \frac{s\pi x}{l} \, dx. \dots \dots (9)$$

If we suppose that $y_0 = 0$ throughout and that $\dot{y}_0$ is finite only in the neighbourhood of $x = \xi$, we have $A_s = 0$, and

$$B_s = \frac{2}{\pi a s} \sin \frac{s\pi \xi}{l} . Y, \dots \dots \dots \dots (10)$$

where $Y = \int \dot{y}_0 dx$. The energy in the various modes being proportional to $B_s{}^2 / \tau_s{}^2$, or to

$$\frac{1}{s^2 \tau_s{}^2} \sin^2 \frac{s\pi \xi}{l},$$

in which $s^2 \tau_s{}^2 = \tau_1{}^2$, is thus independent of s except for the factor $\sin^2 (s\pi\xi/l)$. And even this limited dependence on s disappears if we take the mean with respect to ξ. We may conclude that in the mean the energy of every mode is the same; and since the modes are uniformly spaced with respect to their frequency (proportional to s) and *not* with respect to their period or wavelength, this result corresponds with a constant ordinate of the energy curve when k is taken as abscissa.

It is to be noted that the above corresponds to an arbitrary localized *velocity*. We shall obtain a higher and perhaps objectionable degree of discontinuity, if we make a similar supposition with respect to the *displacement*. Setting in (9) $\dot{y}_0 = 0$ throughout and $y_0 = 0$ except in the neighbourhood of ξ, we get $B_s = 0$ and

$$A_s = \frac{2}{l} \sin \frac{s\pi \xi}{l} . Y_1, \dots \dots \dots \dots (11)$$

where $Y_1 = \int y_0 dx$.   By (8) the mean energy in the various modes is now proportional to $1/\tau_s^2$ or to s^2.   When l is made infinite, so that τ_s may be treated as continuous, we have an energy curve in which the ordinate is proportional to s^2 or k^2, k being abscissa.

We may sum up by saying that if the velocity curve is arbitrary at every point the energy between k and $k + dk$ varies as dk, but if the displacement be arbitrary the energy over the same range varies as $k^2 dk$.

In Schuster's Periodogram, as applied to meteorology, the conception of energy does not necessarily enter, and the definitions may be made at pleasure.   But unless some strong argument should appear to the contrary, it would be well to follow optical (or rather mechanical) analogy, and this, if I understand him, Schuster professes to do.   If the energy associated with the curve $\phi(x)$ to be analysed is represented by $\int \{\phi(x)\}^2 dx$, $\phi(x)$ must be assimilated to the *velocity* and not to the *displacement* of a stretched string.

We have seen that when $\phi(x)$ is arbitrary at all points the ordinate of the energy curve is independent of k.   In the curves with which we are concerned in meteorology the values of $\phi(x)$ at neighbouring points are related, being influenced by the same accidental causes.   But at sufficiently distant points the values of $\phi(x)$ will be independent.   Equation (6) suggests that in such cases the ordinate of the energy curve (k being abscissa) will tend to become constant when k is small enough.

Another illustration of the application of Fourier's theorem to the analysis of irregular curves may be drawn from the optical theory of gratings.   For this purpose we imagine the aperture of a telescope to be reduced to a horizontal strip bounded below by a straight edge and above by the curve to be analysed, such as might be provided by a self-registering tide-gauge. Any periodicities in the curve will then exhibit themselves by bright lines in the image of a source of homogeneous light, corresponding to the usual diffraction spectra of the various orders.   An aperture of the kind required may be obtained by holding the edge of a straight lath against the teeth of a hand-saw.   When the combination is held square in front of the telescope, we have spectra corresponding to the number of teeth.   When the aperture is inclined, not only do the previously existing spectra open out, but new spectra appear in intermediate positions.   These depend upon the fact that the period now involves a sequence of *two* teeth inasmuch as alternate teeth are bent in opposite directions out of the general plane.

The theory of diffraction* shows that the method is rigorous when the source of light is a point and when we consider the illumination at those points of the focal plane which lie upon the horizontal axis (parallel to the straight edge of the aperture).

---

* See, for example, "Wave Theory of Light," *Encyc. Brit.*; *Scientific Papers*, iii. pp. 80, 87. Make $q = 0$.

In order to illustrate the matter further, Mr Gordon constructed an aperture (cut from writing-paper) in which the curved boundary* had the equation

$$y = \sin 2x + \sin (3x + \tfrac{3}{4}\pi).$$

The complete period was about half an inch and the maximum ordinate about one inch. The aperture was placed in front of a 3-inch telescope provided with a high-power eye-piece. When desired, the plane of the aperture could be considerably sloped so as to bring more periods into action and increase the dispersion.

The light employed was from a paraffin lamp†, and it was convenient to limit it by slits. Of these the first was vertical, as in ordinary spectrum work, and it was crossed by another so that at pleasure a linear or a point source could be used. In the latter case the spectrum observed agreed with expectation. Subdued spectra of the first order (corresponding to the complete period) and traces of the fourth and fifth orders were indeed present, as well as the second and third orders alone represented in the aperture-curve. *But along the horizontal axis of the diffraction pattern these subsidiary spectra vanished;* so that the absence of all components, except the second and third, from the aperture-curve could be inferred from the observation.

It will be evident from what has already been said that confusion arises when the point-source is replaced by a linear one; and this is what theory would lead us to expect. In a diffraction-grating, as usually constructed, where all the lines are of equal length, the spectra are of the same character whether the source be elongated, or not, in the vertical direction; but it is otherwise here. The inadmissibility of a linear source and the necessity for limiting the observation to the axis seriously diminish the prospect of making this method a practical one for the discovery of unknown periods in curves registering a meteorological or similar phenomena; but the fact that the analysis can be made at all in this way, without any calculation, is at least curious and instructive.

It may be added that a similar method is applicable when the phenomena to be analysed occur discontinuously. Thus if the occurrence of earthquakes be recorded by ruling fine vertical lines of given length with abscissæ proportional to time, so as to constitute a grating, the positions of the bright places in the resulting spectrum will represent the periodicities that may be present in the time distribution of the earthquakes. And in this case the use of a linear source of light, from which to form the spectrum, is admissible.

[1910. Compare Schuster, *Phil. Mag.* v. p. 344, 1903.]

---

* Figured in Thomson and Tait's *Natural Philosophy*, § 62.
† Doubtless a more powerful source would be better.

286.

CONSIDERATIONS RESPECTING THE COMBUSTION OF MODERN PROPELLANTS IN CLOSED VESSELS AND IN GUNS.

[*Minutes of Explosives Committee*, 1903.]

Closed Vessels.

THE rate of combustion (if the term may be allowed) per unit of surface is assumed to be some function $f(p)$ of the pressure at the moment (t) under consideration. This assumption does not imply that the pressure rather, for instance, than the temperature, is the governing circumstance, but merely that the pressure sufficiently *determines* the state of affairs. To a high pressure will, of necessity, correspond a high temperature.

The case of a tubular Cordite of annular section is the simplest. If r denote the external and r' the internal radius at time t,

$$- dr/dt = dr'/dt = f(p). \quad \ldots\ldots\ldots\ldots\ldots\ldots\ldots\ldots(1)$$

Thus, $r + r'$ remains constant, and with it the *surface* at any time exposed. Accordingly the rate of total combustion depends only upon the pressure, and this simplifies the question considerably.

In addition to (1) another relation is required. It is usual to assume that the pressure is proportional to the quantity of propellant already burned. This supposition will be made in some of the calculations which follow; but it may be remarked that it cannot be accurately true. In ordinary practice, the pressure is atmospheric (and not evanescent) when the combustion begins; and, as the combustion progresses, the temperature rises. Even if it were otherwise, Boyle's law would not apply strictly to the high pressures here involved.

On these accounts the pressure would be greater than it is estimated to be. On the other hand, the increase in space available for the gases as the solid propellant disappears tells in the other direction, so that there is here a tendency to compensation. Upon the whole, the simple law may not be far from the mark, so long as the temperature does not greatly vary.

In the present case it suffices to assume proportionality between the *increment* of pressure and the rate of burning, or

$$dp/dt = Af(p), \dots\dots\dots\dots\dots\dots\dots\dots\dots(2)$$

where A is constant in each experiment, but may vary in different experiments, if the weight of the charge, or the aggregate surface, or the space, be altered.

For (2) the relation between pressure and time is determined by a simple integration. Or conversely, if this relation be known from experiment, (2) determines the form of $f(p)$. It is evident that for this purpose the use of tubular Cordite offers advantages.

If $f(p) = p^n$, where n is positive, (2) gives

$$p^{1-n} = A(1-n)t + C, \dots\dots\dots\dots\dots\dots\dots(3)$$

C being a constant of integration.

If n be less than unity, (3) may be written in the form

$$p^{1-n} = A(1-n)(t-t_0), \dots\dots\dots\dots\dots\dots\dots(4)$$

t_0 being the time at which the pressure is zero. This law obtains so long as the Cordite burns with constant surface. In the ideal case the walls of the tubes become everywhere infinitely thin and disappear simultaneously, the surface exposed falling discontinuously from a constant finite value to zero. From (4), if the time occupied in the rise of pressure from zero to the maximum P be t_1,

$$P^{1-n} = A(1-n)t_1 ;$$

so that if the time be measured from the moment of zero pressure

$$(p/P)^{1-n} = t/t_1. \dots\dots\dots\dots\dots\dots\dots\dots(5)$$

For example, if, as has sometimes been supposed, $n = \frac{1}{2}$,

$$p/P = (t/t_1)^2, \dots\dots\dots\dots\dots\dots\dots\dots\dots(6)$$

the pressure being proportional to the square of the time which has elapsed from the commencement of the burning.

If in (3), n be greater than unity, the pressure cannot be supposed to be zero. The meaning of this is that under such a law the burning could not commence from a zero of pressure. In order to obtain a practical result, we should have to take into account an initial pressure, whether atmospheric or (in virtue of a primer) exceeding atmosphere. These initial pressures being relatively very small, we may expect the commencement of the burning to be slow and uncertain.

The case $n = 1$, probably somewhat closely approached in practice, is critical. Equation (3) is replaced by

$$\log p = At + C, \quad \dots\dots\dots\dots\dots\dots\dots\dots\dots\dots\dots(7)$$

or,

$$p = De^{At}, \quad \dots\dots\dots\dots\dots\dots\dots\dots\dots\dots\dots\dots\dots(8)$$

signifying that the pressure rises according to the law of compound interest. Here again the pressure cannot rise at all from zero, and from an initial atmospheric pressure the rise would be comparatively slow.

If the Cordite be solid instead of tubular, the surface continually diminishes as the combustion proceeds. The conclusion that if n be not less than unity, the pressure is incapable of rising from zero still holds good. If $n = \frac{1}{2}$, the calculation is easily made; and it appears that if t_2 be the total time of burning, (6) is replaced by

$$p/P = \sin^2(\pi t / 2t_2). \quad \dots\dots\dots\dots\dots\dots\dots\dots\dots(9)$$

The problem of finding the relation between R (the initial radius of solid Cordite) and R_1, R_2 (the inner and outer radii of tubular Cordite) in order that the times of burning of equal weights in the same vessel may be equal, has been solved by Lieutenant Wright, R.N. (*O.C. Minute* 46,198/19. 10. 98). The conclusion is

$$\tfrac{1}{2}\pi R = R_2 - R_1. \quad \dots\dots\dots\dots\dots\dots\dots\dots\dots\dots(10)$$

But, as has already been remarked, the law of the rise of pressure is quite different in the two cases. Thus, even though (10) be satisfied, a pressure nearly equal to the common maximum is reached earlier in the case of the solid than in that of the tubular Cordite, notwithstanding that equal times are needed for the final maximum. This is a natural consequence of the circumstances that the surface of the solid Cordite diminishes gradually to zero, while that of the tubular remains constant until the last moment.

Combustion in Guns.

When we consider the case of a gun in place of a closed vessel, the question is further complicated by the increasing space becoming available for the gases of combustion as the shot advances. Of special importance is the relation between the pressure and the travel (x) of the shot. The total work done is represented by $\int p\,dx$, so that if the muzzle-velocity of a given shot from a given gun is prescribed, the *mean* pressure (estimated with regard to *distance*) is thereby determined. A lowered chamber, or maximum, pressure of necessity involves a raised pressure at some other part of the course of the shot.

The muzzle velocity, giving the *mean* pressure, and the *maximum* pressure (as determined by crusher gauges) are two very important data, but they by no means exhaust all that it is desirable to know. More telling are such investigations as those of Sir A. Noble where, by delicate chronoscopes, the shot is timed as it passes various points of the bore. From these times the *accelerations* may be deduced, to which approximately the pressures are proportional. In this way we may find the strength necessary in the various parts of the gun when a given shot is fired with a given propellant.

But, for the purposes of the Explosives Committee, the converse problem seems the more important. Indeed, I am strongly impressed with the feeling that the natural order of procedure is first to determine, either from the strength of the gun, or for other reasons, what pressures it is desired to have—in other words, to define p as a function of x—and then to investigate, by theory and experiment, how nearly the desired law of pressure can be obtained by varying the nature and form of the propellant. Of course, experience may show that an improvement is attainable by an altered design of gun and an altered law of pressure, but at a given time the practical problem is to fit the propellant to the gun.

The question as to what law of pressure is desirable must be decided by experts. Economy of propellant suggests that the combustion should be finished at a comparatively early stage, say before the shot has travelled more than one quarter of its course. But at the present time the object appears rather to be to get the most out of the gun. If the maximum pressure is laid down, the aim would then be to reach this pressure early, and to maintain it without much drop over the strong part of the gun. Subsequently the pressure should fall somewhat rapidly to that considered desirable at the muzzle. From the curves that I have seen, I should suppose that there is not much to complain of in the manner of reaching the maximum, but that the maximum itself is insufficiently maintained. It cannot be right that the high pressure should operate over a very small part only of the travel of the shot.

These considerations suggest the problem of finding whether it is possible to maintain the highest pressure over a finite travel of the shot, and I believe that an answer can be given sufficient to afford practical guidance, although no doubt leaving much to be desired from a theoretical point of view.

Although some deduction for friction and perhaps for other complications would be proper, it may suffice to assume that the momentary acceleration is proportional to the pressure operative behind the shot, so that

$$p = A\, d^2x/dt^2. \quad\dots\dots\dots\dots(11)$$

In (11) the changes of x represent the travel of the shot, but the origin of x may be chosen at convenience.

The rate of total combustion at time t may be equated to $Sf(p)$, when S is the (momentary) surface of the propellant, and $f(p)$ denotes, as in (1), a suitable function of the pressure. The pressure itself depends upon the total amount of the gases, *i.e.*, upon the quantity of propellant already burned, upon the volume, and upon the temperature. The changes of temperature during the course of the combustion are certainly not insignificant, and they would have to be allowed for if the object were a precise quantitative estimate. But so long as the pressure is nearly constant, they must be unimportant, and this suffices for the immediate purpose. In the general problem, especially during the final expansion towards the muzzle of the gun, a different treatment might be necessary.

Neglecting then the change of temperature, we have to consider the effect of volume. To the original volume in the chamber we have, of course, to add that provided by the forward displacement of the shot from its seat. But a further question arises as to what is to be regarded as the chamber-volume. As the propellant burns away, the actual gas-volume is increased independently of the motion of the shot. But in consequence of the deviation from Boyle's law at high pressures, we shall obtain the closest approach to the facts if we reckon the chamber-volume throughout as that part originally unoccupied by propellant. On the whole the volume may be considered to be represented by x, where x is measured, not from the initial position of the base of the shot, but from a point further behind. The initial value of x may be denoted by x_0.

In accordance with the suppositions already detailed, the product (px) of the pressure and volume represents the total quantity of gas, and its differential co-efficient with respect to time is the rate of total combustion. Hence as our second equation we may take—

$$B \frac{d}{dt}(px) = Sf(p) \quad \dotfill (12)$$

when B is some constant; or in a form more telling for our immediate purpose—

$$S = \frac{B}{f(p)} \frac{dx}{dt}\left(p + x\frac{dp}{dx}\right). \quad \dotfill (13)$$

In (13) dx/dt represents the instantaneous velocity of the shot.

We are now in a position to examine effectively what is implied in a pressure-curve such that over a finite range p is independent of x, so that this part of the curve BC reduces to a straight line parallel to the axis of x (fig. 1). If in (13) p is constant, dp/dx is zero, and we see that S must be proportional to dx/dt, independently altogether of the form of f. Hence, in order that the maximum pressure may be maintained, the operative surface of the propellant must *increase*, and that somewhat rapidly, so as to remain proportional to the increasing velocity of the shot. This conclusion appears to be of some importance, and it shows how hopeless must be the attempt to obtain the desired feature in the pressure-curve so long as we limit ourselves to *solid* Cordite, of which the surface is all the time diminishing. Even if we substitute simple tubular Cordite we merely *maintain* the surface, and it still remains impossible, according to our equation, to keep the pressure constant.

Fig. 1.

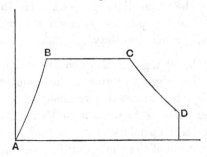

If it be desired to approximate at all closely to the condition of a maintained maximum pressure, other means must be resorted to. We might imagine fresh surfaces brought into play by the removal of an inactive coating; but the most natural device would seem to be the adoption of a multi-tubular form of propellant. When the section is simply annular, the total surface is maintained constant because the internal surface increases at the same rate as the external surface decreases. It is only necessary to provide two or more perforations in order that the gain of internal surface may *exceed* the loss of external surface.

The theory of the gain or loss of surface during combustion of a propellant burning practically in two dimensions, is more simple than might at first be supposed, at any rate so far as the earlier stages are concerned. If the boundary be circular, the rate of change of surface is the same whatever be the radius of the circle. It is on this principle that an inner circular surface always balances an external circular surface. Hence if there be two circular perforations through a round stick, the net gain of surface is at the same rate as the loss of surface which ensues when there are no perforations. And by increasing the number of perforations the gain may be made as rapid as we please.

But it is not necessary to limit these statements to circular boundaries. So long as the boundary remains *oval, i.e.*, of one curvature throughout, the result is the same, whether it be circular or not. For example, the gain at an internal circular perforation is the same as the loss at an external elliptical boundary, and this form of external boundary might, perhaps, be recommended when it is desired to work with *two* perforations. It must not be overlooked, however, that, in the final stage, a distinction will arise between the behaviour of a simple annular and a multi-tubular form. In the former case the surface disappears suddenly, while the latter must involve the separation of portions whose burning will be more like that of threads of unperforated material.

It may be argued that we do not desire an absolute maintenance of the highest pressure over any finite portion of the curve. Perhaps the requirements of the case may be met by a propellant of which the surface remains constant over the space in question. The pressure must indeed fall, but possibly not faster than is admissible. This state of things may be symbolized by the curve (fig. 2), in which BC is shown straight, though no longer horizontal. In this case, as in the former, we have still to consider what is implied in the corners, or places of strong curvature, desirable at B and C, and the required information can be obtained from (13).

Fig. 2.

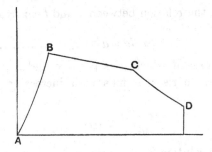

In passing through B the value of dp/dx drops suddenly from a large positive value to zero (fig. 1), or to a negative value (fig. 2), while there is no sudden change in the values of p or x or dx/dt. Hence at the point B there must be a *sudden* drop in the value of S; and again, for the same reason, there must be another sudden drop in S at the point C. The requirements for S are thus somewhat peculiar. At B there is to be a sudden drop. From B to C, S should increase, or at any rate not diminish, and again at C there is to be a sudden drop. The last drop is naturally obtained by the burning out of a simple tubular propellant, so that S falls to zero and remains at zero. But how is the sudden drop in the value of S, required at B, to be secured?

It would appear that the desired features in the curve of pressure can only be obtained by combining in the charge *two* forms of propellant. If we are limited to simple annular sections, there must be two of different times of burning. Of these, one, the thinner in the walls, burns out at B, giving the first drop of operative surface. From B to C the second portion continues to burn with constant surface, until in its turn it disappears at C. From C to D the gases expand without addition to their mass.

It will be understood that sharp corners are spoken of merely for the sake of brevity. In practice it would not be possible, or perhaps desirable, to have them quite sharp. The whole surface of a quantity of annular propellant can never disappear with absolute simultaneity, and the corners would inevitably be rounded off.

If the propellant is entirely burned away at C, the remaining part of the curve CD is out of control, but it can be calculated upon known principles, and is (I believe) well understood. In designing a propellant to suit a given gun, attention would first be given to defining the positions of B and C. The next step would be to find, by calculation, and, if necessary, by special experiment, the forms and proportional quantities of the two kinds of annular propellant necessary. It would seem that in the present problem, *i.e.*, to get the most out of the gun, the *solid* form of propellant has no place.

It may be of interest to notice that over any straight portion, such as BC, of the pressure-curve, the relation between x and t can readily be calculated. For by (11) we have

$$d^2x/dt^2 = a + bx, \quad \dots\dots\dots\dots\dots\dots(14)$$

where a, b are constants, of which a is positive and b positive or negative according as the pressure rises or falls with increasing x. Equation (14) may also be written,

$$\frac{d^2(c+x)}{dt^2} = b(c+x), \quad \dots\dots\dots\dots\dots(15)$$

where $bc = a$; and the solution is

$$c + x = Ge^{\sqrt{b}.t} + He^{-\sqrt{b}.t}, \quad \dots\dots\dots\dots(16)$$

or, $$c + x = G'\cos(\sqrt{b'}.t) + H'\sin(\sqrt{b'}.t). \quad \dots\dots\dots(17)$$

The first form obtains when b is positive, the second when b is negative, b' being equal to $-b$. G, H, G', H' are constants of integration, to be determined by the initial circumstances at B, the commencement of the straight portion of the pressure-curve.

In the case of fig. 1, where $b = 0$, (14) becomes

$$d^2x/dt^2 = a, \quad \dots\dots\dots\dots\dots\dots\dots(18)$$

whence $$x = L + Mt + \tfrac{1}{2}at^2, \quad \dots\dots\dots\dots\dots(19)$$

the motion being that of a falling body.

Another problem, whose solution would be of interest from the theoretical point of view, is that corresponding to $S = $ constant, as in the combustion of a propellant of tubular form. The case where $f(p)$ is taken proportional to p could be dealt with graphically. For from (11), (12), with omission of constant multipliers, we get

$$\frac{d}{dt}(px) = p = \frac{d^2x}{dt^2},$$

or on integration
$$px = \frac{dx}{dt} + p_1 x_1, \quad \dots\dots\dots\dots\dots\dots(20)$$

p_1, x_1 being simultaneous values of p and x when $dx/dt = 0$.

In (20) we may again replace p by d^2x/dt^2, so that

$$\frac{d^2x}{dt^2} = \frac{x_1 p_1 + dx/dt}{x}, \quad \dots\dots\dots\dots\dots\dots(21)$$

an equation by means of which the curve showing the relation between x and t could be constructed. But it may be questioned whether the actual solution would convey anything of practical value.

Although at some points confirmation may be needed, the results of this discussion seem to me to show that the proper course is first to define the pressure-curve to be aimed at. Until this has been done, at least in some degree, it is impossible to say whether one propellant is better or worse than another. When the pressure-curve has been laid down, it will be possible to approximate to it by a suitable choice of propellant. I am disposed to think that the requirements of the case cannot be met by a single form of propellant, but that a combination of two forms of tubular character may suffice.

In conclusion it may be observed that when the power of the gun is pushed by the employment of a propellant capable of giving rather high forward pressures, we can hardly expect to retain the highest standard in respect of uniformity of ballistics. The latter requirement, as well as economy of material, is promoted by an early burning out of the propellant, which may then assume the solid or unperforated form. When nearly the maximum work is obtained from the propellant, a slight irregularity in the manner of burning has but little effect upon the ballistics.

287.

ON THE BENDING OF WAVES ROUND A SPHERICAL OBSTACLE.

[*Proceedings of the Royal Society*, LXXII. pp. 40—41, 1903.]

In the *Proceedings* for January 21, 1903, Mr H. M. Macdonald discusses the effect of a reflecting spherical obstacle upon electrical and aerial waves for the case where the radius of the sphere is large compared with the wave-length (λ) of the vibrations. The remarkable success of Marconi in signalling across the Atlantic suggests a more decided bending or diffraction of the waves round the protuberant earth than had been expected, and it imparts a great interest to the theoretical problem. Mr Macdonald's results, if they can be accepted, certainly explain Marconi's success; but they appear to me to be open to objection.

If C be the source of sound, P a point upon the sphere whose centre is at O, ϕ_1 the velocity-potential at P due to the source (in the absence of the sphere), χ the angle subtended by OC, Mr Macdonald finds for the actual potential at P,

$$\phi = \phi_1 (1 - \cos \chi), \quad\dots\dots\dots\dots\dots\dots\dots\dots\dots(1)$$

so that there is no true shadow near the surface of the sphere. If C be infinitely distant, and μ denote (as usual) the cosine of the angle between OP and OC,

$$\phi = \phi_1 (1 + \mu). \quad\dots\dots\dots\dots\dots\dots\dots\dots\dots(2)$$

That the sound should vanish at the point opposite, and be quadrupled at the point immediately under the source is what would be expected; but that (however large the sphere) the shadow should be so imperfect at, for example, $\mu = -\frac{1}{2}$, is indeed startling.

The first objection that I have to offer is that nothing of this sort is observed in the case of light. The relation of wave-length to diameter of obstacle is about the same in Marconi's phenomenon as when visible light impinges upon a sphere 1 inch (2·54 cm.) in diameter. So far as I am aware no creeping of light into the dark hemisphere through any sensible angle is observed under these conditions even though the sphere is highly polished*.

But I shall doubtless be asked whether I have any complaint against the mathematical argument which leads up to (2).

As in *Theory of Sound*, § 334, the question relates to the ratio between a certain function of c (the radius) and its differential coefficient with respect to c. The function is that which occurs in the representation of a disturbance which travels outwards, and (§ 323) may be denoted by

$$\frac{e^{-ikc}}{c} f_n(ikc), \quad\dots\dots\dots\dots\dots\dots(3)$$

where
$$k = 2\pi/\lambda,$$

and
$$f_n(ikc) = 1 + \frac{n(n+1)}{2 \cdot ikc} + \frac{(n-1)\dots(n+2)}{2 \cdot 4 \cdot (ikc)^2} + \dots \quad\dots\dots\dots(4)$$

The differential coefficient of (3) is

$$-\frac{e^{-ikc}}{c^2} \{(1+ikc)f_n(ikc) - ikc f_n'(ikc)\}, \quad\dots\dots\dots\dots(5)$$

so that the ratio in question takes the form

$$\frac{-c f_n(ikc)}{(1+ikc)f_n(ikc) - ikc f_n'(ikc)}. \quad\dots\dots\dots\dots\dots(6)$$

In these expressions n is the order of the Legendre's function $P_n(\mu)$ which occurs in the series representative of the velocity-potential.

When kc is very great, the ratio expressed in (6) may assume a simplified form. From (4) we see that, if n be finite,

$$f_n(ikc) = 1, \qquad ikc f_n'(ikc) = 0,$$

ultimately, so that

$$(6) = -\frac{1}{ik}, \quad\dots\dots\dots\dots\dots\dots\dots(7)$$

independent of n.

This is the foundation of the simple result reached by Mr Macdonald. Its validity depends, therefore, upon the applicability of (7) to all values of n that need to be regarded. If when kc is infinite, only finite values of n are

---

* It may be remarked that at the centre of the shadow thrown at some distance (say 1 metre) behind, there is a bright spot similar to that seen when a disc is substituted for the sphere. This effect is observed with a magnifying lens. If the eye, situated at the centre of the shadow, be focused upon the sphere, the edge of the obstacle is seen bounded by a very narrow ring of light.

important, (7) is sufficiently established; but (§ 328) it appears that under these conditions the most important terms are of infinite order. I think it will be found that for the most important terms n is approximately equal to kc, and that accordingly (7) is not available. In any case it could not be relied upon without a further examination.

In *Theory of Sound*, § 328, the problem is treated for the case where kc is small, and the calculation is pushed as far as $kc = 2$. The results indicate no definite shadow. I have commenced a calculation for $kc = 10$, about the highest value for which the method is practicable. But it is doubtful whether even this value is high enough to throw light upon what happens when kc is really large*.

* [See Art. 292 below.]

288.

ON THE PROPORTION OF ARGON IN THE VAPOUR RISING FROM LIQUID AIR.

[*Philosophical Magazine*, v. pp. 677—680, 1903.]

THE boiling-point of argon being intermediate between those of nitrogen and oxygen, it may be expected that any operations of evaporation and condensation which increase the oxygen relatively to the nitrogen will at the same time increase the argon relatively to the nitrogen and diminish it relatively to the oxygen. In the experiments about to be detailed the gas analysed was that given off from liquid air, either freshly collected, or after standing (with evaporation) for some time—from a day to a week. The analyses were for oxygen and for argon, and were made upon different, though similar, samples. Thus after an analysis of a sample for oxygen by Hempel's method with copper and ammonia, 4 or 5 litres would be collected in a graduated holder, and then the first analysis confirmed on a third sample. In no case, except one to be specified later, was the quantity of gas withdrawn sufficient to disturb sensibly the composition. The liquid was held in Dewar's vessels, but the evolution of gas from below was always sufficient to keep the mass well mixed.

The examination for argon was made in a large test-tube inverted over alkali, into which the gas was fed intermittently from the holder. The nitrogen was gradually oxidized by the electric discharge from a Ruhmkorff coil in connexion with the public supply of alternating current, the proportion of oxygen being maintained suitably by additions of oxygen or hydrogen as might be required. In the latter case the feed should be very slow, and the electric discharge should be near the top of the test-tube. Great care is required to prevent the hydrogen getting into excess; for if this should occur, the recovery of the normal condition by addition of oxygen is a very risky process. After sufficient gas from the holder, usually about 2 litres, had been introduced, the discharge was continued until no more nitrogen remained, as was evidenced by the cessation of contraction and by the disappearance of the nitrogen line from the spectrum of the discharge when the terminals were connected with a leyden-jar. When it was certain that all nitrogen had been removed, the residual oxygen was taken up by ignition of a piece of phosphorus. On cooling, the residue of argon was measured, and its amount expressed as a percentage of the total gas taken from the holder.

The results are shown in the following table. The oxygen, expressed as a percentage of the whole, varied from 30 to about 98. From 43 to 90 per cent. of oxygen, the argon, as a percentage of the whole, scarcely varied from 2·0.

Percentage of Oxygen	Percentage of Argon	Argon as a percentage of the Nitrogen and Argon
30	1·3	1·9
43	2·0	3·5
64	2·0	5·6
75	2·1	8·4
90	2·0	20·0
98	0·76	33·0
100	0·0	100·0

The experiment entered under the head of 98 per cent. oxygen is not comparable with the others. In this case $5\frac{1}{2}$ litres of gas were collected as the *last* portion coming away from a stock of liquid as it dried up. Nor was the subsequent treatment quite parallel, for the whole of the oxygen was first removed with copper and ammonia leaving 125 c.c. of mixed nitrogen and argon, of which again by subsequent analysis 42 c.c. was found to be argon. The last entry corresponding to 100 per cent. of oxygen is theoretical and does not represent any actual experiment.

It must be clearly understood that these results relate to the *vapour* rising from the liquid, and *not* to the composition of the liquid itself. So far as the oxygen content is concerned, the comparison may be made by means of Mr Baly's observations (*Phil. Mag.* XLIX. p. 517, 1900). It will appear, for example, that when the vapour contains 30 per cent. of oxygen, the liquid will contain about 60 per cent., and that when the vapour contains 90 per cent. the liquid will be of 95 or 96 per cent. At every stage the liquid will be the stronger in the less volatile constituents; so that the proportion of argon to nitrogen, or to nitrogen + argon, will be higher in the liquid than in the vapour.

The constancy of the proportion of argon to the whole over a considerable range may be explained to a certain extent, for it will appear that the proportion must rise to a maximum and thence decrease to zero. To understand this, we must remember that "liquid air" is something of a misnomer. In the usual process the whole of the air concerned is not condensed, but only a part; and the part that is condensed is of course not a sample of the whole. As compared with the atmosphere the liquid contains the less volatile ingredients in increased proportion, and the part not condensed and rejected contains the more volatile ingredients in increased proportion. The vapour coming away from the liquid as first collected has the same composition as the gas rejected in the process of condensation. At the beginning of our table, a point, however, which it would be difficult to reach in actual experiment, we should have an oxygen content much below 20 per cent., a ratio of argon to nitrogen + argon below 1 per cent., and in all probability a ratio of argon to the whole also below 1 per cent.

The object which I had in view was principally to obtain information as to the most advantageous procedure for the preparation of argon. So many laboratories are now provided with apparatus for liquifying air, that it will usually be convenient to start in this way if a sufficient advantage can be gained. The above results show clearly that the advantage that may be gained is great. Something depends upon the procedure to be adopted for eliminating the nitrogen. Upon a moderate scale and where there is a supply of alternating current, the method of oxidation, as in the analyses, is probably the most convenient. In this case it may be an advantage to retain the oxygen. If the oxygen content be about 60 per cent., as in the third experiment, the proportion is about sufficient to oxidize the nitrogen. We may compare this with the mixture of atmospheric air and oxygen which behaves in the same manner. In the latter case the proportion of argon would be reduced from 2·0 per cent. to about ·4 per cent., so that the advantage of using the liquid air amounts to about *five* times. In the arrangement that I described for oxidizing nitrogen upon a large scale* the mixed gases were absorbed at the rate of 20 litres per hour.

In the alternative method the nitrogen is absorbed by magnesium or preferably by *calcium* formed *in situ* by heating a mixture of lime and magnesium as proposed by Maquenne†. In this case it is necessary first to remove the oxygen; but oxygen is so much more easily dealt with than nitrogen that its presence, even in large proportion, is scarcely an objection. On this view, and on the supposition that liquid air is available in large quantities, it is advantageous to allow the evaporation to proceed to great lengths. A 20 per cent. mixture of argon and nitrogen (experiment 5) is easily obtained. Prof. Dewar has shown me a note of an experiment executed in 1899, in which a mixture of argon and nitrogen was obtained containing 25 per cent. of the former. In the 6th experiment 33 per cent. was reached, and there is no theoretical limit.

P.S.—I see that Sir W. Ramsay (*Proc. Roy. Soc.*, March 1903) alludes to an experiment in which the argon content was doubled by starting from liquid air.

---

* *Chem. Soc. Journ.* LXXI. p. 181 (1897); *Scientific Papers*, Vol. IV. p. 270.

† I employed this method successfully in a lecture before the Royal Institution in April 1895 (*Scientific Papers*, Vol. IV. p. 188). In a subsequent use of it I experienced a disagreeable explosion, presumably on account of the lime being insufficiently freed from combined water.

289.

ON THE THEORY OF OPTICAL IMAGES, WITH SPECIAL REFERENCE TO THE MICROSCOPE.

(SUPPLEMENTARY PAPER.)

[*Journ. R. Micr. Soc.* pp. 474—482, 1903.]

IN the memoir, above* reprinted from the *Philosophical Magazine*, I discussed the theories of Abbe and Helmholtz, and endeavoured to show their correlation. It appeared that the method of the former, while ingenious and capable of giving interesting results in certain directions, was inapplicable to many of the problems which it is necessary to attack. As an example of this, it may suffice to mention the case of a *self-luminous* object.

The work of Helmholtz, to which attention has recently been recalled by Mr J. W. Gordon in a lively criticism (p. 381), was founded upon the processes already developed by Airy, Verdet, and others for the performance of the telescope. The theories both of Abbe and Helmholtz pointed to a tolerably definite limit to the powers of the Microscope, dependent, however, upon the wave-length of the light employed and upon the medium in which the object is imbedded. It appeared that two neighbours, whether constituting a single pair of points or forming part of an extended series of equidistant points, could not be properly distinguished if the distance were less than half the wave-length of the light employed. The importance of this conclusion, as imposing a limit upon our powers of direct observation, can hardly be overestimated; but there has been in some quarters a tendency to ascribe to it a more precise character than it can bear, or even to mistake its meaning altogether. A few words upon this subject may not be out of place.

The first point to be emphasised is that nothing whatever is said as to the smallness of a single object that may be made visible. The eye, whether

---

* [In *Journ. R. Micr. Soc.* See this collection, IV. p. 235.]

unaided or armed with a telescope, is able to see as points of light stars subtending no sensible angle. The visibility of the star is a question of brightness simply, and has nothing to do with resolving power. The latter element enters only when it is a question of recognising the duplicity of a double star, or of distinguishing detail upon the surface of a planet. So in the Microscope there is nothing except lack of light to hinder the visibility of an object however small. But if its dimensions be much less than the half wave-length, it can only be seen as a whole, and its parts cannot be distinctly separated, although in cases near the border line some inference may be possible founded upon experience of what appearances are presented in various cases. Thus a practised astronomer may conclude with certainty that a star is double, although its components cannot be properly seen. He knows that a single star would present a round (though false) disc, and any departure from this condition of things he attributes to a complication. A slightly oval disc may suffice not only to prove that the star is double but even to fix the line upon which the components lie, and their probable distance apart.

What has been said about a luminous point applies equally to a luminous *line*. If bright enough, it will be visible, however narrow; but if the real width be much less than the half wave-length, the apparent width will be illusory. The luminous line may be regarded as dividing the otherwise dark field into two portions; and we see that this separation does not require a luminous interval of finite width, but may occur, however narrow the interval, provided that its intrinsic brightness be proportionally increased.

The consideration of a luminous line upon a dark ground is introduced here for comparison with the case, suggested by Mr Gordon, of a dark line upon a (uniformly) bright ground. Calculations to be given later confirm Mr Gordon's conclusion that the line may be visible (but not in its true width), although the actual width fall considerably short of the half wave-length. Although in both these cases there is something that may be described as resolution, what is seen as distinct from the ground is really but a *single* object. So far as I see, there is no escape from the general conclusion, as to the microscopic limit, glimpsed originally by Fraunhofer and afterwards formulated by Abbe and Helmholtz; but it must be remembered that near the limit the question is one of degree, and that the degree may vary with the character of the detail whose visibility is under consideration.

Mr Gordon comments upon the fact that Helmholtz gave no direct proof of his pronouncement that a grating composed of parallel, equidistant, infinitely narrow, luminous lines shows no structure at a certain degree of closeness, and he appears to regard the question as still open. This matter was, however, fully discussed in my paper of 1896, where it is proved that as the grating-interval diminishes, structure finally disappears when the distance between the geometrical images of neighbouring lines falls to equality with half the width of the diffraction pattern due to a single line, reckoned from the first blackness on one side to the first blackness on the other. It is easy to see that the same limit obtains when the lines have a finite width, provided, of course, that the widths and intrinsic luminosities of the lines are equal. If the grating-*interval*, that is the distance between centres or *corresponding* edges of neighbouring lines, be less than the amount above mentioned, no structure can be seen. The microscopic limit occurs when the grating-interval is equal to half the wave-length of the light in operation.

The method employed in 1896 depends upon the use of Fourier's theorem. The critical case, where the structure has *just* disappeared, may be treated in a somewhat more elementary manner as follows. It is required to prove that

$$\frac{\sin^2 u}{u^2} + \frac{\sin^2 (u + \pi)}{(u + \pi)^2} + \frac{\sin^2 (u - \pi)}{(u - \pi)^2} + \frac{\sin^2 (u + 2\pi)}{(u + 2\pi)^2} + \dots, \quad \dots (76)$$

obtained by writing π for v in (22) [Art. 222], is the same for all values of u. In (76) the (sine)^2 have all the same value, so that what has to be proved may be written

$$\frac{1}{\sin^2 u} = \frac{1}{u^2} + \frac{1}{(u + \pi)^2} + \frac{1}{(u - \pi)^2} + \frac{1}{(u + 2\pi)^2} + \dots. \quad \dots (77)$$

This follows readily from the expression for the sine in factors. If we write

$$\sin u = C u \, (u + \pi) \, (u - \pi) \, (u + 2\pi) \dots,$$

or

$$\log \sin u = \log C + \log u + \log (u + \pi) + \dots,$$

we get on differentiation

$$\frac{d \log \sin u}{du} = \frac{1}{u} + \frac{1}{u + \pi} + \frac{1}{u - \pi} + \dots,$$

and again

$$-\frac{d^2 \log \sin u}{du^2} = \frac{1}{u^2} + \frac{1}{(u + \pi)^2} + \frac{1}{(u - \pi)^2} + \dots.$$

In these equations

$$\frac{d \log \sin u}{du} = \cot u, \qquad -\frac{d^2 \log \sin u}{du^2} = \frac{1}{\sin^2 u},$$

from which (77) follows.

We infer that a grating of the degree of closeness in question presents to the eye a uniform field of light and no structure, but it is not proved by this method that structure might not reappear at a greater degree of closeness. If however we take $v = \frac{1}{2}\pi$, that is, suppose the lines to be exactly twice as close as above, a similar method applies. The illumination at the point is now expressed by

$$\frac{\sin^2 u}{u^2} + \frac{\sin^2 (u + \frac{1}{2}\pi)}{(u + \frac{1}{2}\pi)^2} + \frac{\sin^2 (u - \frac{1}{2}\pi)}{(u - \frac{1}{2}\pi)^2} + \ldots,$$

or by

$$\sin^2 u \left\{ \frac{1}{u^2} + \frac{1}{(u + \pi)^2} + \frac{1}{(u - \pi)^2} + \frac{1}{(u + 2\pi)^2} + \ldots \right\}$$

$$+ \cos^2 u \left\{ \frac{1}{(u + \frac{1}{2}\pi)^2} + \frac{1}{(u - \frac{1}{2}\pi)^2} + \frac{1}{(u + \frac{3}{2}\pi)^2} + \ldots \right\}.$$

The value of the first series has above been shown to be unity, and by a like method the same may be proved of the second. The illumination for all values of u is thus equal to 2. That it should be twice as great as before might have been expected.

But my principal object at present is to consider the problem, suggested by Mr Gordon, of a dark line of finite width upon a uniformly bright ground. The problem assumes two forms according as the various parts of the ground are supposed to be self-luminous or to give rise to waves which are all in one phase. The latter is the case of an opaque wire or other linear obstacle upon which impinge plane waves of light in a direction parallel to the axis of the instrument (telescope or microscope), and as it is somewhat the simpler we may consider it first*.

In (28) [Art. 222] we have the expression for the resultant amplitude at any point u due to a series of points or lines, whose geometrical images are situated at $u = 0$, $u = \pm v$, $u = \pm 2v$, &c. If *all* values of u are equally geometrical images of a uniformly bright ground of light, we have to consider

$$\int_{-\infty}^{+\infty} \frac{\sin u}{u} \, du = \pi. \quad\quad\quad\quad\quad\quad\quad\quad (78)$$

At present we suppose that the bright ground is interrupted at points corresponding to $u = \alpha$, $u = -\alpha$, so that 2α represents the width of the geometrical image of the dark obstacle. The amplitude at u is the same for a given numerical value of u, whether u be positive or negative. It will suffice therefore to suppose u positive. If $u < \alpha$, we have

$$A(u) = \int_{-\infty}^{+\infty} \frac{\sin u}{u} \, du - \int_{0}^{\alpha - u} \frac{\sin u}{u} \, du - \int_{0}^{\alpha + u} \frac{\sin u}{u} \, du \quad\quad (79)$$

---

* It should be remarked that in point of fact the field is limited through the operation of a cause not taken into account in the formation of (28) [Art. 222]. It is there assumed that equality of phase in the light emitted from the various points of the object carries with it a like equality of phase at the geometrical images of these points. This will hold good only near the centre of the field. At a moderate distance out the illumination is destroyed by the phase-differences here neglected.

which gives the resultant amplitude at any point u as a function of u and α. If $u > \alpha$, we have

$$A(u) = \int_{-\infty}^{+\infty} \frac{\sin u}{u}\, du + \int_{0}^{u-\alpha} \frac{\sin u}{u}\, du - \int_{0}^{u+\alpha} \frac{\sin u}{u}\, du. \quad \ldots\ldots(80)$$

By (78) the first term is equal to π.

The integral in (79), (80) is known as the *sine-integral*. In the usual notation

$$\int_{0}^{x} \frac{\sin u}{u}\, du = \operatorname{si}(x) \quad \ldots\ldots\ldots\ldots\ldots\ldots\ldots\ldots.(81)$$

so that (79) may be written

$$A(u) = \pi - \operatorname{si}(\alpha - u) - \operatorname{si}(\alpha + u), \quad \ldots\ldots\ldots\ldots.(82)$$

and (80) may be written

$$A(u) = \pi + \operatorname{si}(u - \alpha) - \operatorname{si}(u + \alpha). \quad \ldots\ldots\ldots\ldots.(83)$$

The function si has been tabulated by Dr Glaisher*.

At the centre of the geometrical image of the bar, $u = 0$, and (82) becomes

$$A(0) = \pi - 2\operatorname{si}(\alpha). \quad \ldots\ldots\ldots\ldots\ldots\ldots\ldots.(84)$$

If x is small, (81) gives

$$\operatorname{si}(x) = x - \frac{x^3}{3.1.2.3} + \frac{x^5}{5.1.2.3.4.5} - \ldots; \quad \ldots\ldots\ldots\ldots.(85)$$

so that in (82) if α be small,

$$A(u) = \pi - 2\alpha + \frac{2\alpha(\alpha^2 + 3u^2)}{3.1.2.3} - \ldots. \quad \ldots\ldots\ldots\ldots.(86)$$

From this we see that over the whole geometrical image of the bar the amplitude of vibration is nearly the same. If we write I for the intensity, where $I(u) = \{A(u)\}^2$, and denote by I_0 the value of I corresponding to a uniform ground ($\alpha = 0$), then

$$\frac{I_0 - I}{I_0} = \frac{4\alpha}{\pi}. \quad \ldots\ldots\ldots\ldots\ldots\ldots\ldots\ldots.(87)$$

This gives the proportional loss of illumination over the image of the bar, and it suffices for the information required near the limit of visibility. For example, if the loss of light over the image be one-eighth of the maximum, $2\alpha = \frac{1}{16}\pi$; so that a single bar upon a bright ground might well remain apparent when its width is reduced to $\frac{1}{32}$ of the minimum grating-interval (2π) necessary for visibility.

The above gives the loss of brightness over the region occupied by

---

* *Phil. Trans.*, 1870.

the geometrical image. Outside this region we have from (80), when 2α is small,

$$A\left(u\right)=\pi-\int_{u-a}^{u+a}\frac{\sin u}{u}\,du=\pi-2\alpha\,\frac{\sin u}{u}, \qquad \ldots\ldots\ldots\ldots(88)$$

whence

$$\frac{I_0-I\left(u\right)}{I_0}=\frac{4\alpha}{\pi}\,\frac{\sin u}{u}. \qquad \ldots\ldots\ldots\ldots\ldots\ldots\ldots(89)$$

Here (89) identifies itself with (87) when u is small, and it does not alter greatly until $u=\frac{1}{2}\pi$. The slightly darkened image of the bar has thus a width corresponding to the interval $u=\pm\frac{1}{2}\pi$, exceeding to a great extent the width of the geometrical image when the latter is very small. The conclusion is that, although a very narrow dark bar on a bright ground may make itself visible, the apparent width is quite illusory.

$$A\ (u).$$

$\pm u$	$a=1$	$a=2$	$a=3$
0	$+1 \cdot 520$	$-\ \cdot 068$	$-\ \cdot 556$
1	$1 \cdot 807$	$+\ \cdot 347$	$-\ \cdot 221$
2	$2 \cdot 509$	$1 \cdot 384$	$+\ \cdot 646$
3	$3 \cdot 259$	$2 \cdot 538$	$1 \cdot 717$
4	$3 \cdot 711$	$3 \cdot 322$	$2 \cdot 633$
5	$3 \cdot 745$	$3 \cdot 536$	$3 \cdot 173$
6	$3 \cdot 507$	$3 \cdot 326$	$3 \cdot 326$
7	$3 \cdot 263$	$3 \cdot 027$	$3 \cdot 242$
8	$2 \cdot 932$	$2 \cdot 909$	$3 \cdot 114$

The annexed table gives the values of $A\left(u\right)$ for $\alpha=1,2,3$ for $u=0,1\ldots8$. Corresponding to any value of α,

$$u\left(\infty\right)=\pi=3 \cdot 142.$$

It will be remembered that 2α is the width of the geometrical image of the bar, so that when $\alpha=3$ the width is about the same as the minimum resolvable grating interval (2π).

We now pass to the case of a *self-luminous* ground interrupted by a dark bar. As in (22) [Art. 222], we have for the illumination at any point u within the geometrical image

$$I\left(u\right)=\int_{-\infty}^{+\infty}\frac{\sin^2 u}{u^2}\,du-\int_0^{a-u}\frac{\sin^2 u}{u^2}\,du-\int_0^{a+u}\frac{\sin^2 u}{u^2}\,du\ \ \ldots\ldots(90)$$

and for any point on the positive side beyond the geometrical image

$$I\left(u\right)=\int_{-\infty}^{+\infty}\frac{\sin^2 u}{u^2}\,du+\int_0^{u-a}\frac{\sin^2 u}{u^2}\,du-\int_0^{u+a}\frac{\sin^2 u}{u^2}\,du,\ \ldots\ldots(91)$$

2α denoting as before the width of the geometrical image of the bar, while u is reckoned from the centre of symmetry. If $\alpha = 0$,

$$I(u) = \int_{-\infty}^{+\infty} \frac{\sin^2 u}{u^2} \, du = \pi. \quad \ldots\ldots\ldots\ldots(92)$$

The integrals in (90), (91) may be reduced to dependence upon the sine-integral. It may be proved* that

$$\int_0^x \frac{\sin^2 u}{u^2} \, du = \int_0^{2x} \frac{\sin u}{u} \, du - \frac{\sin^2 x}{x} = \mathrm{si}\,(2x) - \frac{\sin^2 x}{x}. \quad \ldots\ldots(93)$$

Thus, inside the geometrical image,

$$I(u) = \pi - \mathrm{si}\,(2\alpha - 2u) + \frac{\sin^2 (\alpha - u)}{\alpha - u} - \mathrm{si}\,(2\alpha + 2u) + \frac{\sin^2 (\alpha + u)}{\alpha + u}; \quad \ldots(94)$$

and beyond it,

$$I(u) = \pi + \mathrm{si}\,(2u - 2\alpha) - \frac{\sin^2 (u - \alpha)}{u - \alpha} - \mathrm{si}\,(2u + 2\alpha) + \frac{\sin^2 (u + \alpha)}{u + \alpha}. \quad \ldots(95)$$

At the centre ($u = 0$)

$$I(0) = \pi - 2\,\mathrm{si}\,(2\alpha) + \frac{2 \sin^2 \alpha}{\alpha}. \quad \ldots\ldots\ldots\ldots(96)$$

As in the former case an approximate expression (85) for $\mathrm{si}\,(x)$ gives the desired information near the limit of visibility. If α be small, we have for the illumination within the geometrical image from (90)

$$I(u) = \pi - 2\alpha, \quad \ldots\ldots\ldots\ldots\ldots\ldots(97)$$

so that

$$\frac{I_0 - I}{I_0} = \frac{2\alpha}{\pi}. \quad \ldots\ldots\ldots\ldots\ldots\ldots(98)$$

The visibility of a bar of width 2α is thus only half as great as before.

Outside the geometrical image we have approximately, when u considerably exceeds α,

$$I(u) = \pi - \int_{u-\alpha}^{u+\alpha} \frac{\sin^2 u}{u^2} \, du = \pi - 2\alpha \frac{\sin^2 u}{u^2}, \quad \ldots\ldots\ldots(99)$$

whence

$$\frac{I_0 - I(u)}{I_0} = \frac{2\alpha}{\pi} \frac{\sin^2 u}{u^2}. \quad \ldots\ldots\ldots\ldots\ldots(100)$$

The following table gives some values of $I(u)$ calculated from (94), (95).

$$I(u).$$

$\pm u$	$a = \frac{1}{2}$	$a = 1$	$a = 2$
0	2·170	1·349	·453
1	2·442	1·797	·827
2	2·921	2·621	1·711
3	...	3·056	2·565

* E.g. by writing ru for u in the integral to be examined and differentiating with respect to r. Or (93) may be verified by differentiating with respect to x.

The complete value of $I(u)$, when u is great, is π. The width of the geometrical image of the bar is 2α, and the smallest resolvable grating interval is π. The dark bar should be easily recognisable in the first case when its width is but one-third of the minimum grating interval.

In conclusion I may mention the results of a simple experiment conducted almost entirely without apparatus. In front of the naked eye was held a piece of copper foil perforated by a fine needle-hole. Observed through this, the structure of some gauze just disappeared at a distance from the eye equal to 17 in. (inch = 2·54 cm.), the gauze containing 46 meshes to the inch. On the other hand, a single wire ·034 in. in diameter remained fairly visible up to a distance of 20 ft. or 240 in. The ratio between the angles subtended by the periodic structure of the gauze and the diameter of the wire was thus

$$\frac{\cdot022}{\cdot034} \times \frac{240}{17} = 9\cdot1.$$

Using this in (98), we find for the proportional loss of illumination at the centre of the wire

$$\frac{I - I_0}{I_0} = \cdot11,$$

about what might have been expected.

ON THE PRODUCTION AND DISTRIBUTION OF SOUND.

[*Philosophical Magazine*, Vol. VI. pp. 289—305, 1903.]

Theory of Conical Trumpet.

THE theory of small periodic vibrations having their origin at a single point of a gas and thence spreading symmetrically has long been known. The following statement is from *Theory of Sound**, § 280. In it a denotes the velocity of sound, and $k = 2\pi/\lambda$, λ being the wave-length.

" If the velocity-potential be

$$\phi = -\frac{A}{4\pi r} \cos k\,(at - r), \qquad \dotfill (1)$$

we have for the total current crossing a sphere of radius r,

$$4\pi r^2 \frac{d\phi}{dr} = A\,\{\cos k\,(at - r) - kr \sin k\,(at - r)\} = A \cos kat,$$

when r is small enough. If the maximum rate of introduction of fluid be denoted by A, the corresponding potential is given by (1).

" It will be observed that when the source, as measured by A, is finite, the potential and the pressure-variation (proportional to $d\phi/dt$) are infinite at the pole. But this does not, as might for a moment be supposed, imply an infinite emission of energy. If the pressure be divided into two parts, one of which has the same phase as the velocity, and the other the same phase as the acceleration, it will be found that the former part, on which the work depends, is finite. The infinite part of the pressure does no work on the whole, but merely keeps up the vibration of the air immediately round the source, whose effective inertia is indefinitely great.

" We will now investigate the energy emitted from a simple source of given magnitude, supposing for the sake of greater generality that the source

* Macmillan & Co., first edition 1878, second edition 1896.

is situated at the vertex of a rigid cone of solid angle ω.   If the rate of introduction of fluid at the source be $A \cos kat$, we have

$$\omega r^2 d\phi/dr = A \cos kat$$

ultimately, corresponding to

$$\phi = - \frac{A}{\omega r} \cos k\,(at - r), \quad \dots\dots\dots\dots\dots\dots(2)$$

whence

$$\frac{d\phi}{dt} = \frac{kaA}{\omega r} \sin k\,(at - r), \quad \dots\dots\dots\dots\dots\dots(3)$$

and

$$\omega r^2 d\phi/dr = A\,\{\cos k\,(at - r) - kr \sin k\,(at - r)\}. \quad \dots\dots\dots\dots(4)$$

Thus if dW be the work transmitted in time dt, we get, since $\delta p = - \rho\, d\phi/dt$,

$$\frac{dW}{dt} = - \frac{\rho kaA^2}{\omega r} \sin k\,(at - r) \cos k\,(at - r) + \rho\,\frac{k^2 aA^2}{\omega} \sin^2 k\,(at - r).$$

Of the right-hand member the first term is entirely periodic, and in the second the mean value of $\sin^2 k\,(at - r)$ is $\frac{1}{2}$.   Thus in the long run

$$W = \frac{\rho k^2 aA^2}{2\omega} t. \quad \dots\dots\dots\dots\dots\dots\dots\dots(5)$$

"It will be remarked that when the source is given, the amplitude varies inversely as ω, and therefore the intensity inversely as ω^2.   For an acute cone the intensity is greater, not only on account of the diminution in the solid angle through which the sound is distributed, but also because the total energy emitted from the source is itself increased.

"When the source is in the open, we have only to put $\omega = 4\pi$, and when it is close to a rigid plane, $\omega = 2\pi$.

"These results find an interesting application in the theory of the speaking-trumpet, or (by the law of reciprocity, §§ 109, 294) hearing-trumpet. If the diameter of the large open end be small in comparison with the wave-length $(2\pi/k)$, the waves on arrival suffer copious reflexion, and the ultimate result, which must depend largely on the precise relative lengths of the tube and of the wave, requires to be determined by a different process.   But by sufficiently prolonging the cone, this reflexion may be diminished, and it will tend to cease when the diameter of the open end includes a large number of wave-lengths.   Apart from friction it would therefore be possible by diminishing ω to obtain from a given source any desired amount of energy, and at the same time by lengthening the cone to secure the unimpeded transference of this energy from the tube to the surrounding air.

"From the theory of diffraction it appears that the sound will not fall off to any great extent in a lateral direction, unless the diameter at the large end exceed half a wave-length.   The ordinary explanation of the effect of a common trumpet, depending upon a supposed concentration of *rays* in the axial direction, is thus untenable."

Data respecting Fog-Signals.

The above theory should throw light upon the production of sound in "fog-signals," where sirens, or vibrating reeds, are associated with long conical trumpets. In the practice of the Trinity House these are actuated by air compressed to a pressure (above atmosphere) of 25 lbs. per square inch, or 1760 gms. per sq. cm., a pressure which appears rather high. According to Stone the highest pressure used in orchestral wind-instruments is 40 inches (102 cm.) of water.

As might be expected from the high pressure, the energy consumed during the sounding of the signal is very considerable. The high note of the St Catherine's Service signal takes 130 horse-power, and the corresponding note of a Scottish signal (tested at St Catherine's in 1901) requires as much as 600 horse-power. The question obtrudes itself whether these enormous powers are really utilized for the production of sound, or whether from some cause, possibly unavoidable, a large proportion may not be wasted.

Comparison with Musical Instruments, &c.

These statements as to horse-power may be better appreciated if I record for comparison the results of some rough measurements, made in 1901, upon the power absorbed by smaller instruments. In the calculation it will suffice to regard the compressions and rarefactions as *isothermal*. Thus if v_0, p_0 represent the volume and pressure of air in its natural (atmospheric) condition, v, p the corresponding quantities under compression, so that according to Boyle's law $pv = p_0 v_0$, then the work (W) of compression is given by

$$W = p_0 v_0 \log (p/p_0), \quad \dots\dots\dots(6)$$

or, if the compression be small,

$$W = p_0 v_0 \frac{p - p_0}{p_0}. \quad \dots\dots\dots(7)$$

In C.G.S. measure p_0 (the atmospheric pressure) will be 10^6, and if v_0 be measured in cubic centimetres, W will be expressed in ergs. If in (7) v_0 be understood to mean the volume compressed *per second of time*, W will be given in ergs per second, of which $7\cdot46 \times 10^9$ go to the horse-power.

The first example is that of a small horn (without valves) blown by the lips. It resonates to e' of my harmonium, and the pitch when sounded is about e' flat. From one inspiration I can blow it for about 30 seconds with a pressure (in the mouth) of $1\frac{1}{2}$ inch ($3\cdot8$ cm.) of mercury. The contents of the lungs may be taken at 1200 c.c., giving 40 c.c. per second as the wind consumption. This is the value of v_0, and $(p - p_0)/p_0$ is $\frac{1}{20}$. Hence W in ergs per second will be 2×10^6, or in horse-power

$$W = \cdot00027 \text{ H.P.}$$

The sound from this very small horse-power is unpleasantly loud when given in a room of moderate dimensions.

In the case of the harmonium reed e' the wind consumption was 220 c.c. per second, and the pressure 2 inches of water, so that $(p - p_0)/p_0 = \frac{1}{200}$. Hence

$$W = \cdot00015 \text{ H.P.}$$

A small hand fog-signal of Holmes' pattern, known as the "Little Squeaker," consumed a horse-power calculated on the basis of similar measurements to be ·03. For the very effective "Manual" of the Trinity House Service the horse-power was about 3·0.

These examples may all be classed under the head of reeds, the harmonium reed being "free" and probably in consequence less efficient, and the others "striking." To them may be added the case of a whistle of high pitch*, for which the wind consumption represented $1\cdot8 \times 10^6$ ergs per second, or ·00024 horse-power, practically the same as for the small horn above. The latter was certainly the more powerful of the two, considered as a source of audible sound.

It may now be instructive to consider the case of a large siren, such as the 7-inch disk siren experimented upon at St Catherine's in 1901. The wind consumption here was 29 cubic feet, or 810 litres, per second. This average current, for the purposes of a rough calculation, may be analysed into a steady current of the same amount and an alternating current whose extremes are represented by ± 810 litres per second, the latter being alone effective for the production of sound. The first question that arises is to what pressure does this correspond, and is it a reasonable fraction of the actual pressure employed?

The answer must depend upon the other circumstances of the case, such as the character of the cone or other tubular resonator associated with the siren. We shall begin by supposing that there is nothing of this kind, so that the above alternating flow takes place at the surface of a sphere of radius r situated in the open. The velocity-potential and the rate of total flow being given by (1) and (2) with ω equal to 4π, we have for the maximum rate of that flow $A\sqrt{\{1 + k^2r^2\}}$, or with sufficient approximation for our purpose A simply. If s be the "condensation,"

$$\delta p = a^2 \rho s = -\rho\, d\phi/dt,$$

so that by (1)

$$s_{\max} = \frac{kA}{4\pi ar} = \frac{A}{2\lambda ar}. \qquad\qquad\dots\dots\dots\dots\dots\dots(8)$$

---

* *Proc. Roy. Soc.* xxvi. p. 248 (1877); *Scientific Papers*, i. p. 329.

To obtain a numerical result we must make some supposition as to the magnitude of r. Let us take $kr = \frac{1}{4}$. Then

$$s_{max} = \frac{4\pi A}{\lambda^2 a}, \dots\dots\dots\dots\dots\dots\dots\dots\dots\dots\dots\dots(9)$$

in which A is the maximum flow, λ the wave-length, and a the velocity of sound, i.e. 3×10^4 cm. per second. In the experiments referred to, the pitch was low and such that

$$\lambda = 8 \text{ feet} = 240 \text{ cm.},$$

whence with $A = 8\cdot 1 \times 10^5$ c.c. per second we find

$$s_{max} = \frac{1}{180}.$$

The maximum condensation corresponding to the assumed introduction of air is thus only $\frac{1}{180}$ of an atmosphere. The pressure is in the same proportion, and we see that it is but an insignificant fraction of the pressure actually employed (25 lbs. per square inch)*. We infer that no moderate pressure can be utilized in this way, and that some cone or resonating tube is a necessity. It may be remarked that the radius r of the sphere, on which the introduction of air is supposed to take place, is $1/4k$ or $\lambda/8\pi$, that is in the case taken 4 inches or 10 cms.

Cones and Resonators.

The next question is what improvement in the direction of utilizing higher pressures can be attained by the association of cones and resonators? But to this it is at present difficult to give a satisfactory answer. Theory shows that, apart from friction and other complications perhaps not very important, the efficiency of a *small* source may by these means be increased to any extent. Thus, in the case of the cone, if u be the maximum [particle] velocity [in] a progressive wave at a point where the section is σ, conservation of energy requires that σu^2 be constant. The maximum total flow (σu) is therefore proportional to σ , i.e. to the linear dimension of the section. If the vibrations are infinitesimal, we may begin with as small a diameter as we please and end with a large one, and thus obtain any desired multiplication of the source. For it is the total flow at the *open* end of the cone which measures the power of the source for external purposes. If, however, the quantity of air periodically introduced at the small end can no longer be treated as infinitesimal, this argument fails; and it is probable that the advantage derivable from the cone diminishes. In an extreme case we can easily recognize that this must be so. For the most that the cone could do would be to add its own contents to that of the air forcibly introduced. As

* [Atmosphere = 15 lbs. per square inch.]

the latter increases without limit, the addition must at last become relatively unimportant, and then the cone might as well be dispensed with. Similar considerations apply to the use of a resonator.

There is no reason to doubt that great advantage accrues from the use of the conical trumpet in existing fog-signalling apparatus, although probably it falls short of what would be expected according to the theory of infinitely small vibrations. If it be a question of striving to augment still further the force of the sound, we must remember that the application of power has already been carried to great lengths. The utilization of more power might demand an increase in the scale of the apparatus. This in itself would present no particular difficulty, but we must not forget that everything has relation to the wave-length of the sound, and that this is to a great extent fixed for us by the nature of the ear. It may well be that we are trying to do more than the conditions allow, and that further advance would require a different kind of apparatus. As matters stand, it seems to be generally admitted that the instruments using great power are not proportionally effective.

If, as I incline to believe, a large proportion of the power applied to important instruments is not converted into sound, there should be an opening for reducing the very large demands now made. We have to consider what becomes of the power wasted. I have long thought that it is spent in the eddies consequent upon the passage of the air through the comparatively narrow ports of the siren, and in this opinion my friend Sir O. Lodge, with whom I have recently had an opportunity of discussing the matter, concurs. If indeed it were a question of steady flow, one might pronounce with certainty that a great improvement would ensue from a better shaping of the passages, which on the down-stream-side should cone out gradually from the narrowest place. And although the intermittent character of the stream is an important element, this conclusion can hardly be altogether disturbed. The advantage of an enlargement of the ports themselves should also be kept in sight.

The conical trumpets at present employed must act to some extent as resonators, so that the precise relation of the pitch or speed of the siren to the trumpet cannot be a matter of indifference. Although the relation in question is liable to be disturbed by changes of temperature, it would appear that a better adjustment than is feasible with the present governors should be arrived at. To effect this an instrument capable of indicating the vigour of the vibration within the trumpet, as the speed of the siren varies, would be useful.

Vibration Indicator.

Experiments that I have tried appear to prove that the problem above proposed can be solved in a very simple manner. The principle is that of the unsymmetrical formation of jets when an alternating air-current flows through an aperture coned *upon one side.* An experiment given in *Theory of Sound*, § 322, may be quoted in illustration:—"When experimenting with one of König's brass resonators of pitch *c'*, I noticed that when the corresponding fork, strongly excited, was held to the mouth, a wind of considerable force issued from the nipple at the opposite side. This effect may rise to such intensity as to blow out a candle upon whose wick the stream is directed....Closer examination revealed the fact that at the sides of the nipple the outward flowing stream was replaced by one in the opposite direction, so that a tongue of flame from a suitably placed candle appeared to enter the nipple at the same time that another candle situated immediately in front was blown away. The two effects are of course in reality alternating, and only appear to be simultaneous in consequence of the inability of the eye to follow such rapid changes."

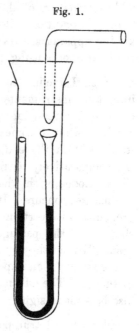

Fig. 1.

The application of the principle for the present purpose is very simple. The candle is replaced by a U pressure-gauge (fig. 1), the jet from a contracted nozzle playing into one limb. The whole is inclosed air-tight in a test-tube, so that no permanent pressure or suction has any effect upon the gauge. The nozzle communicates by means of a flexible tube with the space where the vibration is to be measured. Some throttling to check the vibration of the liquid may be convenient.

The small Holmes apparatus gave an indication of about 3 inches of water, and the Trinity House "Manual" one inch of *mercury.* The sensitiveness may be lessened by contracting the nozzle or probably by insertion of water to diminish the air-space within the test-tube.

Reeds.

[In comparison with sirens] reeds have the advantage of working without a governor, and the pitch once properly fixed is more likely to be maintained. I do not know whether reeds have been tried for very large scale instruments.

In the Barker apparatus* *three* reeds are combined with one trumpet. At first sight it may seem doubtful whether the tongues would vibrate in the same phase, but upon examination I think it will appear that this is the only way in which they could vibrate. On a large scale either the reeds must be multiplied, or an entirely different shape must be adopted, out of all proportion *broader* than at present. Some experiments that I have tried seem to show that the latter alternative is not impracticable.

Trumpets of Elongated Section.

In the trumpets at present employed the section is of circular form and the greater part of the axis is vertical. This disposition has its conveniences, but it entails bending the axis at the wide end of the cone if the mouth is to face horizontally. The effect of such a bending upon the propagation of the wave within the trumpet is hard to estimate. When, as in the case of certain rock-stations, the sound is required to be heard in all directions, a symmetrical form is adopted in the Trinity House Service, the mouth of the trumpet which faces vertically being partially stopped by an obstacle known as the "mushroom." The intention is to cut off the sound in a vertical direction while allowing it to spread in horizontal directions through the annular aperture between the bell-mouth of the trumpet and the mushroom.

Considering the case of the axis horizontal throughout, we may inquire into the probable distribution of the sound. The ratio between the diameter of the mouth and the wave-length is here of essential importance. If the diameter much exceed the half wave-length, the sound is concentrated in the prolongation of the axis. If on the other hand the diameter do not exceed the above-mentioned quantity, we may expect a tolerably equable distribution of sound, at any rate through angles with the axis less than 80°. It follows that the behaviour of the various components of a compound sound may be quite different. The fundamental tone may spread fairly well, while the octave and higher elements are unduly concentrated in the neighbourhood of the axis.

It appears then that a limitation must be imposed upon the size of the mouth, if it be desired that the sound should spread. But since the spreading is required only in the horizontal plane, the limitation applies only to the *horizontal* dimension of the mouth. There is no corresponding limitation upon the vertical diameter. We are thus led to prefer an elongated form of section, the horizontal dimension being limited to the half wave-length, while the vertical dimension may amount if desired to many wave-lengths.

---

* Report of Trinity House Fog-Signal Committee on Experiments conducted at St Catherine's Point, Isle of Wight, 1901.

This subject was explained and illustrated in a lecture before the Royal Institution*, the source of (inaudible) sound being a " bird-call" giving waves of 3 cms. length, which issued from a flattened trumpet whose mouth measured 5 cms. by 1½ cm. The indicator was a high-pressure sensitive flame, and it appeared very clearly that when the long dimension of the section stood vertical the sound was approximately limited to the horizontal plane, but within that plane spread without much loss through all directions less inclined than 80°.

In order to carry the demonstration a little nearer to what would be required in practice, I have lately experimented with the sound from a reed organ-pipe, giving waves of length equal to 8 inches (20 cms.) and thus easily audible. The trumpet is of wood, pyramidal in form, and the section at the mouth is 36 × 4 inches (91 × 10 cms.). The length (OB) is 6 feet (183 cms.). These dimensions were chosen so that OA, OC should exceed OB by ¼λ. A larger difference might entail too great a discrepancy of phase in the waves at A and B; a less difference might lead to an unnecessary prolongation of the cone along the axis. The trumpet was so mounted that its mouth just projected from an open window, and that it could be readily turned round OB as horizontal axis so as to allow the length of the section (AC) to be either horizontal or vertical. The observers took up various positions on a lawn at a moderate distance from the window.

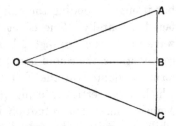

To observers in the line of the axis OB it should make no difference how the trumpet is rotated. On the whole this expectation was confirmed, but a little precaution is required. As usual the phenomenon was complicated by reflected sounds (doubtless from the ground). It was well in every case to make sure, by slightly raising or lowering the head, that the maximum sound was being heard.

When the point of observation deviated (in the horizontal plane) from the axis, the difference due to rotation was soon apparent. At 30° obliquity the sound appears greatly increased as AC passes from the horizontal to the vertical position. At higher obliquities with AC horizontal the sound falls off greatly, but recovers when AC is made vertical. Altogether the effects are very striking, and carry conviction to the mind more fully than experiments with sensitive flames where one is more or less in doubt as to the *magnitude* of the differences indicated by the flame. It will be remarked that to carry out this experiment upon a practical scale will mean a very large structure, the linear magnification being that (6 times) required to pass from an 8-inch wave-length to one (say) of 48 inches.

* Proc. Roy. Inst. Jan. 1902; Nature, 66, p. 42 (1902) [Vol. v. p. 1].

Work done by Detached Sources.

In the case of a single source the pressures to be overcome are proportional to the magnitude of the source, and thus the work done is proportional to the *square* of the magnitude of the source, as is indeed otherwise evident. If, as is usually the case in practice, the object be to emit sound in one (horizontal) plane only, an economy may be effected by *distributing* the source. If sources all in one phase be distributed along a vertical line, the effect is the same at distant points in the horizontal plane as if they were all concentrated in one point, but the *work* required to be done may be much less, the saving corresponding to the fact that in directions other than horizontal the sound is now diminished. We will begin by considering two unit sources in the same phase.

If ϕ, ψ be the potentials of these sources at a point whose distances are r, r', we have, as in (1),

$$\phi = -\frac{\cos k(at - r)}{4\pi r}, \qquad \psi = -\frac{\cos k(at - r')}{4\pi r'} . \ldots\ldots\ldots(10)$$

Thus, when $r = 0$, $4\pi r^2 \dfrac{d\phi}{dr} = \cos kat - kr \sin kat$,

$$\frac{d\phi}{dt} + \frac{d\psi}{dt} = \frac{ka \sin kat}{4\pi r} + \frac{ka}{4\pi D} \sin k(at - D),$$

if D denote the distance between the sources. The work done by the source at $r = 0$ is accordingly proportional to

$$1 + \frac{\sin kD}{kD}, \qquad \ldots\ldots\ldots\ldots\ldots\ldots\ldots\ldots\ldots(11)$$

and an equal work is done by the source at $r' = 0$. If D be infinitely great, the sources act independently, and thus the scale of measurement in (11) is such that unity represents the work done by each source when isolated. If $D = 0$, the work done by *each* source is doubled, and the two sources become equivalent to one of doubled magnitude.

If D be equal to $\frac{1}{2}\lambda$, or to any multiple thereof, $\sin kD = 0$, and we see from (11) that the work done by each source is unaffected by the presence of the other. This conclusion may be generalized. If any number (n) of equal sources in the same phase be arranged in a vertical line so that the distance between immediate neighbours is $\frac{1}{2}\lambda$, the work done by each is the same as if the others did not exist. The whole work accordingly is n, whereas the work to be done at a single source of magnitude n would be n^2. Thus if the sound be only wanted in the horizontal plane, the distribution into n parts effects an economy in the proportion of $n : 1$. It is not necessary that all the possible places between the outer limits be actually occupied.

All that is necessary is that there be n equal sources altogether, and that the distance between any pair of them be a multiple of $\frac{1}{2}\lambda$.

Returning for the moment to the case of two sources only, we may be interested to estimate the work consumed by following the law of emission of sound to a distance in the various directions. If μ be the cosine of the angle between any direction and the vertical, the relative retardation due to the difference of situation is μD. If the potential at any great distance due to one source is $\cos kat$*, that due to the other may be represented by $\cos k\,(at-\mu D)$. For the aggregate potential we have

$$\cos kat\,(1+\cos\mu kD)+\sin kat\,\sin\mu kD,$$

or for the *intensity*

$$2\,(1+\cos\mu kD). \qquad\qquad\dots\dots\dots\dots\dots\dots(12)$$

This is in direction μ. For the total intensity over angular space we must integrate with respect to μ from -1 to $+1$. The *mean* intensity is thus

$$2\int_{0}^{1}(1+\cos\mu kD)\,d\mu = 2\left(1+\frac{\sin kD}{kD}\right). \qquad\dots\dots\dots\dots(13)$$

The scale of measurement is at once recovered by supposing $D=0$, in which case the intensity in various directions would be uniform. The ratio of the mean intensities, which is also that of the work done, is thus

$$\tfrac{1}{2}\left(1+\frac{\sin kD}{kD}\right). \qquad\qquad\dots\dots\dots\dots\dots\dots(14)$$

This is the ratio in which the work done is diminished when a source is divided into two parts and these parts separated to a distance D.

While from the theoretical point of view there is no doubt as to the saving that might arise from the use of a number of separated sources, it is to be noticed that the saving is in the *pressure*. Since at the present time most of the pressure employed with a single source appears to be wasted, we are left in doubt whether with the existing arrangements economy would be attained by breaking up the source.

We will now investigate the expression for the energy radiated from any number of sources of the same pitch situated at finitely distant points in the neighbourhood of the origin O. The velocity-potential ϕ of the motion due to one of the sources at $(x,\,y,\,z)$ is at Q

$$\phi = -\frac{A}{4\pi R}\cos\,(nt+\epsilon-kR), \qquad\dots\dots\dots\dots\dots(15)$$

where R is the distance between Q and $(x,\,y,\,z)$. At a great distance from the origin we may identify R in the denominator with OQ, or ρ; while under the cosine we write

$$R = \rho-(\lambda x+\mu y+\nu z), \qquad\dots\dots\dots\dots\dots(16)$$

---

* It is not necessary to exhibit the dependence on r.

λ, μ, ν being the direction-cosines of OQ. On the whole

$$- 4\pi\rho\phi = \Sigma A \cos\{nt + \epsilon - k\rho + k(\lambda x + \mu y + \nu z)\}, \quad \ldots\ldots(17)$$

in which ρ is a constant for all the sources, but A, ϵ, x, y, z vary from one source to another. The *intensity* in the direction λ, μ, ν is thus represented by

$$[\Sigma A \cos\{\epsilon + k(\lambda x + \mu y + \nu z)\}]^2 + [\Sigma A \sin\{\epsilon + k(\lambda x + \mu y + \nu z)\}]^2,$$

or by

$$\Sigma A^2 + 2\Sigma A_1 A_2 \cos[\epsilon_1 - \epsilon_2 + k\{\lambda(x_1 - x_2) + \mu(y_1 - y_2) + \nu(z_1 - z_2)\}], \quad \ldots(18)$$

the second summation being for every pair of sources of which A_1, A_2 are specimens. We have now to integrate (18) over angular space.

It will suffice if we effect the integration for the specimen term; and this we shall do most easily if we take the line through the points (x_1, y_1, z_1), (x_2, y_2, z_2) as axis of reference, the distance between them being denoted by D. If λ, μ, ν make an angle with D whose cosine is μ,

$$D\mu = \lambda(x_1 - x_2) + \mu(y_1 - y_2) + \nu(z_1 - z_2), \quad \ldots\ldots\ldots\ldots(19)$$

and the mean value of the specimen term is

$$A_1 A_2 \int_{-1}^{+1} \cos\{\epsilon_1 - \epsilon_2 + kD\mu\}\, d\mu,$$

that is

$$\frac{2A_1 A_2}{kD} \sin kD \cos(\epsilon_1 - \epsilon_2). \ldots\ldots\ldots\ldots\ldots\ldots(20)$$

The mean value of (18) over angular space is thus

$$\Sigma A^2 + 2\Sigma \frac{A_1 A_2 \cos(\epsilon_1 - \epsilon_2)\sin kD}{kD}, \quad \ldots\ldots\ldots\ldots(21)$$

where D denotes the distance between the specimen pair of sources. If all the sources are in the same phase $\cos(\epsilon_1 - \epsilon_2) = 1$. If the distance between every pair of sources is a multiple of $\frac{1}{2}\lambda$, $\sin kD = 0$ and (21) reduces to its first term.

We fall back upon a former particular case if we suppose that there are only two sources, that these are units, and are in the same phase. (21) then becomes

$$2 + 2\frac{\sin kD}{kD},$$

agreeing with (11), which represents the work done by each source.

If the question of the phases of the two unit sources be left open (21) gives

$$2 + 2\cos(\epsilon_1 - \epsilon_2)\frac{\sin kD}{kD}. \quad \ldots\ldots\ldots\ldots\ldots(22)$$

If D be small, this reduces to

$$2 + 2\cos(\epsilon_1 - \epsilon_2),$$

which is zero if the sources be in opposite phases, and is equal to 4 if the phases be the same.

If, however, $\sin kD$ be equal to -1, the case is altered. Thus when $D = \frac{3}{2}\lambda$, we get

$$2 - \frac{4}{3\pi} \cos (\epsilon_1 - \epsilon_2),$$

and this is a minimum (and not a maximum) when the phases are the same.

In (22) if the phases are 90° apart, the cosine vanishes. The work done is then simply the double of what would be done by each source acting alone, and this whatever the distance D may be.

Continuous Distributions.

If the distribution of a source be continuous, the sum in (21) is to be replaced by a double integral. As an example, consider the case of a source all in one phase and uniformly distributed over a complete circular arc of radius c. If D be the distance of two elements $d\theta$, $d\theta'$, we have to consider the integral

$$\iint \frac{\sin kD}{kD} \, d\theta \, d\theta', \quad \dots\dots\dots\dots\dots\dots\dots(23)$$

where
$$D = 2c \sin \tfrac{1}{2} (\theta - \theta'). \quad \dots\dots\dots\dots\dots\dots(24)$$

Since every element $d\theta'$ contributes equally, it suffices to take $\theta' = 0$, so that the integral to be evaluated is

$$\int_0^\pi \frac{\sin (2kc \sin \tfrac{1}{2} \theta)}{2kc \sin \tfrac{1}{2} \theta} \, d\theta, \quad \dots\dots\dots\dots\dots\dots(25)$$

or, if $\tfrac{1}{2}\theta = \phi$, $2kc = x$,

$$2 \int_0^{\frac{1}{2}\pi} \frac{\sin (x \sin \phi)}{x \sin \phi} \, d\phi. \quad \dots\dots\dots\dots\dots\dots(26)$$

The integral (26) may be expressed by means of Bessel's function J_0, for

$$J_0 (x) = \frac{2}{\pi} \int_0^{\frac{1}{2}\pi} \cos (x \sin \phi) \, d\phi,$$

so that
$$\int_0^x J_0 (x) \, dx = \frac{2}{\pi} \int_0^{\frac{1}{2}\pi} \frac{\sin (x \sin \phi)}{\sin \phi} \, d\phi.$$

Thus, if constant factors be disregarded, we get

$$\frac{1}{x} \int_0^x J_0 (x) \, dx, \quad \dots\dots\dots\dots\dots\dots(27)$$

in which $x = 2kc$. Since (27) reduces to unity when c, or x, vanishes, it represents the ratio in which the work done is diminished when a source, originally concentrated at the centre, is distributed over a circular arc of radius c.

The case of a source uniformly distributed over a circular *disk* of radius c is investigated in my book on the *Theory of Sound**. According to what is there proved, the factor, analogous to (27), expressing the ratio in which the work done is diminished when a source originally concentrated at the centre is expanded over the disk, has the form

$$\frac{2}{k^2 c^2} \left\{ 1 - \frac{J_1(2kc)}{kc} \right\}, \quad \dots\dots\dots\dots\dots\dots(28)$$

where, as usual, $J_1(z) = \frac{z}{2} - \frac{z^3}{2^2 \cdot 4} + \frac{z^5}{2^2 \cdot 4^2 \cdot 6} - . \quad \dots\dots\dots\dots(29)$

Another case of interest is when the distribution takes place over the surface of a *sphere* of radius c. In (25) we have merely to introduce the factor $\sin\theta$, equal to $2\sin\frac{1}{2}\theta\cos\frac{1}{2}\theta$, so that we get instead of (26)

$$4 \int_0^{\frac{1}{2}\pi} \frac{\sin(x\sin\phi)}{x} \cos\phi \, d\phi = \frac{4}{x^2}(1 - \cos x).$$

The factor, corresponding to (27), is therefore simply

$$\frac{\sin^2 kc}{k^2 c^2} . \quad \dots\dots\dots\dots\dots\dots\dots\dots(30)$$

No work at all is done if c be such that kc is a multiple of π, or $2c$ a multiple of λ.

By the method of the *Theory of Sound* (*loc. cit.*) we might in like manner investigate the effect of distributing a source of sound uniformly throughout the *volume* of a sphere, but the above examples will suffice for our purpose.

Experimental Illustrations.

There is no difficulty in illustrating upon a small scale the results above deduced from theory. The simplest experiment is with an ordinary open organ-pipe, gently blown, so as to exclude overtones as much as may be. The open ends act as two equal sources of sound in the same phase. Connected by a long flexible tube with a well-regulated bellows, the pipe can be held in any position and be observed in the open air from a moderate distance. When the length of the pipe is perpendicular to the line of observation, the two sources are at the same distance and the effects conspire. But if the pipe point toward the observer, the two sources, being at about $\frac{1}{2}\lambda$ apart, are in antagonism and the sound is much enfeebled.

* Macmillan, 1st edition, 1878 ; 2nd edition, 1896, § 302.

In order to exemplify the principle further, a multiple pipe was constructed (fig. 2). This consisted of a straight lead tube 31 inches long and ·35 inch bore*, open at the ends A, B. At four points D, E, F, G, distant $6\frac{1}{2}$ inches from the ends and 6 inches from one another, the tube was perforated, and the holes were blown by four streams of wind from the branched supply-tube C. The whole was cemented to a framework of wood, so that it could be turned round without relative displacement. The intention was that all six apertures A, D, E, F, G, B should act as sources of sound in the same phase, but one could hardly be sure à *priori* that this behaviour would be observed. Would the simultaneous motions of the air-column on the two sides of E (for example) be both towards E or both from E? Might it not rather

Fig. 2.

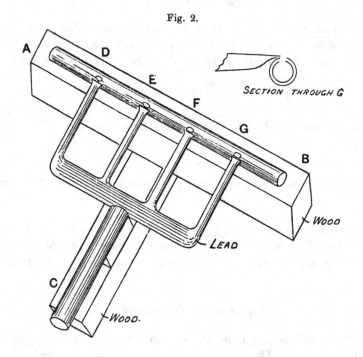

be that the motions would take the same absolute direction, in which case E, F, &c., would fail to act as sources? A little observation, however, sufficed to prove that the apparatus really acted as intended. By listening through a rubber tube whose outer end was brought into proximity with the apertures, it was easy to satisfy oneself that D, E, F, G were effective sources of sound, and were in fact more powerful than the open ends A, B, as was to be expected. The half wave-length of the actual sound was 8 inches, showing that the "openness" of the pipe at D, E, F, G was rather imperfect, owing

* Inch = 2·54 cms.

to the smallness of the holes.   Still the apparatus afforded a combination
of six sources of sound, all in the same phase and at about half a wave-
length apart.

The observations were made upon a lawn; and, as the sound was rather
feeble, a very moderate distance sufficed.   When AB was vertical, the sources
cooperated and the maximum sound was heard.   But when AB was [turned
round], so as no longer to be perpendicular to the (horizontal) line of observation,
the sound was less, a deviation of 30° causing a very great falling off.   The
effect was as if the sound had suddenly gone away to a great distance.   The
success of the experiment no doubt depended a good deal upon the absence
of over-tones, a condition of things favoured by the feebleness of the sound
and also by the high pitch.

291.

ON THE WORK DONE BY FORCES OPERATIVE AT ONE OR MORE POINTS OF AN ELASTIC SOLID.

[*Philosophical Magazine*, Vol. VI. pp. 385—392, 1903.]

AN investigation of the waves generated in an isotropic elastic solid by a periodic impressed force, localized in the neighbourhood of a point, was first given by Stokes *. A simpler treatment of the problem will be found in my paper "On the Light from the Sky, &c.†," and more fully in *Theory of Sound*, § 378. It will be desirable to recapitulate the principal steps.

If α, β, γ be the displacements at any point of the solid, X', Y', Z' the impressed forces reckoned per unit of *mass*, we have equations such as

$$\frac{d^2\alpha}{dt^2} = (a^2 - b^2)\frac{d\delta}{dx} + b^2\nabla^2\alpha + X', \quad \ldots\ldots\ldots\ldots\ldots(1)$$

in which a and b are the velocities of dilatational and transverse waves respectively, and δ represents the dilatation expressed by

$$\delta = \frac{d\alpha}{dx} + \frac{d\beta}{dy} + \frac{d\gamma}{dz}.\ldots\ldots\ldots\ldots\ldots\ldots\ldots\ldots(2)$$

If, as throughout the present paper, α, β, γ, &c. be proportional to e^{ipt}, $d^2\alpha/dt^2 = -p^2\alpha$, and (1) &c. become

$$(a^2 - b^2)\,d\delta/dx + b^2\nabla^2\alpha + p^2\alpha + X' = 0, \quad\ldots\ldots\ldots\ldots(3)$$
$$(a^2 - b^2)\,d\delta/dy + b^2\nabla^2\beta + p^2\beta + Y' = 0, \quad\ldots\ldots\ldots\ldots(4)$$
$$(a^2 - b^2)\,d\delta/dz + b^2\nabla^2\gamma + p^2\gamma + Z' = 0. \quad\ldots\ldots\ldots\ldots(5)$$

These are the fundamental equations. For our purpose we may suppose that X', Y' vanish throughout, and that Z' is finite only in the neighbourhood of the origin. It will be convenient to write

$$k = p/b, \quad h = p/a. \quad\ldots\ldots\ldots\ldots\ldots\ldots\ldots\ldots(6)$$

---

* *Camb. Phil. Trans.* Vol. IX. p. 1 (1849); *Collected Works*, Vol. II. p. 243.

† *Phil. Mag.* XLI. pp. 107, 274 (1871); *Scientific Papers*, I. p. 96.

The *dilatation* δ is readily found. Differentiating (3), (4), (5) with respect to x, y, z and adding, we get

$$\nabla^2\delta + h^2\delta + a^{-2}dZ'/dz = 0. \quad\dots\dots\dots\dots\dots(7)$$

The solution of (7) is

$$\delta = \frac{1}{4\pi a^2}\iiint \frac{dZ'}{dz}\frac{e^{-ihr}}{r}\,dx\,dy\,dz,$$

r denoting the distance between the element at x, y, z near the origin (O) and the point (P) under consideration. If we integrate partially with respect to z, we find

$$\delta = -\frac{1}{4\pi a^2}\iiint Z'\frac{d}{dz}\left(\frac{e^{-ihr}}{r}\right)dx\,dy\,dz, \quad\dots\dots\dots\dots(8)$$

the integrated term vanishing in virtue of the condition that Z' is finite only within a certain space T. Moreover, since the dimensions of T are supposed to be very small in comparison with the wave-length, $d\,(r^{-1}e^{-ihr})/dz$ may be removed from under the integral sign.

It will be convenient also to change the meaning of x, y, z, so that they shall represent as usual the coordinates of P relatively to O. Thus, if $Z_1\,e^{ipt}$ denote the whole force applied at the origin, so that

$$Z_1 = \rho \iiint Z'\,dx\,dy\,dz,$$

in which ρ is the density,

$$\delta = \frac{Z_1}{4\pi a^2\rho}\frac{d}{dz}\left(\frac{e^{-ihr}}{r}\right), \quad\dots\dots\dots\dots\dots\dots(9)$$

giving the dilatation at the point P.

In like manner we may find the *rotations* ϖ', ϖ'', ϖ''', defined by

$$\frac{d\gamma}{dy} - \frac{d\beta}{dz} = 2\varpi', \quad \frac{d\alpha}{dz} - \frac{d\gamma}{dx} = 2\varpi'', \quad \frac{d\beta}{dx} - \frac{d\alpha}{dy} = 2\varpi''' \quad\dots\dots(10)$$

For from (3), (4), (5) we have

$$\nabla^2\varpi' + k^2\varpi' + \tfrac{1}{2}b^{-2}dZ'/dy = 0, \quad\dots\dots\dots\dots\dots(11)$$
$$\nabla^2\varpi'' + k^2\varpi'' - \tfrac{1}{2}b^{-2}dZ'/dx = 0, \quad\dots\dots\dots\dots\dots(12)$$
$$\nabla^2\varpi''' + k^2\varpi''' \qquad\qquad = 0, \quad\dots\dots\dots\dots\dots(13)$$

whence $\varpi''' = 0$, and

$$\varpi' = \frac{Z_1}{8\pi b^2\rho}\frac{d}{dy}\left(\frac{e^{-ikr}}{r}\right), \quad \varpi'' = -\frac{Z_1}{8\pi b^2\rho}\frac{d}{dx}\left(\frac{e^{-ikr}}{r}\right). \quad\dots\dots(14)$$

These are the results given in my paper of 1871.

The values of δ, ϖ', ϖ'', ϖ''' determine those of α, β, γ. If we take

$$\alpha = \frac{d^2\chi}{dx\,dz}, \quad \beta = \frac{d^2\chi}{dy\,dz}, \quad \gamma = \frac{d^2\chi}{dz^2} + w, \quad\dots\dots\dots(15)$$

where
$$w = k^2 A \frac{e^{-ikr}}{r}, \qquad \chi = A \left(\frac{e^{-ikr}}{r} - \frac{e^{-ihr}}{r} \right), \quad \dots\dots\dots(16)$$

and
$$A = \frac{Z_1}{4\pi k^2 b^2 \rho}, \quad \dots\dots\dots\dots\dots\dots\dots(17)$$

it is easy to verify that these forms give the correct values to δ, ϖ', ϖ'', ϖ'''. As regards the dilatation,

$$\delta = \frac{d}{dz} (\nabla^2 \chi + w),$$

in which
$$\nabla^2 \chi = A \left(-k^2 \frac{e^{-ikr}}{r} + h^2 \frac{e^{-ihr}}{r} \right).$$

This reproduces (9).

As regards the rotations, we see that χ does not influence them. In fact

$$\varpi''' = 0, \qquad \varpi' = \tfrac{1}{2} \frac{dw}{dy}, \qquad \varpi'' = -\frac{dw}{dx};$$

and these agree with (14). The solution expressed by (15), (16), (17) is thus verified, and it applies whether the solid be compressible or not.

In the case of incompressibility, $h = 0$. If we restore the time-factor e^{ipt} and throw away the imaginary part of the solution, we get

$$\alpha = \frac{A k^2 xz}{r^3} \left[\left(-1 + \frac{3}{k^2 r^2} \right) \cos (pt - kr) - \frac{3}{kr} \sin (pt - kr) - \frac{3}{k^2 r^2} \cos pt \right],$$
$$\dots\dots(18)$$

$$\gamma = \frac{A k^2}{r} \left[\left(1 - \frac{z^2}{r^2} + \frac{3z^2}{k^2 r^4} - \frac{1}{k^2 r^2} \right) \cos (pt - kr) \right.$$

$$\left. + \left(\frac{1}{kr} - \frac{3z^2}{kr^3} \right) \sin (pt - kr) - \left(\frac{3z^2}{k^2 r^4} - \frac{1}{k^2 r^2} \right) \cos pt \right], \quad \dots(19)$$

the value of β differing from that of α merely by the substitution of y for x. The value of A is given by (17), and $Z_1 \cos pt$ is the whole force operative at the origin at time t.

At a great distance from the origin (18), (19) reduce to

$$\alpha = -\frac{Z_1}{4\pi b^2 \rho} \frac{xz}{r^2} \frac{\cos (pt - kr)}{r}, \qquad \gamma = +\frac{Z_1}{4\pi b^2 \rho} \left(1 - \frac{z^2}{r^2} \right) \frac{\cos (pt - kr)}{r}.$$

$$\dots\dots(20, 21)$$

Upon this (*Theory of Sound*, § 378) I commented:—" W. König (Wied. *Ann.* XXXVII. p. 651, 1889) has remarked upon the non-agreement of (18), (19), first given in a different form by Stokes, with the results of a somewhat similar investigation by Hertz (Wied. *Ann.* XXXVI. p. 1, 1889), in which the terms involving $\cos pt$, $\sin pt$ do not occur, and he seems disposed to regard Stokes's results as affected by error. But the fact is that the problems treated are essentially different, that of Hertz having no relation to elastic solids. The source of the discrepancy is in the first terms of (3) &c., which are omitted by Hertz in his theory of the æther. But assuredly in a theory of elastic solids these terms must be retained. Even when the material is supposed to be incompressible, so that δ vanishes, the retention is still necessary, because, as was fully explained by Stokes in the memoir referred to, the factor $(a^2 - b^2)$ is infinite at the same time."

Although the substance of the above comment appears to be justified, I went too far in saying that Hertz's solution has no relation to elastic solids. It is indeed not permissible to omit the first terms of (3) &c. merely because the solid is incompressible; but if, though the solid is compressible, it be in fact not compressed, these terms disappear. Now Hertz's solution, corresponding to the omission of the second part of χ in (16), makes

$$\alpha = \frac{d^2}{dx\,dz}\left(\frac{e^{-ikr}}{r}\right), \quad \beta = \frac{d^2}{dy\,dz}\left(\frac{e^{-ikr}}{r}\right), \quad \gamma = \left(\frac{d^2}{dz^2}+k^2\right)\left(\frac{e^{-ikr}}{r}\right),$$
$$\dots\dots(22)$$

and accordingly
$$\delta = \frac{d\alpha}{dx}+\frac{d\beta}{dy}+\frac{d\gamma}{dz}=0:$$

values which satisfy $(\nabla^2 + k^2)(\alpha, \beta, \gamma) = 0.$ ....(23)

Thus (3), (4), (5) are satisfied, and the solution applies to an elastic solid upon which no forces act except at the origin. The only question remaining open is as to the character of the forces which must be supposed to act at that place. This is rather a delicate matter; but it is evident at any rate that the forces are not of the simple character contemplated in the preceding investigation. It would appear that they must be double or multiple, and have components parallel to x and y as well as z. By a *double* force is meant the limit of a *couple* of given moment when the components increase and their mutual distance decreases, analogous to the double source of acoustics.

I now propose to calculate the work done by the force Z_1 at the origin as it generates the waves represented by (18), (19). For this purpose we require the part of γ in the neighbourhood of the origin which is in quadrature with the force, *i.e.* is proportional to $\sin pt$. From (19) we get

$$\frac{Ak^2\sin pt}{r}\left[\left(1-\frac{z^2}{r^2}+\frac{3z^2}{k^2r^4}-\frac{1}{k^2r^2}\right)\sin kr + \left(\frac{1}{kr}-\frac{3z^2}{kr^3}\right)\cos kr\right], \dots(24)$$

the last term (in $\cos pt$) not contributing. Expanding $\sin kr$, $\cos kr$ and retaining the terms of order kr, we get for the square bracket

$$\tfrac{2}{3}kr + \frac{z^2}{r^2}(0).$$

Thus $(24) = \tfrac{2}{3}k^3A\sin pt,$

when r is small, so that the part proportional to $\sin pt$ is in the limit finite and independent of z/r. If W be the work done in time dt,

$$\frac{dW}{dt} = Z_1 \cos pt \cdot \tfrac{2}{3} k^3 pA \cos pt \,;$$

and by (6), (17)

$$\text{mean } \frac{dW}{dt} = \frac{k^2 Z_1{}^2}{12\pi b\rho}. \qquad \dots\dots\dots\dots\dots\dots\dots(25)$$

The right-hand member of (25) is thus the work done (on the average) in unit of time.

This result may be confirmed by a calculation of the energy radiated away in unit time, for which purpose we may employ the formulæ (20), (21) applicable when r is great. The energy in question is the double of the kinetic energy to be found in a spherical shell whose thickness $(r_2 - r_1)$ is the distance travelled by transverse waves in the unit of time, viz. b. In the expression for the kinetic energy the resultant (velocity)² at any point x, y, z is by (20), (21) proportional to

$$\frac{x^2 z^2}{r^4} + \frac{y^2 z^2}{r^4} + \frac{(r^2 - z^2)^2}{r^4} = 1 - \frac{z^2}{r^2} = 1 - \mu^2,$$

a quantity symmetrical with respect to the axis. Also $\sin^2(pt - kr)$ is to be replaced by its mean value, viz. $\tfrac{1}{2}$. Thus the kinetic energy is

$$\tfrac{1}{2}\rho \left(\frac{Z_1}{4\pi b^2 \rho}\right)^2 \int_{-1}^{+1} \int_{r_1}^{r_2} \tfrac{1}{2} p^2 (1 - \mu^2)\, 2\pi\, d\mu\, dr,$$

the double of which is identical with (25).

We will now form the expression for the resolved displacement at P due to $Z_1 \cos pt$ acting at O (parallel to OZ), the displacement being resolved in

Fig. 1.

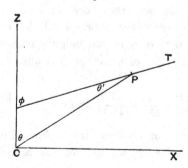

a direction PT in the plane ZOP making an angle θ' with OP (fig. 1). The angle between PT and OZ is denoted by ϕ, so that $\phi = \theta + \theta'$.

The resolved displacement is

$$\gamma \cos \phi + \alpha \sin \phi, \qquad \qquad (26)$$

α and γ being given by (18), (19), in which we write

$$z/r = \cos \theta, \quad x/r = \sin \theta.$$

We find

$$\frac{\gamma \cos \phi + \alpha \sin \phi}{A k^2 / r} = \cos (pt - kr) \left[- \sin \theta \sin \theta' - \frac{\cos \phi}{k^2 r^2} + \frac{3}{k^2 r^2} \cos \theta \cos \theta' \right]$$

$$+ \frac{\sin (pt - kr)}{kr} \left[\cos \phi - 3 \cos \theta \cos \theta' \right] + \frac{\cos pt}{k^2 r^2} \left[\cos \phi - 3 \cos \theta \cos \theta' \right],$$

a symmetrical function of θ and θ' as required by the general principle of reciprocity (*Theory of Sound*, § 108). The value of A is given by (17) in which, however, we will now write $\mathfrak{F}$ for Z_1, so that

$$A = \frac{\mathfrak{F}}{4 \pi k^2 b^2 \rho} . \qquad \qquad (27)$$

The above equation gives the resolved displacement at P in a direction making an angle θ' with r due to a force $\mathfrak{F} \cos pt$ at O acting in a direction inclined to r at angle θ. If we suppose that a force $\mathfrak{F}' \cos pt$ acts at P in direction PT and inquire as to the work done by this force upon the motion due to $\mathfrak{F}$, we have to retain that part of the resolved displacement due to $\mathfrak{F}$ which is in quadrature with $\mathfrak{F}' \cos pt$, viz. the part proportional to $\sin pt$. The mean work is given by the symmetrical expression

$$\frac{k^2 \mathfrak{F} \mathfrak{F}'}{8 \pi b \rho} \left\{ \frac{\cos (\theta + \theta') - 3 \cos \theta \cos \theta'}{k^2 r^2} \left(\cos kr - \frac{\sin kr}{kr} \right) - \frac{\sin kr}{kr} \sin \theta \sin \theta' \right\}.$$

$$\qquad \qquad (28)$$

If the forces are parallel, $\phi = 0$, $\theta' = -\theta$, and (28) becomes

$$\frac{k^2 \mathfrak{F} \mathfrak{F}'}{8 \pi b \rho} \left[\sin^2 \theta \frac{\sin kr}{kr} + \frac{1 - 3 \cos^2 \theta}{k^2 r^2} \left(\cos kr - \frac{\sin kr}{kr} \right) \right]. \qquad (29)$$

If we further suppose that kr is very small, the square bracket reduces to the value $\frac{2}{3}$, and we get

$$\frac{k^2 \mathfrak{F} \mathfrak{F}'}{12 \pi b \rho} . \qquad \qquad (30)$$

Comparing with (25) we see that the work done by $\mathfrak{F}'$ on the motion due to an equal $\mathfrak{F}$ is the same as that done by $\mathfrak{F}$ itself, as should evidently be.

If in (29) $\theta = 90°$, so that the forces are perpendicular to the line joining the two points of application, we get

$$\frac{k^2 \mathfrak{F} \mathfrak{F}'}{8 \pi b \rho} \left[\frac{\sin kr}{kr} + \frac{1}{k^2 r^2} \left(\cos kr - \frac{\sin kr}{kr} \right) \right]. \qquad (31)$$

As we have seen, when kr is small (31) is finite and positive. It vanishes when

$$\tan kr = \frac{kr}{1 - k^2 r^2}, \qquad \dots\dots\dots\dots\dots\dots\dots\dots\dots(32)$$

and this occurs first in the second quadrant.

In general, when there are a number of forces acting at detached points, the whole work done must be obtained by a double summation of (28). If the forces are continuously distributed, the sum becomes a double integral. A particular case, of interest in connexion with the problem of electrical vibrations along a circular wire*, occurs when the forces act tangentially at the various points of a circular arc. Here $\theta' = \theta$, and (28) becomes

$$\frac{k^2 \mathfrak{F} \mathfrak{F}'}{8\pi b \rho} \left\{ \frac{1 + \cos^2 \theta}{k^2 r^2} \left(\frac{\sin kr}{kr} - \cos kr \right) - \frac{\sin kr}{kr} \sin^2 \theta \right\}. \qquad \dots\dots\dots(33)$$

---

* Compare Pocklington, *Proc. Camb. Phil. Soc.* IX. p. 324 (1897); *Nature*, LXII. p. 486 (1903). It would seem that (33) must lead to a more complicated expression for the energy radiated than that in Dr Pocklington's investigation.

292.

ON THE ACOUSTIC SHADOW OF A SPHERE.

WITH AN APPENDIX*, GIVING THE VALUES OF LEGENDRE'S FUNCTIONS FROM P_0 TO P_{20} AT INTERVALS OF 5 DEGREES. BY PROFESSOR A. LODGE.

[*Phil. Trans.* 203 A, pp. 87—110, 1904.]

IN my book on the *Theory of Sound*, § 328, I have discussed the effect upon a source of sound of a rigid sphere whose surface is close to the source.

The question turns upon the relative magnitudes of the wave-length (λ) and the radius (c) of the sphere. If kc be *small*, where $k = 2\pi/\lambda$, the presence of the sphere has but little effect upon the sound to be perceived at a distance.

The following table was given, showing the effect in three principal directions of somewhat larger spheres:

kc	μ	$F^2 + G^2$
$\frac{1}{2}$	1	·294291
	−1	·259729
	0	·231999
1	1	·502961
	−1	·285220
	0	·236828
2	1	·6898
	−1	·3182
	0	·3562

* [1911. Only the Table is reproduced here. In the original paper will be found a description of the process of calculation.]

Here $F^2 + G^2$ represents the intensity of sound at a great distance from the sphere in directions such that μ is the cosine of the angle between them and that radius which passes through the source. Upon the scale of measurement adopted, $F^2 + G^2 = \frac{1}{4}$ for all values of μ, when $kc = 0$, that is, when the propagation is undisturbed by any obstacle. The increased values under $\mu = 1$ show that the sphere is beginning to act as a *reflector*, the intensity in this direction being already more than doubled when $kc = 2$. "In looking at these figures, the first point which attracts attention is the comparatively slight deviation from uniformity in the intensities in different directions. Even when the circumference of the sphere amounts to twice the wavelength, there is scarcely anything to be called a sound shadow. But what is, perhaps, still more unexpected is that in the first two cases the intensity behind the sphere [$\mu = -1$] exceeds that in a transverse direction [$\mu = 0$]. This result depends mainly on the preponderance of the term of the first order, which vanishes with μ. The order of the more important terms increases with kc; when kc is 2, the principal term is of the second order.

"Up to a certain point the augmentation of the sphere will increase the total energy emitted, because a simple source emits twice as much energy when close to a rigid plane as when entirely in the open. Within the limits of the table this effect masks the obstruction due to an increasing sphere, so that when $\mu = -1$, the intensity is greater when the circumference is twice the wave-length than when it is half the wave-length, the source itself remaining constant."

The solution of the problem when kc is very great cannot be obtained by this method, but it is to be expected that when $\mu = 1$ the intensity will be quadrupled, as when the sphere becomes a plane, and that when μ is negative the intensity will tend to vanish. It is of interest to trace somewhat more closely the approach to this state of things—to treat, for example, the case of $kc = 10*$. In every case where it can be carried out the solution has a double interest, since in virtue of the law of reciprocity it applies when the source and point of observation are interchanged, thus giving the intensity at a point on the sphere due to a source situated at a great distance.

But before proceeding to consider a higher value of kc, it will be well to supplement the information already given under the head of $kc = 2$. The original calculation was limited to the principal values of μ, corresponding to the poles and the equator, under the impression that results for other values of μ would show nothing distinctive. The first suggestion to the contrary was from experiment. In observing the shadow of a sphere, by listening

---

* See Rayleigh, *Proc. Roy. Soc.* Vol. LXXII. p. 40 [*Scientific Papers*, Vol. v. p. 114]; also Macdonald, Vol. LXXI. p. 251; Vol. LXXII. p. 59; Poincaré, Vol. LXXII. p. 42.

through a tube whose open end was presented to the sphere, it was found that the somewhat distant source was *more* loudly heard at the anti-pole ($\mu = -1$) than at points 40° or 50° therefrom. This is analogous to Poisson's experiment, where a bright point is seen in the centre of the shadow of a circular disc—an experiment easily imitated acoustically*—and it may be generally explained in the same manner. This led to further calculations for values of μ between 0 and -1, giving numbers in harmony with observation. The complete results for this case ($kc = 2$) are recorded in the annexed table. In obtaining them, terms of Legendre's series, up to and including P_8, were retained. The angles θ are those whose cosine is μ.

$$kc = 2.$$

θ	$F + iG$	$F^2 + G^2$	$4(F^2 + G^2)$
0	$+ \cdot 7968 + \cdot 2342\ i$	$\cdot 6898$	$2\cdot 759$
15	$+ \cdot 8021 + \cdot 1775\ i$	$\cdot 6749$	$2\cdot 700$
30	$+ \cdot 7922 + \cdot 0147\ i$	$\cdot 6278$	$2\cdot 511$
45	$+ \cdot 7139 - \cdot 2287\ i$	$\cdot 5619$	$2\cdot 248$
60	$+ \cdot 5114 - \cdot 4793\ i$	$\cdot 4912$	$1\cdot 965$
75	$+ \cdot 1898 - \cdot 6247\ i$	$\cdot 4263$	$1\cdot 705$
90	$- \cdot 1538 - \cdot 5766\ i$	$\cdot 3562$	$1\cdot 425$
105	$- \cdot 3790 - \cdot 3413\ i$	$\cdot 2601$	$1\cdot 040$
120	$- \cdot 3992 - \cdot 0243\ i$	$\cdot 1600$	$0\cdot 640$
135	$- \cdot 2401 + \cdot 2489\ i$	$\cdot 1196$	$0\cdot 478$
150	$- \cdot 0088 + \cdot 4157\ i$	$\cdot 1729$	$0\cdot 692$
165	$+ \cdot 1781 + \cdot 4883\ i$	$\cdot 2701$	$1\cdot 080$
180	$+ \cdot 2495 + \cdot 5059\ i$	$\cdot 3182$	$1\cdot 273$

A plot of $4(F^2 + G^2)$ against θ is given in fig. 1, curve A.

The investigation for $kc = 10$ could probably be undertaken with success upon the lines explained in *Theory of Sound*; but as it is necessary to include some 20 terms of the expansion in Legendre's series, I considered that it would be advantageous to use certain formulæ of reduction by which the functions of various orders can be deduced from their predecessors, and this involves a change of notation. Formulæ convenient for the purpose have been set out by Professor Lamb†. The velocity-potential ψ is supposed to be proportional throughout to e^{ikat}, but this time-factor is usually omitted. The general differential equation satisfied by ψ is

$$\frac{d^2\psi}{dx^2} + \frac{d^2\psi}{dy^2} + \frac{d^2\psi}{dz^2} + k^2\psi = 0, \quad \text{...........................(1)}$$

of which the solution in polar coordinates applicable to a *divergent* wave of the nth order in Laplace's series may be written

$$\psi_n = S_n r^n \chi_n(kr). \quad \text{...........................(2)}$$

---

* *Phil. Mag.* Vol. IX. p. 278, 1880; *Scientific Papers*, Vol. I. p. 472.

† *Hydrodynamics*, § 267; *Camb. Phil. Trans.* Vol. XVIII. p. 350, 1900.

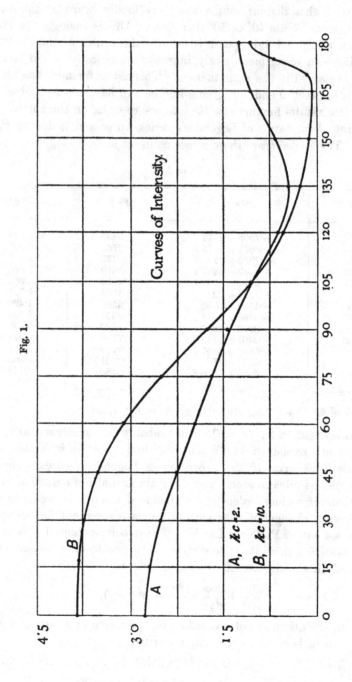

Fig. 1.

Curves of Intensity.

A, $kc = 2$.
B, $kc = 10$.

For the present purpose we may suppose without loss of generality that $k = 1$. The differential equation satisfied by $\chi_n(r)$ is

$$\frac{d^2\chi_n}{dr^2} + \frac{2n+2}{r} \cdot \frac{d\chi_n}{dr} + \chi_n = 0, \quad\text{............(3)}$$

and of this the solution which corresponds to a divergent wave is

$$\chi_n(r) = \left(-\frac{d}{r\,dr}\right)^n \frac{e^{-ir}}{r}. \quad\text{............(4)}$$

Putting $n = 0$ and $n = 1$, we have

$$\chi_0(r) = \frac{e^{-ir}}{r}, \qquad \chi_1(r) = \frac{(1+ir)\,e^{-ir}}{r^3}. \quad\text{............(5)}$$

It is easy to verify that (4) satisfies (3). For if χ_n satisfies (3), $r^{-1}\chi'_n$ satisfies the corresponding equation for χ_{n+1}. And $r^{-1}e^{-ir}$ satisfies (3) when $n = 0$.

From (3) and (4) the following formulæ of reduction may be verified:

$$\chi'_n(r) = -r\chi_{n+1}(r), \quad\text{............(6)}$$

$$r\chi'_n(r) + (2n+1)\chi_n(r) = \chi_{n-1}(r), \quad\text{............(7)}$$

$$\chi_{n+1}(r) = \frac{(2n+1)\chi_n(r) - \chi_{n-1}(r)}{r^2}. \quad\text{............(8)}$$

By means of the last, χ_2, χ_3, &c. may be built up in succession from χ_0 and χ_1.

From (2) $\qquad d\psi_n/dr = S_n(nr^{n-1}\chi_n + r^n\chi'_n),$

or, with use of (7),

$$d\psi_n/dr = r^{n-1}S_n\{\chi_{n-1} - (n+1)\chi_n\}. \quad\text{............(9)}$$

Thus, if U_n be the nth component of the normal velocity at the surface of the sphere $(r = c)$,

$$U_n = c^{n-1}S_n\{\chi_{n-1}(c) - (n+1)\chi_n(c)\}. \quad\text{............(10)}$$

When $n = 0$, $\qquad U_0 = S_0\chi'_0(c) = -S_0c\chi_1(c). \quad\text{............(11)}$

The introduction of S_n from (10), (11) into (2) gives ψ_n in terms of U_n supposed known.

When r is very great in comparison with the wave-length, we get from (4)

$$\chi_n(r) = \frac{i^n e^{-ir}}{r^{n+1}}, \quad\text{............(12)}$$

so that $\qquad \psi_n = S_n \frac{i^n e^{-ir}}{r}. \quad\text{............(13)}$

In order to find the effect at a great distance of a source of sound localised on the surface of the sphere at the point $\mu = 1$, we have only to expand the complete value of U in Legendre's functions. Thus

$$U_n = \tfrac{1}{2}(2n+1) P_n(\mu) \int_{-1}^{+1} U P_n(\mu)\, d\mu$$

$$= \tfrac{1}{2}(2n+1) P_n(\mu) \int_{-1}^{+1} U\, d\mu = \frac{2n+1}{4\pi c^2} P_n(\mu) \iint U\, dS, \quad \ldots(14)$$

in which $\iint U\, dS$ denotes the magnitude of the *source*, i.e., the integrated value of U over the small area where it is sensible. The complete value of ψ may now be written

$$\psi = \frac{\iint U\, dS \cdot e^{i(at-r)}}{4\pi r} \sum \frac{(2n+1)\, i^n P_n(\mu)}{c^{n+1}\{\chi_{n-1}(c) - (n+1)\chi_n(c)\}} \cdot \quad \ldots\ldots(15)$$

When $n = 0$, $\chi_{n-1}(c) - (n+1)\chi_n(c)$ is to be replaced by $-c^2 \chi_1(c)$.

If we compare (15) with the corresponding expression in *Theory of Sound*, (3), § 238, we get

$$c^{n+1}\{\chi_{n-1}(c) - (n+1)\chi_n(c)\} = -i^n e^{-ic} F_n(ic). \quad \ldots\ldots\ldots(16)$$

Another particular case of interest arises when the point of observation, as well as the source, is on the sphere, so that, instead of $r = \infty$, we have $r = c$. Equation (15) is then replaced by

$$\psi = \frac{\iint U\, dS \cdot e^{ikat}}{4\pi c} \sum \frac{(2n+1)\,\chi_n(c)\, P_n(\mu)}{\chi_{n-1}(c) - (n+1)\chi_n(c)} \cdot \quad \ldots\ldots\ldots(17)$$

It may be remarked that, since ψ in (17) is infinite when $\mu = +1$ and accordingly $P_n = 1$, the convergence at other points can only be attained in virtue of the factors P_n. The difficulties in the way of a practical calculation from (17) may be expected to be greater than in the case of (15).

We will now proceed to the actual calculation for the case of $c = 10$, or $kc = 10$. The first step is the formation of the values of the various functions $\chi_n(10)$, starting from $\chi_0(10)$, $\chi_1(10)$. For these we have from (5)

$$10\chi_0(10) = \cos 10 - i \sin 10,$$

$$10^2 \chi_1(10) = \tfrac{1}{10}\cos 10 + \sin 10 + i(\cos 10 - \tfrac{1}{10}\sin 10).$$

The angle (10 radians) $= 540° + 32° 57' \cdot 468$; thus

$$\sin 10 = -\cdot5440210, \quad \cos 10 = -\cdot8390716,$$

and

$$10\chi_0 = -\cdot8390716 + \cdot5440210\, i,$$

$$10^2 \chi_1 = -\cdot6279282 - \cdot7846695\, i.$$

From these, χ_2, χ_3, ... are to be computed in succession from (8), which may be put into the form

$$10^{n+2}\chi_{n+1} = \frac{2n+1}{10}\,10^{n+1}\chi_n - 10^n\chi_{n-1}.$$

For example,

$$10^3\chi_2 = \cdot3\,(10^2\chi_1) - 10\chi_0 = +\cdot6506931 - \cdot7794218\,i.$$

When the various functions $10^{n+1}\chi_n$ have been computed, the next step is the computation of the denominators in (15). We write

$$D_n = 10^{n+1}\{\chi_{n-1} - (n+1)\chi_n\} = 10 \times 10^n\chi_{n-1} - (n+1)\,10^{n+1}\chi_n, \ldots(18)$$

and the values of D_n are given along with $10^{n+1}\chi_n$ in the annexed table.

n	$10^{n+1}\chi_n\,(10)$		D_n	
0	−	$0\cdot83907 + 0\cdot54402\,i$	+	$6\cdot2793 + 7\cdot8467\,i$
1	−	$0\cdot62793 - 0\cdot78467\,i$	−	$7\cdot1349 + 7\cdot0095\,i$
2	+	$0\cdot65069 - 0\cdot77942\,i$	−	$8\cdot2314 - 5\cdot5084\,i$
3	+	$0\cdot95327 + 0\cdot39496\,i$	+	$2\cdot6938 - 9\cdot3741\,i$
4	+	$0\cdot01660 + 1\cdot05589\,i$	+	$9\cdot4498 - 1\cdot3299\,i$
5	−	$0\cdot93834 + 0\cdot55534\,i$	+	$5\cdot7960 + 7\cdot2269\,i$
6	−	$1\cdot04877 - 0\cdot44501\,i$	−	$2\cdot0420 + 8\cdot6685\,i$
7	−	$0\cdot42506 - 1\cdot13386\,i$	−	$7\cdot0872 + 4\cdot6208\,i$
8	+	$0\cdot41117 - 1\cdot25578\,i$	−	$7\cdot9512 - 0\cdot0366\,i$
9	+	$1\cdot12406 - 1\cdot00096\,i$	−	$7\cdot1288 - 2\cdot5482\,i$
10	+	$1\cdot72454 - 0\cdot64605\,i$	−	$7\cdot7293 - 2\cdot9031\,i$
11	+	$2\cdot49747 - 0\cdot35574\,i$	−	$12\cdot7243 - 2\cdot1916\,i$
12	+	$4\cdot01964 - 0\cdot17216\,i$	−	$27\cdot2807 - 1\cdot3194\,i$
13	+	$7\cdot55164 - 0\cdot07465\,i$	−	$65\cdot5265 - 0\cdot6764\,i$
14	+	$16\cdot36978 - 0\cdot02941\,i$	−	$170\cdot030 \ -0\cdot3054\,i$
15	+	$39\cdot92071 - 0\cdot01062\,i$	−	$475\cdot033 \ -0\cdot124\,i$
16	+	$107\cdot3844 \ -0\cdot00353\,i$	−	$1426\cdot33 \ -0\cdot047\,i$
17	+	$314\cdot45 \quad -0\cdot0010\,i$	−	$4586\cdot2 \ -0\cdot017\,i$
18	+	$993\cdot19 \quad -0\cdot000\,i$	−	$15725\cdot0 \ -0\cdot010\,i$
19	+	$3360\cdot3$	−	57274
20	+	12112		-220750
21	+	46299		-897460
22	+	186974		-38374×10^2

It will be seen that the imaginary part of $10^{n+1}\chi_n\,(10)$ tends to zero, as n increases. It is true that if we continue the calculation, having used throughout, say, 5 figures, we find that the terms begin to increase again. This, however, is but an imperfection of calculation, due to the increasing value of $\frac{1}{10}(2n+1)$ in the formula and consequent loss of accuracy, as each term is deduced from the preceding pair. Any doubt that may linger will be removed by reference to (4), according to which the imaginary term in question has the expression

$$-ir^{n+1}\left(-\frac{d}{r\,dr}\right)^n\frac{\sin r}{r}.$$

Now, if we expand $r^{-1} \sin r$ and perform the differentiations, the various terms disappear in order. For example, after the 25th operation we have

$$\left(-\frac{d}{r\,dr}\right)^{25}\frac{\sin r}{r} = \frac{50.48\ldots 4.2}{51!} - \frac{52.50\ldots 6.4}{53!}r^2 + \frac{54\ldots 6}{55!}r^4 - \&c.,$$

the first term being in every case positive and the subsequent terms alternately negative and positive. The series is convergent, since the numerical values of the terms continually diminish, the ratio of consecutive terms being (when $r = 10$)

$$\frac{100}{2.53}, \quad \frac{100}{4.55}, \quad \frac{100}{6.57}, \quad \&c.$$

Accordingly the first term gives a limit to the sum of the series. On introduction of the factor 10^{n+1}, this becomes

$$\frac{10^{26}}{1.3.5\ldots 49.51},$$

i.e., approximately $10^{-8} \times 3\cdot0$. A fortiori, when n is greater than 25, the imaginary part of $10^{n+1}\chi_n(10)$ is wholly negligible.

We can now form the coefficients of P_n under the sign of summation in (15), i.e., the values of

$$i^n(2n+1)D_n^{-1}. \quad\quad\quad\ldots\ldots\ldots\ldots\ldots\ldots\ldots(19)$$

For a reason that will presently appear, it is convenient to separate the odd and even values of n.

n	$i^n(2n+1)D_n^{-1}$	n	$i^n(2n+1)D_n^{-1}$
0	$+0\cdot06217 - 0\cdot07769\,i$	1	$+0\cdot21020 - 0\cdot21396\,i$
2	$+0\cdot41955 - 0\cdot28076\,i$	3	$+0\cdot68978 - 0\cdot19822\,i$
4	$+0\cdot93391 + 0\cdot13143\,i$	5	$+0\cdot92629 + 0\cdot74289\,i$
6	$+0\cdot33469 + 1\cdot42083\,i$	7	$-0\cdot96831 + 1\cdot48517\,i$
8	$-2\cdot13800 + 0\cdot00984\,i$	9	$-0\cdot84474 - 2\cdot36328\,i$
10	$+2\cdot38104 - 0\cdot89430\,i$	11	$+0\cdot30236 + 1\cdot75549\,i$
12	$-0\cdot91426 + 0\cdot04422\,i$	13	$-0\cdot00425 - 0\cdot41200\,i$
14	$+0\cdot17056 - 0\cdot00031\,i$	15	$+0\cdot00002 + 0\cdot06526\,i$
16	$-0\cdot02314 + 0\cdot00000\,i$	17	$-0\cdot00000 - 0\cdot00762\,i$
18	$+0\cdot00235$	19	$+0\cdot00068\,i$
20	$-0\cdot00018$	21	$-0\cdot00005\,i$
22	$+0\cdot00001$		

In the case of $\theta = 0$, or $\mu = +1$, the P's are all equal to $+1$, and we have nothing more to do than to add together all the terms in the above table. When $\theta = 180°$, or $\mu = -1$, the even P's assume (as before) the value $+1$, but now the odd P's have a reversed sign and are equal to -1. If we add together separately the even and odd terms, and so obtain the two partial sums Σ_1 and Σ_2, then $\Sigma_1 + \Sigma_2$ will be the value of Σ for $\theta = 0$, and $\Sigma_1 - \Sigma_2$ will be the value of Σ for $\theta = 180$. And this simplification applies not merely to the special values 0 and 180, but to all intermediate pairs of angles. If $\Sigma_1 + \Sigma_2$ corresponds to θ, $\Sigma_1 - \Sigma_2$ will correspond to $180 - \theta$.

For 0 and 180 we find

$$\Sigma_1 = + 1\cdot22870 + \cdot35326\,i, \quad \Sigma_2 = + 0\cdot31135 + \cdot85436\,i;$$

whence for $\theta = 0$

$$\Sigma = 2\,(F + iG) = + 1\cdot54005 + 1\cdot20762\,i,$$

and for $\theta = 180$

$$\Sigma = 2\,(F + iG) = + 0\cdot91735 - 0\cdot50110\,i.$$

When $\theta = 90°$, the odd P's vanish, and the even ones have the values

$$P_0 = 1, \qquad P_2 = -\tfrac{1}{2}, \qquad P_4 = \frac{1.3}{2.4}, \qquad P_6 = -\frac{1.3.5}{2.4.6}, \quad \&\text{c.}$$

For other values of θ we require tables of $P_n(\theta)$ up to about $n = 20$. That given by Professor Perry* is limited to n less than 7, and the results are expressed only to 4 places of decimals. I have been fortunate enough to interest Professor A. Lodge in this subject, and the Appendix to this paper gives a table calculated by him containing the P's up to $n = 20$ inclusive, and for angles from 0° to 90° at intervals of 5°. As has already been suggested, the range from 0° to 90° practically covers that from 90° to 180°, inasmuch as

$$P_{2n}(90 + \theta) = P_{2n}(90 - \theta), \qquad P_{2n+1}(90 + \theta) = -P_{2n+1}(90 - \theta).$$

In the table of coefficients it will be observed that the highest entry occurs at $n = 10$, in accordance with an anticipation expressed in a former paper.

As will readily be understood, the multiplication by P_n and the summations involve a good deal of arithmetical labour. These operations, as well as most of the preliminary ones, have been carried out in duplicate with the assistance of Mr C. Boutflower, of Trinity College, Cambridge.

$$kc = 10.$$

θ	$2\,(F+iG)$	$4\,(F^2+G^2)$
0	$+1\cdot54005 + 1\cdot20762\,i$	3·8300
5	$+1\cdot58407 + 1\cdot14959\,i$	3·8309
10	$+1\cdot70186 + 0\cdot96603\,i$	3·8295
15	$+1\cdot84773 + 0\cdot63523\,i$	3·8176
30	$+1\cdot52622 - 1\cdot17708\,i$	3·7148
45	$-1\cdot13754 - 1\cdot48453\,i$	3·4978
60	$-0\cdot74695 + 1\cdot59745\,i$	3·1098
75	$+1\cdot45160 - 0\cdot62553\,i$	2·4984
90	$-1\cdot31954 - 0\cdot09924\,i$	1·75104
105	$+0\cdot94204 + 0\cdot41681\,i$	1·06117
120	$-0\cdot57769 - 0\cdot48417\,i$	0·56815
135	$+0\cdot29444 + 0\cdot43841\,i$	0·27890
150	$-0\cdot08146 - 0\cdot35600\,i$	0·13338
165	$-0\cdot12081 + 0\cdot28341\,i$	0·09492
170	$+0\cdot35454 + 0\cdot01457\,i$	0·12591
175	$+0\cdot76023 - 0\cdot34059\,i$	0·69395
180	$+0\cdot91735 - 0\cdot50110\,i$	1·09263

* *Phil. Mag.* Vol. xxxii. p. 516, 1891 ; see also Farr, Vol. xlix. p. 572, 1900.

The results are recorded in the annexed table and in curve B, fig. 1. The intention had been to limit the calculations to intervals of 15°, but the rapid increase in $(F^2 + G^2)$ between 165° and 180° seemed to call for the interpolation of two additional angles. This increase, corresponding to the bright point in Poisson's experiment of the shadow of a circular *disc*, is probably the most interesting feature of the results. A plot is given in fig. 1, showing the relation between the angle θ, measured from the pole, and the *intensity*, proportional to $F^2 + G^2$. It should, perhaps, be emphasised that the effect here dealt with is the intensity of the *pressure* variation, to which some percipients of sound, *e.g.*, sensitive flames, are obtuse. Thus at the antipole a sensitive flame close to the surface would not respond to a distant source, since there is at that place no periodic *motion*, as is evident from the symmetry.

I now proceed to consider the case where the source, as well as the place of observation, are situated upon the sphere; but as this is more difficult than the preceding, I shall not attempt so complete a treatment. It will be supposed still that $kc = 10$.

The analytical solution is expressed in (17), which we may compare with (15). Restricting ourselves for the present to the factors under the sign of summation, we see that the coefficient of P_n in (17) is

$$\frac{(2n+1)\,c^{n+1}\chi_n(c)}{c^{n+1}\{\chi_{n-1}(c)-(n+1)\chi_n(c)\}} = \frac{(2n+1)\,c^{n+1}\chi_n}{D_n},$$

while the corresponding coefficient in (15) is $(2n+1)\,i^n/D_n$.

If these coefficients be called C_n, C'_n respectively, we have

$$C_n = i^{-n}c^{n+1}\chi_n(c)\,.\,C'_n, \quad \dots\dots\dots\dots\dots\dots(20)$$

in which the complex factors $c^{n+1}\chi_n(c)$, C'_n, for $c = 10$, have already been tabulated. We find

n	$\dfrac{(2n+1)\,10^{n+1}\chi_n}{D_n}$	n	$\dfrac{(2n+1)\,10^{n+1}\chi_n}{D_n}$
0	$-0{\cdot}0099 + 0{\cdot}0990\,i$	1	$-0{\cdot}0306 + 0{\cdot}2999\,i$
2	$-0{\cdot}0542 + 0{\cdot}5097\,i$	3	$-0{\cdot}0835 + 0{\cdot}7358\,i$
4	$-0{\cdot}1233 + 0{\cdot}9883\,i$	5	$-0{\cdot}1827 + 1{\cdot}2817\,i$
6	$-0{\cdot}2813 + 1{\cdot}6390\,i$	7	$-0{\cdot}4666 + 2{\cdot}0956\,i$
8	$-0{\cdot}8667 + 2{\cdot}6889\,i$	9	$-1{\cdot}8111 + 3{\cdot}3152\,i$
10	$-3{\cdot}5284 + 3{\cdot}0805\,i$	11	$-4{\cdot}2766 + 1{\cdot}3796\,i$
12	$-3{\cdot}6673 + 0{\cdot}3351\,i$	13	$-3{\cdot}1110 + 0{\cdot}0629\,i$
14	$-2{\cdot}7920 + 0{\cdot}0100\,i$	15	$-2{\cdot}6051 + 0{\cdot}0014\,i$
16	$-2{\cdot}4844 + 0{\cdot}0001\,i$	17	$-2{\cdot}3998$
18	$-2{\cdot}3369$	19	$-2{\cdot}2881$
20	$-2{\cdot}2496$	21	$-2{\cdot}2183$
22	$-2{\cdot}1925$	23	$-2{\cdot}1711$
24	$-2{\cdot}1528$	25	$-2{\cdot}1374$
26	$-2{\cdot}1240$		

The product above tabulated shows marked signs of approaching the limit -2, as n increases; so that the series (17) is divergent when $P_n = 1$, i.e., when $\theta = 0$, as was of course to be expected. The interpretation may be followed further. By the definition of P_n, we have

$$\{1 - 2\alpha . \cos\theta + \alpha^2\}^{-\frac{1}{2}} = 1 + P_1 . \alpha + P_2 . \alpha^2 + \ldots + P_n . \alpha^n + \ldots ; \ldots (21)$$

so that, if we put $\alpha = 1$,

$$1 + P_1 + P_2 + P_3 + \ldots = \frac{1}{2\sin(\frac{1}{2}\theta)}. \qquad \ldots (22)$$

Thus, when θ is small, and the series tends to be divergent, we get from (17)

$$\psi = -\frac{\iint U\, dS . e^{ikat}}{2\pi . 2c \sin(\frac{1}{2}\theta)}; \qquad \ldots (23)$$

and this is the correct value, seeing that $2c \sin(\frac{1}{2}\theta)$ represents the distance between the source and the point of observation, and that on account of the sphere the value of ψ is twice as great in the neighbourhood of the source as it would be were the source situated in the open.

When $\theta = 180°$, i.e., at the point on the sphere immediately opposite to the source, the series converges, since P_n takes alternately the values $+1$ and -1. It will be convenient to re-tabulate continuously these values from $n = 18$ onwards without regard to sign and to exhibit the differences.

n	Function	First difference	Second difference	Third difference
18	2·3369	—	—	—
19	2·2881	−·0488	—	—
20	2·2496	−·0385	+·0103	—
21	2·2183	−·0313	+·0072	−·0031
22	2·1925	−·0258	+·0055	−·0017
23	2·1711	−·0214	+·0044	−·0011
24	2·1528	−·0183	+·0031	−·0013
25	2·1374	−·0154	+·0029	−·0002
26	2·1240	−·0134	+·0020	−·0009

In summing the infinite series, we have to add together the terms as they actually occur up to a certain point and then estimate the value of the remainder. The simple addition is carried as far as $n = 21$ inclusive, and the result is for the even values of n

$$-18·3939 + 9·3506\, i,$$

and for the odd values

$$-19·4734 + 9·1721\, i,$$

or, with signs reversed to correspond with $P_{2n+1}(180) = -1$,

$$+19·4734 - 9·1721\, i.$$

The complete sum up to $n = 21$ inclusive is thus

$$+ 1\cdot0795 + \cdot1785\, i. \quad\dotfill(24)$$

The remainder is to be found by the methods of Finite Differences. The formula applicable to series of this kind may be written

$$\phi(0) - \phi(1) + \phi(2) - \dots = \tfrac{1}{2}\phi(0) - \tfrac{1}{4}\Delta\phi(0) + \tfrac{1}{8}\Delta^2\phi(0) - \dots,$$

in which we may put

$$\phi(0) = 2\cdot1925, \quad \phi(1) = 2\cdot1711, \quad \&c.$$

Thus

$$\phi(0) - \phi(1) + \dots = + 1\cdot0962 + \cdot0054 + \cdot0004 = 1\cdot1020,$$

and for the actual remainder this is to be taken negatively. The sum of the infinite series for $\theta = 180°$ is accordingly

$$- \cdot0225 + \cdot1785\, i, \quad\dotfill(25)$$

from which the intensity, represented by $(\cdot0225)^2 + (\cdot1785)^2$, is proportional to $\cdot03237$. Referring to (17), we see that the amplitude of ψ is in this case

$$\frac{\iint U\, dS}{4\pi c} \times \surd(\cdot03237). \quad\dotfill(26)$$

We may compare this with the amplitude of the vibration which would occur at the same place if the sphere were removed. Here

$$\frac{\iint U\, dS}{4\pi r} = \frac{\iint U\, dS}{4\pi c} \times \surd(\cdot25), \quad\dotfill(27)$$

since $r = 2c$. The effect of the sphere is therefore to reduce the intensity in the ratio of $\cdot25$ to $\cdot03237$.

In like manner we may treat the case of $\theta = 90°$, $i.e.$, when the point of observation is on the equator. The odd P's now vanish and the even P's take signs alternately opposite. The following table gives the values required for the direct summation, $i.e.$, up to $n = 21$ inclusive:

n	$\dfrac{(2n+1)\,10^{n+1}\chi_n \cdot P_n(90)}{D_n}$	n	$\dfrac{(2n+1)\,10^{n+1}\chi_n \cdot P_n(90)}{D_n}$
0	$-\ \cdot0099 + \cdot0990\, i$	2	$+\ \cdot0271 - \cdot2548\, i$
4	$-\ \cdot0462 + \cdot3706\, i$	6	$+\ \cdot0879 - \cdot5122\, i$
8	$-\ \cdot2370 + \cdot7353\, i$	10	$+\ \cdot8683 - \cdot7581\, i$
12	$-\ \cdot8273 + \cdot0756\, i$	14	$+\ \cdot5848 - \cdot0021\, i$
16	$-\ \cdot4879 + \cdot0000\, i$	18	$+\ \cdot4334$
20	$-\ \cdot3964$		
	$-2\cdot0047 + 1\cdot2805\, i$		$+2\cdot0015 - 1\cdot5272\, i$

The next three terms, written without regard to sign, and their differences are as follows :

22	·3688	—	—
24	·3470	− ·0218	—
26	·3292	− ·0178	+ ·0040

The remainder is found, as before, to be

$$+ \tfrac{1}{2}(\cdot 3688) + \tfrac{1}{4}(\cdot 0218) + \tfrac{1}{8}(\cdot 0040) = + \cdot 1903.$$

The sum of the infinite series from the beginning is accordingly

$$+ \cdot 1871 - \cdot 2467\, i, \quad\dots\dots\dots\dots\dots\dots(28)$$

in which $(\cdot 1871)^2 + (\cdot 2467)^2 = \cdot 09588.$

The distance between the source and the point of observation is now $2c \sin 45° = c \sqrt{2}$.

The intensity in the actual case is thus ·09588 as compared with ·5 if the sphere were away.

For other angular positions than those already discussed, not only would the arithmetical work be heavier on account of the factors P_n, but the remainder would demand a more elaborated treatment.

APPENDIX. By Professor A. Lodge.

Table of Zonal Harmonics; *i.e.*, of the Coefficients of the Powers of x as far as P_{20} in the Expansion of $(1 - 2x \cos \theta + x^2)^{-\frac{1}{2}}$ in the form $1 + P_1 x + P_2 x^2 + \ldots + P_n x^n + \ldots$ for 5° Intervals in the Values of θ from 0° to 90°. The Table is calculated to 7 decimal places, and the last figure is approximate.

θ	$P_1 (= \cos \theta)$	P_2	P_3	P_4	P_5
0	1·0000000	1·0000000	1·0000000	1·0000000	1·0000000
5	·9961947	·9886059	·9772766	·9622718	·9436768
10	·9848078	·9547695	·9105688	·8532094	·7839902
15	·9659258	·8995191	·8041639	·6846954	·5471259
20	·9396926	·8245333	·6648847	·4749778	·2714918
25	·9063078	·7320907	·5016273	·2465322	+ ·0008795
30	·8660254	·6250000	·3247595	+ ·0234375	− ·2232722
35	·8191520	·5065151	+ ·1454201	− ·1714242	− ·3690967
40	·7660444	·3802362	− ·0252333	− ·3190044	− ·4196822
45	·7071068	·2500000	− ·1767767	− ·4062500	− ·3756505
50	·6427876	·1197638	− ·3002205	− ·4275344	− ·2544885
55	·5735764	− ·0065151	− ·3886125	− ·3851868	− ·0867913
60	·5000000	− ·1250000	− ·4375000	− ·2890625	+ ·0898437
65	·4226183	− ·2320907	− ·4452218	− ·1552100	·2381072
70	·3420201	− ·3245333	− ·4130083	− ·0038000	·3280672
75	·2588190	− ·3995191	− ·3448846	+ ·1434296	·3427278
80	·1736482	− ·4547695	− ·2473819	+ ·2659016	·2810175
85	·0871557	− ·4886059	− ·1290785	·3467670	·1576637
90	Nil	− ·5000000	Nil	·3750000	Nil

θ	P_6	P_7	P_8	P_9	P_{10}
0	1·0000000	1·0000000	1·0000000	1·0000000	1·0000000
5	·9215975	·8961595	·8675072	·8358030	·8012263
10	·7044712	·6164362	·5218462	+ ·4227908	+ ·3214371
15	·3983060	+ ·2455411	+ ·0961844	− ·0427679	− ·1650562
20	+ ·0719030	− ·1072262	− ·2518395	− ·3516966	− ·4012692
25	− ·2039822	− ·3440850	− ·4062285	− ·3895753	− ·3052371
30	− ·3740235	− ·4101780	− ·3387755	− ·1895752	− ·0070382
35	− ·4114480	− ·3095600	− ·1154393	+ ·0965467	+ ·2541595
40	− ·3235708	− ·1006016	+ ·1386270	·2900130	·2973452
45	− ·1484376	+ ·1270581	·2983398	·2855358	+ ·1151123
50	+ ·0563782	·2854345	·2946824	+ ·1040702	− ·1381136
55	·2297230	·3190966	+ ·1421667	− ·1296151	− ·2692039
60	·3232421	·2231445	− ·0736389	− ·2678985	− ·1882286
65	·3138270	+ ·0422192	− ·2411439	− ·2300283	+ ·0323225
70	·2088770	− ·1485259	− ·2780153	− ·0475854	+ ·2192910
75	+ ·0431002	− ·2730500	− ·1702200	+ ·1594939	·2316302
80	− ·1321214	− ·2834799	+ ·0233080	·2596272	+ ·0646821
85	− ·2637801	− ·1778359	·2017462	·1912893	− ·1498947
90	− ·3125000	Nil	·2734376	Nil	− ·2460938

Table of Zonal Harmonics; *i.e.*, of the Coefficients of the Powers of x as far as P_{20} in the Expansion of $(1 - 2x \cos \theta + x^2)^{-\frac{1}{2}}$ in the form $1 + P_1 x + P_2 x^2 + \ldots + P_n x^n + \ldots$ for 5° Intervals in the Values of θ from 0° to 90°. The Table is calculated to 7 decimal places, and the last figure is approximate—*continued*.

θ	P_{11}	P_{12}	P_{13}	P_{14}	P_{15}
0	1·0000000	1·0000000	1·0000000	1·0000000	1·0000000
5	·7639723	·7242508	·6822849	+ ·6383094 -	+ ·5925694
10	+ ·2199746	+ ·1205620	+ ·0252742	− ·0639478	− ·1453436
15	− ·2654901	− ·3402156	− ·3868998	− ·4048245	− ·3948856
20	− ·4001361	− ·3528461	− ·2682722	− ·1585374	− ·0376336
25	− ·1739692	− ·0223995	+ ·1215469	+ ·2332489	+ ·2952537
30	+ ·1607048	+ ·2732027	+ ·3066580	+ ·2584895	+ ·1465789
35	·3096940	+ ·2532528	+ ·1130760	− ·0565267	− ·1950586
40	+ ·1712040	− ·0211959	− ·1892595	− ·2599246	− ·2083112
45	− ·1041843	− ·2467193	− ·2393239	− ·0972709	+ ·0903925
50	− ·2640939	− ·1987621	− ·0019170	+ ·1821884	+ ·2281988
55	− ·1769491	+ ·0522404	+ ·2209602	+ ·1959135	+ ·0110216
60	+ ·0638713	·2337529	+ ·1658041	− ·0571737	− ·2100185
65	+ ·2351950	+ ·1608831	− ·0863490	− ·2197701	− ·0989734
70	+ ·1864450	− ·0787947	− ·2239288	− ·0745390	+ ·1597121
75	− ·0305439	− ·2274796	− ·0850288	+ ·1687887	+ ·1638193
80	− ·2145820	− ·1307104	+ ·1544264	+ ·1730902	− ·0860215
85	− ·1988401	+ ·1041876	+ ·2010073	− ·0629592	− ·1982155
90	Nil	+ ·2255858	Nil	− ·2094726	Nil

θ	P_{16}	P_{17}	P_{18}	P_{19}	P_{20}
0	1·0000000	1·0000000	1·0000000	1·0000000	1·0000000
5	+ ·5453192	+ ·4968206	+ ·4473403	+ ·3971492	+ ·3465207
10	− ·2173739	− ·2787566	− ·3284945	− ·3658960	− ·3905880
15	− ·3594981	− ·3024136	− ·2284640	− ·1432466	− ·0527721
20	+ ·0801110	+ ·1815511	+ ·2560661	+ ·2965867	+ ·3002029
25	·2997862	+ ·2495290	+ ·1566049	+ ·0399984	− ·0780855
30	+ ·0036143	− ·1318805	− ·2254922	− ·2553464	− ·2169986
35	− ·2565851	− ·2244163	− ·1151189	+ ·0289684	+ ·1556356
40	−. ·0654984	+ ·0986597	+ ·2088162	+ ·2180388	+ ·1273281
45	+ ·2150310	+ ·2100803	+ ·0857609	− ·0809310	− ·1930653
50	+ ·1133974	− ·0732822	− ·1986904	− ·1792842	− ·0359655
55	− ·1714205	− ·2012353	− ·0625381	+ ·1207910	+ ·1945128
60	− ·1498551	+ ·0522168	+ ·1922962	+ ·1377671	− ·0483584
65	+ ·1249926	+ ·1956924	+ ·0427633	− ·1501989	− ·1644051
70	+ ·1757158	− ·0336558	− ·1883363	− ·0935549	+ ·1165241
75	− ·0760903	− ·1924117	− ·0249700	+ ·1696995	+ ·1093683
80	− ·1912133	+ ·0165069	+ ·1861639	+ ·0473145	− ·1608344
85	+ ·0255526	+ ·1908789	+ ·0082151	− ·1794383	− ·0383005
90	+ ·1963808	Nil	− ·1854706	Nil	+ ·1761970

NOTE BY LORD RAYLEIGH.

Professor Lodge's comparison of P_{20} with Laplace's approximate value*
suggests the question whether it is possible to effect an improvement in the
approximate expression without entailing too great a complication. The
following, on the lines of the investigation in Todhunter's *Functions of
Laplace, &c.*†, § 89, shows, I think, that this can be done.

We have

$$P_n = \frac{4}{\pi k (2n+1)} \Big\{ \sin(n+1)\theta$$

$$+ \frac{1.(n+1)}{1.(2n+3)} \sin(n+3)\theta + \frac{1.3.(n+1)(n+2)}{1.2.(2n+3)(2n+5)} \sin(n+5)\theta + \ldots \Big\}, \ldots(\alpha)$$

with

$$\frac{1}{k} = \frac{2.4.6 \ldots 2n}{1.3.5 \ldots (2n-1)} = \sqrt{(\pi n)} . \Big\{ 1 + \frac{1}{8n} + \ldots \Big\}. \quad \ldots\ldots\ldots(\beta)$$

When n is great, approximate values may be used for the coefficients of
the sines in (α). To obtain Laplace's expression it suffices to take

$$1, \quad \frac{1}{2}, \quad \frac{1.3}{2.4}, \quad \frac{1.3.5}{2.4.6}, \quad \&c.;$$

but now we require a closer approximation. Thus

$$\frac{1.(n+1)}{1.(2n+3)} = \frac{1}{2}\Big(1 - \frac{1}{2n+2} \Big),$$

$$\frac{1.3.(n+1).(n+2)}{1.2.(2n+3).(2n+5)} = \frac{1.3}{2.4}\Big(1 - \frac{1}{2n+2} - \frac{1}{2n+4} \Big),$$

and so on. If we write

$$x = 1 - \frac{1}{2n}, \ldots\ldots\ldots\ldots\ldots\ldots\ldots\ldots\ldots\ldots\ldots(\gamma)$$

the coefficients are approximately

$$1, \quad \frac{1}{2}x, \quad \frac{1.3}{2.4}x^2, \quad \frac{1.3.5}{2.4.6}x^3, \quad \&c.,$$

and the series takes actually the form assumed by Todhunter for analytical
convenience. In his notation

$$C = t\cos\theta + \tfrac{1}{2}t^3 \cos 3\theta + \frac{1.3}{2.4} t^5 \cos 5\theta + \ldots,$$

$$S = t\sin\theta + \tfrac{1}{2}t^3 \sin 3\theta + \frac{1.3}{2.4} t^5 \sin 5\theta + \ldots,$$

and

$$P_n = \frac{4}{\pi k (2n+1)} \{ C\sin n\theta + S\cos n\theta \},$$

where ultimately t is to be made equal to unity.

---

* $[P_n(\theta) = \frac{\sqrt{2}}{\sqrt{(n\pi \sin\theta)}} \cos(n\theta + \tfrac{1}{2}\theta - \tfrac{1}{4}\pi).]$

† Macmillan and Co., London, 1875.

By summation of the series $(t < 1)$,

$$C = \frac{t}{\sqrt{\rho}} \cos (\theta + \tfrac{1}{2}\phi), \qquad S = \frac{t}{\sqrt{\rho}} \sin (\theta + \tfrac{1}{2}\phi),$$

where $\qquad \rho^2 = 1 - 2t^2 \cos 2\theta + t^4, \qquad \tan\phi = \dfrac{t^2 \sin 2\theta}{1 - t^2 \cos 2\theta}.$ ....(δ)

For our purpose it is only necessary to write C/t and S/t for C and S respectively, and to identify t^2 with x in (γ). Thus

$$P_n = \frac{4}{\pi k (2n+1) \sqrt{\rho}} \sin (n\theta + \theta + \tfrac{1}{2}\phi), \qquad \ldots\ldots\ldots\ldots(\epsilon)$$

ρ and ϕ being given by (δ). We find, with $t = 1 - 1/4n$,

$$\rho^2 = 4 \sin^2 \theta (1 - 1/2n),$$

so that $\qquad \sqrt{\rho} = \sqrt{(2 \sin \theta)} \cdot (1 - 1/8n);$ ....(ζ)

and $\qquad\qquad \tan\phi = \dfrac{\sin 2\theta}{2 \sin^2 \theta + 1/2n},$

whence $\qquad\qquad \phi = \dfrac{\pi}{2} - \theta - \dfrac{\cot \theta}{4n}.$ ....(η)

Using (ζ), (η), (β) in (ε) we get

$$P_n = \frac{\sqrt{2}}{\sqrt{(\pi n \sin \theta)}} \left\{1 - \frac{1}{4n}\right\} \cdot \cos \left\{n\theta + \frac{\theta}{2} - \frac{\pi}{4} - \frac{\cot\theta}{8n}\right\}, \quad\ldots\ldots\ldots(\theta)$$

which is the expression required.

By this extension, not only is a closer approximation obtained, but the logic of the process is improved.

A comparison of values according to (θ) with the true values may be given in the case of n equal to 20.

Values of P_{20}.

θ	True value	According to (θ)
15°	$-\cdot05277$	$-\cdot05320$
30	$-\cdot21700$	$-\cdot21712$
45	$-\cdot19307$	$-\cdot19306$
60	$-\cdot04836$	$-\cdot04834$
75	$+\cdot10937$	$+\cdot10937$
90	$+\cdot17620$	$+\cdot17618$

[1911. I find that (θ) had been given some years earlier and in a more general form by Hobson, *Phil. Trans.* A, Vol. 187, p. 490, 1896.]

293.

SHADOWS.

[*Royal Institution Proceedings*, Jan. 15, 1904.]

My subject is shadows, in the literal sense of the word—shadows thrown by light, and shadows thrown by sound. The ordinary shadow thrown by light is familiar to all. When a fairly large obstacle is placed between a small source of light and a white screen, a well-defined shadow of the obstacle is thrown on the screen. This is a simple consequence of the approximately rectilinear path of light. Optical shadows may be thrown over great distances, if the light is of sufficient intensity: in a lunar eclipse the shadow of the earth is thrown on the moon: in a solar eclipse the shadow of the moon is thrown on the earth. Acoustic shadows, or shadows thrown by sound, are not so familiar to most people; they are less perfect than optical shadows, although their imperfections are usually over-estimated in ordinary observations. The ear is able to adjust its sensitiveness over a wide range, so that, unless an acoustic shadow is very complete, it often escapes detection by the unaided ear, the sound being sufficiently well heard in all positions. In certain circumstances, however, acoustic shadows may be very pronounced, and capable of easy observation.

The difference between acoustic and optical shadows was considered of so much importance by Newton, that it prevented him from accepting the wave theory of light. How, he argued, can light and sound be essentially similar in their physical characteristics, when light casts definite shadows, while sound shadows are imperfect or non-existent? This difficulty disappears when due weight is given to the consideration that the lengths of light waves and sound waves are of different orders of magnitude. Visible light consists of waves of which the average length is about one forty-thousandth of an inch. Audible sound consists of waves ranging in length from about an inch to nearly forty feet: the wave length corresponding to the middle C of the musical scale is roughly equal to four feet. It is, therefore, no matter for wonder that the effects produced by sound waves and by light waves differ in important particulars.

Moreover, the wave-length is not the only magnitude on which the perfection of the shadow depends; the size of the obstacle, and the distance across which the shadow is thrown, must also be taken into consideration. The optical shadow of a small object, thrown across a considerable distance, partakes of the imperfections generally observed in connection with sound shadows.

It was calculated by the French mathematician, Poisson, that, according to the wave theory of light, there should be a bright spot in the middle of the shadow of a small circular disc—a result that was thought to disprove the wave theory by a *reductio ad absurdum*. Although unknown to Poisson, this very phenomenon had actually been observed some years earlier, and was easily verified when a suitably arranged experiment was made.

Under suitable conditions a bright spot can be observed at the centre of the shadow of a three-penny bit. The coin may be supported by three or four very fine wires, and its shadow thrown by sunlight admitted at

Fig. 1.

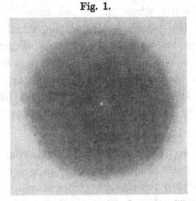

Reproduction of a Photograph of the Shadow of a Silver Penny Piece.

a pin-hole aperture placed in the shutter of a darkened room. The coin may be at a distance of about fifteen feet from the aperture, and the screen at about fifteen feet beyond the coin. To obtain a more convenient illumination, a larger aperture in the shutter may be filled by a short focus lens, which forms a diminutive image of the sun; this image serves as a point source of light. A smaller disc has some advantages. Fig. 1 is reproduced from a photograph of the shadow of a silver *penny* piece, struck at the time of the Coronation. The shadow, formed in the manner just described, was allowed to fall directly on a photographic plate; after development a negative was obtained, in which the dark parts of the shadow were represented by transparent gelatine, while the bright parts were represented by opaque deposits of silver. To obtain a correct representation, a contact print was formed from the negative in the usual way, upon a lantern plate; and from this fig. 1 has been reproduced.

It is at once evident that at the centre of the shadow, where one would expect the darkness to be most complete, there is a distinct bright spot. This result has always been considered a valuable confirmation of the wave theory of light.

I now propose to speak of acoustic shadows—shadows thrown by sound. The most suitable source of sound for the following experiments is the bird-call*, which emits a note of high pitch—so high, indeed, that it is inaudible to most elderly people. The sound emitted has two characteristics, valuable for our purpose—the wave-length is very short; and the sound is thrown forward, without too much tendency to spread, thus differing from sounds produced by most other means.

Since the sound emitted is nearly inaudible, some objective method of observing it is required. For this purpose we may utilize the discovery of Barrett and Tyndall, that a gas flame issuing under somewhat high pressure from a pin-hole burner *flares* when sound waves impinge on it, but recovers and burns steadily when the sound ceases. The sensitiveness of the flame depends on the pressure of the gas, which should be adjusted so that flaring just does not occur in the absence of sound. If the bird-call is directed towards the sensitive flame, the latter flares so long as the call is sounded and no obstacle intervenes. On interposing the hand about midway between the two, the flame recovers and burns steadily. Thus the sound emitted by the bird-call casts a shadow, and to this extent resembles light.

It will now be shown that the sensitive flame flares when it is placed at the centre of the acoustic shadow thrown from a circular disc, but recovers in any other position within the shadow; thus proving that there is sound at the centre of the shadow, although at a small distance from this point there is silence. The part of the flame which is sensitive to sound is that just above the pin-hole orifice, so that it is necessary to arrange the bird-call, the centre of the disc, and the pin-hole orifice in a straight line. For the disc, it is convenient to use a circular plate of glass about 18 inches in diameter with a piece of black paper pasted over its middle portion, a small hole being cut in the *paper* exactly at the centre of the disc. The glass disc is hung by two wires, and the positions of the bird-call and sensitive flame can be adjusted by sighting through the hole in the paper. If the disc is caused to oscillate in its own plane, the flame flares every time that the disc passes through its position of equilibrium, and recovers whenever the disc is not in that position. The analogy between this experiment, and that in which a bright spot is formed at the centre of the optical shadow of a small disc, is sufficiently obvious.

* See *Proc. Roy. Inst.* Jan. 17, 1902. [This Collection, Vol. v. p. 1.]

The approximate theory of the shadow of the circular disc is easily given, and it explains the leading features of the phenomenon. But, even in the simpler case of sound, an exact calculation which shall take full account of the conditions to be satisfied at the edge, has so far baffled the efforts of mathematicians. When the obstacle is a sphere, the problem is more tractable, and, in a recent memoir in the *Philosophical Transactions* a solution is given, embracing the cases where the circumference of the sphere is as great as two or even ten wave-lengths. When the sphere is small relatively to the wave-length, the calculation is easy, but the difficulty rapidly increases as the diameter rises. The diagram* gives the intensity in various positions on the surface of the sphere when plane waves of sound, *i.e.* waves proceeding from a distant source, impinge upon it. The intensity is a maximum at the point 0° nearest to the source, which may be called the pole. From the pole to the equator, distant 90° from it, the intensity falls off, and the fall continues as we enter the hinder hemisphere. But at an angular distance from the pole of about 135° in one case and 165° in the other, the intensity reaches a minimum and thence increases towards the antipole at 180°.

In private experiments the distribution of sound over the surface of the sphere may be explored with the aid of a small Helmholtz resonator and a flexible tube, and in this way evidence may be obtained of the rise of sound in the neighbourhood of the antipole. A more satisfactory demonstration is obtained by the method already employed in the case of the disc, the disc being replaced by a globe (about 12 inches in diameter), or by a croquet-ball of about 3½ inches diameter. In the former case the burner may be situated behind the sphere at such a distance as 5 inches. In the latter a distance of 1½ inches (from the surface) suffices. By a suitable adjustment of the flame, flaring ensues when everything is exactly in line, but the flame recovers when the ball is displaced slightly in a transverse direction. Since the wave-length of the sound is 3 cm. and the circumference of the croquet-ball is about 30 cm., this case corresponds to the curve B of our diagram.

In connection with the mathematical investigation which led to the results represented graphically in fig. 2*, there is a point of interest which I should like to mention. The investigation was carried out upon the supposition that the source of sound is at a considerable distance, so that the waves reaching the sphere are plane; and that the receiver, by which the sound is detected, is situated on the surface of the sphere. At any given position on the surface of the sphere, the receiver will indicate the reception of sound of a certain intensity, which may be read off from fig. 2*. Now the final results assume a form which shows that, if the positions of the

* [This Collection, Vol. v. p. 152.]

source and the receiver are interchanged, the latter will indicate the reception of sound of the same intensity as in the original arrangement. Thus each of the curves in fig. 2* represents the solution of two distinct problems: the intensity of the sound derived from a distant source and detected at any point on the surface of the sphere; and the intensity of the sound derived from a source on the surface of the sphere, and observed at a distant point. This result forms an interesting example of a principle of very wide application, which I have termed the *Principle of Reciprocity*. Some special cases were given many years ago by Helmholtz.

It is a matter of common observation that if one person can see another, either directly or by means of any number of reflections in mirrors, then the second person can equally well see the first. The same law applies to hearing, apparent exceptions being easily explained. For instance, such is the case of a lady sitting in a closed carriage, listening to a gentleman talking to her through the open window. If the street is noisy, the lady can hear what the gentleman says very much more distinctly than he can hear what she replies. This is due to the fact that the gentleman's ears are assailed by noises of the street from which the lady's ears are shielded by the walls of the carriage.

Another instance may be mentioned, which will appeal to electricians. In the arrangement known as Wheatstone's bridge, resistances are joined in the form of a lozenge, a galvanometer being connected between two opposite angles of the lozenge, while a battery is connected between the other two angles. When the resistances are suitably adjusted, no current flows through the galvanometer; but a slight want of adjustment produces a deflection of the galvanometer, thus indicating the passage of a small current. Now, if the positions of the battery and the galvanometer are interchanged, without alteration of resistance, the same current as before will flow through the galvanometer, and therefore the deflection will be the same as before. Thus with a given cell, galvanometer and set of resistances, the sensitiveness of the Wheatstone's bridge arrangement is the same whichever pair of opposite angles of the lozenge are joined by the galvanometer. If a source of alternating E.M.F. is used instead of the battery, and a telephone is substituted for the galvanometer, then the principle of reciprocity still applies, whether the resistances are inductive or non-inductive.

A simple illustration, of a mechanical nature, is now shown. Fig. 3 represents a straightened piece of watch-spring clamped at one end to a firm support. A weight can be hung at either of the points A or B of the spring, when it may be observed that the deflection at B due to the suspension of the weight at A, is exactly equal to the deflection at A due to the

* [This Collection, Vol. v. p. 152.]

suspension of the weight at B. This result is equally true wherever the points A and B may be situated; it applies not only to a loaded spring, which has been chosen as suitable for a simple lantern demonstration, but also to any sort of beam or girder.

It will have become clear, from what has been said, that waves encounter considerable difficulty in passing round the outside of a curved surface. I wish now to refer to a complementary phenomenon—the ease with which

Fig. 3.

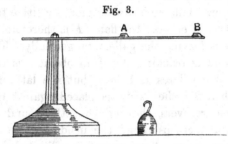

To Illustrate a Simple Mechanical Application of the Principle of Reciprocity.

waves travel round the *inside* of a curved surface. This is the case of the whispering gallery, of which there is a good example in St Paul's Cathedral. The late Sir George Airy considered that the effect could be explained as an instance of concentrated echo, the sound being concentrated by the curved walls, just as light may be brought to a focus by a concave mirror. From my own observations, made in St Paul's Cathedral, I think that Airy's

Fig. 4.

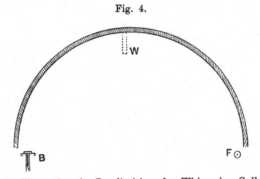

Model Illustrating the Peculiarities of a Whispering Gallery.

explanation is not the true one; for it is not necessary, in order to observe the effect, that the whisperer and the listener should occupy particular positions in the gallery. Any positions will do equally well. Again, whispering is heard more distinctly than ordinary conversation, especially if the whisperer's face is directed along the gallery towards the listener. It is known that a whisper has less tendency to spread than the full-spoken voice; thus a whisper, heard easily in front of the whisperer, is inaudible behind that person's head. These considerations led me to form a fairly satisfactory theory of the whispering gallery, nearly twenty-five years ago*. The phenomenon may be illustrated experimentally by the small scale arrangement represented diagrammatically in fig. 4. A strip of zinc, about 2 feet wide and 12 feet long, is bent into the form of a semicircle ; this forms the model of the whispering gallery. The bird-call B [$\lambda = 2$ cm.] is adjusted so that it throws the sound tangentially against the inner surface of the zinc: it thus takes the place of the whisperer. The sensitive flame F takes the place of the listener. A flame is always more sensitive to sound reaching it in one direction than in others ; the flame F is therefore adjusted so that it is sensitive to sounds leaving the gallery tangentially. The flaring of the flame shows that sound is reaching it: if an obstacle is interposed in the straight line FB the flame flares as before; but if a lath of wood W, which need not be more than 2 inches wide, is placed against the inner surface of the zinc, the flame recovers, showing that the sound has been inter-cepted. Thus the sound creeps round the inside surface of the zinc, and there is no disturbance except at points within a limited distance from that surface†.

* *Theory of Sound*, § 287.

† [1911. For a theoretical treatment of the question see *Phil. Mag.* Vol. xx. p. 1001, 1910.]

294.

SIR GEORGE GABRIEL STOKES, BART., 1819—1903.

[*Royal Society Year-Book*, 1904.]

IN common with so many distinguished men Sir George Stokes was the son of a clergyman. His father, Gabriel Stokes, who was Rector of Skreen, County Sligo, married Elizabeth Haughton, and by her had eight children of whom George was the youngest. The family can be traced back to Gabriel Stokes, born 1680, a well known engineer in Dublin and Deputy Surveyor General of Ireland, who wrote a treatise on Hydrostatics and designed the Pigeon House Wall in Dublin Harbour. This Gabriel Stokes married Elizabeth King in 1711 and among his descendants in collateral branches there are several mathematicians, a Regius Professor of Greek, two Regius Professors of Medicine, and a large sprinkling of scholars of Trinity College, Dublin. In more recent times Margaret Stokes, the Irish Antiquary, and the Celtic scholar, Whitley Stokes, children of the eminent physician, Dr William Stokes, have, among others, shed lustre on the name.

The home at Skreen was a very happy one. In the excellent sea air the children grew up with strong bodies and active minds. Of course great economy had to be practised to meet the educational needs of the family; but in the Arcadian simplicity of a place where chickens cost sixpence and eggs were five or six a penny, it was easy to feed them. They were all deeply attached to their mother, a beautiful and severe woman who made herself feared as well as loved.

Stokes was taught at home; he learnt reading and arithmetic from the parish Clerk, and Latin from his father who had been a scholar of Trinity College, Dublin. The former used to tell with great delight that Master George had made out for himself new ways of doing sums, better than the book. In 1832, at 13 years of age, he was sent to Dr Watts' school in Dublin; and in 1835 for two years to Bristol College, of which Dr Jerrard was Principal. There is a tradition that he did many of the propositions of Euclid, as problems, without looking at the book. He considered that he owed much to the teaching of Francis Newman, brother of the Cardinal, then mathematical master at Bristol College and a man of great charm of character as well as of unusual attainments.

On the first crossing to Bristol the ship nearly foundered; and his brother, who was escorting him, was much impressed by his coolness in face of danger. His habit, often remarked in after life, of answering with a plain "yes" or "no," when something more elaborate was expected, is supposed to date from this time, when his brothers chaffed him and warned him that if he gave "long Irish answers" he would be laughed at by his school-fellows.

It is surprising to learn that as a little boy he was passionate, and liable to violent, if transitory, fits of rage. So completely was this tendency overcome that in after life his temper was remarkably calm and even. He was fond of botany, and when about sixteen or seventeen, collected butterflies and caterpillars. It is narrated that one day while on a walk with a friend he failed to return the salutation of some ladies of his acquaintance, afterwards explaining his conduct by remarking that his hat was full of beetles!

In 1837, the year of Queen Victoria's accession, he commenced residence at Cambridge, where he was to find his home, almost without intermission, for sixty-six years. In those days sports were not the fashion for reading men, but he was a good walker, and astonished his contemporaries by the strength of his swimming. Even at a much later date he enjoyed encounters with wind and waves in his summer holidays on the north coast of Ireland. At Pembroke College his mathematical abilities soon attracted attention, and in 1841 he graduated as Senior Wrangler and first Smith's Prizeman. In the same year he was elected Fellow of his College.

After his degree, Stokes lost little time in applying his mathematical powers to original investigation. During the next three or four years there appeared papers dealing with hydrodynamics, wherein are contained many standard theorems. As an example of these novelties, the use of a stream-function in three dimensions may be cited. It had already been shown by Lagrange and Earnshaw that in the motion of an incompressible fluid in *two dimensions* the component velocities at any point may be expressed by means of a function known as the stream-function, from the property that it remains constant along any line of motion. It was further shown by Stokes that there is a similar function in three dimensions when the motion is symmetrical with respect to an axis. For many years the papers now under consideration were very little known abroad, and some of the results are still attributed by Continental writers to other authors.

A memoir of great importance on the "Friction of Fluids in Motion, etc.," followed a little later (1845). The most general motion of a medium in the neighbourhood of any point is analysed into three constituents—a motion of pure translation, one of pure rotation, and one of pure strain. These results are now very familiar; it may assist us to appreciate their novelty at the time, if we recall that when similar conclusions were put forward by Helmholtz twenty-three years later, their validity was disputed by so acute a critic as Bertrand. The splendid edifice, concerning the theory of inviscid fluids, which Helmholtz raised upon these foundations, is the admiration of all students of Hydrodynamics.

In applying the above purely kinematical analysis to viscous fluids, Stokes lays down the following principle:—"That the difference between the pressure on a plane passing through any point P of a fluid in motion and the pressure which would exist in all directions about P if the fluid in its neighbourhood were in a state of relative equilibrium depends only on the relative motion of the fluid immediately about P; and that the relative motion due to any motion of rotation may be eliminated without affecting the differences of the pressures above mentioned." This leads him to general dynamical equations, such as had already been obtained by Navier and Poisson, starting from more special hypotheses as to the constitution of matter.

Among the varied examples of the application of the general equations two may be noted. In one of these, relating to the motion of fluid between two coaxial revolving cylinders, an error of Newton's is corrected. In the other, the propagation of sound, as influenced by viscosity, is examined. It is shown that the action of viscosity (μ) is to make the intensity of the sound diminish as the time increases, and to render the velocity of propagation less than it would otherwise be. Both effects are greater for high than for low notes; but the former depends on the first power of μ, while the latter depends only on μ^2, and may usually be neglected.

In the same paragraph there occur two lines in which a question, which has recently been discussed on both sides, and treated as a novelty, is disposed of. The words are—"we may represent an arbitrary disturbance of the medium as the aggregate of series of plane waves propagated in all directions."

In the third section of the memoir under consideration, Stokes applies the same principles to find the equations for an elastic solid. In his view the two elastic constants are independent and not reducible to one, as in Poisson's theory of the constitution of matter. He refers to indiarubber as hopelessly violating Poisson's condition. Stokes' position, powerfully supported by Lord Kelvin, seems now to be generally accepted. Otherwise, many familiar materials must be excluded from the category of elastic solids.

In 1846 he communicated to the British Association a Report on Recent Researches in Hydrodynamics. This is a model of what such a survey should be, and the suggestions contained in it have inspired many subsequent investigations. He greatly admired the work of Green, and his comparison of opposite styles may often recur to the reader of mathematical lucubrations. Speaking of the Reflection and Refraction of Sound, he remarks that "this problem had been previously considered by Poisson in an elaborate memoir. Poisson treats the subject with extreme generality, and his analysis is consequently very complicated. Mr Green, on the contrary, restricts himself to the case of plane waves, a case evidently comprising nearly all the phenomena connected with this subject which are of interest in a physical point of view, and thus is enabled to obtain his results by a very simple analysis. Indeed Mr Green's memoirs are very remarkable, both for the elegance and rigour of the analysis, and for the ease with which he arrives at most important results. This arises in a great measure from his divesting the problems he considers of all unnecessary generality; where generality is really of importance he does not shrink from it. In the present instance there is one important respect in which Mr Green's investigation is more general than Poisson's, which is, that Mr Green has taken the case of any two fluids, whereas Poisson considered the case of two elastic fluids, in which equal condensations produce equal increments of pressure. It is curious, that Poisson, forgetting this restriction, applied his formulæ to the case of air and water. Of course his numerical result is quite erroneous. Mr Green easily arrives at the ordinary laws of reflection and refraction. He obtains also a very simple expression for the intensity of its reflected sound...." As regards Poisson's work in general there was no lack of appreciation. Indeed, both Green and Stokes may be regarded as followers of the French school of mathematicians.

The most cursory notice of Stokes' hydrodynamical researches cannot close without allusion to two important memoirs of somewhat later date. In 1847 he investigated anew the theory of oscillatory waves, as on the surface of the sea, pursuing the approximation so as to cover the case where the height is not very small in comparison with the wave-length. To the reprint in *Math. and Phys. Papers* are added valuable appendices pushing the approximation further by a new method, and showing that the slopes which meet at the crest of the highest possible wave (capable of propagation without change of type) enclose an angle of 120°.

The other is the great treatise on the Effect of Internal Friction of Fluids on the Motion of Pendulums. Here are given the solutions of difficult mathematical problems relating to the motion of fluid about vibrating solid masses of spherical or cylindrical form; also, as a limiting case, the motion of a viscous fluid in the neighbourhood of a uniformly advancing solid sphere, and a calculation of the *resistance* experienced by the latter. In the application of the results to actual pendulum observations, Stokes very naturally assumed that the viscosity of air was proportional to density. After Maxwell's great discovery that viscosity is independent of density within wide limits, the question assumed a different aspect; and in the reprint of the memoir Stokes explains how it happened that the comparison with theory was not more prejudiced by the use of an erroneous law.

In 1849 appeared another great memoir on the Dynamical Theory of Diffraction, in which the luminiferous æther is treated as an elastic solid so constituted as to behave as if it were nearly or quite incompressible. Many fundamental propositions respecting the vibration of an elastic solid medium are given here for the first time. For example, there is an investigation of the disturbance due to the operation at one point of the medium of a periodic force. The waves emitted are of course symmetrical with respect to the direction of the force as axis. At a distance, the displacement is transverse to the ray and in the plane which includes the axis, while along the axis itself there is no disturbance. Incidentally a general theorem is formulated connecting the disturbances due to initial displacements and velocities. "If any material system in which the forces acting depend only on the positions of the particles be slightly disturbed from a position of equilibrium, and then left to itself, the part of the subsequent motion which depends on the initial displacements may be obtained from the part which depends upon the initial velocities by replacing the arbitrary functions, or arbitrary constants, which express the initial velocities by those which express the corresponding initial displacements, and differentiating with respect to the time."

One of the principal objects of the memoir was to determine the law of vibration of the secondary waves into which in accordance with Huygens' principle a primary wave may be resolved, and thence by a comparison with phenomena observed with gratings to answer a question then much agitated but now (unless restated) almost destitute of meaning, viz., whether the vibrations of light are parallel or perpendicular to the plane of polarisation. As to the law of the secondary wave Stokes' conclusion is expressed in the following theorem: "Let $\xi = 0$, $\eta = 0$, $\zeta = f(bt - x)$ be the displacements corresponding to the incident light; let O_1 be any point in the plane P, dS an element of that plane adjacent to O_1; and consider the disturbance due to that portion only of the incident disturbance which passes continually across dS. Let O be any point in the medium situated at a distance from the point O_1 which is large in comparison with the length of a wave; let $OO_1 = r$, and let this line make angles θ with the direction of propagation of the incident light, or the axis of x, and ϕ with the direction of vibration, or the axis of z. Then the displacement at O will take place in a direction perpendicular to OO_1, and lying in the plane ZO_1O; and if ζ' be the displacement at O_1 reckoned positive in the direction nearest to that in which the incident vibrations are reckoned positive,

$$\zeta' = \frac{dS}{4\pi r}(1 + \cos\theta)\sin\phi \cdot f'(bt - r).$$

In particular, if

$$f(bt - x) = c \sin \frac{2\pi}{\lambda}(bt - x),$$

we shall have

$$\zeta' = \frac{cdS}{2\lambda r}(1 + \cos\theta)\sin\phi \cdot \cos\frac{2\pi}{\lambda}(bt - r)."$$

Stokes' own experiments on the polarisation of light diffracted by a grating led him to the conclusion that the vibrations of light are perpendicular to the plane of polarisation.

The law of the secondary wave here deduced is doubtless a possible one, but it seems questionable whether the problem is really so definite as Stokes regarded it. A merely mathematical resolution may be effected in an infinite number of ways; and if the problem is regarded as a physical one, it then becomes a question of the character of the obstruction offered by an actual screen.

As regards the application of the phenomena of diffraction to the question of the direction of vibration, Stokes' criterion finds a better subject in the case of diffraction by very small *particles* disturbing an otherwise uniform medium, as when a fine precipitate of sulphur falls from an aqueous solution.

The work already referred to, as well as his general reputation, naturally marked out Stokes for the Lucasian Professorship, which fell vacant at this time (1849). It is characterised throughout by accuracy of thought and lucidity of statement. Analytical results are fully interpreted, and are applied to questions of physical interest. Arithmetic is never shirked.

Among the papers which at this time flowed plentifully from his pen, one "On Attractions, and on Clairaut's Theorem" deserves special mention. In the writings of earlier authors the law of gravity at the various points of the earth's surface had been deduced from more or less doubtful hypotheses as the distribution of matter in the interior. It was reserved for Stokes to point out that, in virtue of a simple theorem relating to the potential, the law of gravity follows immediately from the form of the surface, assumed to be one of equilibrium, and that no conclusion can be drawn concerning the internal distribution of attracting matter.

From an early date he had interested himself in Optics, and especially in the Wave Theory. Although, not long before, Herschel had written ambiguously, and Brewster, the greatest living authority, was distinctly hostile, the magnificent achievements of Fresnel had converted the younger generation; and, in his own University, Airy had made important applications of the theory, *e.g.*, to the explanation of the rainbow, and to the diffraction of object-glasses. There is no sign of any reserve in the attitude of Stokes. He threw himself without misgiving into the discussion of outstanding difficulties, such as those connected with the aberration of light, and by further investigations succeeded in bringing new groups of phenomena within the scope of the theory.

An early example of the latter is the paper "On the Theory of certain Bands seen in the Spectrum." These bands, now known after the name of Talbot, are seen when a spectrum is viewed through an aperture half covered by a thin plate of mica or glass. In Talbot's view the bands are produced by the interference of the two beams which traverse the two halves of the aperture, darkness resulting whenever the relative retardation amounts to an odd number of half-wave lengths. This explanation cannot be accepted as it stands, being open to the same objection as Arago's theory of stellar scintillation. A body emitting homogeneous light would not become invisible on merely covering half the aperture of vision with a half-wave plate. That Talbot's view is insufficient is proved by the remarkable observation of Brewster—that the bands are seen only when the retarding plate is held towards the blue side of the spectrum. By Stokes' theory this polarity is fully explained, and the formation of the bands is shown to be connected with the limitation of the aperture, viz., to be akin to the phenomena of diffraction.

A little later we have an application of the general principle of reversion to explain the perfect blackness of the central spot in Newton's rings, which requires that when light passes from a second medium to a first the coefficient of reflection shall be numerically the same as when the propagation is in the opposite sense, but be affected with the reverse sign—the celebrated "loss of half an undulation." The result is obtained by expressing the conditions that the refracted and reflected rays, due to a given incident ray, shall on reversal reproduce that ray and no other.

It may be remarked that on any mechanical theory the reflection from an infinitely thin plate must tend to vanish, and therefore that a contrary conclusion can only mean that the theory has been applied incorrectly.

A not uncommon defect of the eye, known as astigmatism, was first noticed by Airy. It is due to the eye refracting the light with different power in different planes, so that the eye, regarded as an optical instrument, is not symmetrical about its axis. As a consequence, lines drawn upon a plane perpendicular to the line of vision are differently focussed according to their direction in that plane. It may happen, for example, that vertical lines are well seen under conditions where horizontal lines are wholly confused, and *vice versâ*. Airy had shown that the defect could be cured by cylindrical lenses, such as are now common; but no convenient method of testing had been proposed. For this purpose Stokes introduced a pair of plano-cylindrical lenses of equal cylindrical curvatures, one convex and the other concave, and so mounted as to admit of relative rotation. However the components may be situated, the combination is upon the whole neither convex nor concave. If the cylindrical axes are parallel, the one lens is entirely compensated by the other, but as the axes diverge the combination forms an astigmatic lens of gradually increasing power, reaching a maximum when the axes are perpendicular. With the aid of this instrument, an eye, already focussed as well as possible by means (if necessary) of a suitable spherical lens, convex or concave, may be corrected for any degree or direction of astigmatism; and from the positions of the axes of the cylindrical lenses may be calculated, by a simple rule, the curvatures of a single lens which will produce the same result. It is now known that there are comparatively few eyes whose vision may not be more or less improved by an astigmatic lens.

Passing over other investigations of considerable importance in themselves, especially that on the composition and resolution of streams of polarised light from different sources, we come to the great memoir on what is now called Fluorescence, the most far-reaching of Stokes' experimental discoveries. He "was led into the researches detailed in this paper by considering a very singular phenomenon which Sir J. Herschel had discovered in the case of a weak solution of sulphate of quinine and various other salts of the same alkaloid. This fluid appears colourless and transparent, like water, when viewed by transmitted light, but exhibits in certain aspects a peculiar blue colour. Sir J. Herschel found that when the fluid was illuminated by a beam of ordinary daylight, the blue light was produced only throughout a very thin stratum of fluid adjacent to the surface by which the light entered. It was unpolarised. It passed freely through many inches of the fluid. The incident beam after having passed through the stratum from which the blue light came, was not sensibly enfeebled or coloured, but yet it had lost the power of producing the usual blue colour when admitted into a solution of sulphate of quinine. A beam of light modified in this mysterious manner was called by Sir J. Herschel *epipolised.*

"Several years before, Sir D. Brewster had discovered in the case of an alcoholic solution of the green colouring matter of leaves a very remarkable phenomenon, which he has designated as *internal dispersion.* On admitting into this fluid a beam of sunlight condensed by a lens, he was surprised by finding the path of the rays within the fluid marked by a bright light of a blood-red colour, strangely contrasting with the beautiful green of the fluid itself when seen in moderate thickness. Sir David afterwards observed the same phenomenon in various vegetable solutions and essential oils, and in some solids. He conceived it to be due to coloured particles held in suspension. But there was one circumstance attending the phenomenon which seemed very difficult of explanation on such a supposition, namely, that the whole or a great part of the dispersed beam was unpolarised, whereas a beam reflected from suspended particles might be expected to be polarised by reflection. And such was, in fact, the case with those beams which were plainly due to nothing but particles held in suspension.   From the general identity of the circumstances attending the two phenomena, Sir D. Brewster was led to conclude that epipolic was merely a particular case of internal dispersion, peculiar only in this respect, that the rays capable of dispersion were dispersed with unusual rapidity.   But what rays they were which were capable of affecting a solution of sulphate of quinine, why the active rays were so quickly used up, while the dispersed rays which they produced passed freely through the fluid, why the transmitted light when subjected to prismatic analysis showed no deficiences in those regions to which, with respect to refrangibility, the dispersed rays chiefly belonged, were questions to which the answers appeared to be involved in as much mystery as ever."

Such a situation was well calculated to arouse the curiosity and enthusiasm of a young investigator.   A little consideration showed that it was hardly possible to explain the facts without admitting that in undergoing dispersion the light *changed its refrangibility*, but that if this rather startling supposition were allowed, there was no further difficulty; and experiment soon placed the fact of a change of refrangibility beyond doubt.   "A pure spectrum from sunlight having been formed in air in the usual manner, a glass vessel containing a weak solution of sulphate of quinine was placed in it.   The rays belonging to the greater part of the visible spectrum passed freely through the fluid, just as if it had been water, being merely reflected here and there from motes.   But from a point about halfway between the fixed lines G and H to far beyond the extreme violet, the incident rays gave rise to a light of a sky-blue colour, which emanated in all directions from the portion of the fluid which was under the influence of the incident rays.   The anterior surface of the blue space coincided, of course, with the inner surface of the vessel in which the fluid was contained.   The posterior surface marked the distance to which the incident rays were able to penetrate before they were absorbed.   This distance was at first considerable, greater than the diameter of the vessel, but it decreased with great rapidity as the refrangibility of the incident rays increased, so that from a little beyond the extreme violet to the end, the blue space was reduced to an excessively thin stratum adjacent to the surface by which the incident rays entered.   It appears, therefore, that this fluid, which is so transparent with respect to nearly the whole of the visible rays, is of an inky blackness with respect to the invisible rays, more refrangible than the extreme violet.   The fixed lines belonging to the violet and the invisible region beyond were beautifully represented by dark planes interrupting the blue space.   When the eye was properly placed these planes were, of course, projected into lines."

At a time when photography was of much less convenient application than at present—even wet collodion was then a novelty—the method of investigating the ultra-violet region of the spectrum by means of fluorescence was of great value. The obstacle presented by the imperfect transparency of *glass* soon made itself apparent, and this material was replaced by *quartz* in the lenses and prisms, and in the mirror of the heliostat. When the electric arc was substituted for sunlight a great extension of the spectrum in the direction of shorter waves became manifest.

Among the substances found "active" were the salts of uranium—an observation destined after nearly half a century to become in the hands of Becquerel the starting point of a most interesting scientific advance, of which we can hardly yet foresee the development.

In a great variety of cases the refrangibility of the dispersed light was found to be *less* than that of the incident. That light is always degraded by fluorescence is sometimes referred to as Stokes' law. Its universality has been called in question, and the doubt is perhaps still unresolved. The point is of considerable interest in connection with theories of radiation and the second law of Thermodynamics.

Associated with fluorescence there is frequently seen a "false dispersion," due to suspended particles, sometimes of extreme minuteness. When a horizontal beam of falsely dispersed light was viewed from above in a vertical direction, and analysed, it was found to consist chiefly of light polarised in the plane of reflection. On this fact Stokes founded an important argument as to the direction of vibration of polarised light. For "if the diameters of the (suspended) particles be small compared with the length of a wave of light, it seems plain that the vibrations in a reflected ray cannot be perpendicular to the vibrations in the incident ray." From this it follows that the direction of vibration must be perpendicular to the plane of polarisation, as Fresnel supposed, and the test seems to be simpler and more direct than the analogous test with light diffracted from a grating. It should not be overlooked that the argument involves the supposition that the effect of a particle is to *load* the æther.

It was about this time that Lord Kelvin learned from Stokes "Solar and Stellar Chemistry." "I used always to show [in lectures at Glasgow] a spirit lamp flame with salt on it, behind a slit prolonging the dark line *D* by bright continuation. I always gave your dynamical explanation, always asserted that certainly there was sodium vapour in the sun's atmosphere and in the atmospheres of stars which show presence of the *D*'s, and always pointed out that the way to find other substances besides sodium in the sun and stars was to compare bright lines produced by them in artificial flames with dark lines of the spectra of the lights of the distant bodies*."

---

* Letter to Stokes, published in Edinburgh address, 1871.

Stokes always deprecated the ascription to him of much credit in this matter; but what is certain is that had the scientific world been acquainted with the correspondence of 1854, it could not have greeted the early memoir of Kirchhoff (1859) as a new revelation. This correspondence will appear in Vol. IV of Stokes' collected papers, now being prepared under the editorship of Prof. Larmor. The following is from a letter of Kelvin, dated March 9, 1854: "It was Miller's experiment (which you told me about a long time ago) which first convinced me that there must be a physical connection between agency going on in and near the sun, and in the flame of a spirit lamp with salt on it. I never doubted, after I learned Miller's experiment, that there *must* be such a connection, nor can I conceive of any one knowing Miller's experiment and doubting....If it could only be made out that the bright line D never occurs without soda, I should consider it perfectly certain that there is soda or sodium in some state in or about the sun. If bright lines in any other flames can be traced, as perfectly as Miller did in his case, to agreement with dark lines in the solar spectrum, the connection would be equally certain, to my mind. I quite expect a qualitative analysis of the sun's atmosphere by experiments like Miller's on other flames."

By temperament, Stokes was over-cautious. "We must not go too fast," he wrote. He felt doubts whether the effects might not be due to some constituent of sodium, supposed to be broken up in the electric arc or flame, rather than to sodium itself. But his facts and theories, if insufficient to satisfy himself, were abundantly enough for Kelvin, and would doubtless have convinced others. If Stokes hung back, his correspondent was ready enough to push the application and to formulate the conclusions.

It is difficult to restrain a feeling of regret that these important advances were no further published than in Lord Kelvin's Glasgow lectures. Possibly want of time prevented Stokes from giving his attention to the question. Prof. Larmor significantly remarks that he became Secretary of the Royal Society in 1854. And the reader of the Collected Papers can hardly fail to notice a marked falling off in the speed of production after this time. The reflection suggests itself that scientific men should be kept to scientific work, and should not be tempted to assume heavy administrative duties, at any rate until such time as they have delivered their more important messages to the world.

But if there was less original work, science benefited by the assistance which, in his position as Secretary of the Royal Society, he was ever willing to give to his fellow workers. The pages of the *Proceedings* and *Transactions* abound with grateful recognitions of help thus rendered, and in many cases his suggestions or comments form not the least valuable part of memoirs which appear under the names of others. It is not in human nature for an author to be equally grateful when his mistakes are indicated, but from the point of view of the Society and of science in general, the service may be very great. It is known that in not a few cases the criticism of Stokes was instrumental in suppressing the publication of serious errors.

No one could be more free than he was from anything like an unworthy jealousy of his comrades. Perhaps he would have been the better for a little more wholesome desire for reputation. As happened in the case of Cavendish, too great an indifference in this respect, especially if combined with a morbid dread of mistakes, may easily lead to the withholding of valuable ideas and even to the suppression of elaborate experimental work, which it is often a labour to prepare for publication.

In 1857 he married Miss Robinson, daughter of Dr Romney Robinson, F.R.S., astronomer of Armagh. Their first residence was in rooms over a nursery gardener's in the Trumpington Road, where they received visits from Whewell and Sedgwick. Afterwards they took Lensfield Cottage, where they resided until her death in 1899. Though of an unusually quiet and silent disposition, he did not like being alone. He was often to be seen at parties and public functions, and, indeed, rarely declined invitations. In later life, after he had become President of the Royal Society, the hardihood and impunity with which he attended public dinners were matters of general admiration. The nonsense of fools, or rash statements by men of higher calibre, rarely provoked him to speech; but if directly appealed to, he would often explain his view at length with characteristic moderation and lucidity.

His experimental work was executed with the most modest appliances. Many of his discoveries were made in a narrow passage behind the pantry of his house, into the window of which he had a shutter fixed with a slit in it and a bracket on which to place crystals and prisms. It was much the same in lecture. For many years he gave an annual course on Physical Optics, which was pretty generally attended by candidates for mathematical honours. To some of these, at any rate, it was a delight to be taught by a master of his subject, who was able to introduce into his lectures matter fresh from the anvil. The present writer well remembers the experiments on the spectra of blood, communicated in the same year (1864) to the Royal Society. There was no elaborate apparatus of tanks and "spectroscopes." A test-tube contained the liquid and was held at arm's length behind a slit. The prism was a small one of 60°, and was held to the eye without the intervention of lenses. The blood in a fresh condition showed the characteristic double band in the green. On reduction by ferrous salt, the double band gave place to a single one, to re-assert itself after agitation with air. By such simple means was a fundamental reaction established. The impression left upon the hearer was that Stokes felt himself as much at home in chemical and botanical questions as in Mathematics and Physics.

At this time the scientific world expected from him a systematic treatise on Light, and indeed a book was actually advertised as in preparation. Pressure of work, and perhaps a growing habit of procrastination, interfered. Many years later (1884—1887) the Burnett Lectures were published. Simple and accurate, these lectures are a model of what such lectures should be, but they hardly take the place of the treatise hoped for in the sixties. There was, however, a valuable report on Double Refraction, communicated to the British Association in 1862, in which are correlated the work of Cauchy, MacCullagh and Green. To the theory of MacCullagh, Stokes, imbued with the ideas of the elastic solid theory, did less than justice. Following Green, he took too much for granted that the elasticity of æther must have its origin in *deformation*, and was led to pronounce the incompatibility of MacCullagh's theory with the laws of Mechanics. It has recently been shown at length by Prof. Larmor that MacCullagh's equations may be interpreted on the supposition that what is resisted is not deformation, but *rotation*. It is interesting to note that Stokes here expressed his belief that the true dynamical theory of double refraction was yet to be found.

In 1885 he communicated to the Society his observations upon one of the most curious phenomena in the whole range of Optics— a peculiar internal coloured reflection from certain crystals of chlorate of potash. The seat of the colour was found to be a narrow layer, perhaps one-thousandth of an inch in thickness, apparently constituting a twin stratum. Some of the leading features were described as follows:

(1) If one of the crystalline plates be turned round in its own plane, without alteration of the angle of incidence, the peculiar reflection vanishes twice in a revolution, viz., when the plane of incidence coincides with the plane of symmetry of the crystal.

(2) As the angle of incidence is increased, the reflected light becomes brighter, and rises in refrangibility.

(3) The colours are not due to absorption, the transmitted light being strictly complementary to the reflected.

(4) The coloured light is not polarised.

(5) The spectrum of the reflected light is frequently found to consist almost entirely of a comparatively narrow band. In many cases the reflection appears to be almost total.

Some of these peculiarities, such for example as the evanescence of the reflection at perpendicular incidence, could easily be connected with the properties of a twin plane, but the copiousness of the reflection at moderate angles, as well as the high degree of selection, were highly mysterious. There is reason to think that they depend upon a regular, or nearly regular, alternation of twinning many times repeated.

It is impossible here to give anything more than a rough sketch of Stokes' optical work, and many minor papers must be passed over without even mention. But there are two or three contributions to other subjects as to which a word must be said.

Dating as far back as 1857 there is a short but important discussion on the effect of wind upon the intensity of sound. That sound is usually ill heard up wind is a common observation, but the explanation is less simple than is often supposed. The velocity of moderate winds in comparison with that of sound is too small to be of direct importance. The effect is attributed by Stokes to the fact that winds usually increase overhead, so that the front of a wave proceeding up wind is more retarded above than below. The front is thus tilted; and since a wave is propagated normally to its front, sound proceeding up wind tends to rise, and so to pass over the heads of observers situated at the level of the source, who find themselves, in fact, in a sound shadow.

In a more elaborate memoir (1868) he discusses the important subject of the communication of vibration from a vibrating body to a surrounding gas. In most cases a solid body vibrates without much change of volume, so that the effect is represented by a distribution of sources over the surface, of which the components are as much negative as positive. The resultant is thus largely a question of *interference,* and it would vanish altogether were it not for the different situations and distances of the positive and negative elements. In any case it depends greatly upon the *wave-length* (in the gas) of the vibration in progress. Stokes calculates in detail the theory for vibrating spheres and cylinders, showing that when the wave-length is large relatively to the dimensions of the vibrating segments, the resultant effect is enormously diminished by interference. Thus the vibrations of a piano-string are communicated to the air scarcely at all directly, but only through the intervention of the sounding board*.

On the foundation of these principles he easily explains a curious observation by Leslie, which had much mystified earlier writers. When a bell is sounded in hydrogen, the intensity is greatly reduced. Not only so, but reduction accompanies the actual addition of hydrogen to rarified air. The fact is that the hydrogen increases the wave-length, and so renders more complete the interference between the sounds originating in the positively and negatively vibrating segments.

The determination of the laws of viscosity in gases was much advanced by him. Largely through his assistance and advice, the first decisive determinations at ordinary temperatures and pressures were effected by

---

* It may be worth notice that similar conclusions are more simply reached by considering the particular case of a *plane* vibrating surface.

Tomlinson. At a later period he brilliantly took advantage of Crookes' observations on the decrement of oscillation of a vibrator in a partially exhausted space to prove that Maxwell's law holds up to very high exhaustion and to trace the mode of subsequent departure from it. Throughout the course of Crookes' investigations on the electric discharge in vacuum tubes, in which he was keenly interested and closely concerned, he upheld the British view that the cathode stream consists of projected particles which excite phosphorescence in obstacles by impact: and accordingly, after the discovery of the Röntgen rays, he came forward with the view that they consisted of very concentrated spherical pulses travelling through the æther, but distributed quite fortuitously because excited by the random collisions of the cathode particles.

A complete estimate of Stokes' position in scientific history would need a consideration of his more purely mathematical writings, especially of those on Fourier series and the discontinuity of arbitrary constants in semi-convergent expansions over a plane, but this would demand much space and another pen. The present inadequate survey may close with an allusion to another of those "notes," suggested by the work of others, where Stokes in a few pages illuminated a subject hitherto obscure. By an adaptation of Maxwell's colour diagram he showed (1891) how to represent the results of experiments upon ternary mixtures, with reference to the work of Alder Wright. If three points in the plane represent the pure substances, all associations of them are quantitatively represented by points lying within the triangle so defined. For example, if two points represent water and ether, all points on the intermediate line represent associations of these substances, but only small parts of the line near the two ends correspond to *mixture*. If the proportions be more nearly equal, the association separates into two parts. If a third point (off the line) represents alcohol, which is a solvent for both, the triangle may be divided into two regions, one of which corresponds to single mixtures of the three components, and the other to proportions for which a single mixture is not possible.

A consideration of Stokes' work, even though limited to what has here been touched upon, can lead to no other conclusion than that in many subjects, and especially in Hydrodynamics and Optics, the advances which we owe to him are fundamental. Instinct, amounting to genius, and accuracy of workmanship are everywhere apparent; and in scarcely a single instance can it be said that he has failed to lead in the right direction. But, much as he did, one can hardly repress a feeling that he might have done still more. If the activity in original research of the first fifteen years had been maintained for twenty years longer, much additional harvest might have been gathered in. No doubt distractions of all kinds multiplied, and he was very punctilious in the performance of duties more or less formal. During the sitting of the last Cambridge Commission he interrupted his holiday in Ireland to attend a single meeting, at which however, as was remarked, he scarcely opened his mouth. His many friends and admirers usually took a different view from his of the relative urgency of competing claims. Anything for which a date was not fixed by the nature of the case, stood a poor chance. For example, owing to projected improvements and additions, the third volume of his Collected Works was delayed until eighteen years after the second, and fifty years after the first appearance of any paper it included. Even this measure of promptitude was only achieved under much pressure, private and official.

But his interest in matters scientific never failed. The intelligence of new advances made by others gave him the greatest joy. Notably was this the case in late years with regard to the Röntgen rays. He was delighted at seeing a picture of the arm which he had broken sixty years before, and finding that it showed clearly the united fracture.

Although this is not the place to dilate upon it, no sketch of Stokes can omit to allude to the earnestness of his religious life. In early years he seems to have been oppressed by certain theological difficulties, and was not exactly what was then considered orthodox. Afterwards he saw his way more clearly. In later life he took part in the work of the Victoria Institute: the spirit which actuated him may be judged from the concluding words of an Address on Science and Revelation. "But whether we agree or cannot agree with the conclusions at which a scientific investigator may have arrived, let us, above all things, beware of imputing evil motives to him, of charging him with adopting his conclusions for the purpose of opposing what is revealed. Scientific investigation is eminently truthful. The investigator may be wrong, but it does not follow he is other than truth-loving. If on some subjects which we deem of the highest importance he does not agree with us—and yet he may agree with us more nearly than we suppose—let us, remembering our own imperfections, both of understanding and of practice, bear in mind that caution of the Apostle: 'Who art thou that judgest another man's servant? To his own master he standeth or falleth.'"

Scientific honours were showered upon him. He was Foreign Associate of the French Institute, and Knight of the Prussian Order *Pour le Mérite*. He was awarded the Gauss Medal in 1877; the Arago on the occasion of the Jubilee Celebration in 1899, and the Helmholtz in 1901. In 1889 he was made a Baronet on the recommendation of Lord Salisbury. From 1887 to 1891 he represented the University of Cambridge in Parliament, in this, as in the Presidency of the Society, following the example of his illustrious predecessor in the Lucasian Chair. He was Secretary of the Society from 1854 to 1885, President from 1885 to 1890, received the Rumford medal in 1852, and the Copley in 1893.

But the most remarkable testimony by far to the estimation in which he was held by his scientific contemporaries was the gathering at Cambridge in 1899, in celebration of the Jubilee of his Professorship. Men of renown flocked from all parts of the world to do him homage, and were as much struck by the modesty and simplicity of his demeanour as they had previously been by the brilliancy of his scientific achievements. The beautiful lines by his colleague, Sir R. Jebb, cited below, were written upon this occasion.

There is little more to tell. In 1902 he was chosen Master of Pembroke. But he did not long survive. At the annual dinner of the Cambridge Philosophical Society, held in the College about a month before his death, he managed to attend though very ill, and made an admirable speech, recalling with charming simplicity and courtesy his lifelong intimate connection with the College, to the Mastership of which he had recently been called, and with the Society through which he had published much of his scientific work. Near the end, while conscious that he had not long to live, he retained his faculties unimpaired; only during the last few hours he wandered slightly, and imagined that he was addressing the undergraduates of his College, exhorting them to purity of life. He died on the first of February, 1903.

> Clear mind, strong heart, true servant of the light,
> True to that light within the soul, whose ray
> Pure and serene, hath brightened on thy way,
> Honour and praise now crown thee on the height
> Of tranquil years. Forgetfulness and night
> Shall spare thy fame, when in some larger day
> Of knowledge yet undream'd, Time makes a prey
> Of many a deed and name that once were bright.
>
> Thou, without haste or pause, from youth to age,
> Hast moved with sure steps to thy goal. And thine
> That sure renown which sage confirms to sage,
> Borne from afar. Yet wisdom shows a sign
> Greater, through all thy life, than glory's wage;
> Thy strength hath rested on the Love Divine.

295.

ON THE MEASUREMENT OF CERTAIN VERY SHORT INTERVALS OF TIME.

[*Nature*, Vol. LXIX, pp. 560, 561, 1904.]

ACCORDING to the discovery of Kerr, a layer of bisulphide of carbon, bounded by two parallel plates of metal and thus constituting the dielectric of a condenser or leyden, becomes doubly refracting when the leyden is charged. The plates, situated in vertical planes, may be of such dimensions as 18 cm. long, 3 cm. high, and the interval between them may be 0·3 cm., the line of vision being along the length and horizontal. If the polarising and analysing nicols be set to extinction, with their principal planes at 45° to the horizontal, there is revival of light when the leyden is charged. If the leyden remain charged for some time and be then suddenly discharged, and if the light under observation be sensibly instantaneous, it will be visible if the moment of its occurrence be previous to the discharge; if, however, this moment be subsequent to the discharge, the light will be invisible. The question now suggests itself, what will happen if the instantaneous light be that of the spark by which the leyden is discharged? It is evident that the conditions are of extraordinary delicacy, and involve the duration of the spark, however short this may be. The effect requires the simultaneity of light and double refraction, whereas here, until the double refraction begins to fail, there is no light to take advantage of.

The problem thus presented has been very skilfully treated by MM. Abraham and Lemoine (*Ann. de Chimie*, t. XX, p. 264, 1900). The sparks are those obtained by connecting the leyden with a deflagrator and with the terminals of a large Ruhmkorff coil fed with an alternating current. It is known that if the capacity be not too small, several charges and discharges occur during the course of one alternation in the primary, and that while the charges are gradual, the discharges are sudden in the highest degree. If, as in the present case, the capacity is small, it is necessary to submit the poles of the deflagrator to a blast of air, otherwise the leyden goes out of action and the discharge becomes continuous. Under the blast, the number of sparks may amount to several thousands per second of time. In this way the intensity of the light is much increased and the impression upon the eye becomes continuous, but in other respects the phenomenon is the same as if there were but one spark.

In order to obtain a measure of the double refraction, which is rapidly variable in time, somewhat special arrangements are necessary. At the receiving end the light, after emergence from the trough containing the bisulphide of carbon, falls first upon a double image prism, of somewhat feeble separating power, so held that one of the images is extinguished when the leyden is out of action. The other image would be of full brightness, but this, in its turn, is quenched by an analysing nicol. When there is double refraction to be observed, the nicol is slightly rotated until the two images are of equal brightness. This equality occurs in two positions, and the angle between them may be taken as a measure of the effect. A full discussion is given in the paper referred to.

The finiteness of the angle, which in my experiments amounted to 12°, is a proof that the light on arrival at the CS_2 still finds it in some degree doubly refracting. To obtain the greatest effect the leads from the leyden to the deflagrator should be as short as the case admits, and the course of the light from the sparks to the CS_2 should not be unnecessarily prolonged. The measure of the double refraction, and in an even greater degree the brightness of the light as received, are favoured by connecting a very small leyden directly with the spark terminals, but the advantage is hardly sufficient to justify the complication.

The observations of Abraham and Lemoine bring out the striking fact that if the course of the light be prolonged with the aid of reflectors so as to delay by an infinitesimal time the arrival at the CS_2, the opportunity to pass afforded by the double refraction is in great degree lost, and the angular measure of the effect is largely reduced. There is here no change in the electrical conditions under which the spark occurs, but merely a delay in the arrival of the light.

The optical arrangements which I found most convenient in repeating the above experiment differ somewhat from those of the original authors. The sparks are taken at a short distance from the polarising nicol and somewhat on one side, and in both cases they are focussed upon the analysing nicol. When the course is to be a minimum, the light is reflected obliquely by a narrow strip of mirror situated in the axial line, and focussed by a lens of short focus placed near the first nicol. This lens and mirror are so mounted on stands that they can be quickly withdrawn, and by means of suitable guidance and stops as quickly restored to their positions. In this case the distance travelled by the light from its origin to the middle of the length of CS_2 is about 30 cm.

The arrangements for a more prolonged course are similar, and they remain undisturbed during one set of comparisons. The mirror is larger, and reflects nearly perpendicularly; it is placed upon the axial line at a sufficient distance behind the sparks. The light is rendered nearly parallel by a photographic portrait lens of about 18 cm. focus, the aperture of which suffices to fill up the field of view unless the distance is very long. In all cases the eye of the observer is focussed upon the double image of the interval between the plates of the CS_2 leyden.

The earlier experiments were made at home somewhat under difficulties. For the blast nothing better was available than a glass-blowing foot bellows; but nevertheless the results were fairly satisfactory. Afterwards at the Royal Institution the use of a larger coil in connection with the public supply of electricity, and of an automatic blowing machine, gave steadier sparks and facilitated the readings. An increase of about one metre in the total distance travelled by the light reduced the measured angle from 12° to 6°, so that the time occupied by light in traversing one metre was very conspicuous.

It is principally with the view of directing attention to the remarkable results of Abraham and Lemoine that I describe the above repetition of their experiment, but I have made one variation upon it which is not without interest. In this case the spark is placed directly in the axial line and at some distance behind, which involves the use of longer leads, and therefore probably of a lower degree of instantaneity. The additional retardation is now obtained by the insertion of a 60 cm. long tube containing CS_2 between the sparks and the first nicol, and the comparison relates to the readings obtained with and without this column, all else remaining untouched. The difference is very distinct, and it represents the time taken in traversing the CS_2 over and above that taken in traversing the same length of air. It should be remarked that what we are here concerned with is not the wave-velocity in the CS_2, but the *group*-velocity, which differs from the former on account of the dispersion.

In the above experiments the leyden, where the Kerr effect is produced, is charged comparatively slowly and only suddenly discharged. For some purposes the scope of the method would be extended if the whole duration of the double refraction were made comparable with the above time of discharge. This could be effected somewhat as in Lodge's experiments, where a spark, called the B-spark, occurs between the outer coatings of two jars at the same moment as the A-spark between their inner coatings. The outer coatings remain all the while connected by a feeble conductor, which does not prevent the formation of the B-spark under the violent conditions which attend the passage of the A-spark. The plates of the Kerr leyden would be connected with the outer coatings of the jars, or themselves constitute the "outer" plates of two leydens replacing the jars.

296.

NOTE ON THE APPLICATION OF POISSON'S FORMULA TO DISCONTINUOUS DISTURBANCES.

[*Proceedings of the London Mathematical Society*, Ser. 2, Vol. II. pp. 266—269, 1904.]

In a recent paper* Prof. Love draws attention to "the discovery of an oversight in Stokes's justly famous memoir on the 'Dynamical Theory of Diffraction.'" The dilatation Δ satisfies the partial differential equation $d^2\Delta/dt^2 = a^2\nabla^2\Delta$, and is calculated from it by means of Poisson's integral formula. "According to this formula any function f which satisfies an equation of the form $d^2f/dt^2 = a^2\nabla^2 f$ can be expressed in terms of initial values by the equation

$$f = \frac{t}{4\pi} \iint \dot{f_0}(at)\, d\sigma + \frac{d}{dt}\left\{\frac{t}{4\pi}\iint f_0(at)\, d\sigma\right\}, \quad\ldots\ldots\ldots\ldots(A)$$

in which the integration refers to angular space about the point at which f is estimated, and $f_0(at)$ and $\dot{f_0}(at)$ denote the initial values of f and df/dt on a sphere of radius at with its centre at the point."

"...it will be seen that all Stokes's results depend upon the employment of Poisson's integral formula to express the dilatation and the components of the rotation. In a recent paper I have pointed out that this formula does not in general yield correct expressions for these quantities. In the same paper I identified the formula (A) with one which has been used by Poincaré and others, viz.,

$$f = \frac{1}{4\pi}\iint\left(t\dot{f_0} + f_0 + r\frac{df_0}{dr}\right)_{r=at} d\sigma, \quad\ldots\ldots\ldots\ldots\ldots(B)$$

where r denotes distance from the point at which f is estimated."

"The reason for the failure of such formulæ as (B) to represent the dilatation and the components of rotation is clear from an inspection of (B). When the point at which the disturbance is estimated is near the front of an advancing wave the sphere described about the point penetrates but a little way into the region within which the initial disturbance is confined, and the part of the sphere which is included in the integration is very small. Thus the formula cannot express any quantity which has a value different from zero at the front of an advancing wave. Now there is no kinematical

* *Proc. London Math. Soc.*, Ser. 2, Vol. I. p. 291, 1903.

or dynamical reason why the dilatation and rotation in an elastic solid should be supposed to vanish at the front of an advancing wave, and it appears therefore that Stokes's analysis is adequate to express the effects of particular types of initial disturbance, but not those of an arbitrary initial disturbance confined to a finite portion of the medium."

Having myself on a former occasion* applied Poisson's formula to the forbidden case of a uniform initial condensation limited to the slice bounded by two parallel planes without meeting any difficulty, I was naturally rather taken aback by the above criticism, although it is true that I then contemplated f as representing the *velocity-potential* rather than the dilatation. But the argument for the dilatation assumes much the same form, and it may be desirable to set it out in full.

Let us suppose then that a gaseous medium is initially undisturbed except between the parallel planes A, B, and that within AB there is initially no velocity, but only a uniform dilatation (or condensation). We know, of course, what will happen from the theory of plane waves. The initial state of things is equivalent to the superposition of two progressive waves between which the dilatation is shared. These advance in opposite directions, and in each the particle velocity is uniform and in the direction of propagation. We have now to inquire what account of the matter Poisson's formula will give.

In this f_0 represents the initial dilatation, confined to AB. The initial velocity of dilatation $\dot{f}_0$ is zero both within and without the slice, but this is not of itself sufficient to establish the evanescence of the first integral in (A) after the sphere has reached the slice. We have also to consider what may happen at the boundary planes A and B themselves. Taking the plane A, we see that immediately in front of it the dilatation rises suddenly from 0 to $\frac{1}{2}f_0$, but the effect of this is compensated in the integral by the drop from f_0 to $\frac{1}{2}f_0$ which occurs symmetrically behind. In like manner the boundary B can contribute nothing, and we may equate the first integral to zero.

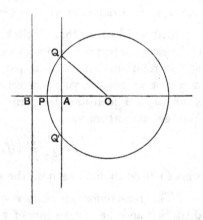

In the second integral, f_0 has a constant value over the portion QPQ' of the sphere and vanishes over the remainder. Also, if θ denote the angle AOQ,

$$\iint d\sigma = 2\pi (1 - \cos \theta) = 2\pi \frac{AP}{OP};$$

* *Theory of Sound*, § 274.

and $OP = at$. Thus

$$f = \tfrac{1}{2} f_0 \frac{d}{dt}\left(t \frac{AP}{at}\right) = \tfrac{1}{2} f_0,$$

since AP increases with velocity a; and accordingly the dilatation is correctly given by Poisson's formula.

When the wave has passed, the sphere cuts completely through the slice, and

$$\iint d\sigma = 2\pi \frac{AB}{OP} = 2\pi \frac{AB}{at};$$

so that $t \iint d\sigma = \text{constant}$,

and consequently $f = 0$. In all respects the passage of the wave is correctly represented.

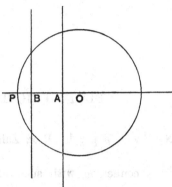

It is clear that the objection is really directed not against (A), but against (B), which, so far as I know, was not used at all by Poisson or by Stokes. And indeed this is recognized by Prof. Love himself in a later passage (p. 321), where he seems to shift the ground and advances against (A) an entirely distinct objection. With this I am not at present concerned, though I have my doubts whether it is more substantial than the first.

Even as regards (B), the charge of failure preferred against it seems to ignore the fact that the integrand becomes infinite in the case supposed of a discontinuous initial condition. Let us apply (B) to the circumstances already defined of a dilatation limited to the slice between two planes A and B. The first term in f_0 vanishes as before. The second term in f_0, also as before, assumes the value $\tfrac{1}{2} f_0 (1 - \cos\theta)$. The only difficulty is in the third term, where df_0/dr, vanishing within and without the slice, becomes infinite on the circle of transition whose diameter is QQ'. If x be a coordinate measured parallel to AB, f_0 is a function of x only, and

$$\frac{df_0}{dr} = \frac{df_0}{dx}\cos\theta.$$

Thus the third integral

$$= \tfrac{1}{2}\int r\frac{df_0}{dx}\cos\theta\sin\theta\,d\theta = \tfrac{1}{2}\int \frac{df_0}{dx}\cos\theta\,dx = \tfrac{1}{2}\cos\theta\,.f_0;$$

and altogether (B) gives $\tfrac{1}{2} f_0$, the correct result. According to Prof. Love's indictment the formula yields zero when the sphere only cuts a little into the slice.

297.

FLUID FRICTION ON EVEN SURFACES.

[Note to a paper by Prof. Zahm, *Phil. Mag.*, Vol. VIII. pp. 66, 67, 1904.]

IN connexion with such experiments as those of Froude and Zahm respectively on flat surfaces moving tangentially through water or air, it is of interest to inquire how much can be inferred from the principle of dynamical similarity. Dynamical similarity includes, of course, geometrical similarity, so that in any comparisons the surfaces must be similar, not only in respect of their boundaries, but also in respect of their roughnesses, at any rate until it is proved that the restriction may be dispensed with. Full geometrical similarity being presupposed, the tangential force per unit area F may be regarded as a function of a the linear dimension of the solid, ρ the density of the fluid, v the velocity, and ν the kinematic coefficient of viscosity. It is assumed that the compressibility of the fluid does not come into account. As in a similar problem relating to the flow of fluid along pipes*, the method of dimensions shows that, if it be a function of the above quantities only,

$$F = \rho v^2 \cdot f(av/\nu), \quad \dots\dots\dots\dots\dots\dots\dots\dots\dots(1)$$

where f denotes an arbitrary function, whose form must be obtained, if at all, from other considerations. If F be independent of ν, f is constant, and F is proportional to the density and the simple square of the velocity. Conversely, if F be not proportional to v^2, f is not constant, and the viscosity and the linear dimension enter.

If the general formula be admitted, several conclusions of importance can be drawn, even though the form of f be entirely unknown. For example, in the case of a given fluid (ρ and ν constant), F is strictly proportional to v^2, provided a be taken inversely as v. Again, if the fluid be varied, we may make comparisons relating to the same surface (a constant). For if v be taken proportional to ν, f remains unaffected; so that F is proportional to ρv^2

---

* *Phil. Mag.* XXXIV. p. 59 (1892); *Scientific Papers*, III. p. 575.

simply. For air the kinematic viscosity is about 10 times greater than for water, so that to obtain comparable cases the velocity through air would need to be 10 times greater than through water, a remaining unchanged. This condition being satisfied, the frictions per unit area would be as ρv^2, or since water is about 1000 times as dense as air, the actual friction would be about 10 times greater for the water than for the air.

According to Dr Zahm's experiments, where the velocity alone is varied, the form of f is within certain limits determined to be

$$f(x) = A x^{-\cdot 15}.$$

Within the same limits (1) gives as the complete expression for F,

$$F = A \rho v^{1\cdot85} (\nu/a)^{\cdot15}. \quad\quad\quad\dots\dots\dots\dots\dots\dots\dots(2)$$

[1911. In a series of experiments designed to determine the form of F, there would be advantage in keeping v (and a) constant, while ν is varied.]

298.

ON THE ELECTRICAL VIBRATIONS ASSOCIATED WITH THIN TERMINATED CONDUCTING RODS.

[*Philosophical Magazine*, Vol. VIII. pp. 105—107, 1904.]

In his discussion of this subject Prof. Pollock* rejects the simple theoretical result of Abraham and others, according to which the wave-length (λ) of the gravest vibration is equal to twice the length (l) of the rod, in favour of the calculation of Macdonald which makes $\lambda = 2\cdot53\ l$. On this I would make a few remarks, entirely from the theoretical point of view.

The investigation of Abraham† is a straightforward one; and though I do not profess to have followed it in detail, I see no reason for distrusting it. It relates to the vibration about a perfect conductor in the form of an elongated *ellipsoid* of revolution; and the above-mentioned conclusion follows when the minor axis ($2b$) of the ellipse vanishes in comparison with the major axis l. As a second approximation Abraham finds

$$\lambda = 2l\,(1 + 5\cdot6\ \epsilon^2), \dots\dots\dots\dots\dots\dots\dots\dots(1)$$

where

$$1/\epsilon = 4\log\,(l/b). \ \dots\dots\dots\dots\dots\dots\dots\dots(2)$$

But a question arises as to whether a result obtained for an infinitely thin ellipsoid can be applied to an infinitely thin rod of *uniform section*. So far as I see, it is not discussed by Abraham, though he refers to his conductor as rod-shaped. The character of the distinction may be illustrated by considering the somewhat analogous case of aerial vibrations within a cavity having the shape of the conductor. If the section be uniform, the wave-length of the longitudinal aerial vibration is exactly twice the length; but if an ellipsoidal cavity of the same length be substituted, then, however narrow it may be, the wave-length will be diminished in a finite ratio on account of the expansion of the section towards the central parts‡. This example may

---

* *Phil. Mag.* VII. p. 635 (1904).

† *Wied. Ann.* LXVI. p. 435 (1898).

‡ See *Theory of Sound*, § 265.

suffice to show that no general extension can be made from the ellipsoidal to the cylindrical shape, however attenuated the section may be.

When we ask whether the extension is justifiable in the present case, we shall find, I think, that the answer is in the affirmative so far as the first approximation is concerned, but in the negative for the second approximation.

Let us commence with the consideration of the known solution for an infinite conducting cylinder of radius r_1 enclosed in a coaxal sheath of radius r_2*. If in Maxwell's notation P, Q, R be the components of electromotive intensity; a, b, c those of magnetic induction; V the velocity of light; we have

$$P,\ Q,\ R = \cos pt \cdot \cos mz\left(\frac{x}{r^2},\ \frac{y}{r^2},\ 0\right),\quad a,\ b,\ c = \cos pt \cdot \cos mz\left(\frac{y}{r^2},\ -\frac{x}{r^2},\ 0\right),$$

in which z is measured along the axis and $m = p/V$. These expressions are independent of r_1 and r_2, and thus the nodal distance corresponding to a given frequency of vibration is the same whatever may be the diameters of the cylinders. But although the relation of m to p is unchanged, as r_1 is supposed to be reduced, the corresponding energies increase and that without limit. Apart from the first two factors, the value of (e.g.) the resultant electromotive intensity varies as $1/r$; so that the integrated square is proportional to $\log(r_2/r_1)$. We see that when r_1 is reduced without limit, the phenomenon is ultimately dominated by the infinite energies associated with the immediate neighbourhood of the attenuated conductor. Moreover, when r_1 is already infinitesimal, its further reduction in a finite ratio causes only a vanishing relative change in the value of the energy of vibration.

In the problem for which the solution is above analytically expressed the section of the rod is circular and uniform, but the considerations already advanced point to the conclusion that when the section is infinitesimal neither the circularity nor the uniformity is essential. So long, at any rate, as all diameters, whether of the same section or of different sections, are in finite ratios to one another, the relation of nodal interval to frequency remains undisturbed.

The same line of argument further indicates that the conclusion may be extended to a terminated rod of infinitesimal section. For the infinite energy associated with the neighbourhood of the conductor is unaffected except at points infinitely near the ends. It appears therefore that the wave-length of the electrical vibration associated with a straight terminated rod of infinitesimal section, is equal to twice the length of the rod, whether the shape

* See, for example, *Phil. Mag.* XLIV. p. 199 (1897); *Scientific Papers*, IV. p. 327.

be cylindrical so that the radius is constant, or ellipsoidal so that the radius varies in a finite ratio at different points of the length. And this conclusion still remains undisturbed, even though the shape be not one of revolution.

Whether the conditions of the limit can be sufficiently attained in experiment is a question upon which I am not prepared to express a decided opinion. From the logarithmic character of the infinity upon which the argument is founded, one would suppose that there might be practical difficulty in reducing the section sufficiently. Even if an adequate reduction were possible mechanically, the conductivity of actual materials might fail. We must remember that in the theory the conductivity is supposed to be *perfect*.

299.

ON THE DENSITY OF NITROUS OXIDE.

[*Proceedings of the Royal Society*, Vol. LXXIV. pp. 181—183, 1904.]

IN the *Proceedings*, Vol. LXXII. p. 204, 1897 *, I have given particulars of weighings of nitrous oxide purified by two distinct methods. In the first procedure, solution in water was employed as a means of separating less soluble impurities, and the result was 3·6356 grammes. In the second method a process of fractional distillation was employed. Gas drawn from the liquid so prepared gave 3·6362. These numbers may be taken to represent the corrected weight of the gas which fills the globe at 0° C. and at the pressure of the gauge (at 15°), and they correspond to 2·6276 for oxygen.

Inasmuch as nitrous oxide is heavier than the impurities likely to be contained in it, the second number was the more probable. But as I thought that the first method should also have given a good result, I contented myself with the mean of the two methods, viz. 3·6359, from which I calculated that referred to air (free from H_2O and CO_2) as unity, the density of nitrous oxide was 1·52951.

The corresponding density found by M. Leduc is 1·5301, appreciably higher than mine; and M. Leduc argues that the gas weighed by me must still have contained one or two thousandths of nitrogen †. According to him the weight of the gas contained in my globe should be 3·6374, or 1·5 milligrammes above the mean of the two methods.

Wishing, if possible, to resolve the question thus raised, I have lately resumed these researches, purifying the nitrous oxide with the aid of liquid air kindly placed at my disposal by Sir J. Dewar, but I have not succeeded in raising the weight of my gas by more than a fraction of the discrepancy (1·5 milligramme). I have experimented with gas carefully prepared in the laboratory from nitrate of ammonia, but as most of the work related to material specially supplied in an iron bottle I will limit myself to it.

---

* Or *Scientific Papers*, Vol. IV. p. 350.

† *Recherches sur les Gaz*, Paris, 1898.

There are two ways in which the gas may be drawn from the supply. When the valve is upwards, the supply comes from the vapourous portion within the bottle, but when the valve is downwards, from the liquid portion. The latter is the more free from relatively volatile impurities, and accordingly gives the higher weight, and the difference between the two affords an indication of the amount of impurity present. After treatment with caustic alkali and sulphuric acid, the gas is conducted through a tap, which is closed when it is desired to make a vacuum over the frozen mass, and thence over phosphoric anhydride to the globe. For the details of apparatus, &c., reference must be made to former papers. The condensing chamber, which can be immersed in liquid air, is in the form of a vertical tube, $2\frac{1}{2}$ cm. in diameter and 22 cm. long, closed below and above connected laterally to the main channel.

The first experiment on July 13 was upon gas from the top of the bottle as supplied, and without treatment by liquid air, with the view of finding out the worst. The weight was 3·6015, about 35 milligrammes too light. The stock of material was then purified, much as in 1896. For this purpose the bottle was cooled in ice and salt* and allowed during about one hour to blow off half its contents, being subjected to violent shaking at frequent intervals. Subsequently three weighings were carried out with gas drawn from the bottom, but without treatment by liquid air. The results stand :—

July 18	3·6368
July 20	3·6360
July 25	3·6362
Mean	3·63633

Next followed experiments in which gas, still drawn from the bottom of the bottle, was further purified by condensation with liquid air. The gas, arriving in a regular stream, was solidified in the condensing chamber. When it was judged that a sufficiency had been collected, the tap behind was turned and a *high* vacuum established over the solid mass with the aid of the Töpler pump. The pump would then be cut off and the gas allowed to evaporate and accumulate in the globe. A reapplication of liquid air caused the gas to desert the globe for the condensing chamber, until in a surprisingly short time a vacuum was re-established. Little or nothing could now be drawn off by the pump, and it was thought that a distinct difference could be perceived between the first and second operations, indicating that in the first condensation a little impurity remained gaseous. If desired, the condensation could be repeated a third time. On one occasion (August 7) the condensed gas was allowed to *liquefy*, for which purpose the

---

* The lower the temperature below the critical point, the more effective is this procedure likely to be.

pressure must rise to not far short of atmospheric, and to blow off part of its contents :—

August 1	....	3·6363
August 3	....	3·6367
August 7	....	3·6366
	Mean	3·63653

The treatment with liquid air raised the weight by only 0·2 milligramme, but the improvement is probably real. That the stock in the bottle still contained appreciable impurity is indicated by a weighing on August 13, in which without liquid air the gas was drawn from the *top* of the bottle. There appeared

August 13........................ 3·6354,

about 1 milligramme short of the proper weight.

It will be seen that the result without liquid air is almost identical with that found by the same method in 1896, and that the further purification by means of liquid air raises the weight only to 3·6365. I find it difficult to believe that so purified the gas still contains appreciable quantities of nitrogen.

The corresponding weight of air being 2·3772*, we find that, referred to air as unity, the density of nitrous oxide is

$$\frac{3·6365}{2·3772} = 1·5297.$$

Again, if oxygen be taken as 16, the density of nitrous oxide will be

$$\frac{3·6365 \times 16}{2·6276} = 22·143.$$

The excess above 22 is doubtless principally due to the departure of nitrous oxide from Boyle's law between atmospheric pressure and a condition of great rarefaction. I hope shortly to be in a position to apply the correction which will allow us to infer what is the ratio of molecular weights according to Avogadro's rule†.

* *Roy. Soc. Proc.* Vol. LIII. p. 134, 1893 ; *Scientific Papers*, Vol. IV. p. 47.

† [1911. See Art. 303 below.]

300.

NOTE TO A PAPER BY PROF. WOOD ON THE ACHROMATIZATION OF APPROXIMATELY MONOCHROMATIC INTERFERENCE FRINGES BY A HIGHLY DISPERSIVE MEDIUM.

[*Phil. Mag.*, Vol. VIII. pp. 330, 331, 1904.]

HAVING had an opportunity of seeing the above paper in proof, I append with Prof. Wood's permission a few remarks.

The remarkable shift of the bands of helium light when a layer of sodium vapour is interposed in the path of one of the interfering pencils, is of the same nature as the displacement of the white centre found by Airy and Stokes to follow the insertion of a thin plate of glass. If D denote the thickness of the plate and μ its refractive index, $(\mu-1)D$ is the retardation due to the insertion of the plate, and if R be the relative retardation due to other causes, the whole relative retardation is

$$R + (\mu - 1)D, \quad \dots\dots\dots\dots\dots\dots\dots(1)$$

in which R and D are supposed to be independent of the wave-length λ, while μ does depend upon it. The order of the band (n) is given by

$$n = \frac{R + (\mu - 1)D}{\lambda}. \quad \dots\dots\dots\dots\dots\dots(2)$$

For the achromatic band in the case of white light, or for the place of greatest distinctness when the bands are formed with light approximately homogeneous, n must be stationary as λ varies, *i.e.*

$$dn/d\lambda = 0. \quad \dots\dots\dots\dots\dots\dots\dots(3)$$

For a small range of wave-length we may write

$$\lambda = \lambda_0 + \delta\lambda,$$

so that

$$n = \frac{R + \left(\mu_0 + \frac{d\mu}{d\lambda_0}\delta\lambda - 1\right)D}{\lambda_0 + \delta\lambda} = \frac{R + (\mu_0 - 1)D}{\lambda_0} + \frac{\delta\lambda}{\lambda_0}\left(D\frac{d\mu}{d\lambda_0} - \frac{R + (\mu_0 - 1)D}{\lambda_0}\right)\dots.(4)$$

The achromatic band occurs, not when the whole relative retardation (1) vanishes, but when

$$R + (\mu_0 - 1) D = D\lambda_0 \frac{d\mu}{d\lambda_0}. \quad \dots\dots\dots\dots(5)$$

If D be great enough, there is no limit to the shift that may be caused by the introduction of the dispersive plate.

As Schuster has especially emphasised, the question here is really one of the *group-velocity*. Approximately homogeneous light consists of a train of waves in which the amplitude and wave-length slowly vary. A local peculiarity of amplitude or wave-length travels in a dispersive medium *with* the *group* and not with the *wave*-velocity; and the relative retardation with which we are concerned is the relative retardation of the groups. From this point of view it is obvious that, what is to be made to vanish is not (1) in which μ is the ratio of wave-velocities V_0/V, but that derived from it by replacing μ by U_0/U, or V_0/U, where U is the group-velocity in the dispersive medium. In vacuum the distinction between U_0 and V_0 disappears, but in the dispersive medium

$$U = d\,(\kappa V)/d\kappa, \quad \dots\dots\dots\dots\dots\dots(6)^*$$

κ being the reciprocal of the wave-length in the *medium*. If we denote as usual the wave-length *in vacuo* by λ,

$$\kappa = \frac{2\pi\mu}{\lambda} = \frac{2\pi V_0}{\lambda V}. \quad \dots\dots\dots\dots\dots(7)$$

Accordingly

$$\frac{V_0}{U} = \frac{V_0 d\kappa}{d\,(\kappa V)} = \frac{d\,(\mu/\lambda)}{d\,(1/\lambda)} = \mu - \lambda\frac{d\mu}{d\lambda}. \quad \dots\dots\dots(8)$$

Substituting this for μ in (1), we see that the position of the most distinct band is given by

$$R + \left(\mu - 1 - \lambda\frac{d\mu}{d\lambda}\right) D = 0, \quad \dots\dots\dots\dots(9)$$

in agreement with (5).

* *Theory of Sound*, § 191, 1877.

301.

ON THE OPEN ORGAN-PIPE PROBLEM IN TWO DIMENSIONS.

[*Philosophical Magazine*, Vol. VIII. pp. 481—487, 1904.]

IN the usual symmetrical organ-pipe of radius R, supposed to be provided at the mouth with an infinite flange, we know that the correction (α) that must be added to the length in order that the open end may be treated as a loop, lies between $\frac{1}{4}\pi R$ and $8R/3\pi$. The wave-length of vibration is here supposed to be very great, so that in the neighbourhood of the mouth the flow follows the electrical law. If we use this analogy and regard the walls of the pipe and the flange as non-conductors, the question is one of the *resistance* of the air-space, measured from a section well inside the pipe to an infinite distance beyond the mouth. And in spite of the extension to infinity this resistance is finite. For if r be the radius of a large sphere whose centre is at the mouth, the resistance between r and $r + dr$ is $dr/2\pi r^2$; and the part corresponding to the passage from a sufficiently great value of r outwards to infinity may be neglected.

A parallel treatment of the problem in *two dimensions*, where inside the mouth the boundary consists of two parallel planes, appears to fail. The resistance to infinity, involving now $\int^{\infty} r^{-1}dr$, instead of $\int^{\infty} r^{-2}dr$, has no finite limit; and we must conclude that when the wave-length (λ) is very great, the correction to the length becomes an infinite multiple of the width of the pipe*. But it remains an open question how the correction to the length compares with λ,—whether, for example, when λ is given it would vanish when the width of the pipe is indefinitely diminished.

The following consideration suggests an affirmative answer to the last question. If we start with a pipe of circular form and suppose the section,

---

* [1911. For a further discussion of the problem which arises when λ is treated as infinitely great see J. J. Thomson's *Recent Researches in Electricity*, § 241, 1893.]

while retaining its area, to become more and more elliptical, it would appear that the correction to the length must continually diminish. But the question has sufficient interest to justify a more detailed treatment.

In *Theory of Sound*, § 302, the problem is considered of the reaction of the air upon the vibratory motion of a circular plate forming part of an infinite and otherwise fixed plane. For our present purpose the circular plate is to be replaced by an infinite rectangular strip extending from $y = -\infty$ to $y = +\infty$, and in the other direction of width x. If $d\phi/dn$ be the given normal velocity of the element dS of the plane and $k = 2\pi/\lambda$,

$$\phi = -\frac{1}{2\pi} \iint \frac{d\phi}{dn} \frac{e^{-ikr}}{r} dS \dots\dots\dots(1)$$

gives the velocity-potential at any point P distant r from dS. In the present case $d\phi/dn$ is constant, where it differs from zero, and accordingly may be removed from under the integral sign. If a denote the velocity of sound, ϕ varies as e^{ikat}, and if σ be the density, we get for the whole variation of pressure acting upon the plate

$$\iint \delta p \, dS = -\sigma \iint \phi \, dS = -ika\sigma \iint \phi \, dS.$$

Thus by (1)

$$\iint \delta p \, dS = \frac{ika\sigma}{\pi} \frac{d\phi}{dn} \Sigma\Sigma \frac{e^{-ikr}}{r} dS dS'. \dots\dots\dots(2)$$

In the double sum

$$\Sigma\Sigma \frac{e^{-ikr}}{r} dS dS', \dots\dots\dots(3)$$

which we have now to evaluate, each pair of elements is to be taken *once* only, and the product is to be summed after multiplication by the factor $r^{-1}e^{-ikr}$, depending on their mutual distance. The best method is that suggested by Maxwell for the common potential. The quantity (3) is regarded as the work that would be consumed in the complete dissociation of the matter composing the plate, that is to say, in the removal of every element from the influence of every other, on the supposition that the potential of two elements is proportional to $r^{-1}e^{-ikr}$. The amount of work required, which depends only on the initial and final states, may be calculated by supposing the operation performed in any way that may be most convenient. For this purpose we suppose that the plate is divided into elementary strips, and that on one side external strips are removed in succession.

To carry out this method we require first an expression for the potential (V) at the edge of a strip. Here,

$$V = \int_0^x \int_{-\infty}^{+\infty} \frac{e^{-ikr}}{r} dx dy, \dots\dots\dots(4)$$

where $r = \sqrt{(x^2 + y^2)}$; and therein

$$\int_{-\infty}^{+\infty} \frac{e^{-ikr}}{r} dy = 2\int_x^\infty \frac{e^{-ikr}}{y} dr = 2\int_1^\infty \frac{e^{-ikxv}}{\sqrt{v^2-1}} dv, \dots\dots\dots(5)$$

representing the potential of a linear source at a point distant x from it. Convergent and semi-convergent series for (5), applicable respectively when x is small and when x is great, are well known.

We have

$$\int_1^\infty \frac{e^{-ikxv}\,dv}{\sqrt{(v^2-1)}} = \left(\frac{\pi}{2ikx}\right)^{\frac{1}{2}} e^{-ikx}\left\{1 - \frac{1^2}{1\,.\,8ikx} + \frac{1^2\,.\,3^2}{1\,.\,2\,.\,(8ikx)^2} - \dots\right\}$$

$$= -\left(\gamma + \log\frac{ikx}{2}\right)\left\{1 - \frac{k^2x^2}{2^2} + \frac{k^4x^4}{2^2\,.\,4^2} - \dots\right\}$$

$$-\frac{k^2x^2}{2^2}S_1 + \frac{k^4x^4}{2^2\,.\,4^2}S_2 - \frac{k^6x^6}{2^2\,.\,4^2\,.\,6^2}S_3 + \dots, \quad \dots\dots\dots(6)$$

where

$$S_m = 1 + \tfrac{1}{2} + \tfrac{1}{3} + \dots + 1/m, \quad \dots\dots\dots\dots\dots(7)$$

and γ is Euler's constant ($\cdot5772\dots$). A simple method of derivation, adequate so far as the leading terms are concerned, will be found in *Phil. Mag.* XLIII. p. 259, 1897 (*Scientific Papers*, IV. p. 290).

Confining ourselves for the present to the case where the total width of the strip is small compared with the wave-length, we have to integrate the second series in (6) with respect to x, for which purpose we have the formula

$$\int x^n \log x\,dx = \frac{x^{n+1}}{n+1}\left\{\log x - \frac{1}{n+1}\right\}. \quad \dots\dots\dots\dots(8)$$

In this way we obtain V, the potential at the edge of the strip of width x. Afterwards we have to integrate again with respect to y and x. The integration with regard to y introduces simply the factor y, representing the (infinite) length of the strip. The integration with respect to x is again to be taken between the limits 0 and x. Thus

$$\iint V\,dx\,dy = -yx^2\{\gamma - \tfrac{3}{2} + \log(\tfrac{1}{2}ikx)\}, \quad \dots\dots\dots\dots(9)$$

terms in x^4 being omitted. This is the equivalent of (3) in the case of an infinite strip of length y and width x. Accordingly by (2)

$$\iint \delta p\,dS = -\frac{ika\sigma}{\pi}\frac{d\phi}{dn}\,.\,yx^2\{\gamma - \tfrac{3}{2} + \log(\tfrac{1}{2}kx) + \tfrac{1}{2}i\pi\}. \quad \dots\dots(10)$$

The reaction of the air upon the plate may be divided into two parts, of which one is proportional to the velocity of the plate and the other to the acceleration. If ξ denote the displacement of the plate at time t, so that $d\xi/dt = d\phi/dn$, we have

$$\frac{d^2\xi}{dt^2} = ika\frac{d\xi}{dt} = ika\frac{d\phi}{dn};$$

and therefore in the equation of motion of the plate, the reaction of the air is represented by a dissipative force

$$xy\,.\,\frac{d\xi}{dt}\,.\,\frac{a\sigma}{2}\,.\,kx \quad \dots\dots\dots\dots\dots\dots\dots(11)$$

retarding the motion, and by an accession to the inertia equal to

$$xy \cdot \frac{\sigma}{\pi} \cdot x\left(\tfrac{3}{2} - \gamma - \log \frac{kx}{2}\right), \quad \dots\dots\dots\dots\dots(12)$$

the first factor in each case denoting the area of the plate. The mass represented by (12) is that of a volume of air having a base equal to the area of the plate and a *thickness*

$$\frac{x}{\pi}\left(\tfrac{3}{2} - \gamma - \log \frac{kx}{2}\right). \quad \dots\dots\dots\dots\dots(13)$$

When x is given (13) increases without limit when $\lambda \ (= 2\pi/k)$ is made infinite, as we found before. But if we regard λ as given, and suppose x (the width of the plate) to diminish without limit, we see that (13) also diminishes without limit.

The application of these results to the problem of the open pipe in two dimensions depends upon the imaginary introduction of a movable piston, itself without mass, at the mouth of the pipe—a variation without influence upon the behaviour of the air at a great distance outside the mouth, with which we are mainly concerned. The conclusion is that if the wave-length or pipe-length be given, the mouth or open end may be treated more and more accurately as a loop as the width is diminished without limit. Both parts of the pressure-variation, corresponding the one to inertia and the other to dissipative escape of energy, ultimately vanish.

So far we have considered the case where in (3), (4) the width of the strip is *small* in comparison with the wave-length. It remains to say something as to the other extreme case; and it may be well to introduce the discussion by a brief statement of the derivation of the semi-convergent series in (6) by the method of Lipschitz*.

Consider the integral $\int \dfrac{e^{-rw}dw}{\sqrt{(1 + w^2)}}$, where w is a complex variable of the form $u + iv$. If we represent, as usual, simultaneous pairs of values of u and v by the coordinates of a point, the integral will vanish when taken round any closed circuit not including the points $w = \pm i$. The circuit at present to be considered is that enclosed by the lines $u = 0$, $v = 1$, and a quadrant at infinity. It is easy to see that along this quadrant the integral ultimately vanishes, so that the result of the integration is the same whether we integrate from $w = i$ to $w = i\infty$, or from $w = i$ to $w = \infty + i$. Accordingly

$$\int_1^\infty \frac{e^{-irv}dv}{\sqrt{(v^2 - 1)}} = \int_0^\infty \frac{e^{-ir}e^{-ru}du}{\sqrt{(2iu + u^2)}} = \frac{e^{-ir}}{\sqrt{(2ir)}} \int_0^\infty \frac{e^{-\beta}\beta^{-\frac{1}{2}}d\beta}{\sqrt{\left(1 + \dfrac{\beta}{2ir}\right)}}$$

$$= \left(\frac{\pi}{2ir}\right)^{\frac{1}{2}} e^{-ir} \left\{ 1 - \frac{1^2}{1 \cdot 8ir} + \frac{1^2 \cdot 3^2}{1 \cdot 2 \cdot (8ir)^2} - \frac{1^2 \cdot 3^2 \cdot 5^2}{1 \cdot 2 \cdot 3 \cdot (8ir)^3} + \dots \right\}, \quad (14)$$

---

* *Crelle*, Bd. LVI. (1859). See also *Proc. Lond. Math. Soc.* XLIX. p. 504 (1888); *Scientific Papers*, III. p. 44.

on expansion and integration by a known formula. This agrees with (6), if kx be written for r.

In arriving at the value of $\iint V\,dx\,dy$, we have to integrate (5) twice with respect to x between the limits 0 and x. Taking first an integration with respect to x, we have

$$\int_0^x dx \int_1^\infty \frac{e^{-ikxv}dv}{\sqrt{(v^2-1)}} = \int_1^\infty \frac{(1-e^{-ikxv})\,dv}{ikv\sqrt{(v^2-1)}};$$

in which

$$\int_1^\infty \frac{dv}{v\sqrt{(v^2-1)}} = \int \frac{dv}{v^2\sqrt{(1-v^{-2})}} = \frac{\pi}{2}.$$

Again, for the second integration with respect to x,

$$\int_0^x dx \int_1^\infty \frac{e^{-ikxv}dv}{ikv\sqrt{(v^2-1)}} = \int_1^\infty \frac{(1-e^{-ikxv})\,dv}{(ikv)^2\sqrt{(v^2-1)}},$$

in which

$$\int_1^\infty \frac{dv}{v^2\sqrt{(v^2-1)}} = \int_0^1 \frac{dz}{\sqrt{(z^{-2}-1)}} = 1.$$

Thus

$$\iint V\,dx\,dy = y\left\{\frac{\pi x}{ik} + \frac{2}{k^2} - \frac{2}{k^2}\int_1^\infty \frac{e^{-ikxv}dv}{v^2\sqrt{(v^2-1)}}\right\}. \quad\ldots\ldots\ldots\ldots(15)$$

When x is great, the outstanding integral in (15) may be treated in the manner already explained. The integral

$$\int \frac{e^{-rw}dw}{w^2\sqrt{(1+w^2)}}$$

will yield the same value, whether taken from $w=i$ to $w=i+i\infty$, or from $w=i$ to $w=\infty+i$. The first gives

$$-\int_1^\infty \frac{e^{-irv}dv}{v^2\sqrt{(v^2-1)}};$$

the second gives

$$\int_0^\infty \frac{e^{-ru}e^{-ir}du}{(u+i)^2\sqrt{(2iu+u^2)}}.$$

Accordingly

$$\int_1^\infty \frac{e^{-irv}dv}{v^2\sqrt{(v^2-1)}} = \frac{e^{-ir}}{\sqrt{(2ir)}}\int_0^\infty \frac{e^{-\beta}\beta^{-\frac{1}{2}}d\beta}{\left(1+\frac{\beta}{ir}\right)^2\cdot\sqrt{\left(1+\frac{\beta}{2ir}\right)}},$$

and the latter may be expanded in inverse powers of r and integrated as before. For the leading term we have

$$\int_0^\infty e^{-\beta}\beta^{-\frac{1}{2}}d\beta = \sqrt{\pi}.$$

Thus approximately when x is great

$$\iint V \, dx \, dy = y \left\{ \frac{\pi x}{ik} + \frac{2}{k^2} - \frac{2\sqrt{\pi} \cdot e^{-ikx}}{k^2 \sqrt{(2ikx)}} \right\} = xy \left\{ -\frac{i\pi}{k} + \frac{2}{k^2 x} \right\}, \quad \dots(16)$$

if we confine ourselves to the leading real and imaginary terms.

From (2) we now get

$$\iint \delta p \, dS = \frac{ika\sigma}{\pi} \frac{d\phi}{dn} \cdot xy \cdot \left\{ -\frac{i\pi}{k} + \frac{2}{k^2 x} \right\} = xy \cdot a\sigma \cdot \frac{d\xi}{dt} + xy \cdot \frac{2\sigma}{\pi k^2 x} \frac{d^2\xi}{dt^2}. \quad (17)$$

When x is large, the inertia term ultimately vanishes in comparison with the area of the plate. The reaction is then reduced to the dissipative term, which is the same as would be obtained from the theory of plane waves of infinite extent.

302.

EXTRACTS FROM NOBEL LECTURE.

[Given before the Royal Academy of Science at Stockholm, 1904.]

THE subject of the densities of gases has engaged a large part of my attention for over 20 years. In 1882 in an address to the British Association I suggested that the time had come for a redetermination of these densities, being interested in the question of Prout's law. At that time the best results were those of Regnault, according to whom the density of oxygen was 15·96 times that of hydrogen. The deviation of this number from the integer 16 seemed not to be outside the limits of experimental error.

In my work, as in the simultaneous work of Cooke, the method of Regnault was followed in that the working globe was counterpoised by a dummy globe (always closed) of the same external volume as itself. Under these conditions we became independent of fluctuations of atmospheric density. The importance of this consideration will be manifest when it is pointed out that in the usual process of weighing against brass or platinum weights, it might make more apparent difference whether the barometer were high or low than whether the working globe were vacuous or charged with hydrogen to atmospheric pressure. Cooke's result, as at first announced, was practically identical with that of Regnault, but in the calculations of both these experimenters a correction of considerable importance had been overlooked. It was assumed that the external volume of the working globe was the same whether vacuous or charged to atmospheric pressure, whereas of course the volume must be greater in the latter case. The introduction of the correction reduced Cooke's result to the same as that which I had in the meantime announced, viz. 15·88. In this case therefore the discrepancy from Prout's law was increased, and not diminished, by the new determination.

Turning my attention to *nitrogen*, I made a series of determinations, using a method of preparation devised originally by Harcourt, and recommended to me by Ramsay. Air bubbled through liquid ammonia is passed through a tube containing copper at a red heat where the oxygen of the air is consumed by the hydrogen of the ammonia, the excess of the ammonia being subsequently removed with sulphuric acid. In this case the copper serves merely to increase the surface and to act as an indicator. As long as it remains bright, we have security that the ammonia has done its work.

Having obtained a series of concordant observations on gas thus prepared I was at first disposed to consider the work on nitrogen as finished. Afterwards, however, I reflected that the method which I had used was not that of Regnault and that in any case it was desirable to multiply methods, so that I fell back upon the more orthodox procedure according to which, ammonia being dispensed with, air passes directly over red hot copper. Again a series in good agreement with itself resulted, but to my surprise and disgust the densities obtained by the two methods differed by a thousandth part—a difference small in itself but entirely beyond the experimental errors. The ammonia method gave the smaller density, and the question arose whether the difference could be attributed to recognized impurities. Somewhat prolonged inquiry having answered this question in the negative, I was rather at a loss how to proceed. It is a good rule in experimental work to seek to magnify a discrepancy when it first presents itself, rather than to follow the natural instinct of trying to get quit of it. What was the difference between the two kinds of nitrogen? The one was wholly derived from air; the other partially, to the extent of about one-fifth part, from ammonia. The most promising course for magnifying the discrepancy appeared to be the substitution of oxygen for air in the ammonia method, so that *all* the nitrogen should in that case be derived from ammonia. Success was at once attained, the nitrogen from the ammonia being now 1/200 part lighter than that from air, a difference upon which it was possible to work with satisfaction. Among the explanations which suggested themselves were the presence of a gas heavier than nitrogen in the air, or (what was at first rather favoured by chemical friends) the existence in the ammonia-prepared gas of nitrogen in a dissociated state. Since such dissociated nitrogen would probably be unstable, the experiment was tried of keeping a sample for eight months, but the density was found to be unaltered.

On the supposition that the air-derived gas was heavier than the "chemical" nitrogen on account of the existence in the atmosphere of an unknown ingredient, the next step was the isolation of this ingredient by absorption of nitrogen. This was a task of considerable difficulty; and it was undertaken by Ramsay and myself working at first independently but afterwards in concert. Two methods were available,—the first that by which Cavendish had originally established the identity of the principal component of the atmosphere with the nitrogen of nitre and consisting in the oxidation of the nitrogen under the influence of electric sparks with absorption of the acid compounds by alkali. The other method was to absorb the nitrogen by means of magnesium at a full red heat. In both these ways a gas was isolated of amount equal to about one per cent. of the atmosphere by volume and having a density about half as great again as that of nitrogen. From the manner of its preparation it was proved to be non-oxidisable and to refuse absorption by magnesium at a red heat, and further varied attempts to induce chemical combination were without result. On this account the name *argon* was given to it. The most remarkable feature of the gas was the ratio of its specific heats, which proved to be the highest possible, viz. 1·67, indicating that sensibly the whole of the energy of molecular motion is translational.

Argon must not be deemed rare. A large hall may easily contain a greater weight of it than a man can carry.

In subsequent investigations Ramsay and Travers discovered small quantities of new gases contained in the aggregate at first named argon. Helium, originally obtained by Ramsay from clevite, is also present in minute quantity.

Experiments upon the refractivity and viscosity of argon revealed nothing specially remarkable, but the refractivity of helium proved to be unexpectedly low, not attaining one-third of that of hydrogen—the lowest then known.

As regards the preparation of argon, it is advantageous to begin with *liquid air*, for preparing which a plant is now to be found in many laboratories. [See Art. 288, p. 115.]

Although the preparation of a considerable quantity of argon is rather an undertaking, there is no difficulty in demonstrating its existence with the most ordinary appliances. By the use of a specially shaped tube and an ordinary induction-coil actuated by a small Grove battery, I was able to show the characteristic spectrum of argon at atmospheric pressure, starting with 5 c.c. only of air.

Another question relating to the gases of the atmosphere has occupied my attention—namely the amount of free hydrogen. [See Art. 277, p. 49.]

Another branch of my work upon gases has relation to the law of pressure, especially at low pressures. Under these circumstances the usual methods are deficient in accuracy. Thus Amagat considers that under the best conditions it is not possible to answer for anything less than 0·01 mm. of mercury. By the use of a special manometer I was able to carry the accuracy at least 50 times further than Amagat's standard, and thus to investigate with fair accuracy the effect of pressures not exceeding 0·01 mm. in total amount. Boyle's law was fully verified, even in the case of oxygen, for which C. Baur had found anomalies, especially in the neighbourhood of 0·7 mm. pressure.

More recently I have made determinations of the compressibility of gases
between one atmosphere and half an atmosphere of pressure. For this
purpose two manometric gauges, each capable of measuring half an atmosphere, were employed. The equality of the gauges could be tested by using
them *in parallel,* to borrow an electrical term. One of the gauges alone would
thus serve for half an atmosphere, while the two combined *in series* gave the
whole atmosphere. In combination with these gauges volumes in the ratio
of 2 : 1 were needed. Here again the desired result was arrived at by the
use of two equal volumes, either alone or in combination. Any question as
to the precise equality of the two volumes is eliminated in each set of
observations by using the two single volumes alternately. The mean result
then *necessarily* corresponds to the half of the total volume, except in so far
as the capacities of the vessels may be altered by change of pressure.

The annexed table gives a summary of results for the various gases.

Gas	B	Temperature
O_2	1·00038	11·2
H_2	0·99974	10·7
N_2	1·00015	14·9
CO	1·00026	13·8
Air	1·00023	11·4
CO_2	1·00279	15·0
N_2O	1·00327	11·0

Here
$$B = \frac{pv \text{ at } \frac{1}{2} \text{ atmos.}}{pv \text{ at } 1 \text{ atmos.}},$$

the temperature being the same at both pressures and having the value
recorded. That B should be less than unity in the case of hydrogen and
exceed that value for the other gases, is what was to be expected from the
known behaviour at higher pressures.

The principal interest of these results is perhaps to calculate corrections
to ratios of densities, as found at atmospheric pressure, so as to infer what
the ratios would be in a state of great rarefaction. It is only under this
condition that Avogadro's law can be expected to apply accurately, as
I pointed out in 1892 in connection with oxygen and hydrogen.

In the case of nitrogen and oxygen, the correction is not important, and
the original comparison of densities* is sensibly unaffected. According to
this method the atomic weight of nitrogen is 14·01, in opposition to the
14·05 found by Stas.

* Rayleigh and Ramsay, *Phil. Trans.* 1895.

303.

ON THE COMPRESSIBILITY OF GASES BETWEEN ONE ATMOSPHERE AND HALF AN ATMOSPHERE OF PRESSURE.

[*Phil. Trans.* A, Vol. CCIV. pp. 351—372, 1905.]

THE present* is the third of a series of memoirs in which are detailed observations upon the compressibility of the principal gases at pressures from one atmosphere downwards. In the first† of these the pressures dealt with were exceedingly low, ranging from 1·5 millims. to 0·01 millim. of mercury, and the use of a special and extraordinarily delicate manometer allowed the verification of Boyle's law to be pushed to about $\frac{1}{2000}$ of a millimetre of mercury.

In the second‡ memoir the products of pressure and volume at constant temperature (that of the room) were compared when the pressure was changed from 75 millims. to 150 millims. of mercury—in the ratio of 2 : 1. The ratio of the products (denoted by *B*) would be unity according to Boyle's law; for the more condensable gases, *e.g.* nitrous oxide, it exceeds unity. The following were the final mean values :—

Nature of gas	B
Air	·99997
Hydrogen	·99997
Oxygen	1·00024
Nitrous oxide	1·00066
Argon	1·00021
Carbonic oxide§	1·00005

* A Preliminary Notice containing many of the results now recorded in greater detail was published in *Roy. Soc. Proc.*, February, 1904.

† *Phil. Trans.* A, CXCVI. pp. 205—223, 1901 ; *Scientific Papers*, Vol. IV. p. 511.

‡ *Phil. Trans.* A, CXCVIII. pp. 417—430, 1902. On p. 428, line 8 from bottom, read $1+m\tau$ instead of $1-m\tau$. [*Scientific Papers*, v. p. 27.]

§ The number for carbonic oxide was obtained subsequently to the publication of the memoir. It is the mean of two sets of observations, giving severally 1·00003 and 1·00008. The gas was prepared from ferrocyanide of potassium (see *Roy. Soc. Proc.* Vol. LXII. p. 204, 1897 ; *Scientific Papers*, Vol. IV. p. 347).

The deviations from unity in the cases of oxygen and argon were thought to exceed the errors of observation. The results presently to be given for oxygen render it probable that the larger half of the deviation was, in fact, error. At any rate, Boyle's law was sensibly observed by air, hydrogen and carbonic oxide.

The method employed in this research appeared to be satisfactory, and I was desirous of extending it to higher pressures, still, however, below the atmospheric, as to which there seemed to be a great dearth of information. I could find only some incidental observations by Amagat* on air and carbonic acid, and these it may be well to quote:—

Air		Acide carbonique	
Pression initiale en centimètres	$\dfrac{pv}{p'v'}$	Pression initiale en centimètres	$\dfrac{pv}{p'v'}$
37·314	1·0003	37·760	1·0026
37·318	1·0002	37·724	1·0022
31·523	1·0000	37·561	1·0025
30·613	1·0027	24·970	1·0015
25·393	1·0010	24·974	1·0020
25·376	1·0005	24·948	1·0015
24·830	·9997	24·947	1·0021
24·841	1·0002		

The pressures were as 2 : 1, and the "initial pressure" p was the smaller. The temperature was from 17° C. to 19° C. The ratio $pv/p'v'$ is what I have denoted by B. It will be seen that the numbers for air exhibit considerable discrepancies.

The earlier entries in Amagat's table correspond pretty closely with the observations that I proposed to undertake. Besides the general elucidation of the behaviour of gases at reduced pressures, the object in view was to obtain material for comparing the densities of various gases at great rarefactions. In the actual weighings of gases the pressure in the containing vessel is usually atmospheric, but the ratios of densities so obtained are not immediately available for inferring molecular weights according to Avogadro's rule. This rule can only be supposed to apply with rigour when the gases are so far rarefied as to come within the range of Boyle's law†. For this

---

* *Ann. de Chimie*, tome xxviii. 1883.

† The application of this idea to oxygen and hydrogen was made in my paper "On the Relative Densities of Oxygen and Hydrogen," *Roy. Soc. Proc.* Vol. L. p. 448, 1892; *Scientific Papers*, Vol. iii. p. 525. My hesitation then and later to push the investigation further, so as to obtain corrections to the relative densities observed at atmospheric pressure, arose from the uncertainties in which the anomalous observations of Mendeleef and Siljerström had enveloped the behaviour of gases at low pressures.

purpose it is advisable that the range of pressure employed should be sufficient to give accuracy, but not so high that the application to Avogadro's rule involves too much extrapolation. The comparison of volumes at pressures atmospheric and half atmospheric seems to meet these requirements, though we must not forget that (apart from theory) the result is still of the nature of an extrapolation. On this subject reference may be made to a paper by Sir W. Ramsay and Dr B. Steele*.

The guiding idea in the present apparatus, as in that of 1902, is the use of two manometric gauges combined in a special manner. "The object is to test whether when the volume of a gas is halved its pressure is doubled, and its attainment requires two gauges indicating pressures which are in the ratio of 2 : 1. To this end we may employ a pair of independent gauges as nearly as possible similar to one another, the similarity being tested by combination *in parallel*, to borrow an electrical term. When connected below with one reservoir of air and above with another reservoir, or with a vacuum, the two gauges should reach their settings simultaneously, or at least so nearly that a suitable connection can readily be applied. For brevity we may for the present assume precise similarity. If now the two gauges be combined *in series*, so that the low-pressure chamber of the first communicates with the high-pressure chamber of the second, the combination constitutes a gauge suitable for measuring a doubled pressure."

The Manometers.

The construction of the gauges is modelled upon that used extensively in my researches upon the density of gases†. An iron measuring rod, AB, is actually applied to the two mercury surfaces, arranged so as to be vertically superposed. This rod is of about 7 millims. diameter and is pointed below, A. At the upper end, B, it divides at the level of the mercury into a sort of fork, and terminates in a point similar to that at A, and, like it, directed downwards. The coincidence of these points with their images reflected at the mercury surfaces is observed with the aid of lenses of 20 millims. focus suitably held in position. It is, of course, independent of any irregular refractions which the walls of the tube may exercise. In each manometer the distance between the points is 15 inches or 381 millims.

The internal diameter of the tubes, constituting the upper and lower chambers of the manometers, is 22 millims. This is the diameter at the level of the "points" to which the mercury surfaces are set. At the places where the iron rod emerges above and below into the open, the glass is contracted until it becomes an approximate fit, and air-tightness is secured with the aid of cement.

* "On the Vapour-Densities of some Carbon Compounds; an attempt to determine their Molecular Weights," *Phil. Mag.* Oct. 1893.
† *Roy. Soc. Proc.* Vol. LIII. pp. 134—149, 1893; *Scientific Papers*, Vol. IV. p. 41.

Fig. 1.

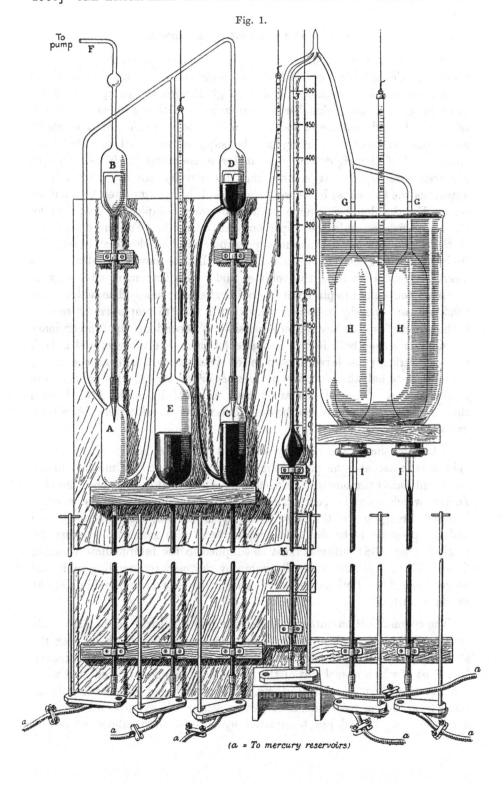

(a = To mercury reservoirs)

General Arrangement of Apparatus.

With one important difference to be explained presently the general scheme is the same as in 1902 and is sketched in fig. 1. The left-hand manometer can be connected above, through F, with the pump or with the gas supply. The lower chamber A communicates with the upper chamber D of the right-hand manometer and with an intermediate reservoir E, to which, as to the manometers, mercury can be supplied from below. The lower chamber C of the right-hand manometer is connected with the principal gas reservoirs H, H. It is here that the novelty enters. In the 1902 apparatus the two equal bulbs were superposed, being connected by a narrow neck. For the doubled volume, both bulbs were occupied by gas; but for the single volume, only the upper one was available. For the comparison of the single and double volume, a principal factor of the final result, preliminary gauging had to be relied upon. In the new apparatus it was desired largely to increase the volumes, and it was both more advantageous and more convenient to place the two bulbs H side by side. The temperature conditions are thereby improved; but what I wish to emphasize at present is the elimination, thereby rendered possible, of dependence upon preliminary gauging, for either bulb is now available for the single volume; and if both are symmetrically employed in each set of observations, the mean necessarily corresponds to half the total volume, *whether or not the two single volumes are precisely equal.* The volumes are defined, as usual, by marks GG, II, upon the associated tubes above and below. The use of the side tube JK will be explained presently.

When, as shown, the mercury stands at the lower marks I, the double volume is in use, and the pressure is such as will balance the mercury in one (the right-hand) manometer. A vacuum is established in the upper chamber D, from which a way is open through ABF to the pump. When the mercury is raised through one of the bulbs to the upper mark G, the volume is halved and the pressure to be dealt with is doubled. Gas sufficient to exert the single pressure (381 millims.) must be supplied to the intermediate chamber E, which is now isolated from the pump by the mercury standing up in the curved tube AB. Both manometers can now be set and the doubling of the pressure verified.

The communication through to the pump is unobstructed, but on a side tube a three-way tap is provided communicating on the one hand with the gas supply and on the other with a vertical tube delivering under mercury, by means of which a wash-out of the generating vessels can be effected when it is not convenient to evacuate them. The six tubes of glass leading downwards from the gas reservoirs, manometers, &c., are all well over a barometer-height in length, and are terminated by suitable indiarubber hoses and reservoirs for the supply of mercury*. By this precaution the internal pressure on the hoses is guaranteed to exceed the external atmospheric pressure, and under this condition the use of indiarubber seems to be free from serious objection. If, however, the external pressure be allowed to be in excess, there are soon signs of the percolation of air and probably of moisture.

When settings are actually in progress, the mercury in the hoses is isolated from that in the reservoirs by pinch-cocks, and the adjustment of the supply is effected by squeezing the hoses. As explained in my first paper, the final adjustment must be made by squeezers which operate upon parts of the hoses which lie flat upon the large wooden tray underlying the whole.

The Side Apparatus.

The use of this was fully explained in my former paper. By the employment of manometric gauges we are enabled to dispense with scales and cathetometers; but since (save as to a small temperature correction) the pressures are defined beforehand, the adjustment is thrown upon the volume. The variable volume is introduced at the side tube JK, which, with its associated bulb, allows of the elimination from the results of the volume which cannot be directly gauged, including that over the mercury in the lower chamber of the right-hand manometer when set. The tubes above and below the bulb were calibrated in the usual manner†. It should be remarked that the diagram shows the mercury in the side apparatus in a position suitable for a measurement at the *doubled* pressure, while in the rest of the apparatus the position of the mercury corresponds to the *single* pressure.

General Sketch of Theory.

It will be convenient to repeat this, nearly as given in the former paper. To save complication, it will be supposed that the temperature is constant, not only throughout the whole apparatus at one time, but also at the four different times concerned.

$V_1 =$ volume of two large bulbs H, H together between GG, II (about 633 cub. centims.).

$V_3 =$ volume between $CJGG$ (the ungauged space).

$V_4 =$ measured volume on upper part of JK from highest mark J downwards.

---

* These reservoirs were protected from external moisture by tubes of chloride of calcium.

† The whole of the apparatus was made under my instructions by the late Mr Gordon, who also took a large part in the observations. I take this opportunity of recording my indebtedness to his faithful assistance over a long series of years.

V_5 = measured volume, including bulb of side apparatus, from highest mark J downwards.

P_1 = smaller pressure (height of mercury in right-hand manometer).

P_2 = larger pressure (sum of heights of mercury in two manometers).

In the first pair of operations, when the large bulbs are in use, the pressure P_1 corresponds to the volume $(V_1 + V_3 + V_5)$, and the pressure P_2 corresponds to $(\tfrac{1}{2}V_1 + V_3 + V_4)$, *the quantity of gas being the same.* Hence the equation

$$P_1(V_1 + V_3 + V_5) = BP_2(\tfrac{1}{2}V_1 + V_3 + V_4),$$

B being a numerical quantity which would be unity according to Boyle's law. In the second pair of operations with the *same nature* but with a *different quantity* of gas, and with the *same pressures*, the mercury stands at GG throughout, and we have

$$P_1(V_3 + V'_5) = BP_2(V_3 + V'_4);$$

whence by subtraction

$$P_1(V_1 + V_5 - V'_5) = BP_2(\tfrac{1}{2}V_1 + V_4 - V'_4).$$

From this equation V_3 has been eliminated, and B is expressed by means of P_1/P_2 and the actually gauged volumes $V_5 - V'_5$, $V_4 - V'_4$. It is important to remark that only the *differences* $V_5 - V'_5$, $V_4 - V'_4$ are involved. The first is measured on the lower part of the side apparatus, and the second on the upper part; while the capacity of the intervening bulb *does not appear.*

If Boyle's law be closely followed, there is nothing to prevent both $V_5 - V'_5$ and $V_4 - V'_4$ from being very small. Except the preliminary comparison of the manometers, the whole of the data required for the verification are then contained in the observations of each set.

When the temperature-changes are taken into account, V_3, V_4, V_5 are not fully eliminated, but they appear with coefficients which are very small if the temperature conditions are good.

Thermometers.

Of these four were employed. The first gave the temperature (τ) of the manometric columns; the second gave the temperature (T) of V_3; the third that of the bulb of the side apparatus (t). The temperature of the water-bath, in which are contained the principal bulbs, is of course the most important. The water was stirred continuously by a stream of air, and the temperature was taken by a thermometer that could be read to $\frac{1}{100}$ of a degree C. No observations were begun until it had been ascertained that the temperature of the water was slowly rising. It is important to understand what really are the demands made upon this thermometer. It was arranged that the mean temperature θ_1 when the double volume was in use should be almost the same as for the single volumes θ_2. The difference was usually less than $\frac{1}{100}$ of a degree and rarely exceeded $\frac{2}{100}$ or $\frac{3}{100}$. Under these circumstances the use of the thermometer was practically only to identify the same temperature on different days, and the actual error of its readings and even of its scale of temperature were of but secondary importance. Comparisons with other thermometers showed that there were no errors which could possibly become sensible. The precautions necessary, in order that the other thermometers should do their work satisfactorily, were indicated in the former paper. In the present work the number of intervening screens was increased.

It is desirable to emphasise that most of the errors that could arise from imperfect action of the thermometers is eliminated in the actual results, which depend only upon a comparison between operations with and without the large bulbs. For example, suppose that there is an error in the rather ill-defined temperature of the space V_3. The conditions are the same whether the large bulbs are in use or not; and thus whatever error occurs in the one case may be expected to repeat itself in the other. So far as this repetition is complete, the error disappears in the comparison. Again, it might happen that one of the large bulbs tended to be warmer than the other or than the thermometer. But this, so far as it is constant, could lead to no error, the effect when the bulb is used alone being compensated by the effect when it forms one of the pair. Purely accidental errors are, in any case, eliminated when the mean is taken of a number of observations.

The Large Reservoirs.

The tubes forming the principal parts are of glass, 25 centims. in length, 41 millims. in internal diameter, and about $2\frac{1}{4}$ millims. thickness in the walls. There are prolongations above and below of narrow bore, upon which are placed the marks defining the volumes.

As has been explained, the accurate comparison of these volumes is unnecessary. As it happens, the actual volumes between the marks are so nearly equal that it is difficult to say which is the larger. The total volume V_1, required only to be roughly known for the sake of the subsidiary terms, is 632·6 cub. centims.

But there is another question to be considered. The single bulbs are used under an internal pressure of an atmosphere. Under the same pressure the combined volume of both bulbs would of course be exactly double the mean of those of the bulbs used separately. But when the bulbs are in combination, the internal pressure is reduced to half an atmosphere, and the bulbs contract. A correction is thus necessary which runs similarly through all the results calculated on the supposition that the ratio of volumes is exactly 2 : 1.

The amount of the correction has been determined in two ways. Direct observation of the change of level of water filling the bulb and standing in the small upper prolongation, when the internal pressure was changed from one atmosphere to half atmosphere, gave a total relative alteration of 4.4×10^{-5} per half atmosphere, of which 2.3×10^{-5} would be due to the contraction of the water. The difference, viz. 2.1×10^{-5}, represents the relative contraction or expansion of the volume per half atmosphere of pressure.

A calculation founded upon the measured dimensions of the tubes, including the thickness of the walls, combined with estimates of the elasticity of the glass, gave 2.0×10^{-5} per half atmosphere, in better agreement than could have been expected.

The real ratio of volumes with which we are concerned in these experiments is thus not 2 exactly, but

$$2 (1 - {\cdot}000021).$$

The value of B calculated without allowance for this correction would be too large, so that a gas which really obeyed Boyle's law exactly would appear to be too condensable, like CO_2. From a value of B so calculated we are to subtract $\cdot000021$.

Comparison of Manometers.

This comparison is effected by combining the manometers in parallel so that the mercury at the lower levels is subject to the pressure of one continuous quantity of gas, while the mercury at the upper levels is *in vacuo*, or at any rate under the pressure of the same very rare gas. Any difference that may manifest itself may be estimated by finding what change of gas-volume is required in order to pass from the pressure appropriate to one manometer to that appropriate to the other.

The first matter requiring attention is the verticality of the measuring rods, or rather of the lines joining the *points* actually applied to the mercury. The points were visually projected upon a plumb-line, hung a few centimetres away, and were observed through a hole of 2 millims. or 3 millims. diameter perforated in a black card. If the adjustment is perfect the same position of the card allows accurate projection upon the thread of the upper and lower points; if not, the necessary motion of the card, perpendicular to the line of sight, gives data for estimating the amount of the error. If x and y be the linear horizontal deviations from true adjustment thus determined in *any* two perpendicular planes, l the length of the rod, the angular error is $\sqrt{(x^2 + y^2)} \div l$, and the proportional error of height for the present purpose is $(x^2 + y^2)/2l^2$. When the manometers were compared, no value of

x or y exceeded $\frac{1}{2}$ millim., so that with $l = 380$ millims. the error of verticality could be neglected.

In effecting the comparison of the manometers, both mercury levels must be set *below* (in order to make the gas volume definite), while the settings above are made alternately. It was at once apparent, when the right-hand manometer was set, that the rod on the left was a little too long, a perceptible interval being manifest between the upper point and the mercury. In these experiments the total gas volume was about 2845 cub. centims., the principal part being the volume of a large bottle protected from rapid changes of temperature by a packing of sawdust. The necessary changes were produced by causing mercury to rise and fall in a vertical tube of small bore, the position of the meniscus being noted at the moment when a setting was judged to be good. The settings of the two manometers must be made alternately in order to eliminate temperature changes; and the result of each set of observations was derived from the means of four settings of one manometer, and of five of the other. The lowering of the mercury in the auxiliary tube required to pass from a setting of the left-hand manometer to a setting of the right-hand manometer, was found on three separate days to be 50·3 millims., 51·3 millims., 49·9 millims., or, as a mean, 50·5 millims. As regards the section of the auxiliary tube, it was found that a mercury thread occupying 85 millims. of it weighed 5·335 grammes. The proportional difference of volume is thus

$$\frac{50 \cdot 5 \times 5 \cdot 335}{85 \times 13 \cdot 54 \times 2845} = \cdot 000082 \, ;$$

and the same fraction represents the proportion by which the left-hand measuring rod exceeds in length its fellow on the right. It would seem that by this procedure the lengths of the rods are compared to about a millionth part.

In the notation employed in the calculations

$$H_1 = 1 \cdot 000082 H_2, \quad \text{whence} \quad \frac{H_1 + H_2}{2 H_2} = 1 \cdot 000041.$$

It may be observed that an error in the comparison of H_1 and H_2 enters to only half its amount into the final result.

The Observations.

The manipulation necessary to imprison the right quantity of gas was described in the former paper. When this has once been secured the observations are straightforward. On each occasion six readings were taken, extending over about an hour, during which time the temperature always rose, and the means were combined into what was treated as one observation.

A complete set usually included eight observations at the high pressure, in four of which one large bulb was in use, and in the second four the other bulb. Interpolated in the middle of these were the observations (usually six in number) of the low pressure where both large bulbs were occupied by gas. Further, each set included eight observations relating to the side apparatus, in which the large bulbs stood charged with mercury. In this way each set contained within itself complete material for the elimination of V_3, which might possibly vary from time to time with the character of the contact between mercury and glass in the lower chamber of the right-hand manometer. Finally the means were taken of all the corresponding observations, no further distinction being maintained between the two large bulbs.

The following table shows in the notation employed the correspondence of volumes and temperatures :—

I.	V_1	θ_1	V_3	T_1	V_5	t_1
II.	$\frac{1}{2}V_1$	θ_2	V_3	T_2	V_4	T_2
III.	—	—	V_3	T_3	V'_5	t_3
IV.	—	—	V_3	T_4	V'_4	T_4

In the first observation V_1 is the volume of the two large bulbs together and θ_1 the temperature of the water-bath, reckoned from some convenient neighbouring temperature as standard. V_3 is the ungauged volume already discussed whose temperature T_1 is given by the upper thermometer. V_5 is the (larger) volume in the side apparatus whose temperature t_1 is that of the associated thermometer. In the second observation $\frac{1}{2}V_1$ is the (mean) volume of a single bulb and θ_2 its temperature. V_4 is the volume in the side apparatus whose temperature, as well as that of V_3, is taken to be T_2. III. and IV. represent the corresponding observations when the large bulbs are not used. The temperatures of the mercury in the manometric columns are represented by τ_1, τ_2, τ_3, τ_4.

As an example of the actual quantities, the observations on hydrogen, April 9—24, 1903, may be taken. The values of V_1 and V_3 are approximate.

$$V_1 = 632\cdot6, \quad V_3 = 11\cdot02, \quad V_5 = 13\cdot978, \quad V_4 = 1\cdot504,$$
$$V_5 - V'_5 = -\cdot650, \quad V_4 - V'_4 = -\cdot245.$$
$$\theta_1 = -\cdot007, \quad \theta_2 = +\cdot001; \quad t_1 = +\cdot31, \quad t_3 = -\cdot01.$$
$$T_1 = +\cdot69, \quad T_2 = +\cdot78, \quad T_3 = +\cdot23, \quad T_4 = +\cdot25.$$
$$\tau_1 = +\cdot43, \quad \tau_2 = +\cdot54, \quad \tau_3 = +\cdot05, \quad \tau_4 = +\cdot11.$$

The volumes are in cubic centimetres and the temperatures are in Centigrade degrees, reckoned from 11°.

The Reductions.

The simple theory has already been given, but the actual reductions are rather troublesome, on account of the numerous temperature corrections. These, however, are but small.

We have first to deal with the expansion of the mercury and of the iron in the manometers. If the actual heights of the mercury (at the same temperature) be H_1, H_2, we have for the relative *pressures* $H/(1 + m\tau)$, where $m = \cdot 00017$. Thus in the notation already employed

$$P_1 = \frac{H_2}{1 + m\tau_1}, \quad \text{or} \quad \frac{H_2}{1 + m\tau_3};$$

and
$$P_2 = \frac{H_1 + H_2}{1 + m\tau_2}, \quad \text{or} \quad \frac{H_1 + H_2}{1 + m\tau_4}.$$

The quantity of gas at a given pressure occupying a known volume is to be found by dividing the volume by the absolute temperature. Hence each volume is to be divided by $1 + \beta\theta$, $1 + \beta T$, $1 + \beta t$, as the case may be, where β is the reciprocal of the absolute temperature chosen as a standard for the set. Thus in the above example for hydrogen, $\beta = 1/(273 + 11) = 1/284$.

Our equations expressing that the quantities of gas are the same at the single and at the doubled pressures accordingly take the form

$$\frac{H_2}{1 + m\tau_1}\left\{\frac{V_1}{1 + \beta\theta_1} + \frac{V_3}{1 + \beta T_1} + \frac{V_5}{1 + \beta t_1}\right\} = \frac{B(H_1 + H_2)}{1 + m\tau_2}\left\{\frac{\frac{1}{2}V_1}{1 + \beta\theta_2} + \frac{V_3 + V_4}{1 + \beta T_2}\right\},$$

$$\frac{H_2}{1 + m\tau_3}\left\{\frac{V_3}{1 + \beta T_3} + \frac{V'_5}{1 + \beta t_3}\right\} = \frac{B(H_1 + H_2)}{1 + m\tau_4}\frac{V_3 + V'_4}{1 + \beta T'_4},$$

where B is the numerical quantity to be determined—according to Boyle's law identical with unity.

By subtraction and neglect in the small terms of the squares of the small temperature differences, we obtain

$$\frac{1}{(1 + m\tau_1)(1 + \beta\theta_1)} - \frac{B(H_1 + H_2)}{2H_2(1 + m\tau_2)(1 + \beta\theta_2)}$$

$$= \frac{V_3}{V_1}\{m(\tau_1 - \tau_3 - 2\tau_2 + 2\tau_4) + \beta(T_1 - T_3 - 2T_2 + 2T_4)\}$$

$$- \frac{2V_4}{V_1}\{m(\tau_2 - \tau_4) + \beta(T_2 - T_4)\}$$

$$+ \frac{V_5}{V_1}\{m(\tau_1 - \tau_3) + \beta(t_1 - t_3)\}$$

$$+ \frac{2(V_4 - V'_4)}{V_1}\{1 - m\tau_4 - \beta T_4\}$$

$$- \frac{V_5 - V'_5}{V_1}\{1 - m\tau_3 - \beta t_3\}.$$

The first three terms on the right, viz., those in V_3, V_4, V_5, vanish if $\tau_1 = \tau_3$, $\tau_2 = \tau_4$, $T_1 = T_3$, $T_2 = T_4$, $t_1 = t_3$. · If in general R denote the sum of the five terms, we may write with sufficient approximation for the actual experiments

$$B = \frac{2H_2}{H_1 + H_2} + m(\tau_2 - \tau_1) + \beta(\theta_2 - \theta_1) - R.$$

It may be well to exhibit further the steps of the reduction in the case of hydrogen above detailed. The five terms composing R are

$$\text{Term in } V_3 = -\cdot000038$$
$$\text{,,} \qquad V_4 = -\cdot000009$$
$$\text{,,} \qquad V_5 = +\cdot000026$$
$$\text{Term in } V_4 - V'_4 = -\cdot000775$$
$$\text{,,} \qquad V_5 - V'_5 = +\cdot001027$$
$$R = +\cdot000231$$

Thus
$$2H_2/(H_1 + H_2) = 1 - \cdot000041$$
$$m(\tau_2 - \tau_1) = +\cdot000019$$
$$\beta(\theta_2 - \theta_1) = +\cdot000028$$
$$-R = -\cdot000231$$
$$B = \quad \cdot999775$$

In the above calculation the volumes of the principal capacities at the two pressures have been assumed to be as $2:1$ exactly. As has already been explained, the value of B so obtained is subject to correction of $\cdot000021$ to be subtracted.

Hence $\qquad\qquad B = \cdot99975$,

a result in strictness applicable to the temperature $11\cdot0°\,$C. The hydrogen is somewhat less compressible than according to Boyle's law, as was to be expected from its known behaviour at pressures above atmosphere.

The Results.

After the above explanation it will suffice to record the final results of the various sets of observations.

HYDROGEN.

Date	Source	Temperature	B
January 17 to 27, 1903.....................	Palladium	10·6	·99974
,, 30 to February 10, 1903	,,	10·6	·99974
April 9 to 24, 1903	,,	11·0	·99975
Mean........................		10·7	·99974

In the case of hydrogen the agreement of single results is remarkably good. This gas, as well as all the others, was carefully dried with phosphoric anhydride.

CARBONIC OXIDE.

Date	Source	Temperature	B
May 2 to 14, 1903	Ferrocyanide	11·7	1·00033
„ 19 to 29, 1903	„	12·5	1·00024
June 17 to 29, 1903	„	14·5	1·00024
July 1 to 15, 1903	„	16·6	1·00023
Mean............................		13·8	1·00026

The gas was prepared from ferrocyanide of potassium and sulphuric acid*, and purified from CO_2 by a long tube of alkali. It is barely possible that the abnormally high number which stands first in the table may be due to imperfect purification on that occasion; on principle, however, it is retained, as no suspicion suggested itself at the time.

NITROGEN.

Date	Source	Temperature	B
July 23 to August 5, 1903	Chemical	16·5	1·00012
August 7 to 19, 1903.......................	„	16·4	1·00022
October 5 to 20, 1903	„	15·4	1·00011
„ 28 to November 9, 1903	From air	13·2	1·00015
November 10 to 23, 1903..................	„	12·9	1·00016
Mean............................		14·9	1·00015

The "chemical" nitrogen was from potassium nitrite and ammonium chloride. That "from air" was prepared by bubbling air through ammonia and passing over red-hot copper and sulphuric acid, with the usual precautions. It contained about 1 per cent. of argon; but this could hardly influence the observed numbers.

OXYGEN.

Date	Source	Temperature	B
November 25 to December 5, 1903......	Perman-ganate	12·4	1·00034
December 8 to 19, 1903		10·9	1·00042
December 21, 1903, to January 5, 1904	„	10·2	1·00038
Mean............................		11·2	1·00038

* See *Roy. Soc. Proc.* Vol. LXII. p. 204, 1897; *Scientific Papers*, Vol. IV. p. 347.

It remains to record certain results with air (free from H_2O and CO_2). It is curious that the greatest discrepancies show themselves here.

The earlier observations, at the end of 1902, were made before the apparatus was perfected, and gave as a mean $B = 1\cdot00022$. Subsequently, return was made to air.

AIR.

Date	Temperature	B
April 1 to 11, 1904	9·5	1·00035
April 12 to 26, 1904	11·3	1·00016
May 14 to 26, 1904	13·5	1·00018
Mean...........................	11·4	1·00023

In partial explanation of the high number which stands first, it should perhaps be mentioned that the set of observations in question was incomplete. Owing to an accident, it was impossible to return from the lower pressure to the higher pressure, as had been intended.

It may be well to repeat here that

$$B = \frac{pv \text{ at } \frac{1}{2} \text{ atmosphere}}{pv \text{ at } 1 \text{ atmosphere}},$$

the temperature being constant and having the values recorded in each case.

Although the accordance of results seems to surpass considerably anything attained in observations below atmosphere at the time this work was undertaken, I must confess that, except in the case of hydrogen, it is not so good as I had expected in view of the design of the apparatus and of the care with which the observations were made. I had supposed that an error of 3 parts in 100,000 (at the outside), corresponding to $\frac{1}{100}°$ C., was as much as was to be feared. As it is, I do not believe that the discrepancies can be explained as due to errors of temperature, or of pressure, or of volume so far as the readings are concerned. But it is possible that a variable contact between mercury and glass in the lower chamber of the first manometer may have affected the volume in an uncertain manner, though care was taken to obviate this as far as could be. It is to be remembered, however, that, except as to the comparison of the two manometers, all sources of error enter independently in each set of observations; and that a mere repetition of the readings without a change in the gas or in the pressure (from half atmosphere to one atmosphere, or *vice versâ*) gave very much closer agreements.

The values of B here discussed are the same as those given in the Preliminary Notice*, except that no account was there taken of the small deviation in the ratio of volumes from $2:1$ in consequence of the yielding to pressure. If we measure p in atmospheres and assume, as has been usually done, e.g., by Regnault and Van der Waals, that at small pressures the equation of an isothermal is

$$pv = PV(1 + \alpha p),$$

where PV is the value of the product in a state of infinite rarefaction, then

$$B = \frac{1 + \tfrac{1}{2}\alpha}{1 + \alpha} = 1 - \tfrac{1}{2}\alpha,$$

or

$$\alpha = \frac{1}{pv}\frac{d(pv)}{dp} = 2(1 - B).$$

In applying α to correct the observed densities of gases at atmospheric pressure, we are met with the consideration that α is itself a function of temperature, and that the value of α really required for our purpose is that corresponding to $0°$ C., at which temperature the weighing vessels are charged with gas†. In the case of the principal gases α is so small that its correction for temperature was not likely to be important for the purpose in hand, but when we come to CO_2 and N_2O the situation might well be altered. If we know the pressure-equation of the gas, there is no difficulty in calculating a correction to α in terms of the critical constants. I had, in fact, calculated such a correction for carbonic acid after Van der Waals, when I became acquainted with the memoir of D. Berthelot‡, in which this and related questions are admirably discussed. The "reduced" form of Van der Waals' equation is

$$\left(\pi + \frac{3}{v^2}\right)(v - \tfrac{1}{3}) = \tfrac{8}{3}\theta;$$

pressure, volume and temperature being expressed in terms of the critical values. From this we find

$$\frac{1}{\pi v}\frac{d(\pi v)}{d\pi} = \frac{1}{8\theta}\left(1 - \tfrac{27}{8}\frac{1}{\theta}\right);$$

and the effect of a change of temperature upon the value of α is readily deduced. Indeed, if the pressure-equation and the critical values could be thoroughly trusted, there would be no need for experiments upon the value of α at all. The object of such experiments is to test a proposed pressure-equation, or to find materials for a new one; but consistently with this we may use a form, known to represent the facts approximately, to supply a subordinate correction.

* Roy. Soc. Proc. Vol. LXXIII. p. 153, February, 1904.

† Except in the comparison of hydrogen and oxygen.

‡ "Sur les thermomètres à gaz," Travaux et Mémoires du Bureau International, tome XIII. I am indebted to the Director for an early copy of this memoir, and of that of Chappuis presently to be referred to.

A careful discussion of the available data relating to various gases has led D. Berthelot to the conclusion that the facts at low pressure are not to be reconciled with Van der Waals' equation, either in its original form or as modified by Clausius (*i.e.*, with the insertion of the absolute temperature in the denominator of the cohesive term).

The equation which best represents the relation of $d(\pi v)/d\pi$ to temperature is

$$\frac{d(\pi v)}{d\pi} = \tfrac{1}{4}\left(1 - \frac{6}{\theta^2}\right),$$

corresponding to the pressure-equation for *low densities*

$$\left(\pi + \tfrac{16}{3}\frac{1}{\theta v^2}\right)(v - \tfrac{1}{4}) = \tfrac{32}{9}\theta.$$

Also
$$\frac{1}{\pi v}\frac{d(\pi v)}{d\pi} = \frac{9}{128\theta}\left(1 - \frac{6}{\theta^2}\right).$$

As an example, let us apply this formula to find for oxygen what change must be made in α in order to pass from the temperature of the observations (11·2° C.) to 0° C. If θ_0 be the value of θ for 0° C., we have

$$\theta_0 = \frac{273}{273 - 118} = \frac{273}{155}, \qquad \theta = \frac{273 + 11\cdot2}{155} = \frac{284\cdot2}{155}.$$

The factor by which the observed value of α must be multiplied is thus

$$\frac{\theta_0^{-1} - 6\theta_0^{-3}}{\theta^{-1} - 6\theta^{-3}},$$

being the same whether the pressures be reckoned in terms of the critical pressure or in atmospheres. In the case of oxygen the factor is

$$\frac{\cdot568 - 1\cdot097}{\cdot545 - \cdot973} = \frac{\cdot529}{\cdot428} = 1\cdot236.$$

The observed value of α is $-\cdot00076$, corresponding to 11·2° C. Hence at 0° C. we should have

$$\alpha = -\cdot00076 \times 1\cdot236 = -\cdot00094.$$

It will be seen that the correction has a considerable relative effect; but α is so small that the calculated atomic weights are not much influenced. It must be admitted, however, that observations for the present purpose would be best made at 0° C.; to this, however, my apparatus does not lend itself.

The following table embodies the results thus obtained.

Gas	B	α	Temperature	α corrected to 0° C.
			° C.	
Oxygen	1·00038	− ·00076	11·2	− ·00094
Hydrogen	·99974	+ ·00052	10·7	+ ·00053
Nitrogen..............	1·00015	− ·00030	14·9	− ·00056
Carbonic oxide	1·00026	− ·00052	13·8	− ·00081
Air	1·00023	− ·00046	11·4	—
Carbon dioxide	1·00279	− ·00558	15·0	− ·00668
Nitrous oxide.........	1·00327	− ·00654	11·0	− ·00747

The experiments on carbonic anhydride and on nitrous oxide were of later date, having been postponed until the apparatus had been well tested on other gases.

In both these cases it was found that the readings were less constant than usual, signs being apparent of condensation upon the walls of the containing vessels, or possibly upon the cement in the manometer. Under these circumstances it seemed desirable to avoid protracted observations and to concentrate effort upon reproducing the conditions (especially as regards time) as closely as possible with and without the use of the large bulbs. In this way, for example, the question of the cement is eliminated. Condensation upon the walls of the large bulbs themselves, if it occurs, cannot be eliminated from the results; all that we can do is not unnecessarily to increase the opportunity for it by allowing too long a time. It is certain that, unless by chance, these results are less accurate than for the other gases, *i.e.*, less accurate absolutely, but the value of α is so much larger that in a sense the loss of accuracy is less important. Two entirely independent results for nitrous oxide agreed well. They were:—

November 22, 26, 1904	$B = 1·003295$
„ 23, 29, 1904	$B = 1·003252$
Mean............................	$B = 1·00327$

The gas was from the same supply as had been used for density determinations.

In applying these results to correct the ratios of densities as observed at atmospheric pressure to what would correspond to infinite rarefaction, we have, taking oxygen as a standard, to introduce the factor $(1 + \alpha)/(1 + \alpha_0)$, α_0 being the value for oxygen. Taking α from the third column, which may be considered without much error to correspond to a temperature of 13° C. throughout, and also from the fifth column, we have:—

Gas	Correcting factor for about 13° C.	Correcting factor for 0° C.
Hydrogen..................	1·00128	1·00147
Nitrogen	1·00046	1·00038
Carbonic oxide	1·00024	1·00013
Carbon dioxide	·99518	·99426
Nitrous oxide	·99422	·99347

The double of the first number in the second column, viz., 2·00256, represents, according to Avogadro's law, the volume of hydrogen which combines with one volume of oxygen to form water, the pressure being atmospheric and the temperature 13° C. Scott gave 2·00245 for 16° C. In his later work Morley found 2·0027, but this appears to correspond to 0° C. The third column in the above table gives for this temperature 2·0029. The agreement here may be regarded as very good.

In correcting the densities directly observed at 0° C., in order to deduce molecular weights, we must use the third column of the above table*. Oxygen being taken as 32, the densities of the various gases at 0° C., and at atmospheric and very small pressures, as deduced from my own observations†, are :—

Gas	Atmospheric pressure	Very small pressure
H_2	2·0149 (16° C.)	2·0173 = 2 × 1·0086
N_2	28·005	28·016 = 2 × 14·008
CO	28·000	28·003
CO_2	44·268	44·014
N_2O	44·285	43·996

From the researches of M. Leduc and Professor Morley it is probable that the above numbers for hydrogen are a little, perhaps nearly one thousandth part, too high. The correction to very small pressures has to be made in a different manner for hydrogen and for the other gases, in consequence of the fact that the observed ratio of densities corresponded not to 0° C., but to 16° C.‡ Now the observed values of B for hydrogen and oxygen relate to about 11° C., so that if we correct a to 0° C., we are, in fact, altering it in the wrong direction. I have employed as the correcting factor to 16° C. the value 1·00118.

It may be noticed that the discrepancy between my ratio of hydrogen to oxygen and that of M. Leduc is partially explained by the fact that my comparisons were at 16° C., and his at 0° C.

* The correction for temperature was neglected in the Preliminary Notice.

† *Roy. Soc. Proc.* Vol. LIII. p. 134, 1893; Vol. LXII. p. 204, 1897. *Scientific Papers*, Vol. IV. pp. 39, 352. Also for Nitrous Oxide, *Roy. Soc. Proc.* Vol. LXXIV. p. 181, 1904.

‡ See *Roy. Soc. Proc.* Vol. L. p. 448, 1892; *Scientific Papers*, Vol. III. p. 533.

The uncorrected number for nitrogen ($14\cdot003$ corresponding to $O = 16$) has already been given[*], and contrasted with the $14\cdot05$ obtained by Stas. This question deserves the attention of chemists. If Avogadro's law be strictly true, it seems impossible that the atomic weight of nitrogen can be $14\cdot05$.

The atomic weight of carbon can be derived in three ways from these results. First from CO and O :—

$$CO = 28\cdot003$$
$$O = 16\cdot000$$
$$\overline{C = 12\cdot003}$$

Secondly from CO_2 and O :—

$$CO_2 = 44\cdot014$$
$$O_2 = 32\cdot000$$
$$\overline{C = 12\cdot014}$$

Thirdly from CO_2 and CO. This method is independent of the density and compressibility of oxygen :—

$$2CO = 56\cdot006$$
$$CO_2 = 44\cdot014$$
$$\overline{C = 11\cdot992}$$

It will be seen that the number for CO_2 is too high to give the best agreement. Were we to suppose that the true number for CO_2 was $44\cdot004$, instead of $44\cdot014$, we should get by the second method $12\cdot004$ and by the third $12\cdot002$, in agreement with one another and with the result of the first method. The alteration required is less than one part in 4000, and is probably within the limits of error for the compressibility (as reduced to $0°$ C.), and perhaps even for the density. The truth is that the second and third methods are not very advantageous for the calculation of the atomic weight of carbon, and would perhaps be best used conversely.

Finally the molecular weight of N_2O allows of another estimate of that of nitrogen. Thus

$$N_2O = 43\cdot996$$
$$O = 16\cdot000$$
$$\overline{N_2 = 27\cdot996}$$

whence $N = 13\cdot998$.

It should be remarked that these results relative to CO_2 and N_2O depend very sensibly upon the correction of α from about $13°$ C. to $0°$ C., and that this depends upon the discussion of M. D. Berthelot. M. Berthelot has himself deduced molecular weights in a similar manner, founded upon Leduc's measures of densities.

---

[*] Rayleigh and Ramsay, *Phil. Trans.* A, Vol. CLXXXVI. p. 187, 1895; *Scientific Papers*, Vol. IV. p. 133.

It remains to refer to some memoirs which have appeared since the greater part of the present work was finished. Foremost among these is that of M. Chappuis*, who has investigated in a very thorough manner the compressibilities of hydrogen, nitrogen, and carbonic anhydride at various temperatures and at pressures in the neighbourhood of the atmospheric. For hydrogen M. Chappuis finds at 0° C. $\alpha = +\cdot00057$ per atmosphere, and for nitrogen $\alpha = -\cdot00043$. In the case of carbonic anhydride pv departs sensibly from a linear function of p. M. Berthelot gives as a more accurate expression

$$pv = 1 - \frac{\cdot00670}{v},$$

founded on Chappuis' measures and applicable at 0° C. The unit of pressure is here the atmosphere. According to this, for $p = 1$ we get $pv = 1 - \cdot00665$, as compared with $pv = 1$ for $v = \infty$, in close agreement with my value recorded above.

A comparison of Chappuis' method and apparatus with mine may not be without interest. On his side lay a very considerable advantage in respect of the "nocuous space" V_3, inasmuch as this was reduced to as little as 1·1 cub. centims., whereas mine was ten times as great. The advantage would be important when working at temperatures other than that of the room. Otherwise, the influence of V_3 did not appear to prejudice my results, except in so far as V_3 might be *uncertain* from capillarity in the manometer; and this cause of error would operate equally in Chappuis' apparatus.

So far as a 2:1 ratio suffices, my method of varying the volume seems the better, and, indeed, not to admit of improvement.

In the manometric arrangements it would seem that both methods are abundantly accurate, so far as the readings are concerned. I am disposed, however, to favour a continual verification of the vacuum by the Töpler pump, and, what is more important, a method of reading which is independent of possible errors arising from irregular refraction at the walls of the manometric tubes.

It may be remarked that, with a partial exception in the case of CO_2, M. Chappuis' work relates to pressures *above* atmosphere.

Other papers which have appeared since my Preliminary Notice are those of M. Guye, working both alone and with the assistance of collaborators†. Several of these relate to the atomic weight of nitrogen and insist on the discrepancy between the number resulting from density and that of Stas. Among the methods employed is that of decomposing nitrous oxide by an incandescent iron wire and comparing the original volume with the residual

* "Nouvelles Études sur les Thermomètres à gaz," *Extrait du tome* XIII. *des Travaux et Mémoires du Bureau International des Poids et Mesures.*

† *C. R.* April 25, May 16, June 13, July 4, October 31, 1904.

nitrogen.   In my hands* this method failed to give good results, in consequence, apparently, of the formation of higher oxides of nitrogen.

P.S., *March* 6.—Some observations upon Ammonia may here be appended.
The gas was evolved (almost without warmth) from the solution in water,
and was dried by very slow passage over fragments of caustic potash.   The
precautions mentioned under nitrous oxide were here followed with more
minute care.   The glass surfaces were in contact with the gas for weeks,
either at half atmospheric or whole atmospheric pressure, and the observations
at full pressure were not commenced until that pressure had prevailed for
a day or more.   On the reduction of pressure to the half atmospheric,
ammonia was sensibly liberated from the walls, and perhaps from the cement
in the manometer.   In the observations to be compared, the *same interval*
was allowed to elapse between the reduction of pressure and the corresponding
readings, whether the big bulbs were in use or not.   Any anomalies not
dependent upon the walls of the big bulbs themselves are thus eliminated.
In addition to commercial ammonia, a special sample prepared by Dr Scott
in accordance with Stas' directions was employed.   It will be seen that there
is no certain difference between the results from the two kinds.

The departure from Boyle's law, in this case, is almost more than can be
provided for in the side apparatus.   It became therefore necessary to allow
a small difference of temperature between the high- and low-pressure
observations such as would somewhat prejudice the accuracy of the results,
were it possible to expect the attainment of the same high degree of accuracy
as for the less condensable gases.   Under the actual circumstances the
variation of temperature was of no importance.

AMMONIA.

Date	Source	Temperature	B
		°C.	
December 31, 1904, January 4, 1905 ...	Commercial	10·3	1·00647
„ 28, 1904, „ 5, 1905 ...	„	9·2	1·00631
Mean...........................	Commercial	9·7	1·00639
January 23, 25, 1905........................	Scott's	9·6	1·00630
„ 24, 27, 1905........................	„	9·8	1·00617
Mean...........................	Scott's	9·7	1·00624
Mean of all		9·7	1·00632

* *Roy. Soc. Proc.* Vol. LXII. p. 204, 1897; *Scientific Papers*, Vol. IV. p. 350.

304.

ON THE PRESSURE OF GASES AND THE EQUATION
OF VIRIAL.

[*Philosophical Magazine*, Vol. IX. pp. 494—505, 1905.]

IF m be the mass of a particle, V its velocity, p the pressure and v the volume of the body composed of the particles, the virial equation is

$$\tfrac{1}{2}\Sigma m V^2 = \tfrac{3}{2}pv + \tfrac{1}{2}\Sigma \rho \phi (\rho), \quad \dots\dots\dots\dots\dots\dots(1)$$

where further ρ denotes the distance between two particles at the moment under consideration, and $\phi(\rho)$ the mutual force, assumed to depend upon ρ only. If the mutual forces can be neglected, either because they are non-existent or for some other reason, (1) coincides with Boyle's law, since the kinetic energy is supposed to represent temperature (T).

According to some experimenters, among whom may be especially mentioned Ramsay and Young, the relation between pressure and temperature at constant volume is in fact linear, or

$$p = T\psi (v) + \chi (v); \quad \dots\dots\dots\dots\dots\dots\dots(2)$$

and it is of interest to inquire whether such a form is to be expected on theoretical grounds, when $\phi (\rho)$ can no longer be neglected. It has indeed been maintained* that (2) is a rigorous consequence of the general laws of thermodynamics and of the hypothesis that the forces between molecules are functions of the distance only. The argument proceeded upon the assumption that the distances of the particles, and therefore the mutual forces between them, remain constant when the temperature changes, provided only that the volume of the body is maintained unaltered. According to this the virial term in (1) is a function of volume only, so that (1) reduces to (2), with $\psi (v)$ proportional to v^{-1}. But, as Boltzmann pointed out, the assumption is unfounded, and in fact inconsistent with the fundamental principles of the molecular theory. The molecules are not at rest but in motion; and when the temperature varies there is nothing to hinder the virial from varying with it.

---

* M. Levy, *C. R.* t. LXXXVII. pp. 449, 488, 554, 649, 676, 826 (1878).

The readiest proof of this assertion is by reference to the case where the molecules are treated as " hard elastic spheres," that is where the force is zero so long as ρ exceeds a certain value (the diameter of the spheres), and then becomes infinite. From the researches of Van der Waals, Lorentz, and Tait it is known that in that case

$$\tfrac{1}{2}\Sigma\rho\,\phi\,(\rho) = -\tfrac{1}{2}\Sigma m V^2 \,.\, b/v, \qquad\qquad\dots\dots\dots\dots\dots(3)$$

where b, denoting four times the total volume of the spheres, is supposed to be small in relation to v. So far from the virial being necessarily independent of temperature, it is here directly proportional to temperature. The introduction of the special value (3) into (1) gives the well-known form

$$p\,(v-b) = \tfrac{1}{3}\Sigma m V^2 = RT, \qquad\qquad\dots\dots\dots\dots\dots(4)$$

in which b is still regarded as small in comparison with v. It is worthy of note that this particular case, although of course sufficient to upset the general argument that the virial is independent of temperature, nevertheless itself conforms to (2), proportionality to T being for this purpose as good as independence of T.

Not only is the linear relation maintained in spite of the forces of collision of elastic spheres when no other forces operate, but it remains undisturbed even when we introduce such forces, provided that they be of the character considered in the theory of capillarity, that is extending to a range which is a large multiple of molecular distances and not increasing so fast with diminishing distance as to make the total effect sensibly dependent upon the positions occupied by neighbours. Under these restrictions symmetry ensures that the resultant force upon a sphere, situated in the interior and not undergoing collision, is zero; and the whole effect of such forces is represented (Young, Laplace, Van der Waals) by an addition to the pressure of a quantity independent of the temperature and inversely proportional to the square of the volume. In Van der Waals' well-known form

$$\left(p + \frac{a}{v^2}\right)(v - b) = RT, \qquad\qquad\dots\dots\dots\dots\dots(5)$$

the relation between p and T is still linear. Even if the particles depart from the spherical form, the virial of collisional and cohesive forces remains a linear function of the temperature[*].

The forces above considered are partly repulsive and partly attractive. Repulsion at a certain degree of proximity seems to be demanded in order to preserve the individuality of molecules and to prevent infinite condensation. It will be remembered that Maxwell proposed a repulsion inversely as the *fifth* power of the distance, partly as the consequence of some faulty experi-

---

[*] "On the Virial of a System of Hard Colliding Bodies," *Nature*, XLV. pp. 80—82 (1891); *Scientific Papers*, III. p. 469.

ments upon the relation of viscosity to temperature and partly no doubt on account of a special facility of calculation upon the basis of this law. So far as viscosity (η) is concerned, its relation to temperature (T) when the force of repulsion varies as ρ^{-n} is readily obtained by the method 'of dimensions*. It appears that

$$\eta \propto T^{\frac{n+3}{2n-2}}. \quad \dots\dots\dots\dots\dots\dots\dots\dots(6)$$

The case of sudden collisions may be represented by taking $n = \infty$, so that

$$\eta \propto T^{\frac{1}{2}}; \quad \dots\dots\dots\dots\dots\dots\dots\dots(7)$$

while if $n = 5$ $\qquad \eta \propto T. \quad \dots\dots\dots\dots\dots\dots\dots\dots(8)$

According to experiments on the more permanent gases n would vary from ·68 for hydrogen to ·81 for argon; but Sutherland's law †

$$\eta \propto \frac{T^{\frac{1}{2}}}{1 + C/T} \quad \dots\dots\dots\dots\dots\dots\dots\dots(9)$$

probably represents the facts better than (6), whatever value may be assigned to n. According to the theory of corresponding states, C should be proportional to the critical temperature when we pass from one gas to another.

A similar application of the method of dimensions will give interesting information respecting the virial, when the force of repulsion is

$$\phi(\rho) = -\mu\rho^{-n}. \quad \dots\dots\dots\dots\dots\dots\dots(10)$$

The virial is a definite function of N the number of molecules, m the mass of each molecule, V the velocity of mean square on which the temperature depends, μ the force at unit distance, and v the volume of the containing vessel. Of these quantities the virial is of the dimensions of energy, N has none, m is a mass simply, V is a velocity, v a volume, while μ has the dimensions

$$\text{mass} \times (\text{length})^{n+1} \times (\text{time})^{-2}.$$

Hence if we suppose that the virial varies as v^{-s}, we find that it must be proportional to

$$(mV^2)^{\frac{n-3s-1}{n-1}} . \mu^{\frac{3s}{n-1}} . v^{-s}; \quad \dots\dots\dots\dots\dots(11)$$

or since mV^2 represents temperature,

$$T^{\frac{n-3s-1}{n-1}} . \mu^{\frac{3s}{n-1}} . v^{-s}. \quad \dots\dots\dots\dots\dots(12)$$

For example, if $s = 0$,

$$\Sigma\rho\phi(\rho) \propto T, \quad \dots\dots\dots\dots\dots\dots\dots(13)$$

---

* *Proceedings Royal Society*, LXVI. p. 68 (1900); *Scientific Papers*, IV. p. 453.
† *Phil. Mag.* Vol. XXXVI. p. 513 (1893).

whatever n may be. Hence a term in the virial equation independent of volume must be proportional to temperature, as in (1). Again, if $s = 1$,

$$\Sigma \rho \phi(\rho) \propto v^{-1} \cdot T^{\frac{n-4}{n-1}}. \quad \dots\dots\dots\dots\dots\dots(14)$$

Of this we have already had examples, both the virial terms in Van der Waals' equation being proportional to v^{-1}. The first, representing the virial of collisional forces, corresponds in (14) to $n = \infty$, giving proportionality to T. The second is independent of T and can be reconciled with (14) only by supposing $n = 4$. It might seem that in a rare gas, whenever the virial depends sensibly upon what occurs during the encounters of simple pairs of molecules, there must be proportionality to v^{-1}, so that (14) would apply. If, as Maxwell supposed, $n = 5$,

$$\Sigma \rho \phi(\rho) \propto v^{-1} \cdot T^{\frac{1}{2}}, \quad \dots\dots\dots\dots\dots\dots(15)$$

in agreement with a result obtained by Boltzmann for this case. If we retain $n = 5$, but leave the relation to v open, we get from (12)

$$\Sigma \rho \phi(\rho) \propto v^{-s} \cdot T^{1 - \frac{3}{4}s}. \quad \dots\dots\dots\dots\dots(16)$$

If we now discard the supposition that the dependence upon v follows the law of v^{-s}, we may interpret (16) to mean that considered as a function of v and T, the virial is limited to the form

$$\Sigma \rho \phi(\rho) = T \cdot F(v T^{\frac{3}{4}}), \quad \dots\dots\dots\dots\dots(17)$$

F denoting an arbitrary function of the *single* variable $vT^{\frac{3}{4}}$.

And more generally, whatever n may be, we find from (12) that the virial is limited to the form

$$\Sigma \rho \phi(\rho) = T \cdot F\left(\frac{v T^{\frac{3}{n-1}}}{\mu^{\frac{3}{n-1}}}\right). \quad \dots\dots\dots\dots(18)$$

A further generalization may be made by discarding altogether the position that $\phi(\rho)$ is represented by any power of ρ. In this case it is convenient to write

$$\phi(\rho) = -\mu' f(\rho/a), \quad \dots\dots\dots\dots\dots\dots(19)$$

where a is a linear quantity. Here f itself may be supposed to be of no dimensions, while μ' has the dimensions of a force. The virial is a function of μ', a, m, V, v; and since its dimensions are those of energy, *i.e.* of mV^2 or T, we may write

$$\Sigma \rho \phi(\rho) = T \cdot F(\mu', a, m, V, v),$$

where F is of no dimensions. It is easy to see that μ', m, and V^2 can occur only in the combination μ'/mV^2 or μ'/T. To make this of no dimensions, we introduce the factor a. Thus F becomes a function of a, v, and $\mu'a/T$, in which again v can occur only in the form a^3/v. Accordingly

$$\Sigma\rho\phi(\rho) = T \cdot F\left(\frac{a^3}{v},\ \frac{\mu'a}{T}\right), \quad \dots\dots\dots\dots\dots(20)$$

F being in general an arbitrary function of *two* variables.

From (20) we may fall back on (18) by the consideration that in accordance with (10) μ' and a can occur only in the combination $\mu'a^n$.

It may be well to remark that the method of dimensions does not tell us whether or no an available solution can be deduced from particular assumptions. What it teaches us is the form which an available solution must assume. For example, equation (14) gives the form of the term in the virial proportional to v^{-1}, under the law of force (10); and nothing has been said as to any restriction upon the value of n. But it is easy to see that n must in fact be greater than 4. Otherwise the integral representing the virial relating to a given particle would not be convergent. We have to consider

$$\int \rho\phi(\rho)\,\rho^2 d\rho$$

with infinity for the upper limit, and this diverges unless n exceed 4.

It is not to be expected that any law included under (10) could represent with completeness the mutual action of the particles of a gas. Under it no provision can be made for repulsion at small distances and attraction at greater ones. And when $n > 4$, the aggregate virial depends too much upon the encounters which take place at exceedingly small distances.

If, as for both the virial terms in Van der Waals' formula, there be proportionality to v^{-1}, (20) becomes

$$\Sigma\rho\phi(\rho) = \frac{a^3 T}{v}\ F\left(\frac{\mu'a}{T}\right), \quad \dots\dots\dots\dots\dots(21)$$

or, if we prefer it,

$$\Sigma\rho\phi(\rho) = \frac{\mu'a^4}{v}\ F\left(\frac{\mu'a}{T}\right), \quad \dots\dots\dots\dots\dots(22)$$

F in both cases denoting an arbitrary function. According to Van der Waals F in (21) is a linear function, the constant part giving the collisional virial and the second term the cohesional virial which is independent of T. Except for one consideration to be mentioned presently, there would appear to be good reason for supposing the virial of a rare gas to be proportional to v^{-1}; but on the other hand it is doubtful whether the cohesional forces are altogether of the kind supposed by Laplace and Van der Waals. We should expect the cohesional virial to be more directly influenced by the approaches of molecules during an encounter; and on the experimental side D. Berthelot has shown cause for preferring to that of Van der Waals the Rankine and

Clausius form, in which a factor T is introduced in the denominator. The most natural extension of the formula would be by substituting a quadratic for a linear form of F in (21). We should then write

$$\tfrac{1}{2}\Sigma\rho\,\phi\,(\rho) = \frac{3a^3}{2v}\left(-AT + B\mu'a + C\,\frac{\mu'^2a^2}{T}\right),\quad\ldots\ldots\ldots\ldots(23)$$

A, B, C being arbitrary constants; and the pressure equation, when written after Van der Waals' manner with neglect of v^{-2}, becomes

$$\left\{p + \frac{a^3}{v^2}\left(B\mu'a + \frac{C\mu'^2a^2}{T}\right)\right\}\left\{v - \frac{a^3A}{R}\right\} = RT.\quad\ldots\ldots\ldots\ldots(24)$$

As has already been said, Van der Waals' form corresponds to $C = 0$. On the other hand, the Rankine and Clausius form requires that $B = 0$, while C remains finite. It will be evident that the two alternatives differ fundamentally. According to the latter the cohesional terms tend to vanish when T is sufficiently increased.

If the cohesional terms are to vanish when T is infinite, the forces concerned must be of an entirely different character from that contemplated in Laplace's and Van der Waals' theory. It has been suggested by Sutherland[*] that the forces may be of electric origin and in themselves (except during actual collision) as much repulsive as attractive. This is not inconsistent with the preponderance of attraction in the final result. "There is this fundamental distinction in the effects of attractive and repulsive forces whose strength decreases with increasing distance, that the attractive forces by their own operation tend to increase themselves, while the repulsive forces tend to decrease themselves." The forces contemplated by Sutherland are such as are due to electric or magnetic doublets, but a rather simpler illustration may be arrived at by retaining the single character of the centres of force, and supposing them to be as much positive as negative, under the usual electrical law that similars repel while opposites attract one another. When T is infinite, so that the paths are not influenced by the forces, the cohesional virial will disappear, but it may become finite as the temperature falls and room is given for the attractive forces to assert their advantage. There is nothing in the argument upon which (21) was founded which is interfered with by the occurrence of the two kinds of particles, and it would seem that F must then become an *even* function of μ', so that in (23) $B = 0$.

As stated, the above argument is probably not quite legitimate, inasmuch as according to (19) a reversal of μ' would imply a reversal of the collisional forces as well as of those which operate at greater distances. The introduction of the two sorts of particles is not supposed to alter the repulsive forces called into play during actual collision. I believe, however, that the

* *Phil. Mag.* Vol. IV. p. 625 (1902).

instantaneous collisional forces may be omitted from (19). The effect of the collisions may be defined without reference to any datum having dimensions other than a, representing the radius of a sphere. The collisions being thus, as it were, already provided for, the argument remains that the virial must be a definite function of N, m, V, μ', a, v, of which N need not be regarded, the force (outside actual collision) being given by (19). Equation (21) then follows as before with its approximate form (23). If we now suppose that the particles are repellent as much as attractive, (19) may be written

$$\phi(\rho) = \pm\, \mu' f(\rho/a); \quad \dots\dots\dots\dots\dots\dots\dots\dots(25)$$

and, since odd powers of μ' are now excluded, $B = 0$ in (23), (24).

We have thus discovered a possible theoretical foundation for the empirical conclusion that T should be introduced into the denominator of the cohesional virial, and it would seem to follow conversely that, if the empirical conclusion is correct, the forces must be intrinsically as much repellent as attractive. This argument may be regarded as a strong confirmation of Sutherland's idea, though a question remains as to how the attraction asserts its superiority over repulsion.

In the above argument the particles are regarded as simple centres of force, half of them being "positive" and half "negative." The advantage is that the form may still be treated as spherical, so that the collisions may be assimilated to those of "elastic spheres." But a polar constitution, such that the positive and negative elements are combined in every particle, is certainly more probable. This will introduce, as another linear datum, the distance between the poles, and the collisions will admit of greater variety. Moreover, there is now kinetic energy of rotation as well as of translation. However, since the kinetic energies are proportional, the argument remains unaffected, so far as it relates to the dependence of the virial of a given gas upon volume and temperature, and the Rankine-Clausius form (24) with $B = 0$ still obtains.

As to the preponderance of attractive over repulsive virial, I think that the conclusion is correct, although Sutherland's argument, quoted above, omits reference to the essential consideration of the *time* for which any particular value of the virial prevails. If we fix our attention upon a pair of particles, acting as simple centres of force, which encounter one another, the corresponding virial varies from moment to moment, but the mean contribution to the total may be represented by

$$\int \phi(\rho)\, \rho\, dt,$$

the integration being taken over the whole range for which $\rho\phi(\rho)$ is sensible. Since only relative motion is in question, the centre of gravity of the two particles may be supposed to be at rest and the problem becomes one of "central forces." In the usual notation we have

$$\frac{d^2r}{dt^2} - r\left(\frac{d\theta}{dt}\right)^2 = P, \quad r^2\frac{d\theta}{dt} = h, \quad\text{.................(26)}$$

so that

$$\int P.r.dt = \left[r\frac{dr}{dt}\right] - \int \left(\frac{dr}{dt}\right)^2 dt - \int r^2\left(\frac{d\theta}{dt}\right)^2 dt$$

$$= \left[r\frac{dr}{dt}\right] - \int v^2 dt = \left[r\frac{dr}{dt}\right] - \int v\,ds, \quad\text{...........(27)}$$

v denoting the resultant velocity. At the upper limit dr/dt is equal to the velocity at ∞, say V, and at the lower limit $dr/dt = -V$. Hence

$$\int P.r.dt = 2rV - \int v\,ds, \quad\text{....................(28)}$$

so that the mean virial is closely connected with the "action" in the orbit.

For a simple illustration it will be more convenient to make θ the independent variable. Thus by (26)

$$\int P.r.\,dt = \frac{1}{h}\int P.r^3.d\theta. \quad\text{....................(29)}$$

Suppose for example that

$$P = \mu r^{-3}. \quad\text{....................(30)}$$

Then

$$\int P.r.\,dt = \frac{\mu}{h}\int d\theta = \frac{\mu}{h}\theta, \quad\text{....................(31)}$$

where θ represents twice the vectorial angle between the initial asymptote and the apse. If h be given, a comparison between repellent and attractive forces (μ given in magnitude but variable in sign) shows that (31) is greater in the case of attraction (fig. 1), so that if attractive and repellent forces occur indifferently the average effect corresponds to attraction. In the case of the particular law (30) we can carry out the calculation. If, as usual, $u = r^{-1}$, the equation of the orbit is

$$\frac{d^2u}{d\theta^2} + u = \frac{\mu}{h^2}u, \quad\text{....................(32)}$$

μ being positive in the case of attraction; whence, if μ be small,

$$u = U\sin\sqrt{(1 - \mu h^{-2})}\,\theta. \quad\text{....................(33)}$$

In (33) $u = 0$, or $r = \infty$, when $\theta = 0$ and when

$$\theta = \pi \div \sqrt{(1 - \mu h^{-2})}; \quad\text{....................(34)}$$

so that from (31)

$$\int P.r.dt = \frac{\mu\pi}{\sqrt{(h^2 - \mu)}}. \quad\text{....................(35)}$$

The solutions (33), (35) hold if μ be numerically less than h^2, and (35) shows that when μ changes sign the virial of attraction preponderates. This conclusion is accentuated by the consideration of what occurs if μ exceed h^2 numerically. Equations (33), (35) still hold if μ be negative, *i.e.* if the force be repulsive. But when μ is positive, the form changes. Thus if $\mu = h^2$, we have

$$u = U\theta, \quad \dots\dots\dots\dots\dots\dots\dots\dots\dots\dots(36)$$

and neither θ in (31) nor the virial has a finite value. The like remains true when $\mu > h^2$.

In the above example Pr^3 remains constant, and the preponderance of attraction over repulsion depends upon the greater vectorial angle in the former case. If Pr^3, instead of remaining constant, continually increases with diminishing r, the preponderance of attraction follows *a fortiori*.

Fig. 1.   Fig. 2.

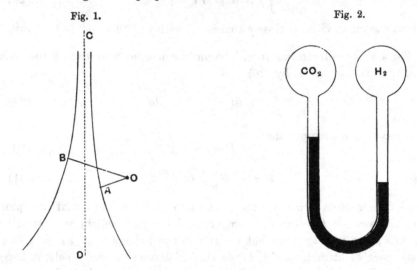

O, centre of force; CD, asymptote; A, B, apses.

A particular case of (32) which arises when $\mu = h^2$ should be singled out for especial notice, *i.e.* the case of circular motion for which $u = $ constant. The attracting particles then revolve round one another in perpetuity, and the virial is infinite in comparison with that of an ordinary encounter. It is this possible occurrence of re-entrant orbits which causes hesitation as to the accuracy with which we may assume the virial of a rare gas to be inversely as the volume. It seems to be generally supposed (see, for example, Meyer's *Kinetic Theory of Gases*, § 4) that if a gas be rare enough no appreciable pairing can occur. But the question is not as to the frequency with which new pairs may form, but as to the relative number of them in existence at any time. It is easy to recognize that the coupling or the severance of a pair of particles cannot occur of itself, but requires always the cooperation of a third particle. If the gas is very rare, no doubt there are few opportunities for the formation of fresh pairs, but for the same reason those already formed have a higher degree of permanence. On the whole it would appear that the number of pairs in existence at any moment is independent of the volume v of a rare gas, and the same would be true of the corresponding virial. At this rate we should have terms in the virial which by (20) come under the form

$$T . F \left(\frac{\mu' a}{T} \right). \dots\dots\dots\dots\dots\dots\dots\dots\dots\dots\dots(37)$$

It will be remarked that if these terms in the virial, independent of v, are sensible, the density of the gas will depart from Avogadro's rule, however greatly it may be rarefied. In the case of elastic spheres, which come into collision when their centres approach to a certain distance, there is naturally a limit to the magnitude of the attraction, and then pairing becomes impossible if the velocity be sufficiently great. Any departure from Avogadro's rule at high rarefactions would thus tend to disappear as the temperature rises.

The behaviour of mere centres of force, which may approach one another without limit, appears to follow a different course. Taking for example the power law of (10), we see from (18) that for any part of the virial which is independent of v, the function F must be constant, so that the virial is proportional to T and independent of μ.

To return to the question with which we started, there seems good reason to doubt that the relation of pressure to temperature with volume constant is accurately linear, even at high rarefactions. On the other hand, it is clear that this relation is approximately satisfied; and the natural course would be to take it as a foundation, determining the functions χ and ψ in (2), as well as the function of v and T jointly which may be required in supplement. As regards the latter part of the question, a differential arrangement in which two gases, say CO_2 and H_2, are balanced against one another at the same temperature, would appear to offer advantages. This is shown diagrammatically in fig. 2, where the two gas-reservoirs are connected by a U-tube containing mercury. According to Boyle's law, even as modified by the introduction of a co-volume, the mercury may stand in the U-tube at fixed marks at the *same level*, in spite of variations of temperature affecting both bulbs alike. And under the more general law (2) the same fixity of the mercury thread can be attained, though now with the extremities at different levels. With such an arrangement the departure from (2) becomes a matter of direct observation, and so long as *uniformity* of temperature is secured, a precise measurement of it, or of the *total* pressure, is of secondary importance. Useful results would probably require a total pressure of four or five atmospheres.

305.

THE DYNAMICAL THEORY OF GASES AND OF RADIATION.

[*Nature*, Vol. LXXI. p. 559; Vol. LXXII. pp. 54, 55; pp. 243, 244, 1905*.]

IN Mr Jeans' valuable work upon this subject† he attacks the celebrated difficulty of reconciling the "law of equipartition of energy" with what is known respecting the specific heats of gases. Considering a gas the molecules of which radiate into empty space, he shows that in an approximately steady state the energy of vibrational modes may bear a negligible ratio to that of translational and rotational modes.

I have myself speculated in this direction; but it seems that the difficulty revives when we consider a gas, not radiating into empty space, but bounded by a perfectly reflecting enclosure. There is then nothing of the nature of dissipation; and, indeed, the only effect of the appeal to the æther is to bring in an infinitude of new modes of vibration, each of which, according to the law, should have its full share of the total energy. I cannot give the reference, but I believe that this view of the matter was somewhere‡ expressed, or hinted, by Maxwell.

We know that the energy of æthereal vibrations, corresponding to a given volume and temperature, is not infinite or even proportional to the temperature. For some reason the higher modes fail to assert themselves§. A full comprehension here would probably carry with it a solution of the specific heat difficulty.

I am glad to have elicited the very clear statement of his view which Mr Jeans gives in *Nature* of April 27. In general outline it corresponds

* The reader interested in this subject should refer to Mr Jeans' letters, *Nature*, Vol. LXXI. p. 607; Vol. LXXII. pp. 101, 102; pp. 293, 294.

† *The Dynamical Theory of Gases*, Camb. Univ. Press, 1904.

‡ Mr Jeans refers to Maxwell's *Collected Works*, Vol. II. p. 433.

§ Compare "Remarks upon the Law of Complete Radiation," *Phil. Mag.* Vol. XLIX. p. 539, 1900. [*Scientific Papers*, Vol. IV. p. 483.]

pretty closely with that expressed by O. Reynolds in a British Association discussion at Aberdeen (*Nature*, Vol. XXXII. p. 534, 1885). The various modes of molecular motion are divided into two sharply separated groups. Within one group, including the translatory modes, equipartition of energy is supposed to establish itself within a small fraction of a second; but between the modes of this group and those of vibration included in the other group, equipartition may require, Mr Jeans thinks, millions of years. Even if minutes were substituted for years, we must admit, I think, that the law of equipartition is reconciled with all that is absolutely proved by our experiments upon specific heat, which are, indeed, somewhat rough in all cases, and especially imperfect in so far as they relate to what may happen over long intervals of time.

As I have already suggested, it is when we extend the application of the law of equipartition to the modes of æthereal vibration that the difficulties thicken, and this extension we are bound to make. The first question is as to the consequences of the law, considered to be applicable, after which, if necessary, we may inquire whether any of these consequences can be evaded by supposing the equipartition to require a long time for its complete establishment. As regards the first question, two things are at once evident. The energy in any particular mode must be proportional to θ, the absolute temperature. And the number of modes corresponding to any finite space occupied by the radiation, is infinite. Although this is enough to show that the law of equipartition cannot apply in its integrity, it will be of interest to follow out its consequences a little further. Some of them were discussed in a former paper*, the argument of which will now be repeated with an extension designed to determine the coefficient as well as the law of radiation.

As an introduction, we consider the motion of a stretched string of length l, vibrating transversely in one plane. If a be the velocity of propagation, ξ the number of subdivisions in any mode of vibration, the frequency f is given by

$$f = a\xi/2l. \dots\dots\dots\dots\dots\dots\dots(1)$$

A passage from any mode to the next in order involves a change of unity in the value of ξ, or of $2lf/a$. Hence if e denote the kinetic energy of a single mode, the law of equipartition requires that the kinetic energy corresponding to the interval df shall be

$$2le/a \cdot df. \dots\dots\dots\dots\dots\dots(2)$$

In terms of λ the wave-length, (2) becomes

$$2le/\lambda^2 \cdot d\lambda. \dots\dots\dots\dots\dots\dots(3)$$

---

* "Remarks upon the Law of Complete Radiation," *Phil. Mag.* Vol. XLIX. p. 539, June, 1900. [*Scientific Papers*, Vol. IV. p. 483.]

This is for the whole length of the string. The longitudinal *density* of the kinetic energy is accordingly

$$2e/\lambda^2 . d\lambda \dots\dots\dots\dots(4)$$

In each mode the potential energy is (on the average) equal to the kinetic, so that if we wish to reckon the whole energy, (4) must be doubled. Another doubling ensues when we abandon the restriction to one plane of vibration; and finally for the total energy corresponding to the interval from λ to $\lambda + d\lambda$ we have

$$8e/\lambda^2 . d\lambda \dots\dots\dots\dots(5)$$

When we proceed to three dimensions, and consider the vibrations within a cube of side l, subdivisions may occur in three directions. In place of (1)

$$f = a/2l . \sqrt{(\xi^2 + \eta^2 + \zeta^2)}, \dots\dots\dots(6)$$

where ξ, η, ζ may assume any integral values. The next step is to ascertain what is the number of modes which corresponds to an assigned variation of f.

If the integral values of ξ, η, ζ be regarded as the coordinates of a point, the whole system of points constitutes a cubic array of volume-density unity. If R be the distance of any point from the origin,

$$R^2 = \xi^2 + \eta^2 + \zeta^2 ;$$

and the number of points between R and $R + dR$, equal to the included volume, is

$$4\pi R^2 dR.$$

Hence the number of modes corresponding to df is

$$4\pi (2l/a)^3 f^2 df,$$

or in terms of λ

$$4\pi . 8l^3 . \lambda^{-4} d\lambda \dots\dots\dots(7)$$

If e be the kinetic energy in each mode, then the kinetic energy corresponding to $d\lambda$ and to unit of volume is

$$32 . \pi . e . \lambda^{-4} d\lambda . \dots\dots\dots(8)$$

Since, as in the case of the string, we are dealing with transverse vibrations, and since the whole energy is the double of the kinetic energy, we have finally

$$128 . \pi . e . \lambda^{-4} d\lambda \dots\dots\dots(9)$$

as the total energy of radiation per unit of volume corresponding to the interval from λ to $\lambda + d\lambda$, and in (9) e is proportional to the absolute temperature θ.

Apart from the numerical coefficient, this is the formula which I gave in the paper referred to as probably representing the truth when λ is large, in place of the quite different form then generally accepted. The suggestion was soon confirmed by Rubens and Kurlbaum, and a little later Planck (*Drude Ann.* Vol. IV. p. 553, 1901) put forward his theoretical formula, which seems to agree very well with the experimental facts. This contains two constants, h and k, besides c, the velocity of light. In terms of λ it is

$$E\,d\lambda = \frac{8\pi ch}{\lambda^5}\,\frac{d\lambda}{e^{ch/k\lambda\theta}-1}, \quad\ldots\ldots\ldots\ldots\ldots(10)$$

reducing when λ is great to

$$E\,d\lambda = 8\pi k\theta\lambda^{-4}d\lambda, \quad\ldots\ldots\ldots\ldots\ldots\ldots(11)$$

in agreement with (9). $E\,d\lambda$ here denotes the volume-density of the energy of radiation corresponding to $d\lambda$.

A very remarkable feature in Planck's work is the connection which he finds between radiation and molecular constants. If N be the number of gaseous molecules in a cubic centimetre at $0°$ C. and under a pressure of one atmosphere,

$$k = \frac{1\text{·}013 \times 10^6}{273\,N}.\quad\ldots\ldots\ldots\ldots\ldots\ldots(12)$$

Though I failed to notice it in the earlier paper, it is evident that (9) leads to a similar connection. For e, representing the kinetic energy of a single mode at temperature θ, may be identified with one-third of the average kinetic energy of a gaseous molecule at that temperature. In the virial equation, if N be the total number of molecules,

$$\tfrac{3}{2}pv = \Sigma\tfrac{1}{2}mV^2 = 3Ne,$$

so that

$$e = pv/2N. \quad\ldots\ldots\ldots\ldots\ldots\ldots\ldots\ldots\ldots(13)$$

If we apply this to one cubic centimetre of a gas under standard conditions, N has the meaning above specified, $v = 1$, and $p = 1\text{·}013 \times 10^6$ C.G.S. Accordingly, at $0°$ C.

$$e = 1\text{·}013 \times 10^6/2N,$$

and at $\theta°$

$$e = \frac{1\text{·}013 \times 10^6 \times \theta}{2 \times 273\,N}. \quad\ldots\ldots\ldots\ldots\ldots(14)$$

Introducing this into (9), we get as the number of ergs per cubic centimetre of radiation

$$\frac{64.\pi.1\text{·}013.10^6.\theta.d\lambda}{273.N.\lambda^4}, \quad\ldots\ldots\ldots\ldots\ldots(15)$$

θ being measured in centigrade degrees. This result is eight times as large as that found by Planck. If we retain the estimate of radiation used in his calculations, we should deduce a value of N eight times as great as his, and probably greater than can be accepted. [See below.]

A critical comparison of the two processes would be of interest, but not having succeeded in following Planck's reasoning I am unable to undertake it. As applying to all wave-lengths, his formula would have the greater value if satisfactorily established. On the other hand, the reasoning which leads to (15) is very simple, and this formula appears to me to be a necessary consequence of the law of equipartition as laid down by Boltzmann and Maxwell. My difficulty is to understand how another process, also based upon Boltzmann's ideas, can lead to a different result.

According to (15), if it were applicable to all wave-lengths, the total energy of radiation at a given temperature would be infinite, and this is an inevitable consequence of applying the law of equipartition to a uniform structureless medium. If we were dealing with elastic solid balls colliding with one another and with the walls of a containing vessel of similar constitution, energy, initially wholly translational, would be slowly converted into vibrational forms of continually higher and higher degrees of subdivision. If the solid were structureless, this process would have no limit; but on an atomic theory a limit might be reached when the subdivisions no longer included more than a single molecule. The energy, originally mechanical, would then have become entirely thermal.

Can we escape from the difficulties, into which we have been led, by appealing to the slowness with which equipartition may establish itself? According to this view, the energy of radiation within an enclosure at given temperature would, indeed, increase without limit, but the rate of increase after a short time would be very slow. If a small aperture is suddenly made, the escaping radiation depends at first upon how long the enclosure has been complete. In this case we lose the advantage formerly available of dividing the modes into two sharply separated groups. Here, on the contrary, we have always to consider vibrations of such wave-lengths as to bear an intermediate character. The kind of radiation escaping from a small perforation must depend upon the size of the perforation.

Again, does the postulated slowness of transformation really obtain? Red light falling upon the blackened face of a thermopile is absorbed, and the instrument rapidly indicates a rise of temperature. Vibrational energy is readily converted into translational energy. Why, then, does the thermopile not itself shine in the dark?

It seems to me that we must admit the failure of the law of equipartition in these extreme cases. If this is so, it is obviously of great importance to ascertain the reason. I have on a former occasion (*Phil. Mag.* Vol. XLIX. p. 118, 1900)* expressed my dissatisfaction with the way in which great potential energy is dealt with in the general theory leading to the law of equipartition.

* [This Collection, Vol. IV. p. 451.]

In *Nature*, May 18, I gave a calculation of the co-efficient of complete radiation at a given absolute temperature for waves of great length on principles laid down in 1900, and it appeared that the result was eight times as great as that deduced from Planck's formula for this case. In connection with similar work of his own Mr Jeans (*Phil. Mag.* July [1905]) has just pointed out that I have introduced a redundant factor 8 by counting negative as well as positive values of my integers ξ, η, ζ.

I hasten to admit the justice of this correction. But while the precise agreement of results in the case of very long waves is satisfactory so far as it goes, it does not satisfy the wish expressed in my former letter for a comparison of processes. In the application to waves that are not long, there must be some limitation on the principle of equipartition. Is there any affinity in this respect between the ideas of Prof. Planck and those of Mr Jeans?

[1911. Since the date of these letters further valuable work has been done by Planck, Jeans, Lorentz, Larmor, Einstein and others. But I suppose the question can hardly yet be considered settled.]

306.

AN OPTICAL PARADOX.

[*Philosophical Magazine*, Vol. IX. pp. 779—781, 1905.]

CONSIDER the following combination:—A point-source A of approximately homogeneous light (λ) is focused by the lens LL upon the object-glass of a telescope T. In its turn the telescope is focused upon L. According to geometrical optics the margin of the lens L should be seen sharp by an eye applied to the telescope; but when we consider the limitation of aperture at the object-glass of the telescope, we come to the conclusion that the definition must be very bad. The image of A at C constitutes the usual diffraction pattern of which most of the light is concentrated in the central disc. The diameter of this disc is of the order $\lambda \cdot LC/LL$. If this be regarded as the

effective aperture of T, the angular resolving power will be found by dividing λ by the above quantity, giving LL/LC; so that the entire angular magnitude of the lens LL is on the limits of resolving power.

If this be admitted, we may consider next the effect of enlarging the source A, hitherto supposed to be infinitely small. If the process be carried far enough, the object-glass of T will become filled with light, and we may expect the natural resolving power to be recovered. But here we must distinguish. If the enlarged source at A be a self-luminous body, such as a piece of white-hot metal or the carbon of an electric arc, no such conclusion will follow. There is no phase-relation between the lights which act at different parts of the object-glass, and therefore no possibility of bringing into play the interferences upon which the advantage of a large aperture depends. It appears, therefore, that however large the self-luminous source at A may be, the definition is not improved, but remains at the miserably low level already specified. If, however, the source at A be not a real one, but merely an aperture through which light from real sources passes, the case may be altered.

Returning to the extended self-luminous source, we see that the inefficiency depends upon the action of the lens L. If the glass be removed from its seat, so that A is no longer focused upon the object-glass, the definition must recover.

I do not know how far the above reasoning will seem plausible to the reader, but I may confess that I was at first puzzled by it. I doubt whether any experimenter would willingly accept the suggested conclusion, though he might be unable to point out a weak place in the argument. He would probably wish to try the experiment; and this is easily done. The lens L may be the collimating-lens of an ordinary spectroscope whose slit is backed by a flame. The telescope is removed from its usual place to a distance of say 10 feet and is focused upon L. The slit is at the same time focused upon the object-glass of the telescope. Although the image of the slit is very narrow, the definition of L as seen in the telescope does not appear to suffer, the vertical parts of the circular edge (parallel to the slit) being as well defined as the horizontal parts. If, however, at the object-glass a material screen be interposed provided with a slit through which the image of the first slit can pass, the definition at the expected places falls off greatly, even although a considerable margin be allowed in the width of the second slit.

This experiment gives the clue to the weak place in the theoretical argument. It is true that the greater part of the light ultimately reaching the eye passes through a very small area of the object-glass; but it does not follow that the remainder may be blocked out without prejudice to the definition of the boundary of the field. In fact, a closer theoretical discussion of the diffraction phenomena leads to conclusions in harmony with experiment.

In the case of a point-source and the complete circular aperture LL, the question turns upon the integral

$$\int_0^\infty J_0(\alpha x)\, J_1(\beta x)\, dx,$$

J_0, J_1 being the Bessel's functions usually so denoted. The integral passes from 0 to $1/\beta$, as α passes through the value β *.

If the aperture of LL be reduced to a narrow annulus, the integral to be considered is

$$\int_0^\infty J_0(\alpha x)\, J_0(\beta x)\, x\, dx.$$

This assumes an infinite value when $\alpha = \beta$ †.

If the apertures be rectangular, the integrals take still simpler forms.

* A theorem attributed to Weber. See Gray and Matthews' *Bessel's Functions*, p. 228.
† See *Theory of Sound*, § 203, equations (14), (16).

307.

THE PROBLEM OF THE RANDOM WALK.

[*Nature*, Vol. LXXII. p. 318, 1905.]

THIS problem, proposed by Prof. Karl Pearson in the current number of *Nature**, is the same as that of the composition of n iso-periodic vibrations of unit amplitude and of phases distributed at random, considered in *Philosophical Magazine*, X. p. 73, 1880; XLVII. p. 246, 1889 (*Scientific Papers*, I. p. 491; IV. p. 370). If n be very great, the probability sought is

$$2n^{-1}e^{-r^2/n}r\,dr.$$

Probably methods similar to those employed in the papers referred to would avail for the development of an approximate expression applicable when n is only moderately great.

---

* [Vol. LXXII. p. 294. "A man starts from a point O and walks l yards in a straight line; he then turns through any angle whatever and walks another l yards in a second straight line. He repeats this process n times. I require the probability that after these n stretches he is at a distance between r and $r+\delta r$ from his starting-point O. The problem is one of considerable interest, but I have only succeeded in obtaining an integrated solution for two stretches. I think, however, that a solution ought to be found, if only in the form of a series in powers of $1/n$, when n is large."]

308.

ON THE INFLUENCE OF COLLISIONS AND OF THE MOTION OF MOLECULES IN THE LINE OF SIGHT, UPON THE CONSTITUTION OF A SPECTRUM LINE.

[Proceedings of the Royal Society, Vol. LXXVI. (A), pp. 440—444, 1905.]

APART from the above and other causes of disturbance, a line in the spectrum of a radiating gas would be infinitely narrow. A good many years ago*, in connection with some estimates by Ebert, I investigated the widening of a line in consequence of the motion of molecules in the line of sight, taking as a basis Maxwell's well-known law respecting the distribution of velocities among colliding molecules, and I· calculated the number of interference-bands to be expected, upon a certain supposition as to the degree of contrast between dark and bright parts necessary for visibility. In this investigation no regard was paid to the collisions; the vibrations issuing from each molecule being supposed to be maintained with complete regularity for an indefinite time.

Although little is known with certainty respecting the genesis of radiation, it has long been thought that collisions act as another source of disturbance. The vibrations of a molecule are supposed to remain undisturbed while a free path is described, but to be liable to sudden and arbitrary alteration of phase and amplitude when another molecule is encountered. A limitation in the number of vibrations executed with regularity necessarily implies a certain indeterminateness in the frequency, that is a dilatation of the spectrum line. In its nature this effect is independent of the Doppler effect—for example, it will be diminished relatively to the latter if the molecules are smaller; but the problem naturally arises of calculating the conjoint action of both causes upon the constitution of a spectrum line. This is the question considered by Mr C. Godfrey in an interesting paper†, upon which it is the principal object

---

* *Phil. Mag.* Vol. xxvii. p. 298, 1889; *Scientific Papers*, Vol. iii. p. 258.

† "On the Application of Fourier's Double Integrals to Optical Problems," *Phil. Trans.* A, Vol. cxcv. p. 329, 1899.

of the present note to comment. The formulæ at which he arrives are some-what complicated, and they are discussed only in the case in which the density of the gas is reduced without limit. According to my view this should cause the influence of the collisions to disappear, so that the results should coincide with those already referred to where the collisions were disregarded from the outset. Nevertheless, the results of the two calculations differ by 10 per cent., that of Mr Godfrey giving a *narrower* spectrum line than the other.

The difference of 10 per cent. is not of much importance in itself, but a discrepancy of this kind involves a subject in a cloud of doubt, which it is desirable, if possible, to dissipate. Mr Godfrey himself characterises the discrepancy as paradoxical, and advances some considerations towards the elucidation of it. I have a strong feeling, which I think I expressed at the time, that the 10-per-cent. correction is inadmissible, and that there should be no ambiguity or discontinuity in passing to the limit of free paths infinitely long. In connection with some other work I have recently resumed the consideration of the question, and I am disposed to think that Mr Godfrey's calculation involves an error respecting the way in which the various free paths are averaged.

The first question is as to the character of the spectrum line corresponding to a regular vibration which extends over a *finite* interval of time. As the energy lying between the limits n and $n + dn$ of frequency (or rather inverse wave-length), Mr Godfrey finds from Fourier's theorem

$$\frac{\sin^2 \pi r n}{n^2}\, dn, \quad \dots\dots\dots\dots\dots\dots\dots\dots\dots\dots\dots(1)$$

r denoting the finite length of the train of waves, and n being measured from that value which would be dominant if r were infinitely long. For the total energy of all wave-lengths we have

$$\int_{-\infty}^{+\infty} \frac{\sin^2 \pi r n}{n^2}\, dn = \pi^2 r. \quad \dots\dots\dots\dots\dots\dots\dots(2)$$

That the total energy should be proportional to r is what we would expect. The maximum coefficient in (1) occurs, of course, when $n = 0$, and is propor-tional to r^2; once proportional to r on account of the greater total energy as given in (2), and again on account of the greater *condensation* of the spectrum as r increases. Expression (1) may be taken to represent the spectrum of the radiation from a single molecule which describes in a given direction and with a given velocity a free path proportional to r. If there be N independent molecules answering to this description, N may be intro-duced as a factor into (1). From this expression Mr Godfrey proceeds to investigate the spectrum corresponding to the aggregate radiation of the gas, integrating first for the different lengths (r) of parallel free paths described with constant velocity, and afterwards for the various component velocities across and in the line of sight, the latter giving rise to the Doppler effect. It is with the first of these integrations that I am more particularly concerned.

In order to effect it, we need to know the probabilities of the various lengths of free path described with given velocity. "Now, Tait has shown that, of all atoms moving with velocity v, a fraction $e^{-f\rho}$ penetrates unchecked to distance ρ where f is [a function of v and of the permanent data of the gas]. From this we see that, of molecules moving with velocity v, a fraction $fe^{-f\rho}d\rho$ have free paths between ρ and $\rho + d\rho$. Now, such a molecule will emit an undisturbed train of waves of length between r and $r + dr$, where $r = \rho \cdot V/v$, and V is the velocity of light. Hence, of all molecules moving with velocity v, a fraction $(fv/V)\,e^{-vfr/V}$ will give free paths between r and $r + dr$. Returning to the expression (1) for the energy of a single train of length r, we see that with the aggregates of molecules now under consideration (definite thwart and line-of-sight velocities) we have for n a proportion of energy

$$\frac{fv}{n^3 V} \int_0^\infty e^{-vfr/V} \sin^2 n\pi r \cdot dr,"$$

or, on effecting the integration,

$$\frac{2\pi^2}{4\pi^2 n^2 + v^2 f^2 / V^2}. \quad \dots\dots\dots\dots\dots\dots(3)^*$$

The next steps are integrations over the various velocities, but it is not necessary to follow them here in detail, inasmuch as the objection which I have to take arises already. It appears to me that what we are concerned with is not the momentary distribution of free paths among the molecules which are describing them, but rather the statistics of the various free paths (described with velocity v) which occur in a relatively long *time*. During this time various free paths occur with frequencies dependent on the lengths. Fix the attention on two of these, one long and one short. They present themselves in certain relative numbers, or say in a certain proportion, and it is with this proportion that we have to do. The other procedure takes, as it were, an instantaneous view of the system and, surveying the molecules, inquires what proportions of them are pursuing free paths of the two lengths under contemplation. It is not difficult to recognize that this is a different question. Of the paths which are described in a given period of time, an instantaneous survey is more likely to hit upon a long one than upon a short one. Thus Mr Godfrey's integration favours unduly the long paths.

---

* Mr Godfrey's expression (iii) differs somewhat from (3). A $4\pi^2$ appears to have been temporarily dropped, but this is not material for my present purpose.

The above consideration indicates that we ought to divide by r previously to integration, that is, evaluate

$$\frac{fv}{n^2 V} \int_0^\infty r^{-1} e^{-vfr/V} \sin^2 n\pi r \, . \, dr. \quad \dots\dots\dots\dots\dots(4)$$

If we write

$$\int_0^\infty r^{-1} e^{-cr} \sin^2 ar \, dr = y,$$

we find

$$\frac{dy}{da} = \frac{2a}{4a^2 + c^2}, \quad \text{and} \quad y = \tfrac{1}{4} \log \frac{4a^2 + c^2}{c^2}.$$

Hence

$$(4) = \frac{fv}{4n^2 V} \log \left\{ 1 + \frac{4n^2 \pi^2 V^2}{v^2 f^2} \right\}, \quad \dots\dots\dots\dots\dots(5)$$

in place of (3).

It must be remarked, however, that an over-valuation of long paths relatively to shorter ones which all correspond to the same velocity would not of itself explain the 10-per-cent. discrepancy; for, when the gas is infinitely rare, all the paths must be considered to be infinitely long, and then the proportion of relatively longer and shorter paths becomes a matter of indifference. In fact (3) should give the correct result in the limit ($f = 0$), even though it be of erroneous form as respects n, provided a suitable function of f and v be introduced as a factor. If we integrate (3) as it stands with respect to n between the limits $-\infty$ and $+\infty$, we obtain

$$\frac{2\pi^2 V}{vf}. \quad \dots\dots\dots\dots\dots\dots\dots\dots(6)$$

But this should certainly be independent of f. I think that if we introduce the factor vf into (3), Mr Godfrey's analysis would then lead to the same result as is obtained by neglecting the influence of collisions *ab initio*.

It may be convenient to recite the constitution and visibility of a spectrum line according to the simple theory, where the Doppler effect is alone regarded. If ξ be the velocity of a molecule in the line of sight, the number of molecules whose velocities in this direction lie between ξ and $\xi + d\xi$ is, by Maxwell's theory,

$$e^{-h\xi^2} d\xi. \quad \dots\dots\dots\dots\dots\dots(7)$$

According to Doppler's principle the reciprocal wave-length of the light received from these molecules is changed from Λ^{-1}, corresponding to $\xi = 0$, to $\Lambda^{-1}(1 + \xi/V)$, V being the velocity of light. If x denote the variation of reciprocal wave-length, $x = \Lambda^{-1} \xi/V$, and the distribution of light in the dilated spectrum line may be taken to be

$$e^{-h V^2 \Lambda^2 x^2} dx. \quad \dots\dots\dots\dots\dots(8)$$

When this light forms interference-bands with relative retardation D, the "visibility" accorded to Michelson's reckoning is expressed by

$$\int_{-\infty}^{+\infty} e^{-hV^2\Lambda^2 x^2} \cos(2\pi Dx)\, dx \div \int_{-\infty}^{+\infty} e^{-hV^2\Lambda^2 x^2}\, dx,$$

that is

$$e^{-\frac{\pi^2 D^2}{hV^2\Lambda^2}}. \quad\dots\dots\dots\dots\dots\dots\dots\dots\dots\dots\dots(9)$$

If v be the velocity of mean square, on which the pressure of the gas depends,

$$v^2 = \frac{3}{2h}. \quad\dots\dots\dots\dots\dots\dots\dots\dots\dots(10)$$

In terms of v the exponent in (8) is

$$-\frac{3V^2\Lambda^2 x^2}{2v^2}, \quad\dots\dots\dots\dots\dots\dots\dots\dots(11)$$

and that in (9) is

$$-\frac{2\pi^2 v^2 D^2}{3V^2\Lambda^2}. \quad\dots\dots\dots\dots\dots\dots\dots\dots(12)$$

309.

ON THE MOMENTUM AND PRESSURE OF GASEOUS VIBRA-TIONS, AND ON THE CONNEXION WITH THE VIRIAL THEOREM.

[*Philosophical Magazine*, Vol. x. pp. 364—374, 1905.]

IN a paper on the Pressure of Vibrations (*Phil. Mag.* III. p. 338, 1902; *Scientific Papers*, v. p. 41) I considered the case of a gas obeying Boyle's law and vibrating within a cylinder in one dimension. It appeared that in consequence of the vibrations a piston closing the cylinder is subject to an additional pressure whose amount is measured by the volume-density of the total energy of vibration. More recently, in an interesting paper (*Phil. Mag.* IX. p. 393, 1905) Prof. Poynting has treated certain aspects of the question, especially the momentum associated with the propagation of progressive waves. Thus prompted, I have returned to the consideration of the subject, and have arrived at some more general results, which however do not in all respects fulfil the anticipations of Prof. Poynting. I commence with a calculation similar to that before given, but applicable to a gas in which the pressure is any arbitrary function of the density.

By the general hydrodynamical equation (*Theory of Sound*, § 253 *a*),

$$\varpi = \int \frac{dp}{\rho} = -\frac{d\phi}{dt} - \tfrac{1}{2} U^2, \qquad \dots\dots\dots\dots\dots(1)$$

where p denotes the pressure, ρ the density, ϕ the velocity-potential, and U the resultant velocity at any point. If we integrate over a long period of time, ϕ disappears, and we see that

$$\int \varpi \, dt + \tfrac{1}{2} \int U^2 dt \qquad \dots\dots\dots\dots\dots\dots(2)$$

retains a constant value at all points of the cylinder. The value at the piston is accordingly the same as the mean value taken over the length of the cylinder.

If p_1, ρ_1 denote the pressure and density at the piston, and p_0, ρ_0 the pressure and density that would prevail throughout were there no vibrations, we have

$$p = f(\rho) = f(\rho_0 + \rho - \rho_0), \quad \dots \dots \dots \dots \dots (3)$$

and approximately

$$\varpi = \int \rho^{-1} f'(\rho)\, d\rho = \int \frac{f'(\rho_0 + \rho - \rho_0)}{\rho_0 + \rho - \rho_0}\, d(\rho - \rho_0)$$

$$= \frac{\rho - \rho_0}{\rho_0} f'(\rho_0) + \frac{(\rho - \rho_0)^2}{2\rho_0{}^2} \{\rho_0 f''(\rho_0) - f'(\rho_0)\}. \quad \dots \dots \dots (4)$$

For the mean value of ϖ at the piston we have only to write ρ_1 for ρ in (4) and integrate with respect to t. And at the piston $U = 0$.

For the mean of the whole length l of the cylinder (parallel to x), we have to integrate with respect to x as well as with respect to t. And in the integration with respect to x the first term of (4) disappears, inasmuch as the mean density remains the same as if there were no vibrations. Accordingly

$$f'(\rho_0) \int \frac{\rho_1 - \rho_0}{\rho_0}\, dt = \tfrac{1}{2} \iint \frac{U^2\, dx\, dt}{l}$$

$$+ \{\rho_0 f''(\rho_0) - f'(\rho_0)\} \left\{ \iint \frac{(\rho - \rho_0)^2}{2\rho_0{}^2} \frac{dx\, dt}{l} - \int \frac{(\rho_1 - \rho_0)^2}{2\rho_0{}^2}\, dt \right\}, \quad \dots \dots (5)$$

the terms on the right being of the second order in the quantities which express the vibration.

Again, $\displaystyle \int (p_1 - p_0)\, dt = \int \{f(\rho_1) - f(\rho_0)\}\, dt$

$$= \rho_0 f'(\rho_0) \int \frac{\rho_1 - \rho_0}{\rho_0}\, dt + \rho_0{}^2 f''(\rho_0) \int \frac{(\rho_1 - \rho_0)^2}{2\rho_0{}^2}\, dt\,;$$

so that by (5) $\displaystyle \int (p_1 - p_0)\, dt = \frac{\rho_0}{2} \iint \frac{U^2\, dx\, dt}{l}$

$$+ \{\rho_0{}^2 f''(\rho_0) - \rho_0 f'(\rho_0)\} \iint \frac{(\rho - \rho_0)^2}{2\rho_0{}^2} \frac{dx\, dt}{l} + \rho_0 f'(\rho_0) \int \frac{(\rho_1 - \rho_0)^2}{2\rho_0{}^2}\, dt. \quad \dots (6)$$

The three integrals on the right in (6) are related in a way which we may deduce from the theory of infinitely small vibrations. If the velocity of propagation of such vibrations be denoted by a, then $f'(\rho_0) = a^2$. By the usual theory we have

$$U = \frac{d\phi}{dx}, \qquad \frac{\rho - \rho_0}{\rho_0} = -\frac{1}{a^2} \frac{d\phi}{dt}. \quad \dots \dots \dots \dots \dots (7)$$

If we suppose that the cylinder is closed at $x = 0$ and at $x = l$, a normal vibration is expressed by

$$\phi = \cos \frac{s\pi x}{l} \cdot \cos \frac{s\pi a t}{l}, \quad \dots \dots \dots \dots \dots \dots (8)$$

where s is any integer, giving

$$\frac{a^2}{2} \int \frac{(\rho_1 - \rho_0)^2}{\rho_0^2}\, dt = a^2 \iint \frac{(\rho - \rho_0)^2}{\rho_0^2} \frac{dx\,dt}{l} = \iint \frac{U^2 dx\,dt}{l}, \quad \dots\dots(9)$$

the integrations with respect to x in (9) being taken from 0 to l, that is over the length of the cylinder.

The same conclusions (9) follow in the general case where ϕ is expressed by a sum of terms derived from (8) by attributing an integral value to s. The latter part expresses the equality of the mean potential and kinetic energies.

Introducing the relations (9) into (6), so as to express the mean pressure upon the piston in terms of the mean kinetic energy, we get as the final formula

$$\int (p_1 - p_0)\, dt = \left\{ \rho_0 + \frac{\rho_0^2 f''(\rho_0)}{2 f'(\rho_0)} \right\} \iint \frac{U^2 dx\,dt}{l}. \quad \dots\dots\dots(10)$$

Among special cases let us first take that of Boyle's law, where $p = a^2\rho$, so that

$$f'(\rho_0) = a^2, \qquad f''(\rho_0) = 0.$$

We have at once

$$\int (p_1 - p_0)\, dt = \rho_0 \iint \frac{U^2 dx\,dt}{l}. \quad \dots\dots\dots(11)$$

The expression on the right represents double the volume-density of the kinetic energy, or the volume-density of the whole energy, and we recover the result of the former investigation.

According to the adiabatic law

$$p/p_0 = (\rho/\rho_0)^\gamma; \quad \dots\dots\dots\dots\dots(12)$$

so that

$$f'(\rho_0) = \frac{p_0 \gamma}{\rho_0}, \qquad f''(\rho_0) = \frac{p_0 \gamma (\gamma - 1)}{\rho_0^2}. \quad \dots\dots\dots(13)$$

Hence from (10)

$$\int (p_1 - p_0)\, dt = \tfrac{1}{2} (\gamma + 1) \rho_0 \iint \frac{U^2 dx\,dt}{l}. \quad \dots\dots\dots(14)$$

The mean pressure upon the piston is now $\tfrac{1}{2}(\gamma + 1)$ of the volume-density of the total energy. We fall back on Boyle's law by taking $\gamma = 1$.

It appears therefore that the result is altered when Boyle's law is departed from. Still more striking is the alteration when we take the case treated in *Theory of Sound*, § 250 of the law of pressure

$$p = \text{Const.} - a^2 \rho_0^2 / \rho. \quad \dots\dots\dots\dots(15)$$

According to this

$$f'(\rho_0) = a^2, \qquad f''(\rho_0) = -2a^2 / \rho_0, \quad \dots\dots\dots(16)$$

and (10) gives

$$\int (p_1 - p_0)\, dt = 0. \quad \dots\dots\dots\dots(17)$$

The law of pressure (15) is that under which waves of finite condensation can be propagated without change of type.

In (17) the mean additional pressure vanishes, and the question arises whether it can be negative. It would appear so. If, for example,

$$p = \text{Const.} - \frac{a^2 \rho_0^3}{2\rho^2}, \quad \dots\dots\dots\dots\dots\dots(18)$$

$$f'(\rho_0) = a^2, \qquad f''(\rho_0) = -3a^2/\rho_0,$$

and
$$\int (p_1 - p_0)\, dt = -\tfrac{1}{2}\rho_0 \iint \frac{U^2\, dx\, dt}{l}. \quad \dots\dots\dots\dots(19)$$

I now pass on to the question of the *momentum* of a progressive train of waves. This question is connected with that already considered; for, as Prof. Poynting explains, if the reflexion of a train of waves exercises a pressure upon the reflector, it can only be because the train of waves itself involves momentum. From this argument we may infer already that momentum is not a necessary accompaniment of a train of waves. If the law were that of (15), no pressure would be exercised in reflexion. But it may be convenient to give a direct calculation of the momentum.

For this purpose we must know the relation which obtains in a progressive wave between the forward particle velocity u (not distinguished in one-dimensional motion from U) and the condensation $(\rho - \rho_0)/\rho_0$, usually denoted by s. When the disturbance is infinitely small, this relation is well known to be $u = as$, in the case of a positive wave. Thus

$$u : s = \sqrt{(dp/d\rho)}. \quad \dots\dots\dots\dots\dots\dots\dots(20)$$

The following is the method adopted in *Theory of Sound*, § 351 :—" If the above solution be violated at any point a wave will emerge, travelling in the negative direction. Let us now picture to ourselves the case of a positive progressive wave in which the changes of velocity and density are very gradual but become important by accumulation, and let us inquire what conditions must be satisfied in order to prevent the formation of a negative wave. It is clear that the answer to the question whether, or not, a negative wave will be generated at any point will depend upon the state of things in the immediate neighbourhood of the point, and not upon the state of things at a distance from it, and will therefore be determined by the criterion applicable to small disturbances. In applying this criterion we are to consider the velocities and condensations not absolutely, but relatively, to those prevailing in the neighbouring parts of the medium, so that the form of (20) proper for the present purpose is

$$du = \sqrt{\left(\frac{dp}{d\rho}\right)} \cdot \frac{d\rho}{\rho}, \quad \dots\dots\dots\dots\dots\dots(21)$$

whence
$$u = \int \sqrt{\left(\frac{dp}{d\rho}\right)} \cdot \frac{d\rho}{\rho}, \quad \dots\dots\dots\dots\dots(22)$$

which is the relation between u and ρ necessary for a positive progressive wave. Equation (22) was obtained analytically by Earnshaw (*Phil. Trans.* 1859, p. 146).

In the case of Boyle's law, $\sqrt{(dp/d\rho)}$ is constant, and the relation between velocity and density, given first, I believe, by Helmholtz, is

$$u = a \log (\rho/\rho_0),$$

if ρ_0 be the density corresponding to $u = 0$."

In our previous notation

$$dp/d\rho = f'(\rho) = a^2 + f''(\rho_0) \cdot (\rho - \rho_0),$$

a being the velocity of infinitely small waves, equal to $\sqrt{\{f'(\rho_0)\}}$; and by (22)

$$u = a\frac{\rho - \rho_0}{\rho_0} + \frac{a}{\rho_0}\left(\frac{f''(\rho_0)}{2a^2} - \frac{1}{\rho_0}\right)\frac{(\rho - \rho_0)^2}{2}, \quad\ldots\ldots\ldots\ldots(23)$$

the first term giving the usual approximate formula.

The momentum, reckoned per unit area of cross section,

$$= \int \rho u\, dx = \rho_0 \int \left(1 + \frac{\rho - \rho_0}{\rho_0}\right) u\, dx.$$

Introducing the value of u from (23) and assuming that the mean density is unaltered by the vibrations, we get

$$\rho_0\left\{\frac{\rho_0 f''(\rho_0)}{4a} + \frac{a}{2}\right\}\int \frac{(\rho - \rho_0)^2}{\rho_0^2}\, dx, \quad\ldots\ldots\ldots\ldots\ldots(24)$$

or, if we prefer it, $\quad\dfrac{\rho_0}{a^2}\left\{\dfrac{\rho_0 f''(\rho_0)}{4a} + \dfrac{a}{2}\right\}\displaystyle\int u^2 dx. \quad\ldots\ldots\ldots\ldots\ldots(25)$

The total energy of the length considered is $\rho_0\displaystyle\int u^2 dx$; and the result may be thus stated

$$\text{momentum} = \left\{\frac{\rho_0 f''(\rho_0)}{4a^3} + \frac{1}{2a}\right\} \times \text{total energy.} \quad\ldots\ldots\ldots(26)$$

This may be compared with (10). If we suppose the long cylinder of length l to be occupied by a train of progressive waves moving towards the piston, the integrated pressure upon the piston during a time t, equal to l/a, should be equal to twice the momentum of the whole initial motion. The two formulæ are thus in accordance, and it is unnecessary to discuss (26) at length. It may suffice to call attention to Boyle's law, where $f''(\rho_0) = 0$, and to the law of pressure (15) under which progressive waves have *no momentum*. It would seem that pressure and momentum are here associated with the tendency of waves to alter their form as they proceed on their course.

The above reasoning is perhaps as simple as could be expected; but an argument to be given later, relating to the kinetic theory of gases, led me to recognize, what is indeed tolerably obvious when once remarked, that there is here a close relation with the virial theorem of Clausius. If x, y, z be the coordinates; v_x, v_y, v_z the component velocities of a material particle of mass m, then

$$\tfrac{1}{2}\Sigma m v^2 = -\tfrac{1}{2}\Sigma X x + \tfrac{1}{4}\frac{d^2 \Sigma m x^2}{dt^2}$$

with two similar equations, X being the impressed force in the direction of x operative upon m. If the motion be what is called stationary, and if we understand the symbols to represent always the mean values with respect to time, the last term disappears, and

$$\tfrac{1}{2}\Sigma m v_x^2 = -\tfrac{1}{2}\Sigma X x. \quad\dots\dots\dots\dots\dots\dots\dots(27)$$

The mean kinetic energy of the system relative to any direction is equal to the virial relative to the same direction.

Let us apply (27) to our problem of the one-dimensional motion of a gas within a cylinder provided with closed ends. As in other applications of the virial theorem, the forces X are divided into two groups, internal and external. The latter reduces to the forces between the ends (pistons) and the gas. If p_1 be the pressure on the pistons—it will be the same on the average at both ends—the external virial is per unit of area $\tfrac{1}{2}p_1 l$ simply. As regards the internal virial, I do not remember to have seen its value stated, probably because in hydrodynamics the mechanical properties of a fluid are not usually traced to forces acting between the particles. There can be no doubt, however, what the value is. If we suppose that the whole mass of gas in (27) is at rest, the left-hand member vanishes, so that the sum of the internal and external virial must vanish. Under a uniform pressure p_0, the internal virial is therefore $\tfrac{1}{2}p_0 l$. In an actual gas the virial for any part can depend only on the local density, so that whether the gas be in motion or not, the value of the internal virial is

$$-\tfrac{1}{2}\int_0^l p\,dx. \quad\dots\dots\dots\dots\dots\dots\dots(28)$$

Hence (27) gives

$$\text{kinetic energy} = \tfrac{1}{2}p_1 l - \tfrac{1}{2}\int_0^l p\,dx = \tfrac{1}{2}(p_1 - p_0)l - \tfrac{1}{2}\int_0^l (p - p_0)\,dx. \quad (29)$$

If the gas be subject to Boyle's law, pressure is proportional to density, and the last term in (29) disappears. The additional pressure on the ends $(p_1 - p_0)$ is thus equal to twice the density of the kinetic energy.

In general, $p - p_0 = a^2(\rho - \rho_0) + \tfrac{1}{2}f''(\rho_0)\cdot(\rho - \rho_0)^2,$

and $\int_0^l (p - p_0)\,dx = \tfrac{1}{2}f''(\rho_0)\int(\rho - \rho_0)^2\,dx.$

If we introduce expressly the integration with respect to t already implied, (29) gives

$$l \int (p_1 - p_0) \, dt = \rho_0 \iint U^2 \, dx \, dt + \tfrac{1}{2} f'' (\rho_0) \iint (\rho - \rho_0)^2 \, dx \, dt$$

$$= \left\{ \rho_0 + \frac{\rho_0^2 f'' (\rho_0)}{2a^2} \right\} \iint U^2 \, dx \, dt,$$

regard being paid to (9). Equation (10) is thus derived very simply from the virial.

In all that precedes, the motion of the gas has been in one dimension, and even when we supposed the gas to be confined in a cylinder, we were able to avoid the consideration of lateral pressures upon the walls of the cylinder by applying the virial equation in its one-dimensional form. We now pass on to the case of three dimensions, and the first question which arises is as to the value of the virial. In place of (27) we have now

$$\tfrac{1}{2} \Sigma m U^2 = - \tfrac{1}{2} \Sigma (Xx + Yy + Zz), \dots \dots \dots (30)$$

U being the resultant velocity, Y, Z impressed forces parallel to the axes of y and z. Let us first apply this to a gas at rest under pressure p_0. The total virial, represented by the right-hand member of (30), is now zero; that is, the internal and external virial balance one another. As is well known and as we may verify at once by considering the case of a rectangular chamber, the external virial is $\tfrac{3}{2} p_0 v$, v denoting the volume of gas. The internal virial is accordingly $- \tfrac{3}{2} p_0 v$; and from this we may infer that whether the pressure be uniform or not, the internal virial is expressed by

$$- \tfrac{3}{2} \iiint p \, dx \, dy \, dz. \dots \dots \dots \dots (31)$$

The difference between the internal virial of the gas in motion and in equilibrium is

$$- \tfrac{3}{2} \iiint (p - p_0) \, dx \, dy \, dz. \dots \dots \dots (31^*)$$

According to the law of Boyle, (31^*) must vanish, since the mean density of the whole mass cannot be altered. The internal virial is therefore the same whether the gas be at rest or in motion.

A question arises here as to whether a particular law of pressure may not be fundamentally inconsistent with the statical Boscovitchian theory of the constitution of a gas upon which the application of the virial theorem is based. If, indeed, we assume Boyle's law in its integrity, the inconsistency does exist. For Maxwell has shown (Maxwell's *Scientific Papers*, vol. II. p. 422) that on a statical theory Boyle's law involves between the molecules of a gas a repulsion inversely as the distance. This makes the internal virial for any pair of molecules independent of their mutual distance, and thus the virial for the whole mass independent of the distribution of the parts. But such an explanation of Boyle's law violates the principle upon which (31) was deduced, making the pressure dependent upon the total quantity of the mass and not merely upon the local density; from which Maxwell concluded that all statical theories are to be rejected. It is to be remarked, however, that our calculations involve the law of pressure only as far as the term involving the square of the variation of density, and that a law agreeing with Boyle's to this degree of approximation may perhaps not be inconsistent with a statical Boscovitchian theory*.

Passing over this point, we find in general from (30)

$$\tfrac{1}{2}\Sigma m\,U^2 = \tfrac{3}{2}\,(p_1 - p_0)\,v - \tfrac{3}{2}\iiint (p - p_0)\,dx\,dy\,dz, \ldots\ldots\ldots\ldots(32)$$

whenever the character of the motion is such that the mean pressure (p_1) is the same at all points of the walls of the chamber. Further, as before,

$$\iiint (p - p_0)\,dx\,dy\,dz = \tfrac{1}{2}f''\,(\rho_0) \iiint (\rho - \rho_0)^2\,dx\,dy\,dz,$$

and finally, regard being paid to (9) as extended to three dimensions,

$$(p_1 - p_0)\,v = \left(\frac{1}{3} + \frac{\rho_0 f''\,(\rho_0)}{2a^2}\right) \times \text{total energy.} \ldots\ldots\ldots\ldots(33)$$

In the case of Boyle's law $f'' = 0$, and we see that the mean pressure upon the walls of the chamber is measured by one-third of the volume-density of the total energy.

For the adiabatic law (12), (13) gives

$$(p_1 - p_0)\,v = \left(\frac{1}{3} + \frac{\gamma - 1}{2}\right) \times \text{total energy.} \ldots\ldots\ldots\ldots(34)$$

In the case of certain gases called *monatomic*, $\gamma = 1\tfrac{2}{3}$, and (34) becomes

$$(p_1 - p_0)\,v = \tfrac{2}{3} \times \text{total energy.} \ldots\ldots\ldots\ldots\ldots(35)$$

Thirdly, in the case of the law (15) for the relation between pressure and density,

$$(p_1 - p_0)\,v = -\tfrac{2}{3} \times \text{total energy,} \ldots\ldots\ldots\ldots\ldots(36)$$

the mean pressure upon the walls being *less* than if there were no motion.

So far we have treated the question on the usual hydrodynamical basis, reckoning the energy of compression or rarefaction as *potential*. It was,

---

* I think the difficulty may be turned by supposing the force, inversely as the distance, to operate only between particles whose mutual distance is small, and that outside a certain small distance the force is zero. All that is necessary is that a pair of particles once within the range of the force should always remain within it—a condition easily satisfied so long as small disturbances alone are considered.

however, on the lines of the kinetic theory that I first applied the virial theorem to the question of the pressure of vibrations. In the form of this theory which regards the collisions of molecules as instantaneous, there is practically no potential, but only kinetic, energy. And if the gas be monatomic, the whole of this energy is translational. If V be the resultant velocity of the molecule whose mass is m, the virial equation gives

$$\tfrac{3}{2}p_1 v = \tfrac{1}{2}\Sigma m V^2, \quad \dots\dots\dots\dots\dots\dots\dots(37)$$

p_1 denoting, as before, the pressure upon the walls, assumed to be the same over the whole area. If necessary, p_1 and $\Sigma m V^2$ are to be averaged with respect to time.

It is usually to a gas in equilibrium that (37) is applied, but this restriction is not necessary. Whether there be vibrations or not, p_1 is equal to $\tfrac{2}{3}$ of the volume-density of the whole energy of the molecules. Consider a given chamber whose walls are perfectly reflecting, and let it be occupied by a gas in equilibrium. The pressure is given by (37). Suppose now that additional energy (which can only be kinetic) is communicated. We learn from (37) that the additional pressure is measured by $\tfrac{2}{3}$ of the volume-density of the additional energy, whether this additional energy be in the form of heat, equally or unequally distributed, or whether it take the form of mechanical vibrations, i.e. of coordinated velocities and density differences. Under the influence of heat-conduction and viscosity the mechanical vibrations gradually die down, but the pressure undergoes no change.

The above is the case of the adiabatic law with $\gamma = 1\tfrac{2}{3}$ already considered in (35), and a comparison of the two methods of treatment, in one of which potential energy plays a large part, while in the other all the energy is regarded as kinetic, suggests interesting reflexions as to what is really involved in the distinction of the two kinds of energy.

If we abandon the restriction to monatomic molecules, the question naturally becomes more complicated. We have first to consider in what form the virial equation should be stated. In the case of a diatomic molecule we have, in the first instance, not only the kinetic energy of the molecule as a whole, but also the kinetic energy of rotation, and in addition the internal virial of the force by which the union of the two atoms is maintained. It is easy to see, however, that the two latter terms balance one another, so that we are left with the kinetic energy of the molecule as a whole. For general purposes a theorem is required for which I have not met a complete statement. For any part of a wider system for which we wish to form the virial equation, we may omit the kinetic energy of the motion relative to the centre of gravity of the part, if at the same time we omit the virial of the internal forces operative in this part and treat the forces acting from outside upon the part, whether from the remainder of the system or wholly from outside, as acting at the centre of gravity of the part. In applying (37) to a gas regarded as composed of molecules, we are therefore to include on the right only the kinetic energy of translation of the molecules. If a gas originally at rest be set into vibration, we have

$$\tfrac{3}{2}\,(p_1 - p_0)\,v = \text{additional energy of translation.} \quad \dots\dots\dots(38)$$

The pressure p_1 does not now, as in the case of monatomic gases, remain constant. Under the influence of viscosity and heat-conduction, part of the energy at first translational becomes converted into other forms.

A complete discussion here would carry us into the inner shrine of the kinetic theory. We will only pursue the subject so far as to consider briefly the case of rigid molecules for which the energy is still entirely kinetic— partly that of the translatory motion of the molecules as wholes and partly rotatory. Of the additional energy E representing the vibrations, half may be regarded as wholly translational. Of the other half, the fraction which is translational is $3/m$, where m is the whole number of modes. The translational part of E is therefore $\tfrac{1}{2}E(1 + 3/m)$; so that

$$(p_1 - p_0)\,v = E\left(\frac{1}{3} + \frac{1}{m}\right). \quad \dots\dots\dots\dots\dots\dots(39)$$

If $m = 3$, as for monatomic molecules, we recover the former result; otherwise $p_1 - p_0$ is less. In terms of γ we have

$$\gamma = 1 + 2/m, \quad \dots\dots\dots\dots\dots\dots\dots(40)$$

and accordingly
$$(p_1 - p_0)\,v = E\left(\frac{1}{3} - \frac{\gamma - 1}{2}\right), \quad \dots\dots\dots\dots\dots(41)$$

in agreement with (34) where what was there called the total energy is now regarded as the additional energy of vibration. In the case of a diatomic gas, $m = 5$, $\gamma = 1\tfrac{2}{5}$.

310.

THE ORIGIN OF THE PRISMATIC COLOURS.

[*Philosophical Magazine*, Vol. x. pp. 401—407, 1905.]

THE fact that by the aid of a spectroscope interferences may be observed with light originally white used to be regarded as a proof of the existence of periodicities in the original radiation; but it seems now to be generally agreed that these periodicities are due to the spectroscope. When a *pulse* strikes a grating, it is obvious that the periodicity and its variation in different directions are the work of the grating. The assertion that Newton's experiments prove the colours to be already existent in white light, is usually made in too unqualified a form.

When a prism, which has no periodicities of figure, is substituted for a grating, the *modus operandi* is much less obvious. This question has been especially considered by Schuster (*Phil. Mag.* XXXVII. p. 509, 1894; VII. p. 1, 1904), and quite recently Ames has given an "Elementary Discussion of the Action of a Prism upon White Light" (*Astrophysical Journal*, July 1905). The aim of the present note is merely to illustrate the matter further.

I commence by remarking that, so far as I see, there is nothing faulty or specially obscure in the traditional treatment founded upon the consideration of simple, and accordingly infinite, trains of waves. By Fourier's theorem any arbitrary disturbance may be thus compounded; and the method suffices to answer any question that may be raised, so long at least as we are content to take for granted the character of the dispersive medium—the relation of velocity to wave-length—without enquiring further as to its constitution. For example, we find the resolving-power of a prism to be given by

$$\frac{\lambda}{d\lambda} = T\frac{d\mu}{d\lambda}, \quad \dots\dots\dots\dots\dots\dots\dots\dots\dots(1)$$

in which λ denotes the wave-length *in vacuo*, T the "thickness" of the prism, μ the refractive index, and $d\lambda$ the smallest difference of wave-length that can be resolved. A comparison with the corresponding formula for a grating shows that (1) gives the number of waves (λ) which travel in the prescribed direction as the result of the action of the prism upon an incident pulse.

But, although reasoning on the above lines may be quite conclusive, a desire is naturally felt for a better understanding of the genesis of the sequence of waves, which seems often to be regarded as paradoxical. Probably I have been less sensible of this difficulty from my familiarity with the analogous phenomena described by Scott Russel and Kelvin, of which I have given a calculation*. "When a small obstacle, such as fishing-line, is moved forward slowly through still water, or (which, of course, comes to the same thing) is held stationary in moving water, the surface is covered with a beautiful wave-pattern, fixed relatively to the obstacle. On the up-stream side the wave-length is short, and, as Thomson has shown, the force governing the vibrations is principally cohesion. On the down-stream side the waves are longer and are governed principally by gravity. Both sets of waves move with the same velocity relatively to the water, namely, that required in order that they may maintain a fixed position relatively to the obstacle. The same condition governs the velocity, and therefore the wave-length, of those parts of the wave-pattern where the fronts are oblique to the direction of motion. If the angle between this direction and the normal to the wave-front be called θ, the velocity of propagation of the waves must be equal to $v_0 \cos \theta$, where v_0 represents the velocity of the water relatively to the (fixed) obstacle." In the laboratory the experiment may be made upon water contained in a large sponge-bath and mounted upon a revolving turn-table. The fishing-line is represented by the impact of a small jet of wind. In this phenomenon the action of a prism is somewhat closely imitated. Not only are there sequences of waves, unrepresented (as would appear) either in the structure of the medium or in the character of the force, but the wave-length and velocity are variable according to the direction considered.

For the purposes of Scott Russel's phenomenon the localized pressure is regarded as permanent; but here it will be more instructive if we suppose it applied for a finite time only. Although the method is general, we may fix our ideas upon deep water, subject to gravity (cohesion neglected), upon which operates a pressure localized in a line and moving transversely with velocity V. In the general two-dimensional problem thus presented, the effect of the travelling pressure is insignificant unless V is a possible wave-velocity; but where this condition is satisfied, a corresponding train of waves is generated. In the case of deep water under gravity the

* "The Form of Standing Waves on the Surface of Running Water," *Proc. Lond. Math. Soc.* Vol. xv. p. 69 (1883); *Scientific Papers*, Vol. ii. p. 258.

condition is always satisfied, for the wave-velocities vary from zero to infinity.

The limitation to a wave-train of velocity V is complete only when the time of application of the pressure is infinitely extended. Otherwise, besides the train of velocity V we have to deal with other trains, of velocities differing so little from V that during the time in question they remain sensibly in step with the first. As is known*, the behaviour of such aggregates is largely a matter of the *group-velocity* U, whose value is given by

$$U = \frac{d(kV)}{dk}, \quad\dotsb\dotsb\dotsb\dotsb\dotsb\dotsb\dotsb(2)$$

k being proportional to the reciprocal of the wave-length *in the medium*. In the particular case of deep-water waves $U = \frac{1}{2}V$.

From this point of view it is easy to recognize that the total length of the train of waves generated in time t' is $\pm (V - U) t'$. If τ be the periodic time of these waves, the wave-length in the medium is $V\tau$, and the number of waves is therefore

$$\pm \frac{V - U}{V} \frac{t'}{\tau}. \quad\dotsb\dotsb\dotsb\dotsb\dotsb\dotsb\dotsb(3)$$

But for our present purpose of establishing an analogy with prisms and their resolving-power, what we are concerned with is not the number of waves at any time in the dispersive medium itself, but rather the number after emergence of the train into a medium which is non-dispersive; and here a curious modification ensues. During the emergence the relative motion of the waves and of the group still continues, and thus we have to introduce the factor V/U, obtaining for the number N of waves outside

$$N = \frac{V - U}{U} \frac{t'}{\tau}. \quad\dotsb\dotsb\dotsb\dotsb\dotsb\dotsb(4)$$

If X be the distance through which the pressure travels, $X = Vt'$; and if V_0 be the (constant) velocity outside and λ the wave-length outside, $\lambda = V_0\tau$. Thus

$$N = \left(\frac{V_0}{U} - \frac{V_0}{V}\right) \frac{X}{\lambda}. \quad\dotsb\dotsb\dotsb\dotsb\dotsb(5)$$

To introduce optical notation, let $\mu = V_0/V$, so that μ is the refractive index. In terms of μ

$$\frac{V_0}{U} = \mu - \lambda \frac{d\mu}{d\lambda}, \quad\dotsb\dotsb\dotsb\dotsb\dotsb\dotsb(6)$$

so that finally

$$N = -X \frac{d\mu}{d\lambda}, \quad\dotsb\dotsb\dotsb\dotsb\dotsb\dotsb\dotsb(7)$$

* See, for example, *Nature*, Vol. xxv. p. 51 (1881); *Scientific Papers*, Vol. I. p. 540.

in close correspondence with (1). A very simple formula thus expresses the number of waves (after emergence) generated by the travel of the pressure over a distance X of a dispersive medium.

The above calculation has the advantage of being clear of the complication due to obliquity; but a very little modification will adapt it to the case of a prism, especially if we suppose that the waves considered are emergent at the second face of the prism without refraction. In the figure, AC represents an incident plane pulse whose trace runs along the first face of the prism from A to B. AF, BE is the direction of propagation of the refracted waves under consideration, to which the second face of the prism is perpendicular. As before, if τ be the period, V the wave-velocity of the waves propagated in direction BE, U the corresponding group-velocity, t' the time of travel of the pulse from A to B, the number of waves within the medium is

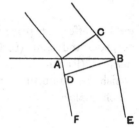

$$\frac{V-U}{V}\frac{t'}{\tau},$$

giving on emergence the number of waves expressed in (4). If V_0 be the velocity in vacuum, $\tau = \lambda/V_0$, and

$$t' = \frac{BC}{V_0} = \frac{AD}{V};$$

so that

$$\frac{t'}{\tau} = \frac{AD}{\lambda}\frac{V_0}{V}.$$

Thus, as in (5), (6), (7),

$$N = \left(\frac{V_0}{U} - \frac{V_0}{V}\right)\frac{AD}{\lambda} = -AD\frac{d\mu}{d\lambda}, \quad\dots\dots\dots\dots\dots(8)$$

in agreement with (1).

Although the process is less easy to follow, the construction of a train of waves from an incident pulse is as definite in the case of a prism as is that of a grating; and its essential features are presented to the eye in Scott Russel's phenomenon.

The above treatment suffices for a general view, but it may be instructive to give an analytical statement; and this I am the more inclined to do as affording an opportunity of calling attention to a rather neglected paper by Lord Kelvin entitled "On the Waves produced by a Single Impulse in Water of any Depth, or in a Dispersive Medium"[*]. When we know the

* *Proc. Roy. Soc.* Vol. XLII. p. 80 (1887).

effect of an impulse, that of a uniform force applied for a finite time can be deduced by integration. It may be convenient to recite the leading steps of Kelvin's investigation.

Let $f(k)$ denote the velocity of propagation corresponding to wave-length (in the medium) $2\pi/k$. The Fourier-Cauchy-Poisson synthesis gives

$$u = \int_0^\infty dk \cos k\,[x - t f(k)] \quad\dots\dots\dots\dots\dots(9)$$

for the effect at place and time (x, t) of an infinitely intense disturbance at place and time $(0, 0)$. When $x - t f(k)$ is very large, the parts of the integral (9) which lie on the two sides of a small range, $\kappa - \alpha$ to $\kappa + \alpha$, vanish by annulling interference; κ being a value, or the value, of k, which makes

$$\frac{d}{dk}\{k\,[x - t f(k)]\} = 0 \quad\dots\dots\dots\dots\dots\dots(10)$$

or $$x = t\{f(\kappa) + \kappa f'(\kappa)\} = Ut, \quad\dots\dots\dots\dots\dots(11)$$

U being the group-velocity. By Taylor's theorem when $k - \kappa$ is very small,

$$k\,[x - t f(k)] = \kappa^2 t f'(\kappa) + \tfrac{1}{2}t\,(k - \kappa)^2\{-\kappa f''(\kappa) - 2f'(\kappa)\}.$$

Using this in (9) and integrating with the aid of

$$\int_{-\infty}^{+\infty} d\sigma \cos \sigma^2 = \int_{-\infty}^{+\infty} d\sigma \sin \sigma^2 = \surd(\tfrac{1}{2}\pi),$$

we find as an approximate value

$$u = \frac{\surd(2\pi)\,.\,\cos\{t\kappa^2 f'(\kappa) + \tfrac{1}{4}\pi\}}{\surd t\,.\,\surd\{-\kappa f''(\kappa) - 2f'(\kappa)\}}\,. \quad\dots\dots\dots(12)$$

As a particular case, for deep-water gravity waves

$$f(k) = \surd(g/k), \qquad f'(k) = -\tfrac{1}{2}g^{\frac{1}{2}}k^{-\frac{3}{2}}, \qquad -k f''(k) - 2f'(k) = \tfrac{1}{4}g^{\frac{1}{2}}k^{-\frac{3}{2}},$$

and finally with use of (11)

$$u = \pi^{\frac{1}{2}}g^{\frac{1}{2}}t x^{-\frac{3}{2}} \cos\left(\frac{gt^2}{4x} - \frac{\pi}{4}\right). \quad\dots\dots\dots(13)*$$

This gives the effect of the impulse at $(0, 0)$. If the impulse be at x', t', we are to write $x - x'$ for x and $t - t'$ for t. For our purpose of finding the effect of a travelling force, we are to make $x' = Vt'$ and integrate with respect to t' from 0 to t', t' being the duration of the force. The integral will depend mainly upon the part where

$$\frac{(t - t')^2}{x - Vt'},$$

---

* An almost equally simple formula applies when more generally $f(k) \propto k^n$.

under the cosine, is stationary.   This occurs when

$$2x = V(t + t'), \quad \dots\dots\dots\dots\dots\dots\dots(14)$$

and then

$$\frac{g(t - t')^2}{4(x - Vt')} = \frac{g(Vt - x)}{V^2}. \quad \dots\dots\dots\dots\dots(15)$$

Omitting the variation of the other factors as less important, we see that, when sensible, the effect is proportional to

$$\cos\left\{\frac{g(Vt - x)}{V^2} - \frac{\pi}{4}\right\}, \quad \dots\dots\dots\dots\dots\dots(16)$$

representing simple waves of velocity V.   But this is limited to such values of x and t as make t' in (14) lie between 0 and t'.   Thus if t be given, the range of x is from $\frac{1}{2}Vt$ to $\frac{1}{2}Vt + \frac{1}{2}Vt'$; so that the train of waves covers a length $\frac{1}{2}Vt'$, agreeing with the general value given before, since here $U = \frac{1}{2}V$.   If, as would be more convenient in order to find the length of the train after emergence into a non-dispersive medium, we regard x as given, we find that t ranges from $2x/V$ to $2x/V + t'$.

I have taken the particular case first, as the reasoning is rather simpler when we have, as in (13), an explicit expression in terms of x and t*.   In general κ cannot be eliminated between (11) and (12), and we must proceed rather differently.   The question is when will

$$k[x - tf(k)] \quad \dots\dots\dots\dots\dots\dots\dots(17)$$

with

$$x = t\{f(k) + kf'(k)\} \quad \dots\dots\dots\dots\dots \dots(18)$$

be stationary with respect to t', $x - Vt'$ being substituted for x and $t - t'$ for t in (17), (18)?   Now

$$\left(V\frac{d}{dx} + \frac{d}{dt}\right)(17) = k\{V - f(k)\} + \left(V\frac{dk}{dx} + \frac{dk}{dt}\right)\left(x - t\frac{d\{kf(k)\}}{dk}\right),$$

of which the second term on the right vanishes by (18).   The variation of (17) vanishes when $V = f(k)$.   Accordingly

$$k[x - Vt' - (t - t')f(k)]$$

with

$$x - Vt' = (t - t')\{f(k) + kf'(k)\}$$

is stationary with respect to t', if $V = f(k)$, and then assumes the form

$$k[x - Vt].$$

Here t' must lie between 0 and t'.   Thus if t be constant, x has a range

$$Vt' - t'\{f(k) + kf'(k)\} = (V - U)t'.$$

---

* For an admirable discussion of the general problem of deep-water waves arising from a localized disturbance, see Lamb, *Proc. Lond. Math. Soc.* Vol. II. p. 371 (1904).

And if x be given, t has a range

$$t' - \frac{Vt'}{f(k) + kf'(k)} = \frac{t'(U - V)}{U} . \qquad \dots\dots\dots\dots(19)$$

These are the limits over which the waves of velocity V extend. And (19) shows that the number of waves which pass a fixed point, either within the dispersive medium or on emergence from it, has the expression

$$\frac{t'}{\tau} \frac{U - V}{U},$$

where τ is the periodic time, in agreement with (4).

311.

ON THE CONSTITUTION OF NATURAL RADIATION.

[*Philosophical Magazine*, Vol. XI. pp. 123—127, 1906.]

THE expression of Prof. Larmor's views in his paper* "On the Constitution of Natural Radiation" is very welcome. Although it may be true that there has been no direct contradiction, public and private communications have given me an uneasy feeling that our views are not wholly in harmony; nor is this impression even now removed. It may conduce to a better understanding of some of these important and difficult questions if without dogmatism I endeavour to define more clearly the position which I am disposed to favour on one or two of the matters concerned.

On p. 580, in comparing white light and Röntgen radiation, Prof. Larmor writes: "Both kinds of disturbance are resolvable by Fourier's principle into trains of simple waves. But if we consider the constituent train having wave-length variable between λ and $\lambda + \delta\lambda$, *i.e.* varying irregularly from part to part of the train within these limits, a difference exists between the two cases. In the case of the white light the vibration-curve of this approximately simple train is in appearance steady; it is a curve of practically constant amplitude, but of wave-length slightly erratic within the limits $\delta\lambda$ and therefore of phase at each point entirely erratic. In the Fourier analysis of the Röntgen radiation the amplitude is not regular, but on the contrary may be as erratic as the phase." This raises the question as to the general character of the resultant of a large number of simple trains of approximately equal wave-length. In what manner will the resultant amplitude and phase vary? In several papers† I have considered particular cases of approximately simple waves, showing how they may be resolved into absolutely simple trains of approximately equal wave-lengths. But now the question presents itself in

---

* *Phil. Mag.* Vol. x. p. 574 (1905).

† See especially *Phil. Mag.* Vol. L. p. 135 (1900). [*Scientific Papers*, Vol. IV. p. 486.]

the converse form.   What are we to expect from the composition of simple trains, severally represented by

$$a_1 \cos \{(n + \delta n_1) t + \epsilon_1\}, \qquad \ldots\ldots\ldots\ldots\ldots\ldots\ldots\ldots(1)$$

where δn_1 is small, while the amplitude a_1 and the initial phase ϵ_1 vary from one train to another?

In virtue of the smallness of δn_1 we may appropriately regard (1) as a vibration of speed n and of phase $\epsilon_1 + \delta n_1 t$, variable therefore with the time. The amplitude and phase may be represented in the usual way by the polar coordinates of a point; and the point representing (1) accordingly lies on the circle of radius a_1 and revolves uniformly with small angular velocity. For the present at any rate I suppose that the amplitudes a_1, a_2, &c. are all equal (1), in which case the points lie all upon the same circle.   The radius from the centre O to any of the points P upon the circumference is a vector fully representative of the vibration, and the resultant of the vectors represents the resultant of the vibrations.

After the lapse of a time t the points have moved from their initial positions P to other positions Q, and the aggregate of the vectors OP is replaced by the aggregate of OQ.   The *difference* is the aggregate of PQ. Now we suppose that t is so related to the greatest δn that all the arcs PQ are small fractions of the quadrant, and the question before us is the amount of the difference between the resultants of the OP's and the OQ's, *i.e.* of the PQ's.   There are certain cases where we can say at once that the difference of resultants is small, small that is relatively to the whole.   This happens when all the P's are rather close together, *i.e.* when the component vibrations have initially nearly the same phase.   It is then certain that at the end of the time t the amplitude and phase are but little altered from what they were at the beginning.   Over this range the vibration is approximately simple, and the range is inversely as the greatest departure from the mean frequency n.

But in general the distribution of initial phases ϵ causes the resultant to be much less than if the phases were in agreement, and it may even happen that the initial resultant is zero.   At the end of the time t the resultant will probably not be zero, so that in this case the change is relatively large. The proposition that small changes in the phases of the components can lead only to relatively small changes in the resultant is thus not universally true; and we must inquire further as to the conditions under which the conclusion is probable.

The most important case for our purpose is when the initial phases are distributed *at random*, as they would presumably be when Röntgen radiation is concerned.   If the components are very numerous (and of equal amplitude unity), the problem is one which I have considered on former occasions[*].

It appears that the probability of a resultant amplitude lying between r and $r + dr$ is

$$\frac{2}{m} e^{-r^2/m} r\, dr, \dots\dots\dots\dots\dots\dots\dots(2)$$

where m is the number of components.   Or the probability of an amplitude exceeding r is $e^{-r^2/m}$.   The *mean* intensity (when the phases are redistributed at random a great many times) is m, corresponding to the amplitude $\sqrt{m}$.

When r is great compared with $\sqrt{m}$, the probability of an amplitude exceeding r becomes vanishingly small.   When on the other hand r is small, the probability of a resultant less than r is approximately r^2/m.   It appears that the chance of the resultant lying outside the range from say $\frac{1}{4}\sqrt{m}$ to $2\sqrt{m}$ is comparatively small.

We have next to consider the resultant of the components PQ.   Here again the phases are distributed in all directions.   The amplitudes, however, are no longer equal, but they are small relatively to unity.   Although the contrary is not impossible, it would seem that in all probability the resultant amplitude of the PQ's is small in comparison with that of the OP's, from which it follows that, exceptional cases apart, the amplitude and phase of the resultant remain but little changed at the end of a time t, such that the changes of phase of the individual components are small.

From the above discussion I am disposed to infer that a Fourier element of radiation necessarily possesses in large degree the characteristic which (if I rightly understand him) Prof. Larmor associates with white light in contrast to Röntgen radiation.   Of course, after the lapse of a sufficient time the final phases of the components lose all simple relation to the initial phases.   The final phase of the resultant is then without relation to the initial phase, and the amplitudes may differ finitely, but in all probability within somewhat restricted limits.   From this variation it seems to me white light cannot be exempt.

In the above and, so far as I remember, in what I have written previously, the question is purely kinematical.   In saying that Fourier's theorem is competent to answer any question that may be raised respecting the action of a dispersive medium, I take for granted that the law of dispersion is given in its entirety.   I quite admit that if there are any wave-lengths for which the behaviour of the medium is unknown, a corresponding uncertainty must attach to the fate of any aggregate in which these are included.   Doubtless

---

[*] *Phil. Mag.* Vol. x. p. 73 (1880); *Scientific Papers*, Vol. I. p. 491; *Theory of Sound*, 2nd ed. Vol. I. § 42 a.

a complete statement of the law of dispersion may involve the case of wave-lengths for which the medium is not transparent.

As regards the passage quoted from Sir G. Stokes, his object was, I think, to explain the absence of refraction when Röntgen rays traverse matter. Taking light of ordinary and absolutely definite wave-length incident upon transparent matter, he contemplates the lapse of 10,000 periods before harmony is established between the ætherial and molecular vibrations, that is, as I understand it, before regular refraction is possible. At this rate the light from a soda flame would be incapable of regular refraction, for the vibrations are certainly not regular for more than 500 periods. Indeed Stokes's argument appears better adapted to prove that Röntgen rays could not traverse material media at all in a regular manner, than that they would do so without change of wave-velocity.

I must confess that I have never fully understood Stokes's position in this matter. A medium is non-refractive and nearly transparent for the pulses constituting Röntgen rays*. What reception would it give to simple waves of half wave-length equal to the thickness of the pulses? I should suppose that it would be non-refractive and transparent for these also, but Stokes's argument seems to imply the contrary. The paradox would then have to be met that the medium treats simple waves less simply than compound ones.

* [1911. The question whether Röntgen radiation is really of this character at all seems still to be an open one.]

312.

ON AN INSTRUMENT FOR COMPOUNDING VIBRATIONS, WITH APPLICATION TO THE DRAWING OF CURVES SUCH AS MIGHT REPRESENT WHITE LIGHT.

[*Philosophical Magazine*, Vol. XI. pp. 127—130, 1906.]

IN discussions respecting the character of the curve by which the vibrations of white light may be expressed, I have often felt the want of some ready, even if rough, method of compounding several prescribed simple harmonic motions. Any number of points on the resultant can of course always be calculated and laid down as ordinates; but the labour involved in this process is considerable. The arrangement about to be described was exhibited early in the year during lectures at the Royal Institution. As it is inexpensive to construct and easily visible to an audience, I have thought that such a description might be useful, accompanied with a few specimens of curves actually drawn with its aid.

A wooden batten, say 1 inch square and 5 feet long, is so mounted horizontally as to be capable of movement only along its length. For this purpose it suffices to connect two points near the two ends, each by means of two thin metallic wires, with four points symmetrically situated in the roof overhead. This mounting, involving four constraints only, allows also of a rotatory or rolling motion, which could be excluded, if necessary, by means of a fifth wire attached to a lateral arm. In practice, however, this provision was not used or needed. The movement of the batten along its length is controlled by a piece of spring-steel against which the pointed extremity of the batten is held by rubber bands. Any force acting in the direction of the length of the batten produces a displacement proportional to the force*. The tracing point, by which the movements are recorded, is at the other end,

---

* In strictness this presupposes the fulfilment of a condition involving the period of the force and that of free vibration under the influence of the spring, which it is scarcely necessary to enter upon.

as nearly as possible in the line joining the two points of attachment of the four suspending wires.

The longitudinal forces are due to the vibrations of pendulums hanging from horizontal cross-pieces attached to the batten at their centres. The two ends of a wire or cord are attached to the extremities of a cross-piece, the bob of the pendulum being a mass of lead (perhaps half a pound) carried at the middle of the cord. When set swinging the movements of the pendulums are thus parallel to the batten and tend to displace it along its length. In my apparatus the length of the longest pendulum is $3\frac{1}{2}$ feet.

Under the influence of one pendulum the tracing point describes a small simple harmonic motion along the length of the batten. In order to draw a curve of sines the smoked glass destined to receive the record should move vertically in its own plane. I found it more convenient and sufficient for my purpose to substitute a movement of rotation. A disk (like the face-plate of a lathe) revolves freely in a vertical plane round a horizontal axis. To this disk a piece of smoked glass is cemented and the tracing is taken near the circumference, the axis of rotation being at the same level as the tracing point, so that the movement of vibration is radial.

The disk must be made to revolve slowly and with uniform angular velocity. To effect this I employed a sand-clock, a device which works better than would be expected*. The sand, carefully sifted and dried, is contained in a vertical metal tube of about 1 inch diameter, and escapes below through a small aperture of size to be determined by trial. On the sand rests a weight, of such diameter as to fit the tube easily; and this in its descent rotates the disk by means of a thread, of which the free part is vertical while the remainder engages a circumferential groove. The descent of the weight is practically independent of the quantity of sand remaining at any time†. It is scarcely necessary to say that the revolving parts must be so weighted as to keep the thread tight.

The advantages of the apparatus depend of course upon the facility with which a number of vibratory movements can be combined‡. It is as easy to record the effect of a number of pendulums as of a single one, the contribution in each case being proportional to the amplitude of vibration. In my instrument there are six pendulums, the shortest of such length as to vibrate about twice as quickly as the longest. The frequencies are in fact somewhat as the numbers 5, 6, 7, 8, 9, 10. No precise adjustment was attempted, the object being in fact rather to avoid anything specially simple.

The lengths of the pendulums were chosen so as to afford an illustration of the vibrations constituting white light. Of course a complete physical

* It was used by H. Draper to drive an equatorially mounted telescope.
† See Note at end of paper.
‡ The principle of mechanical addition is employed in an instrument devised by Michelson.

representation of light from the sun or from the electric arc would need a much larger range of frequency. But we may suppose this light filtered through media capable of sensibly absorbing the ultra-red and ultra-violet, while still remaining white so far as the eye could tell, even with the aid of a prism. The range of an *octave*, for which provision is made, then amply suffices.

The number of pendulums may seem, and perhaps is, rather small. The frequency, *e.g.* 7, given by one of the pendulums must be taken to represent a range from $6\frac{1}{2}$ to $7\frac{1}{2}$, with an error therefore up to 1 in 14. Such an error will be serious after 7 vibrations, but not so for 3 or 4 vibrations. Hence if we limit ourselves to sequences of 3 or 4 waves, the representation is about good enough.

Connected with the above is the question what amplitudes of vibration are to be assigned to the various pendulums. It would not be difficult to give effect to an assigned law of spectrum intensity whether suggested by theory or found in observation. It is to be remembered, however, that such laws relate to averages, and do not give the relative amplitudes at any particular time, which will indeed vary fortuitously over a rather large range. I thought it therefore unnecessary to be very particular in this respect. The vibrations of the shorter pendulums die down more rapidly than the slower ones. By giving the former an advantage at starting a somewhat wide range is covered.

The tracings presented no general features that might not have been anticipated. A few specimens are reproduced—one showing the operation of the longest and shortest pendulums alone, the others the effect of all the pendulums.

Note on the Principle of the Sand-Clock.

The difficulty of propelling a column of sand, occupying a tube, by forces pushing at one end is well known; but I do not remember to have seen any discussion of the question on mechanical principles. A similar phenomenon occurs in the storage of grain, the weight of which, when contained in tall bins, is found to be taken mainly on the sides and but little on the bottom of the bin*.

The unexpectedness of these effects depends upon a half unconscious comparison with fluids which in a state of rest are exempt from friction. In the present case, when the sand is moving, the tangential force at the wall is reckoned at μ times the normal force. We may suppose, as a rough approximation, that there is something like a fluid pressure p. If a be the radius of

---

* I. Roberts, *Proc. Roy. Soc.* Vol. xxxvi. p. 226, 1884. "In any cell which has parallel sides, the pressure of wheat upon the bottom ceases when it is charged up to twice the diameter of the inscribed circle."

the tube and dx an element of length along the axis, the tangential force acting upon the surface $2\pi a\,dx$ is $\mu p.2\pi a.dx$. This is to be equated to the difference of the forces upon the two faces of the slice, viz. $-\pi a^2 dp$. Accordingly

$$\frac{dp}{p\,dx} = -\frac{2\mu}{a},$$

or
$$p = p_0 e^{-2\mu x/a},$$

the pressure diminishing as a increases.   Hence a powerful pressure at $x = 0$ is unable to overcome a very feeble one acting in the opposite direction at a section many diameters away.   The case is similar to that of a rope coiled round a post, as used to check the motion of steamers coming up to a pier.

As regards numbers, it will not be out of the way to suppose $\mu = \frac{1}{10}$. When $x = 10a$, $p/p_0 = e^{-2} = {\cdot}14$.

Fig. 1.

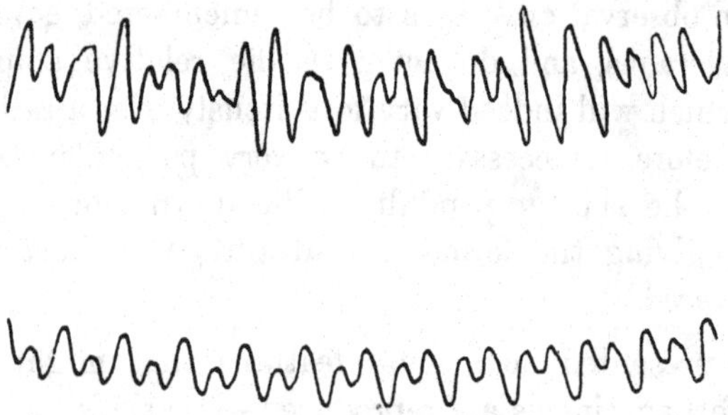

Fig. 2.

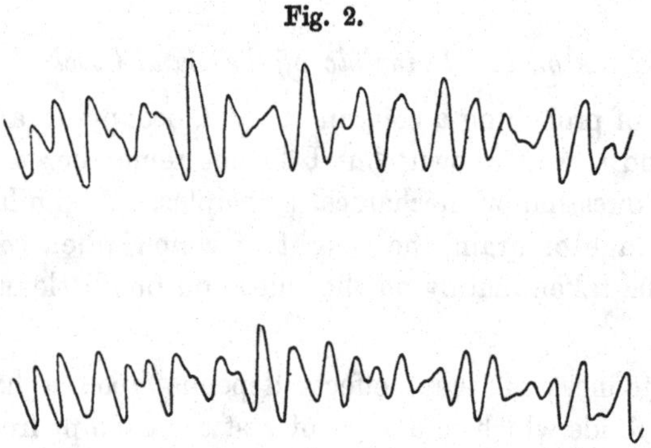

313.

ON ELECTRICAL VIBRATIONS AND THE CONSTITUTION OF THE ATOM.

[*Philosophical Magazine*, Vol. XI. pp. 117—123, 1906.]

IN illustration of the view, suggested by Lord Kelvin, that an atom may be represented by a number of negative electrons, or negatively charged corpuscles, enclosed in a sphere of uniform positive electrification, Prof. J. J. Thomson has given some valuable calculations* of the stability of a ring of such electrons, uniformly spaced, and either at rest or revolving about a central axis. The corpuscles are supposed to repel one another according to the law of inverse square of distance and to be endowed with inertia, which may, however, be the inertia of æther in the immediate neighbourhood of each corpuscle. The effect of the sphere of positive electrification is merely to produce a field of force directly as the distance from the centre of the sphere. The artificiality of this hypothesis is partly justified by the necessity, in order to meet the facts, of introducing from the beginning some essential difference, other than of mere sign, between positive and negative.

Some of the most interesting of Prof. Thomson's results depend essentially upon the finiteness of the number of electrons; but since the experimental evidence requires that in any case the number should be very large, I have thought it worth while to consider what becomes of the theory when the number is infinite. The cloud of electrons may then be assimilated to a fluid whose properties, however, must differ in many respects from those with which we are most familiar. We suppose that the whole quantities of positive and negative are equal. The difference between them is that the positive is constrained to remain undisplaced, while the negative is free to move. In equilibrium the negative distributes itself with uniformity throughout the sphere occupied by the positive, so that the total density is everywhere zero. There is then no force at any point; but if the negative

* *Phil. Mag.* Vol. VII. p. 237 (1904).

be displaced, a force is usually called into existence. We may denote the density of the negative at any time and place by ρ, that of the positive and of the negative, when in equilibrium, being ρ_0. The repulsion between two elements of negative $\rho\,dV$, $\rho'\,dV'$ at distance r is denoted by

$$\gamma . r^{-2} . \rho\,dV . \rho'\,dV'. \quad\dots\dots\dots\dots\dots\dots\dots(1)$$

The negative fluid is supposed to move without circulation, so that a velocity-potential (ϕ) exists; and the first question which presents itself, is as to whether there is "condensation." If this be denoted by s, the equation of continuity is, as usual*,

$$\frac{ds}{dt} + \nabla^2\phi = 0. \quad\dots\dots\dots\dots\dots\dots\dots(2)$$

Again, since there is no outstanding *pressure* to be taken into account, the dynamical equation assumes the form

$$\frac{d\phi}{dt} = R, \quad\dots\dots\dots\dots\dots\dots\dots(3)$$

where R is the potential of the attractive and repulsive forces. Eliminating ϕ, we get

$$\frac{d^2s}{dt^2} = -\nabla^2 R. \quad\dots\dots\dots\dots\dots\dots\dots(4)$$

In equilibrium R is zero, and the actual value depends upon the displacements, which are supposed to be small. By Poisson's formula

$$\nabla^2 R = 4\pi\gamma\rho_0 s, \quad\dots\dots\dots\dots\dots\dots\dots(5)$$

so that

$$\frac{d^2s}{dt^2} + 4\pi\gamma\rho_0 s = 0. \quad\dots\dots\dots\dots\dots\dots\dots(6)$$

This applies to the interior of the sphere; and it appears that any departure from a uniform distribution brings into play forces giving stability, and further that the times of oscillation are the same whatever be the character of the disturbance. It is worthy of note that the constant ($\gamma\rho_0$) of itself determines a *time*.

In considering the significance of the vibrations expressed by (6), we must remember that when s is uniform no external forces having a potential are capable of disturbing the uniformity.

We now pass on to vibrations not involving a variable s, that is of such a kind that the fluid behaves as if incompressible. An irrotational displacement now requires that some of the negative fluid should traverse the surface of the positive sphere (a). In the interior $\nabla^2 R = 0$.

To represent simple vibrations we suppose that ϕ, &c. are proportional to e^{ipt}. By (3) $\nabla^2\phi = 0$; and we take (at any rate for trial)

$$\phi = e^{ipt} r^n S_n, \quad\dots\dots\dots\dots\dots\dots\dots(7)$$

* *Theory of Sound*, § 244.

where S_n is a spherical surface harmonic of the nth order. The velocity across the surface of the sphere at $r = a$ is

$$d\phi/dr = na^{n-1}e^{ipt}S_n;$$

and thus the quantity of fluid which has passed the element of area $d\sigma$ at time t is

$$\rho \int \frac{d\phi}{dr} \, dt \cdot d\sigma = \frac{\rho_0 na^{n-1}}{ip} e^{ipt}S_n d\sigma. \quad \dots\dots\dots\dots\dots(8)$$

The next step is to form the expression for R, the potential of all the forces. In equilibrium the positive and negative densities everywhere neutralize one another, and thus in the displaced condition R may be regarded as due to the surface distribution (8). By a well-known theorem in Attractions we have

$$R = -\frac{4\pi\gamma\rho_0 nr^n S_n e^{ipt}}{ip\,(2n+1)}. \quad \dots\dots\dots\dots\dots\dots(9)$$

But by (3) this is equal to $d\phi/dt$, or $ipe^{ipt}r^n S_n$. The recovery of $r^n S_n$ proves that the form assumed is correct; and we find further that

$$p^2 = \frac{4\pi\gamma\rho_0 \cdot n}{2n+1}. \quad \dots\dots\dots\dots\dots\dots(10)$$

This formula for the frequencies of vibration gives rise to two remarks. The frequency depends upon the density ρ_0, but not upon the radius (a) of the sphere. Again, as n increases, the pitch rises indeed, but approaches a finite limit given by $p^2 = 2\pi\gamma\rho_0$. The approach to a finite limit as we advance along the series is characteristic of the series of spectrum-lines found for hydrogen and the alkali metals, but in other respects the analogy fails. It is p^2, rather than p, which is simply expressed; and if we ignore this consideration and take the square root, supposing n large, we find

$$p \propto 1 - 1/2n,$$

whereas according to observation n^2 should replace n. Further, it is to be remarked that we have found only one series of frequencies. The different kinds of harmonics which are all of one order n do not give rise to different frequencies. Probably the simplicity of this result would be departed from if the number of electrons was treated merely as great but not infinite.

The principles which have led us to (10) seem to have affinity rather with the older views as to the behaviour of electricity upon a conductor than with those which we associate with the name of Maxwell. It is true that the vibrations above considered would be subject to dissipation in consequence of radiation, and that this dissipation would be very rapid, at any rate in the case of n equal to unity*. But this hardly explains the difference between the two views.

* In this case we should have to consider how the positive sphere is to be held at rest.

[1911. Some paragraphs dealing with the question of electrical vibrations outside a conducting sphere (J. J. Thomson, *Proc. Lond. Math. Soc.* Vol. xv. p. 197, 1884; *Recent Researches*, § 312, 1893), or of sonorous vibrations outside a rigid and fixed sphere, are omitted as involving a misconception. The matter had already been satisfactorily treated by Lamb (*Proc. Lond. Math. Soc.* Vol. xxxii. p. 208, 1900) and by Love (*Ibid.* Vol. ii. p. 88, 1904).]

In the calculation of frequencies given above for a cloud of electrons the undisturbed condition is one of equilibrium, and the frequencies of radiation are those of vibration about this condition of equilibrium. Almost every theory of this kind is open to the objection that I put forward some years ago*, viz. that p^2, and not p, is given in the first instance. It is difficult to explain on this basis the simple expressions found for p, and the constant differences manifested in the formulae of Rydberg and of Kayser and Runge. There are, of course, particular cases where the square root can be taken without complication, and Ritz† has derived a differential equation leading to a formula of this description and capable of being identified with that of Rydberg. Apart from the question whether it corresponds with anything mechanically possible, this theory has too artificial an appearance to inspire much confidence.

A partial escape from these difficulties might be found in regarding actual spectrum lines as due to *difference tones* arising from primaries of much higher pitch,—a suggestion already put forward in a somewhat different form by Julius.

In recent years theories of atomic structure have found favour in which the electrons are regarded as describing orbits, probably with great rapidity. If the electrons are sufficiently numerous, there may be an approach to steady motion. In case of disturbance, oscillations about this steady motion may ensue, and these oscillations are regarded as the origin of luminous waves of the same frequency. But in view of the discrete character of electrons such a motion can never be fully steady, and the system must tend to radiate even when undisturbed‡. In particular cases, such as some considered by Prof. Thomson, the radiation in the undisturbed state may be very feeble. After disturbance oscillations about the normal motion will ensue, but it does not follow that the frequencies of these oscillations will be manifested in the spectrum of the radiation. The spectrum may rather be due to the upsetting of the balance by which before disturbance radiation was prevented, and the frequencies will correspond (with modification) rather to the original distribution of electrons than to the oscillations. For example, if four equally spaced electrons revolve in a ring, the radiation is feeble and

---

* *Phil. Mag.* xliv. p. 362, 1897 ; *Scientific Papers*, iv. p. 345.

† Drude, *Ann.* Bd. xii. p. 264, 1903.

‡ Confer Larmor, *Matter and Æther.*

its frequency is four times that of revolution. If the disposition of equal spacing be disturbed, there must be a tendency to recovery and to oscillations about this disposition. These oscillations may be extremely slow; but nevertheless frequencies will enter into the radiation once, twice, and thrice as great as that of revolution, and with intensities which may be much greater than the original radiation of fourfold frequency.

An apparently formidable difficulty, emphasised by Jeans, stands in the way of all theories of this character. How can the atom have the definiteness which the spectroscope demands? It would seem that variations must exist in (say) hydrogen atoms which would be fatal to the sharpness of the observed radiation; and indeed the gradual change of an atom is directly contemplated in view of the phenomena of radioactivity. It seems an absolute necessity that the large majority of hydrogen atoms should be alike in a very high degree. Either the number undergoing change must be very small or else the changes must be sudden, so that at any time only a few deviate from one or more definite conditions.

It is possible, however, that the conditions of stability or of exemption from radiation may after all really demand this definiteness, notwithstanding that in the comparatively simple cases treated by Thomson the angular velocity is open to variation. According to this view the frequencies observed in the spectrum may not be frequencies of disturbance or of oscillations in the ordinary sense at all, but rather form an essential part of the original constitution of the atom as determined by conditions of stability.

314.

ON THE PRODUCTION OF VIBRATIONS BY FORCES OF RELATIVELY LONG DURATION, WITH APPLICATION TO THE THEORY OF COLLISIONS.

[*Philosophical Magazine*, Vol. XI. pp. 283—291, 1906.]

THE problem of the collision of elastic solid bodies has been treated theoretically in two distinct cases. The first is that of the longitudinal impact of elongated bars, which for simplicity may be supposed to be of the same material and thickness. Saint-Venant[*] showed that, except when the lengths are equal, a considerable fraction of the original energy takes the form of vibrations in the longer bar, so that the translational velocities after impact are less than those calculated by Newton for bodies which he called perfectly elastic. It will be understood that in Saint-Venant's theory the *material* is regarded as perfectly elastic, and that the total mechanical energy is conserved. The duration of the impact is equal to the period of the slowest vibration of the longer bar.

The experiments of Voigt[†], undertaken to test this theory, have led to the conclusion that it is inapplicable when the bars differ markedly in length. The observations agree much more nearly with the Newtonian law, in which all the energy remains translational. Further, Hamburger[‡] found that the duration of impact was much greater than according to theory, though it diminished somewhat as the relative velocity increased. I do not think that these discrepancies need cause surprise when we bear in mind that the theory presupposes a condition of affairs impossible to realise in practice. Thus it is assumed that the pressure during collision is uniform over the whole of the contiguous faces. But, however accurately the faces may be prepared, the pressure, at any rate in its earlier and later stages, must certainly be local and be connected with the approach by a law altogether

---

[*] *Liouville's Journal,* XII. (1867). See also Love's *Treatise on the Theory of Elasticity,* Vol. II. p. 137 (1893).

[†] Wied. *Ann.* XIX. (1883).

[‡] Wied. *Ann.* XXVIII. (1886).

different from that assumed in the calculation. Since the region of first contact would yield with relative ease, we may expect a prolongation of the impact, and in consequence, as we shall see more in detail presently, a diminished development of vibrations. Possibly with higher velocities and longer bars a nearer approach might be attained to the theoretical conditions.

But it is with Hertz's[*] solution, under certain conditions, of the problem of impinging curved bodies with which I am now more concerned. He commences with the purely statical problem of contact under pressure. Thus if two equal spheres of similar material be pressed together with a given force P_0, the surfaces of contact are moulded to a plane; and it is required to find the radius of the circle of contact, and more especially the distance (α) through which the centres (or other points remote from the place of contact) approach one another. It appears that the relation between P_0 and α is simply

$$P_0 = k_2 a^{\frac{3}{2}}, \quad \dots\dots\dots\dots\dots\dots\dots\dots\dots(1)$$

where k_2 depends only on the forms and materials of the two bodies. In the particular case above-mentioned,

$$k_2 = \frac{\sqrt{2r} \cdot E}{3(1-\sigma^2)}, \quad \dots\dots\dots\dots\dots\dots\dots(2)$$

where r is the radius of the spheres, E Young's modulus, and σ Poisson's ratio.

In applying this result to impacts Hertz proceeds:—"It follows both from existing observations and from the results of the following considerations, that the time of impact, *i.e.* the time during which the impinging bodies remain in contact, is very small in absolute value; yet it is very large compared with the time taken by waves of elastic deformation in the bodies in question to traverse distances of the order of magnitude of that part of their surfaces which is common to the two bodies when in closest contact, and which we shall call the surface of impact. It follows that the elastic state of the two bodies near the point of impact during the whole duration of impact is very nearly the same as the state of equilibrium which would be produced by the total pressure subsisting at any instant between the two bodies, supposing it to act for a long time. If, then, we determine the pressure between the two bodies by means of the relation which we previously found to hold between this pressure and the distance of approach along the common normal of two bodies at rest, and also throughout the volume of each body make use of the equations of motion of elastic solids, we can trace the progress of the phenomenon very exactly. We cannot in this way expect

[*] *Journal für reine und angewandte Mathematik*, xcii. p. 156 (1881); Hertz's *Miscellaneous Papers*, English edition, p. 146. A good account is given by Love, *loc. cit.*

to obtain general laws; but we may obtain a number of such if we make the further assumption that the time of impact is also large compared with the time taken by elastic waves to traverse the impinging bodies from end to end. When this condition is fulfilled, all parts of the impinging bodies, except those infinitely close to the point of impact, will move as parts of rigid bodies; we shall show from our results that the condition in question may be realised in the case of actual bodies." The above-mentioned condition may in fact always be satisfied by taking the relative velocity of impact to be sufficiently small.

The solution of the problem, thus limited, is now easily found. For the case of two spheres the relative acceleration $\ddot{a}$ is connected with P_0 by the equation

$$P_0 = - \ddot{a}/k_1, \quad \dots\dots\dots\dots\dots\dots\dots\dots(3)$$

where
$$k_1 = (m_1 + m_2)/m_1 m_2,$$

and m_1, m_2 are the masses of the spheres. Eliminating P_0 between (1) and (3), we get

$$\ddot{a} + k_1 k_2 a^{\frac{3}{2}} = 0, \quad \dots\dots\dots\dots\dots\dots\dots\dots(4)$$

and on integration as the equation of energy

$$\dot{a}^2 - \dot{a}_0^2 + \tfrac{4}{5} k_1 k_2 a^{\frac{5}{2}} = 0, \quad \dots\dots\dots\dots\dots\dots(5)$$

a_0 being the relative velocity before impact.

"The greatest compression takes place when $\dot{a}$ vanishes, and if a_1 be the value of a at this instant

$$a_1 = \left[\frac{5\dot{a}_0^2}{4 k_1 k_2} \right]^{\frac{2}{5}}. \quad \dots\dots\dots\dots\dots\dots\dots(6)$$

Before the instant of greatest compression the quantity a increases from zero to a maximum a_1, and $\dot{a}$ diminishes from a maximum $\dot{a}_0$ to zero. After the instant of greatest compression a diminishes from a_1 to zero and $\dot{a}$ increases to $\dot{a}_0$. The bodies then separate, and the velocity with which they rebound is equal to that with which they approach. This result is in accord with Newton's Theory. It might have been predicted from the character of the fundamental assumptions."

"The duration of the impact is

$$2 \int_0^{a_1} \frac{da}{\sqrt{(\dot{a}_0^2 - \tfrac{4}{5} k_1 k_2 a^{\frac{5}{2}})}} = \frac{2a_1}{\dot{a}_0} \int_0^1 \frac{dx}{\sqrt{(1 - x^{\frac{5}{2}})}}$$

$$= \frac{4}{5} \frac{a_1}{\dot{a}_0} \frac{\sqrt{\pi} \cdot \Gamma\left(\frac{2}{5}\right)}{\Gamma\left(\frac{9}{10}\right)} = \frac{a_1}{\dot{a}_0} (2 \cdot 9432), \quad \dots\dots\dots(7)$$

where a_1 is given by (6).

The duration of impact, therefore, varies inversely as the fifth root of the initial relative velocity*."

So long as the condition is satisfied that the duration of the impact is very long in comparison with the free periods, vibrations will not be excited in a sensible degree, the energy remains translational, and Newton's laws find application. It would be of great interest if we could enfranchise ourselves from this restriction. It is hardly to be expected that a complete solution of the problem will prove feasible, but I have thought that it would be worth while to inquire into the circumstances of the *first* appearance of sensible vibrations. We should then be in a better position to appreciate at least the range over which Newton's laws may be expected to hold.

In the case of spheres the vibrations to be considered are those of the "second class" investigated by Lamb†. They involve spherical harmonic functions of the various orders, limited in the present case to the *zonal* kind. But for each order there are an infinite number of modes corresponding to greater or less degrees of subdivision along the radius. The first appearance of vibrations will be confined to those of longest period, of which the most important is of the second order. In this mode the sphere vibrates symmetrically with respect both to a polar axis and to the equatorial plane, the greatest compression along the axis synchronizing with the greatest expansion at the equator. In what follows we shall denote by ϕ_1, ϕ_2, &c. the radial displacement at the pole (point of contact) corresponding to the several modes, the first ϕ_1 being appropriated to that mode in which the sphere moves as a rigid body (spherical harmonic of order 1), the next ϕ_2 to the mode of the second order above described which gives the principal vibration.

Since there is no force of restitution corresponding to ϕ_1, the equation for it takes the simple form

$$a_1 \ddot{\phi}_1 = P_0, \dotfill (8)$$

P_0 being as before the total pressure between the spheres at any time, and a_1 a coefficient of inertia—in this case the simple mass of a sphere. On the other hand, the equations for ϕ_2 &c. are of the form

$$a_2 \ddot{\phi}_2 + c_2 \phi_2 = P_0, \dotfill (9)$$

c_2 &c. being coefficients of stability to be treated as large. This form applies to all the *lower* modes, for which the force of collision operating at any moment may be treated as a whole. By equation (1) of Hertz's theory $P_0 = k_2 \alpha^{\frac{3}{2}}$, but now that we are admitting the possibility of vibrations α must be reckoned no longer from the centre, but from a point which is at

* Love, *loc. cit.* p. 154.
† *Proc. Lond. Math. Soc.* Vol. XIII. p. 189 (1882).

once near the surface and yet distant from it by an amount large in comparison with the diameter of the circle of contact.   We may write

$$\tfrac{1}{2}a = \phi_1 + \phi_2 + \dots, \quad \dots\dots\dots\dots\dots\dots\dots(10)$$

inclusion being made of the coordinates of the lower modes only.   The sum of *all* the coordinates would be zero, since (in the case of equal spheres) the pole does not move.   Thus

$$P_0 = k_2 (2\phi_1 + 2\phi_2 + \dots)^{\frac{3}{2}}. \quad \dots\dots\dots\dots\dots(11)$$

In the first approximation c_2 &c. are regarded as infinite, so that ϕ_2 &c. vanish. P_0 reduces to $k_2 (2\phi_1)^{\frac{3}{2}}$, and so from (8)

$$a_1 \ddot{\phi}_1 = k_2 (2\phi_1)^{\frac{3}{2}}, \quad \dots\dots\dots\dots\dots\dots(12)$$

the solution of which gives ϕ_1 as in Hertz's theory.   If P_0 be regarded as a known function of the time, ϕ_2 is determined by (9); but it may be well at this stage to ascertain how far P_0 is modified in a second approximation. Retaining for brevity ϕ_2 only, we have approximately $\phi_2 = P_0/c_2$.   Hence

$$a_1 \ddot{\phi}_1 = k_2 (2\phi_1)^{\frac{3}{2}} \left\{ 1 + \frac{3}{2} \frac{k_2}{c_2} (2\phi_1)^{\frac{1}{2}} \right\}, \quad \dots\dots\dots\dots(13)$$

and we infer that P_0 is changed by a term of the order c_2^{-1}.

We will now pass on to consider the general problem of a vibrator whose natural vibrations are very rapid in comparison with the force which operates. We write (9) in the form

$$\ddot{\phi} + n^2 \phi = P_0/a_2 = \Phi, \dots\dots\dots\dots\dots(14)$$

where $n^2 = c_2/a_2$, and is to be treated as very great.   If ϕ and $\dot{\phi}$ vanish when $t = 0$, the solution of (14) is*

$$\phi = \frac{1}{n} \int_0^t \sin n\,(t - t')\, \Phi_{t=t'}\, dt'. \quad \dots\dots\dots\dots(15)$$

If the force operates only between $t = 0$ and $t = \tau$ and we require the value of ϕ at a time t greater than τ, that is after the operation of the force has ceased, we may write

$$\phi = \frac{1}{n} \int_0^\tau \sin n\,(t - t') \Phi_{t=t'}\, dt'. \quad \dots\dots\dots\dots(16)$$

If τ be infinitely small, the force reduces to an impulse, and we get

$$\phi = n^{-1} \sin nt \,.\, \int \Phi dt; \quad \dots\dots\dots\dots(17)$$

but it is the other extreme which concerns us at present.

In many cases, especially when $\Phi = 0$ at the limits, we may advantageously integrate (16) by parts.   Thus

$$\phi = n^{-2} \Phi_{t=\tau} \cos n\,(t - \tau) - n^{-2} \Phi_{t=0} \cos nt - \frac{1}{n^2} \int_0^\tau \cos n\,(t - t') \frac{d\Phi}{dt'}\, dt'. \ \dots(18)$$

---

* *Theory of Sound*, § 66.

Again

$$-\frac{1}{n^2}\int_0^\tau \cos n(t-t')\frac{d\Phi}{dt'}dt' = \frac{1}{n^3}\frac{d\Phi_{t=\tau}}{dt}\sin n(t-\tau)$$

$$-\frac{1}{n^3}\frac{d\Phi_{t=0}}{dt}\sin nt - \frac{1}{n^3}\int_0^\tau \sin n(t-t')\frac{d^2\Phi}{dt'^2}dt', \quad \ldots(19)$$

and so on if required. In this way we obtain a series proceeding by descending powers of n, and thus presumably advantageous when n is great.

As an example, let $\Phi = t$, so that $d^2\Phi/dt^2 = 0$. The force rises from zero at $t=0$ to a greatest value at $t=\tau$ and then suddenly drops to zero. From (18), (19) we find

$$\phi = n^{-2}\tau\cos n(t-\tau) + n^{-3}\sin n(t-\tau) - n^{-3}\sin nt. \ldots\ldots(20)$$

Again, take the parabolic law

$$\Phi = t\tau - t^2, \ldots\ldots\ldots\ldots\ldots\ldots\ldots(21)$$

so that $\Phi = 0$ at both limits,

$$d\Phi/dt = \tau - 2t, \qquad d^2\Phi/dt^2 = -2.$$

From (18), (19)

$$\phi = -\tau n^{-3}\sin n(t-\tau) - \tau n^{-3}\sin nt + 2n^{-4}\cos n(t-\tau) - 2n^{-4}\cos nt$$

$$= -2n^{-3}\tau\cos\tfrac{1}{2}n\tau.\sin n(t-\tfrac{1}{2}\tau) + 4n^{-4}\sin\tfrac{1}{2}n\tau.\sin n(t-\tfrac{1}{2}\tau). \ldots(22)$$

If Φ and its differential coefficients up to a high order are continuous within the range of integration and vanish at the limits, the leading term in the development of (16) is of high inverse power in n. An extreme case of this kind is considered by Mr Jeans*, who takes

$$\Phi = \frac{c}{\pi(c^2+t^2)}.$$

In this case the solution involves the factor e^{-nc}, smaller when n is great than any inverse power of n. But the force is not here limited to a finite range of time.

The application of these results to the problem of the collision of equal elastic spheres is not quite so straightforward as had been expected. In (9),

$$\Phi = P_0/a_2 = a_2^{-1}k_2\alpha^{\frac{3}{2}}, \ldots\ldots\ldots\ldots\ldots(23)$$

α denoting, as in (4), (5), (6), (7), the approach of the spheres. The terminal values of α and of $\alpha^{\frac{3}{2}}$ are zero. Again,

$$\frac{d}{dt}(\alpha^{\frac{3}{2}}) = \tfrac{3}{2}\alpha^{\frac{1}{2}}\frac{d\alpha}{dt}, \ldots\ldots\ldots\ldots\ldots(24)$$

---

* *Dynamical Theory of Gases*, Cambridge, 1904, § 241. I should perhaps mention that most of the results of the present paper were obtained before I was acquainted with Mr Jeans' work.

so that $d\Phi/dt$ vanishes at the limits of the range.   But

$$\frac{d^2}{dt^2}(\alpha^{\frac{3}{2}}) = \tfrac{3}{4}\alpha^{-\frac{1}{2}}\left(\frac{d\alpha}{dt}\right)^2 + \tfrac{3}{2}\alpha^{\frac{1}{2}}\frac{d^2\alpha}{dt^2} = \tfrac{3}{4}\dot\alpha_0^2\alpha^{-\frac{1}{2}} - \tfrac{21}{16}k_1k_2\alpha^2, \quad ......(25)$$

use being made of (4), (5); and the first part of this becomes *infinite* at the limits where $\alpha = 0$.

Equations (18), (19) give

$$\phi = -\frac{1}{n^3}\int_0^\tau \sin n\,(t - t')\,\frac{d^2\Phi}{dt'^2}\,dt', \quad ....(26)$$

and in this we have now to consider the two parts

$$\int \sin n\,(t - t')\,\alpha^{-\frac{1}{2}}dt' \quad \text{and} \quad \int \sin n\,(t - t')\,\alpha^2 dt'.$$

For the second we get on integration by parts, since α^2 vanishes at both limits,

$$-\frac{1}{n}\int \cos n\,(t - t')\,\frac{d\alpha^2}{dt'}\,dt',$$

which is of order n^{-1}, or less.   In the first part the relation between dt' and $d\alpha$ is, as in (7),

$$dt' = \frac{d\alpha}{\sqrt{(\dot\alpha_0^2 - \tfrac{4}{5}k_1k_2\alpha^{\frac{5}{2}})}}. \quad ....(27)$$

If we exclude the terminal parts of the range, the integral would be of order n^{-1}, or less, so that it is only the terminal parts that contribute to the leading term.   For the beginning we see from (27), or independently, that

$$\alpha = \dot\alpha_0 t, \quad ....(28)$$

nearly, so that for this terminal region

$$\int \sin n\,(t - t')\,\alpha^{-\frac{1}{2}}dt' = \frac{\dot\alpha_0^{-\frac{1}{2}}}{n^{\frac{1}{2}}}\left\{\sin nt\int\frac{\cos nt'd\,(nt')}{(nt')^{\frac{1}{2}}} - \cos nt\int\frac{\sin nt'd\,(nt')}{(nt')^{\frac{1}{2}}}\right\}.$$

When we suppose n very great, the limits of integration may be identified with zero and infinity; and further by a known theorem

$$\int_0^\infty \frac{\sin x\,dx}{\sqrt{x}} = \int_0^\infty \frac{\cos x\,dx}{\sqrt{x}} = \sqrt{(\tfrac{1}{2}\pi)}.$$

Thus, so far as it depends upon the early part of the collision,

$$\phi = \frac{3\sqrt{\pi}}{4}\frac{k_2\dot\alpha_0^{\frac{3}{2}}}{a_2 n^{\frac{7}{2}}}\cos{(nt + \tfrac{1}{4}\pi)}. \quad ....(29)$$

There will be a similar term due to the end of the collision, derivable from (29) by replacing nt with $n\,(t - \tau)$.

If, as I think must be the case, (29) gives the leading term in the expression for a vibration, the next question is as to the order of magnitude of the

corresponding energy in comparison with the energy before collision, viz.
$\frac{1}{2}$ Mass $\times \dot{a}_0^2$, or $\frac{2}{3}\pi\rho r^3 \dot{a}_0^2$.

The maximum kinetic energy of the vibration is given by

$$\tfrac{1}{2} a_2 \phi^2_{max.} = \frac{9\pi}{32} \frac{k_2^2 \dot{a}_0^3}{a_2 n^5},$$

and the ratio (R) of this to the energy before collision is

$$R = \frac{27}{64} \frac{k_2^2 \dot{a}_0}{a_2 n^5 \rho r^3} = \frac{3}{32} \frac{E^2 \dot{a}_0}{(1-\sigma^2)^2 a_2 n^5 \rho r^2}, \quad \dots\dots\dots\dots(30)$$

if we introduce the value of k_2 from (2).

The precise value of a_2 would have to be calculated from Lamb's theory. It is easy to see that it is decidedly smaller than, but of the same order of magnitude as, the mass of the sphere, viz. $\frac{4}{3}\pi\rho r^3$.

The precise value of R would depend also upon σ, but for our purpose it will suffice to make $\sigma = \frac{1}{2}$. Thus we take

$$R = \frac{E^2}{5\rho^2} \frac{\dot{a}_0}{n^5 r^5}. \quad \dots\dots\dots\dots\dots\dots\dots(31)$$

According to Lamb's calculation * for the principal vibration of the second order in spherical harmonics ($\sigma = \frac{1}{2}$)

$$n = \sqrt{\left(\frac{E}{\rho}\right)} \cdot \frac{\cdot 85\,\pi}{\sqrt{3} \cdot r}, \quad \dots\dots\dots\dots\dots(32)$$

so that approximately $$R = \frac{1}{50} \frac{\dot{a}_0}{\sqrt{(E/\rho)}}. \quad \dots\dots\dots\dots\dots(33)$$

In (33) $\sqrt{(E/\rho)}$ is the velocity of longitudinal vibrations along a bar of the material in question, and the comparison is between this velocity and the velocity of approach before collision. In steel the velocity of longitudinal vibrations is about 500,000 cms. per second, or about 16 times that of sound in air. It will be seen that in most cases of collision R is an exceedingly small ratio.

The general result of our calculation is to show that Hertz's theory of collisions has a wider application than might have been supposed, and that under ordinary conditions vibrations should not be generated in appreciable degree. So far as this conclusion holds, the energy of colliding spheres remains translational, and the velocities after impact are governed by Newton's laws, as deducible from the principles of energy and momentum.

* *Loc. cit.* p. 206.

315.

ON THE DILATATIONAL STABILITY OF THE EARTH.

[*Proceedings of the Royal Society*, A. Vol. LXXVII. pp. 486—499, 1906.]

THE theory of elastic solids usually proceeds upon the assumption that the body is initially in a state of ease, free from stress and strain. Displacements from this condition, due to given forces, or vibrations about it, are then investigated, and they are subject to the limitation that Hooke's law shall be applicable throughout and that the strain shall everywhere be small. When we come to the case of the earth, supposed to be displaced from a state of ease by the mutual gravitation of its parts, these limits are transgressed; and several writers* who have adopted this point of view have indicated the obstacles which inevitably present themselves. In his interesting paper† Professor Jeans, in order to attain mathematical definiteness, goes the length of introducing forces to counteract the self-gravitation: "That is to say, we must artificially annul gravitation in the equilibrium configuration, so that this equilibrium configuration may be completely unstressed, and each element of matter be in its normal state." How wide a departure from actuality is here implied will be understood if we reflect that under such forces the interior of the earth would probably be as mobile as water.

It appears to me that a satisfactory treatment of these problems must start from the condition of the earth as actually stressed by its self-gravitation, and that the difficulties to be faced in following such a course may not be so great as has been supposed. The stress, which is so enormous as to transcend all ordinary experience, is of the nature of a purely hydrostatic pressure, and as to this surely there can be no serious difficulty. After great compression the response to further compressing stress is admittedly less than at first, but there is no reason to doubt that the reaction is purely elastic and that the material preserves its integrity. At this point it may be well to remark, in passing, upon the confusion often met with in geological

---

* See, for example, Love, *Theory of Elasticity*, § 127; Chree, *Phil. Mag.* Vol. XXXII. p. 233, 1891; Jeans, *Phil. Trans.* A. Vol. CCI. p. 157, 1903.

† *Loc. cit.* p. 161.

and engineering writings arising from the failure to distinguish between
a one-dimensional and a three-dimensional, or hydrostatic, pressure. When
rock or cast iron is said to be *crushed* by such and such a pressure, it is the
former kind of pressure which is, or ought to be, meant. There is no evidence
of crushing under purely hydrostatic pressure, however great.

Not only is the integrity of a body unimpaired by hydrostatic pressure,
but there is reason to think that the superaddition of such a pressure may
preserve a body from rupture under stresses that would otherwise be fatal.
FitzGerald raises this question in a review* of Hertz's *Miscellaneous Papers*.
He writes: " In considering the cracking of a material like glass, Hertz seems
to think its cracking will depend only on the tension; that it will crack
where the tension exceeds a certain limit. He does not seem to consider
whether it might not crack by shearing with hardly any tension. It is
doubtful whether a material in which there were sufficient general compression to prevent any tension at all, would crack. Rocks seem capable of being
bent about and distorted to almost any extent without cracking, and this
might very well be expected if they were at a sufficient depth under other
rocks to prevent their parts being under tension. It is an interesting
question whether a piece of glass could be bent without breaking if it were
strained at the bottom of a sufficiently deep ocean. On the other hand,
there seems very little doubt that the parts of a body might slide past one
another under the action of a shear, and would certainly crack unless there
were a sufficiently great compressional stress to prevent the crack; and that
consequently a body might crack, even though the tensions were not by
themselves sufficiently great to cause separation, and might crack where the
shear was greatest, and not where the tensions were greatest."

When we reflect that pieces of lead may be made to unite under pressure
when the surfaces are clean, and upon what is implied when insufficiently
lubricated journals, or slabs of glass under polish, *seize*, we may well doubt
whether it is possible to disintegrate a material at all when subjected to
enormous hydrostatic pressure. In the words of Dr Chree†: " The conditions
under which the deep-seated materials of the earth exist are fundamentally
different from those we are familiar with at the surface. The enormous
pressure, and the presumably high temperature, very likely combine to
produce a state to which the terms solid, viscous, liquid, as we understand
them, are alike inapplicable."

A study of the mechanical operations of coining and of stamping (in
recent years, I believe, much developed) would probably throw light upon
this question. We know that rod or tube may be "squirted" from hot (but
solid) lead. Is the obstacle to a similar treatment of harder material purely

* *Nature*, November 5, 1896; *Scientific Writings*, p. 433.
† *Phil. Mag.* Vol. XLIII. p. 173, 1897.

practical? In the laboratory I have experimented upon jellies of various degrees of stiffness, on the principle of suiting the material to the appliances rather than the appliances to the material. In the simplest arrangement a leaden bullet is imbedded in jelly contained in a strong glass tube which the bullet somewhat nearly fits. Although the tube stand vertical for several days, there is no appreciable descent. But if by numerous longitudinal impacts against a suitable pad the inertia of the bullet be brought into play, movements through several inches may be obtained. Here, although the deformations are very violent, there is no rupture visible, either before or behind the bullet.

When an elastic body is slightly displaced from the condition of ease, the potential energy (V) is expressed by terms involving the squares and products of the displacements. If, however, we suppose given finite forces to be constantly imposed, so that the initial condition is one of strain, the case is somewhat, though not essentially, altered. It may be convenient to make a statement, once for all, in terms of generalised co-ordinates. If under the action of the forces Φ, Θ, ... the co-ordinates assume the values ϕ, θ, ... we have in terms of the potential energy of strain V,

$$\Phi = \frac{dV}{d\phi}, \qquad \Theta = \frac{dV}{d\theta}, \text{ etc. } \dots\dots\dots\dots\dots(1)$$

If the forces permanently imposed be distinguished by the suffix (0), they are connected with the corresponding values of the co-ordinates, ϕ_0, etc., by the equations

$$\Phi_0 = \frac{dV}{d\phi_0}, \qquad \Theta_0 = \frac{dV}{d\theta_0}, \text{ etc. } \dots\dots\dots\dots\dots(2)$$

This strained condition is now to be regarded as initial, and displacements from it are denoted by ascribing to the co-ordinates the slightly altered values $\phi_0 + \delta\phi$, $\theta_0 + \delta\theta$, etc. For the potential energy of strain we have

$$V - V_0 = \frac{dV}{d\phi_0}\delta\phi + \frac{dV}{d\theta_0}\delta\theta + \tfrac{1}{2}\frac{d^2V}{d\phi_0^2}(\delta\phi)^2 + \frac{d^2V}{d\phi_0 d\theta_0}\delta\phi . \delta\theta + \dots, \dots(3)$$

which is of the *first* order of the small quantities $\delta\phi$, etc. But $V - V_0$ is not now the whole potential energy. In addition to the potential energy of strain we have to include that of the steadily imposed forces, represented by the terms

$$- \Phi_0\delta\phi - \Theta_0\delta\theta - \dots \dots\dots\dots\dots\dots(4)$$

The whole potential energy is thus

$$V - V_0 - \Phi_0\delta\phi - \Theta_0\delta\theta - \dots$$

$$= \tfrac{1}{2}\frac{d^2V}{d\phi_0^2}(\delta\phi)^2 + \frac{d^2V}{d\phi_0 d\theta_0}\delta\phi . \delta\theta + \tfrac{1}{2}\frac{d^2V}{d\theta_0^2}(\delta\theta)^2 + \dots, \dots(5)$$

regard being paid to (2). The total potential energy, as given by (5), is now of the second order in $\delta\phi$, &c., as is obviously required by the circumstance that the strained condition ϕ_0, &c., is one of equilibrium under the proposed forces. The coefficients of stability are $d^2V/d\phi_0{}^2$, $d^2V/d\phi_0 d\theta_0$, &c., and they may differ finitely from the values which obtained previously to the application of the forces Φ_0, &c.

As an example having an immediate bearing upon the matter in hand, let us consider the case of a uniform non-gravitating body originally in a state of ease. If a small hydrostatic pressure Φ act upon it, the volume ϕ changes proportionally, and the ratio gives the "compressibility" of the body in this condition. Under the action of a finite pressure Φ_0 the volume may be greatly altered, especially if the body be gaseous, but the new condition is still one of equilibrium and may be regarded as initial. The compressibility now may be quite different from before, but it may be treated in the same way as depending upon the small *change* of volume $\delta\phi$ accompanying the imposition of a small *additional* pressure $\delta\Phi$.

To those who, while accepting the usual elastic theory for bodies in a state of ease, repudiate the application to bodies subject to great hydrostatic pressure, I would suggest that liquids and solids, as we know them, are not really free from stress. In virtue of cohesional forces, there is every reason to believe, the interior of a drop of water is under pressure not insignificant even in comparison with those prevailing inside the earth, and the same may be said of a piece of steel.

The conclusion that I draw is that the usual equations may be applied to matter in a state of stress, provided that we allow for altered values of the elasticities. In general, these elasticities will not only vary from point to point, but be æolotropic in character. If, however, we suppose that the body is naturally isotropic, and that the imposed stress is everywhere merely a hydrostatic pressure, so that by pure expansion a state of ease could be attained, the case is much simpler and probably suffices for an approximate view of the condition of the earth. But although the initial state is one free of shear, we are not to conclude that the rigidity is the same as it would be without the imposed pressure. On the contrary, there is much reason to think that the rigidity would be increased. If there is any analogy to be found in a pile of mutually repellant hard spheres, it will follow that an infinite pressure will entail infinite rigidity as well as infinite incompressibility.

In the original draft of this paper I had supposed that it would be possible upon these lines to find another and a more practical basis for Professor Jeans' analysis. A correspondence with Professor Love[*] has, however, convinced me that this hope is destined to disappointment, and the remainder of

---

[*] To whom I am indebted also for other corrections.

the paper loses accordingly much of the interest which at first I felt for it. In Professor Jeans' theory, if Δ be the dilatation, so that the altered density is $\rho(1 - \Delta)$, U the radial outward displacement, E the potential of a volume-distribution of density $\rho\Delta$, and a surface-distribution of density $-\rho U$, the displacements ξ, η, ζ are subject to

$$\rho\frac{d^2\xi}{dt^2} = (\lambda + \mu)\frac{d\Delta}{dx} + \mu\nabla^2\xi - \rho\frac{dE}{dx}, \quad \ldots\ldots\ldots\ldots(6)$$

and two similar equations relating to η, ζ. In (6) λ and μ are the elastic constants of Lamé's notation, and they relate to displacements from the *compressed* initial condition. From equations (6) we obtain, as usual,

$$\rho\frac{d^2\Delta}{dt^2} = (\lambda + 2\mu)\nabla^2\Delta - \rho\nabla^2 E; \quad \ldots\ldots\ldots\ldots(7)$$

and by Poisson's equation $\nabla^2 E = -4\pi\gamma\rho\Delta, \quad \ldots\ldots\ldots\ldots\ldots\ldots(8)$

γ being the constant of gravitation.   Thus

$$\rho\frac{d^2\Delta}{dt^2} = (\lambda + 2\mu)\nabla^2\Delta + 4\pi\gamma\rho^2\Delta, \quad \ldots\ldots\ldots\ldots(9)$$

which is Professor Jeans' equation *.

The solution of these equations is developed by Professor Jeans with the view of determining at what point instability sets in.   Attention is concentrated mainly upon the solution of (9) expressed by a spherical function of order *one*, as being that which bears upon the question of the evolution of the moon.

I had intended merely to indicate a somewhat simpler treatment, following more closely the notation and method of Lamb's memoir, " On the Vibrations of an Elastic Sphere "†; but as the results so obtained do not agree with those of Jeans, it appears necessary to set forth the argument in fuller detail, so as to facilitate criticism.

If in (9) we assume that Δ is proportional to $\cos pt$, we get

$$(\nabla^2 + h^2)\Delta = 0, \quad \ldots\ldots\ldots\ldots\ldots\ldots(10)$$

where $h^2 = \dfrac{\rho(p^2 + 4\pi\gamma\rho)}{(\lambda + 2\mu)}: \quad \ldots\ldots\ldots\ldots\ldots(11)$

and the solution of (10), subject to the condition of finiteness at the centre, is

$$\Delta = (hr)^{-\frac{1}{2}} J_{n+\frac{1}{2}}(hr) . S_n . \cos pt, \quad \ldots\ldots\ldots\ldots(12)$$

J being the symbol of Bessel's functions, and S_n a spherical surface function of order n.   As is well known, $J_{n+\frac{1}{2}}$ is expressible in finite terms ; in the case of $n = 0$, $(hr)^{\frac{1}{2}} J_{\frac{1}{2}}(hr)$ may be replaced in (12) by $\sin hr/hr$, a constant factor being disregarded.

---

* *Loc. cit.* p. 162.

† *Proceedings London Mathematical Society*, Vol. XIII. p. 192, 1882.

Before going further it may be well to consider the particular case of a *fluid* for which $\mu = 0$. Here the solution for Δ already given suffices to solve the problem, and the condition of no pressure at the surface $(r = a)$ gives at once

$$J_{n+\frac{1}{2}}(ha) = 0, \quad \dots (13)$$

which with (11) determines p^2 in terms γ, ρ, a and the elastic constant λ. The criterion of stability follows by setting $p = 0$. In the case of $n = 0$, where the displacements are symmetrical, $ha = m\pi$, m being an integer; and we see that equilibrium is unstable for symmetrical displacements if

$$a^2\rho^2\gamma > \tfrac{1}{4}\pi\lambda. \quad \dots (14)$$

In general, by (8) and (10)

$$\nabla^2 E = \frac{4\pi\gamma\rho}{h^2}\nabla^2\Delta, \quad \dots (15)$$

so that

$$E = \frac{4\pi\rho\gamma}{h^2}\Delta + e, \quad \dots (16)$$

where e satisfies throughout the sphere

$$\nabla^2 e = 0. \quad \dots (17)$$

Substituting the value of E in (6), we get with regard to (11)

$$(\nabla^2 + k^2)\,\xi = \left(1 - \frac{k^2}{h^2}\right)\frac{d\Delta}{dx} + \frac{\rho}{\mu}\frac{de}{dx}, \quad \dots (18)$$

where

$$k^2 = p^2\rho/\mu. \quad \dots (19)$$

Equation (18) and its companions may be treated as in Lamb's classical paper. A solution is

$$\xi = -\frac{1}{h^2}\frac{d\Delta}{dx} + \frac{1}{p^2}\frac{de}{dx}, \text{ etc.,} \quad \dots (20)$$

where Δ satisfies (10). In virtue of (17) these values satisfy the relation

$$\frac{d\xi}{dx} + \frac{d\eta}{dy} + \frac{d\zeta}{dz} = \Delta;$$

and the solution may be completed by the addition of terms u, v, w, satisfying $(\nabla^2 + k^2)\,u = 0$, etc., as well as the relation

$$\frac{du}{dx} + \frac{dv}{dy} + \frac{dw}{dz} = 0.$$

Professor Lamb gives the general values of u, v, w. For our present purpose, and with limitation to one order of spherical harmonics, it suffices to take

$$u = \psi_{n-1}(kr)\frac{d\phi_n}{dx} - \frac{n}{n+1}\frac{k^2 r^{2n+3}}{2n+1 \cdot 2n+3}\psi_{n+1}(kr)\frac{d}{dx}\frac{\phi_n}{r^{2n+1}}, \quad \dots (21)$$

and two similar equations, where ϕ_n is a solid harmonic of degree n; $r = \sqrt{(x^2 + y^2 + z^2)}$; and ψ_n is defined by the equation

$$\psi_n(\theta) = 1 - \frac{\theta^2}{2 \cdot 2n+3} + \frac{\theta^4}{2 \cdot 4 \cdot 2n+3 \cdot 2n+5} - \dots \quad \dots (22)$$

Save as to a constant multiplier $\theta^n \psi_n(\theta)$ is identical with $\theta^{-\frac{1}{2}} J_{n+\frac{1}{2}}(\theta)$, as employed in (12). ψ_n is thus associated with *solid* in place of *surface* harmonics. The function possesses the following properties

$$\psi_n'(\theta) = -\frac{\theta}{2n+3}\psi_{n+1}(\theta); \quad\quad\quad\quad\quad\text{......................(23)}$$

$$\psi_n(\theta) + \frac{\theta}{2n+1}\psi_n'(\theta) = \psi_{n-1}(\theta); \quad\quad\quad\quad\text{.................(24)}$$

$$\psi_n(\theta) - \psi_{n-1}(\theta) = \frac{\theta^2}{2n+1 \cdot 2n+3}\psi_{n+1}(\theta). \quad\quad\text{...........(25)}$$

A formula in spherical harmonics frequently required is

$$x\phi_n = \frac{r^2}{2n+1}\left\{ \frac{d\phi_n}{dx} - r^{2n+1}\frac{d}{dx}\frac{\phi_n}{r^{2n+1}}\right\}. \quad\quad\text{..............(26)}$$

The term of the nth order in Δ is thus

$$\Delta_n = \psi_n(hr).\,\omega_n \quad\quad\quad\quad\quad\quad\text{............................(27)}$$

and corresponding thereto

$$\xi = -\frac{1}{h^2}\frac{d\Delta_n}{dx} + \frac{1}{p^2}\frac{de_n}{dx} + u, \quad\quad\quad\text{........................(28)}$$

where u is defined as above, and e_n, as well as ϕ_n and ω_n, is a solid harmonic of degree n.

The formation of the boundary conditions to be satisfied at the free surface of the sphere $(r = a)$ proceeds almost exactly as in Lamb's investigation (p. 199), the only difference arising from the fact that h has now a different value. The first of the three symmetrical surface conditions may be written

$$\frac{\lambda}{\mu}x\Delta_n + \left(r\frac{d}{dr} - 1\right)\xi + \frac{d}{dx}(x\xi + y\eta + z\zeta) = 0. \quad\text{............(29)}$$

The terms in (29) depending on the parts of ξ, η, ζ which involve Δ_n are found to be

$$A_n\frac{d\omega_n}{dx} + B_n\frac{d}{dx}\frac{\omega_n}{r^{2n+1}}, \quad\quad\quad\text{..........................(30)}$$

where $\quad A_n = \frac{\lambda + 2\mu}{\mu}\frac{a^2\psi_n(ha)}{2n+1} - \frac{2n-2}{h^2}\psi_{n-1}(ha); \quad\quad\text{....................(31)}$

$$B_n = -\frac{\lambda + 2\mu}{\mu}\frac{a^{2n+3}}{2n+1}\psi_n(ha) + \frac{2(n+2)a^{2n+3}}{2n+1 \cdot 2n+3}\psi_{n+1}(ha). \quad\text{...(32)}$$

In like manner Lamb finds for the terms in (29) arising from u, v, w,

$$C_n\frac{d\phi_n}{dx} + D_n\frac{d}{dx}\frac{\phi_n}{r^{2n+1}}, \quad\quad\quad\text{.........................(33)}$$

where $\quad C_n = -\left\{\frac{k^2a^2}{2n+1}\psi_n(ka) - 2(n-1)\psi_{n-1}(ka)\right\}; \quad\text{.................(34)}$

$$D_n = -\frac{n}{n+1}\frac{k^2a^{2n+3}}{2n+1}\left\{\psi_n(ka) + \frac{2(n+2)}{k^2a^2}ka\,\psi_n'(ka)\right\}. \quad\text{.........(35)}$$

We have now further an additional part arising from e_n, which, it should be observed, makes no contribution to Δ_n. In this

$$\left(r\frac{d}{dr} - 1 \right)\frac{de_n}{dx} = (n-2)\frac{de_n}{dx}, \qquad x\frac{de_n}{dx} + y\frac{de_n}{dy} + z\frac{de_n}{dz} = ne_n;$$

so that the additional part is

$$\frac{2n-2}{p^2}\frac{de_n}{dx} \dots\dots\dots\dots\dots\dots\dots\dots(36)$$

The two most important cases where $n = 0$ and $n = 1$ are also especially simple, in that (36) disappears. It will be convenient to consider them first. When $n = 0$, u, v, w vanish: also, since e_0 is constant, (28) reduces to

$$\xi = -\frac{1}{h^2}\frac{d\Delta_0}{dx}, \qquad \eta = -\frac{1}{h^2}\frac{d\Delta_0}{dy}, \qquad \zeta = -\frac{1}{h^2}\frac{d\Delta_0}{dz}, \qquad \dots\dots(37)$$

where Δ_0 is proportional to $\psi_0(hr)$. The motion is everywhere purely radial. Exactly as in Lamb's investigation of vibrations without gravity, the expression (30) reduces to

$$B_0\frac{d}{dx}\frac{\omega_0}{r},$$

where ω_0 is a constant, so that the surface conditions yield simply $B_0 = 0$, or from (32)

$$\frac{\lambda + 2\mu}{\mu}\psi_0(ha) = \tfrac{4}{3}\psi_1(ha). \dots\dots\dots\dots\dots\dots(38)$$

Writing θ for ha, and for ψ_0 and $\psi_1 (= -3\theta^{-1}\psi_0')$ their values, we get

$$\tan\theta = \frac{4\theta}{4 - \theta^2\dfrac{\lambda + 2\mu}{\mu}}. \dots\dots\dots\dots\dots\dots(39)$$

Except for a slight difference of notation, this is the same as Lamb's equation, and his results are therefore available. They are expressed by means of Poisson's elastic constant σ, and they exhibit ha/π as dependent on σ and on the order of the root. To adapt them it is only necessary to remember that h^2, as given by (11), has here a different value from that which obtains when there is no gravitation ($\gamma = 0$). On the other hand, although γ be finite, ha/π may still be equated to T_1/τ, where T_1 is the time occupied by a plane wave of longitudinal vibration in traversing a space equal to the diameter of the sphere, and τ denotes the time of complete oscillation. The following are the smallest values of ha/π corresponding to selected values of σ, as given by Lamb:—

$\sigma = 0$.	$\sigma = \tfrac{1}{4}$.	$\sigma = \tfrac{3}{10}$.	$\sigma = \tfrac{1}{3}$.
0·6626	0·8160	0·8500	0·8733

For example, if $\sigma = \tfrac{1}{4}$ (Poisson's value), the criterion of stability is

$$\frac{4\pi\gamma\rho^2 a^2}{\lambda + 2\mu} < (0\cdot 8160)^2.$$

If $\sigma = \frac{1}{2}$, the material is incompressible, and motion of the kind now under contemplation is excluded.

When $n = 1$, (36) again vanishes, though for a different reason from before *. The form of the solution is accordingly the same as if there were no gravitation. We have from (31), (32)

$$A_1 = \frac{\lambda + 2\mu}{\mu}\frac{a^2\psi_1(ha)}{3}, \qquad B_1 = -\frac{\lambda + 2\mu}{\mu}\frac{a^5\psi_1(ha)}{3} + \frac{2a^5\psi_2(ha)}{5}; \quad (40, 41)$$

and from (34), (35) omitting some common factors which have no effect,

$$C_1 = \psi_1(ka); \qquad \dots\dots\dots\dots\dots\dots\dots\dots\dots(42)$$

$$D_1 = \frac{a^3}{2}\left\{\psi_1(ka) + \frac{6}{k^2a^2}ka\psi_1'(ka)\right\} = \frac{a^3}{2}\left\{\psi_1(ka) - \tfrac{6}{5}\psi_2(ka)\right\}. \quad (43)$$

The surface conditions (29) are of the form

$$A_1\frac{d\omega_1}{dx} + B_1\frac{d}{dx}\frac{\omega_1}{r^3} + C_1\frac{d\phi_1}{dx} + D_1\frac{d}{dx}\frac{\phi_1}{r^3} = 0,$$

and, as Professor Lamb shows, they require that

$$A_1\omega_1 + C_1\phi_1 = 0; \qquad B_1\omega_1 + D_1\phi_1 = 0. \quad \dots\dots\dots\dots(44)$$

It follows that ϕ_1 and ω_1 must be of the same form, and also that

$$B_1/A_1 = D_1/C_1, \qquad \dots\dots\dots\dots\dots\dots\dots\dots(45)$$

in which the values A_1, B_1, C_1, D_1 are to be substituted from (40), (41), (42), (43). We find

$$\frac{4\mu}{\lambda + 2\mu}\frac{\psi_2(ha)}{\psi_1(ha)} = 5 - 2\frac{\psi_2(ka)}{\psi_1(ka)}; \qquad \dots\dots\dots\dots(46)$$

or if, in accordance with (23), we replace $\psi_2(\theta)$ by $-5\theta^{-1}\psi_1'(\theta)$,

$$-\frac{4\mu}{\lambda + 2\mu}\frac{\psi_1'(ha)}{ha . \psi_1(ha)} = 1 + \frac{2\psi_1'(ka)}{ka . \psi_1(ka)}. \qquad \dots\dots\dots\dots(47)$$

Except for the different value of h, this agrees with Professor Lamb's equation (87).

In equation (46) there is no limitation upon the value of p. If to find the criterion of stability we put p or k equal to zero,

$$\psi_2(ka) = \psi_1(ka) = 1,$$

and the equation reduces to

$$\psi_2(ha) = \frac{3(\lambda + 2\mu)}{4\mu}\psi_1(ha). \qquad \dots\dots\dots\dots(48)$$

The equation may also be written in terms of the Bessel functions. The relation between J and ψ is

$$J_{\frac{1}{2}}(x) \times \surd(\tfrac{1}{2}\pi x) = x\psi_0(x) \quad = \sin x,$$
$$J_{\frac{3}{2}}(x) \times \surd(\tfrac{1}{2}\pi x) = \tfrac{1}{3}x^2\psi_1(x) \quad = x^{-1}\sin x - \cos x,$$
$$J_{\frac{5}{2}}(x) \times \surd(\tfrac{1}{2}\pi x) = \tfrac{1}{15}x^3\psi_2(x) = (3x^{-2} - 1)\sin x - 3x^{-1}\cos x;$$

---

* The terms de_1/dx, etc., in (28) denote in this case a uniform displacement, as of a rigid body, and naturally contribute nothing to the surface condition.

so that in terms of J's (48) becomes

$$\frac{\lambda + 2\mu}{\mu} J_{\frac{3}{2}}(x) = \frac{20}{3x} J_{\frac{5}{2}}(x) ; \dots\dots\dots\dots\dots\dots(49)$$

or, if we introduce the circular functions,

$$\frac{(3 - x^2)\tan x - 3x}{x^2 \tan x - x^3} = \tfrac{3}{20} \frac{\lambda + 2\mu}{\mu} \dots\dots\dots\dots\dots\dots(50)$$

Unfortunately (49) does not agree with the result given by Professor Jeans. In his notation, when $n = 1$,

$$y_1 \text{ (from 59)} = - \tfrac{13}{15}\frac{\rho a^2}{\mu}, \qquad y_2 \text{ (from 60)} = + \tfrac{1}{30}\frac{\rho a^2}{\mu} ;$$

and (54), (57), (58) give

$$\frac{\lambda + 2\mu}{\mu} J_{\frac{3}{2}}(x) = \frac{156}{25x} J_{\frac{5}{2}}(x). \dots\dots\dots\dots\dots\dots(51)$$

The comparison of processes is rendered difficult by the occurrence of several errors (possibly misprints) in Professor Jeans' paper. Thus (33) does not seem correct, and (41), (42) do not follow from (38), (39). Starting from Professor Jeans' equations just mentioned and making use of his (30), (43), (44), (45), (48), (49), I have obtained a result in harmony with my equation (49).

From (50) it is easy to calculate the value of λ/μ corresponding to any value of x. When $\mu = 0$, $\tan x = x$, of which the first root is

$$x = 1\cdot 4303\pi = 4\cdot 493.$$

This gives an angle of $77\frac{1}{2}°$ ($+ 180°$). Calculating for angles of 60°, 50°, 40°, we find

λ/μ	∞	3·840	2·056	1·221
x	4·493	4·189	4·014	3·840

It seems that the value of x is not very sensitive to variation of μ, and for such values of the ratio of λ/μ as are likely to occur, especially under high pressure, we might almost content ourselves with the fluid solution ($\mu = 0$).

The simplicity of the cases so far considered, viz., $n = 0$ and $n = 1$, depends upon our having escaped the necessity of determining the value of e_n. For values of n greater than unity this function remains in the equations, which now demand a more elaborate treatment. From (20), (21)

$$\xi = -\frac{1}{h^2}\frac{d\Delta_n}{dx} + \frac{1}{p^2}\frac{de_n}{dx} + u, \dots\dots\dots\dots\dots\dots(52)$$

in which the second term becomes infinite when $p = 0$. In order to balance this, ϕ_n in (21) must be made infinite of the order p^{-2} or k^{-2}. Thus writing $k^{-2}\Phi_n = \phi_n$, we have

$$\frac{1}{p^2}\frac{de_n}{dx} + u = \frac{\rho}{\mu k^2}\frac{de_n}{dx} + \frac{1}{k^2}\left\{1 - \frac{k^2 r^2}{2 \cdot 2n+1}\right\}\frac{d\Phi_n}{dx}$$

$$- \frac{n}{n+1}\frac{r^{2n+3}}{2n+1 \cdot 2n+3}\left\{1 - \frac{k^2 r^2}{2 \cdot 2n+5} + \ldots\right\}\frac{d}{dx}\frac{\Phi_n}{r^{2n+1}}.$$

Thus, as in the theory of differential equations with equal roots, we have when $k^2 = 0$,

$$\frac{1}{p^2}\frac{de_n}{dx} + u = \frac{df_n}{dx} + \frac{\rho}{\mu}\frac{r^2}{2 \cdot 2n+1}\frac{de_n}{dx} + \frac{\rho}{\mu}\frac{n}{n+1}\frac{r^{2n+3}}{2n+1 \cdot 2n+3}\frac{d}{dx}\frac{e_n}{r^{2n+1}}, \quad \ldots(53)$$

with two similar equations, f_n denoting again a solid harmonic of degree n. From (52), (53) we find for the radial displacement

$$U = \frac{x\xi + y\eta + z\zeta}{r} = -\frac{1}{h^2}\frac{d\Delta_n}{dr} + \frac{nf_n}{r} + \frac{\rho}{\mu}\frac{nre_n}{2 \cdot 2n+3}$$

$$= \frac{\omega_n}{h}\left\{-\frac{n}{ha}\psi_n(ha) + \frac{ha}{2n+3}\psi_{n+1}(ha)\right\} + \frac{nf_n}{a} + \frac{\rho}{\mu}\frac{na e_n}{2 \cdot 2n+3} \quad \ldots(54)$$

at the surface, where $r = a$.

The boundary condition (29) requires a parallel treatment. The terms depending on Δ_n remain as in (30), (31), (32). From (34), (35), we get as appropriate for the present purpose

$$C_n = (2n-2)\left\{1 - \frac{k^2 a^2}{2 \cdot 2n+1}\right\} - \frac{k^2 a^2}{2n+1} = 2n - 2 - \frac{n k^2 a^2}{2n+1},$$

$$D_n = \frac{n k^2 a^{2n+3}}{n+1 \cdot 2n+1 \cdot 2n+3}.$$

Equations (33), (36) now give, Φ being written as before for $k^2\phi$,

$$(2n-2)\left\{\frac{\rho}{\mu k^2}\frac{de_n}{dx} + \frac{1}{k^2}\frac{d\Phi_n}{dx}\right\} - \frac{na^2}{2n+1}\frac{d\Phi_n}{dx}$$

$$+ \frac{n a^{2n+3}}{n+1 \cdot 2n+1 \cdot 2n+3}\frac{d}{dx}\frac{\Phi_n}{r^{2n+1}}, \quad \ldots\ldots(55)$$

which, when k^2 is made to vanish, is to be replaced by

$$(2n-2)\frac{df_n}{dx} + \frac{\rho}{\mu}\frac{na^2}{2n+1}\frac{de_n}{dx} - \frac{\rho}{\mu}\frac{na^{2n+3}}{n+1 \cdot 2n+1 \cdot 2n+3}\frac{d}{dx}\frac{e_n}{r^{2n+1}}. \quad (56)$$

This is additional to (30). The equations to be satisfied at the surface are thus

$$A_n\omega_n + (2n-2)f_n + \frac{\rho}{\mu}\frac{na^2}{2n+1}e_n = 0. \quad \ldots\ldots\ldots\ldots(57)$$

$$B_n\omega_n - \frac{\rho}{\mu}\frac{na^{2n+3}}{n+1 \cdot 2n+1 \cdot 2n+3}e_n = 0. \quad \ldots\ldots\ldots(58)$$

When $n = 1$, f_n disappears, and the final condition is found by eliminating the ratio $\omega_n : e_n$ from (57), (58). This would conduct us again to the results already arrived at for that case. In general we require another equation connecting e_n with ω_n and f_n.

For this purpose we must recur to the definition (16) of e_n and of E. A calculation is made by Jeans on the basis of the expression (12) of Δ by means of Bessel's functions. We have at the surface

$$e_n = \frac{4\pi\rho\gamma a}{2n+1}\left\{- U - (n + \tfrac{1}{2}) h^{-1} (ha)^{-\frac{3}{2}} J_{n+\frac{1}{2}}(ha) . S_n\right.$$

$$\left. - h^{-1} (ha)^{-\frac{1}{2}} J'_{n+\frac{1}{2}}(ha) . S_n\right\}. \quad (59)*$$

In order to express this in our present notation (by means of ψ's), we see by comparison of (12) and (27) that

$$J_{n+\frac{1}{2}}(ha) . S_n = (ha)^{\frac{1}{2}} \psi_n(ha) . \omega_n, \quad\dots\dots\dots\dots(60)$$

so that with use of (24)

$$e_n = \frac{4\pi\rho\gamma a}{2n+1}\left\{- U - \frac{2n+1}{h^2 a} \psi_{n-1}(ha) . \omega_n\right\}. \quad\dots\dots\dots(61)$$

Eliminating U between this and (54), we find

$$f_n = \frac{\rho a^2 e_n}{\mu}\left\{- \frac{(2n+1)\mu}{n(\lambda + 2\mu)h^2 a^2} - \frac{1}{2 \cdot 2n + 3}\right\}$$

$$+ \frac{\omega_n}{h^2}\left\{- \frac{2n+1}{n}\psi_{n-1} + \psi_n - \frac{h^2 a^2}{n \cdot 2n + 3}\psi_{n+1}\right\}. \quad (62)$$

The substitution for f_n in (57) now gives

$$\omega_n\left[A_n + \frac{2n-2}{h^2}\left\{- \frac{2n+1}{n}\psi_{n-1} + \psi_n - \frac{h^2 a^2}{n \cdot 2n + 3}\psi_{n+1}\right\}\right]$$

$$+ \frac{\rho a^2 e_n}{\mu}\left[- \frac{(2n-2)(2n+1)\mu}{n(\lambda + 2\mu)h^2 a^2} - \frac{2n-2}{2 \cdot 2n + 3} + \frac{n}{2n+1}\right] = 0. \quad (63)$$

This equation and (58) determine two values of $\omega_n : e_n$, and the elimination of this ratio gives the required final result. We will write (63) for brevity as

$$Fa^2\omega_n + G(\rho a^2/\mu) e_n = 0, \quad\dots\dots\dots\dots\dots(64)$$

where by (31), and reduction with the aid of (25),

$$F = \frac{\lambda + 2\mu}{\mu}\frac{\psi_n}{2n+1} + \frac{2n-2}{h^2 a^2}\left\{- \frac{2n+1}{n}\psi_n + \frac{h^2 a^2}{2n+1 \cdot 2n + 3}\psi_{n+1}\right\}, \quad(65)$$

$$G = - \frac{(2n-2)(2n+1)\mu}{n(\lambda + 2\mu)h^2 a^2} - \frac{2n-2}{2 \cdot 2n + 3} + \frac{n}{2n+1}. \quad\dots\dots\dots\dots(66)$$

---

* In Professor Jeans' equations (5), (23) the sign of U is positive, but this appears to be an error.

Similarly, if (58) be written

$$Ha^2\omega_n + K\,(\rho a^2/\mu)\,e_n = 0, \quad \ldots\ldots\ldots\ldots\ldots\ldots\ldots(67)$$

we may take $\quad H = -\dfrac{\lambda + 2\mu}{\mu}\,\dfrac{\psi_n}{2n+1} + \dfrac{2\,(n+2)}{2n+1\,.\,2n+3}\,\psi_{n+1} \quad \ldots\ldots\ldots(68)$

$$K = -\dfrac{n}{n+1\,.\,2n+1\,.\,2n+3}\,; \quad \ldots\ldots\ldots\ldots\ldots\ldots\ldots(69)$$

and the final result is $\quad FK - GH = 0, \quad \ldots\ldots\ldots\ldots\ldots\ldots\ldots(70)$

given the ratio $\psi_{n+1}\,(ha) : \psi_n\,(ha)$ in terms of n, ha, and $(\lambda + 2\mu)/\mu$.

In applying results of calculation based upon the assumption of a uniform compressibility to the case of the earth where the variation is likely to be very considerable, we must have regard to the character of the function (12) by which the dilatation is expressed. When $n = 1$ or a greater number, (12) vanishes at the centre and (when $\mu = 0$) at the surface. The values to be ascribed to the elasticities are those proper to an intermediate position, such as half-way between the centre and the surface. For a more complete treatment we might calculate the balance of the elastic and gravitational potential energies on the basis of a displacement still following the same law as has been found to apply to a uniform sphere. In accordance with a general principle the result, so calculated, will be correct as far as the first powers of the variations from uniformity.

Another question, interesting to geologists, upon which our results have a bearing is as to the effect of denudation in altering the surface level. The immediate effect of the removal of material is, of course, to lower the level; but if the material removed is heavy and the substratum very compressible, the springing up of the foundation may more than neutralize the first effect, and leave the new surface higher than the old one. So far as I am aware discussions have been based upon the elastic quality merely of the interior without regard to self-gravitation; but, as is easy to see, if the condition be one approaching instability, the effect of a pressure applied to the surface may be immensely increased.

316.

SOME MEASUREMENTS OF WAVE-LENGTHS WITH A MODIFIED APPARATUS.

[*Philosophical Magazine*, Vol. XI. pp. 685—703, 1906.]

As the result of discussions held during the last three or four years, it seems to be pretty generally agreed that the use of the diffraction-grating in fundamental work must be limited to interpolation between standard wave-lengths determined by other means. Even under the advantageous conditions rendered possible by Rowland's invention of the concave grating, allowing collimators and object-glasses to be dispensed with, the accuracy attained in comparisons of considerably differing wave-lengths is found to fall short of what had been hoped. I think that this disappointment is partly the result of exaggerated expectations, against which in 1888[*] I gave what was intended to be a warning. Quite recently, Michelson[†] has shown in detail how particular errors of ruling may interfere with results obtained by the method of coincidences; but we must admit that the discrepancies found by Kayser[‡] in experiments specially designed to test this question, are greater than would have been anticipated.

Under these circumstances, attention has naturally been directed to interference methods, and especially to that so skilfully worked out by Fabry and Perot. In using an accepted phrase it may be well to say definitely that these methods have no more claim to the title than has the method which employs the grating. The difference between the grating and the parallel plates of Fabry and Perot is not that the latter depends more upon interference than the former, but that in virtue of simplicity the parallel plates allow of a more accurate construction. In Fabry and Perot's work the wave-lengths are directly compared with the green and red of *cadmium*; and they have obtained numbers, apparently of great accuracy, for artificial lights from vacuum-tubes containing various substances, *e.g.* mercury, for numerous

---

[*] Wave Theory, *Enc. Brit.*; *Scientific Papers*, III. p. 111, footnote.
[†] *Astro-physical Journal*, XVIII. p. 278 (1903).
[‡] *Zeitschrift für wiss. Photographie*, Bd. II. p. 49 (1904).

lines from an iron arc, and also for various rays of the solar spectrum. While, so far as I can judge, there has been every disposition to receive with favour work which not only bears the marks of care but is explained with great discrimination, it must still be felt that, in accordance with an almost universal rule, confirmation by other hands is necessary to complete satisfaction. It was with this feeling that about a year ago I commenced some observations of which I now present a preliminary account. I was not without hope that I might be able to introduce some variations which would turn out to be improvements, and which would, at any rate, promote the independence of my results.

In this method the interference rings utilized are of the kind first observed by Haidinger, dependent upon obliquity. Their theory is contained in the usual formulæ for the reflexion and transmission of parallel light by a "thin plate." Thus, if λ be the wave-length of monochromatic light, $\kappa = 2\pi/\lambda$, δ the retardation, e the reflecting power of the surface, we have, in the usual notation for the intensity of reflected light*,

$$R = \frac{4e^2 \sin^2\left(\frac{1}{2}\kappa\delta\right)}{1 - 2e^2\cos\kappa\delta + e^4}, \quad\dots\dots\dots\dots\dots\dots(1)$$

and
$$\delta = 2\mu t \cos\alpha', \quad\dots\dots\dots\dots\dots\dots\dots(2)$$

where t denotes the thickness of the plate, μ the refractive index, and α' the obliquity of the rays within the plate.

Another form of (1) is

$$\frac{1}{R} = 1 + \frac{(1-e^2)^2}{4e^2\sin^2\left(\frac{1}{2}\kappa\delta\right)}, \quad\dots\dots\dots\dots\dots\dots(3)$$

and from this we see that if $e = 1$ absolutely,

$$1/R = R = 1$$

for all values of δ. If $e = 1$ very nearly, $R = 1$ nearly for all values of δ for which $\sin\left(\frac{1}{2}\kappa\delta\right)$ is not very small. In the light reflected from an extended source, the ground will be of the full brightness corresponding to the source, but it will be traversed by *narrow* dark lines. By transmitted light the ground, corresponding to general values of the obliquity, will be dark, but will be interrupted by narrow bright rings whose position is determined by $\sin\left(\frac{1}{2}\kappa\delta\right) = 0$. In permitting for certain directions a complete transmission in spite of a high reflecting power (e) of the surfaces, the plate acts the part of a resonator.

There is no transparent material for which, unless at high obliquity, e approaches unity. In Fabry and Perot's apparatus the reflexions at nearly perpendicular incidence are enhanced by lightly silvering the surfaces. In this way the advantage of narrowing the bright rings is attained in great

* See, for example, Wave Theory, *Enc. Brit.*; *Scientific Papers*, III. pp. 64, 65.

measure without too great a sacrifice of light. The plate in the optical sense is one of air, and is bounded by plates of glass whose inner silvered surfaces are accurately flat and parallel*. The outer surfaces need only ordinary flatness, and it is best that they be not quite parallel to the inner ones.

It will be seen that the optical parts are themselves of extreme simplicity; but they require accuracy of construction and adjustment, and the demand in these respects is the more severe the further the ideal is pursued of narrowing the rings by increase of reflecting power. Two forms of mounting are employed. In one instrument, called the interferometer, the distance between the surfaces—the thickness of the plate—is adjustable over a wide range. In its complete development this instrument is elaborate and costly. The actual measurements of wave-lengths by Fabry and Perot were for the most part effected by another form of instrument called an *étalon* or interference-gauge. The thickness of the optical plate is here fixed; the glasses are held up to metal knobs, acting as distance-pieces, by adjustable springs, and the final adjustment to parallelism is effected by regulating the pressure exerted by these springs.

The theory of the comparison of wave-lengths by means of this apparatus is very simple, and it may be well to give it, following closely the statement of Fabry and Perot†. Consider first the cadmium radiation λ. It gives a system of rings. Let P be the ordinal number of one of these rings, for example the first counting from the centre. This integer is supposed known. The order of interference at the centre will be $p = P + \epsilon$. We have to determine this number ϵ, lying ordinarily between 0 and 1. The diameter of the ring under consideration increases with ϵ; so that a measure of the diameter allows us to determine the latter. Let e‡ be the thickness of the plate of air. The order of interference at the centre is $p = 2e/\lambda$. This corresponds to normal passage. At an obliquity i the order of interference is $p \cos i$. Thus if x be the angular diameter of the ring P, $p \cos \frac{1}{2}x = P$; or since x is small,

$$p = P(1 + \tfrac{1}{8}x^2).$$

In like manner, from observations upon another radiation λ' to be compared with λ, we have

$$p' = P'(1 + \tfrac{1}{8}x'^2);$$

whence if e be treated as an absolute constant,

$$\frac{\lambda'}{\lambda} = \frac{P}{P'}\left(1 + \frac{x^2}{8} - \frac{x'^2}{8}\right). \quad\quad\quad\quad\dots\dots\dots\dots\dots\dots(4)$$

The ratio λ/λ' is thus determined as a function of the angular diameters x, x' and of the integers P, P'.

* The most important requirement is the equidistance of the surfaces, and would not be inconsistent with equal and opposite finite curvatures.

† *Ann. de Chimie*, xxv. p. 110 (1902). A good account is given in Baly's *Spectroscopy*.

‡ Now with an altered meaning.

One of the principal variations in my procedure relates to the manner in which P is determined. MM. Fabry and Perot* say:—"L'étalon, une fois réglé, est mesuré en fonction des longueurs d'onde du cadmium, par les méthodes que nous avons précédement décrites; l'emploi de interféromètre est nécessaire pour cela." I wished to dispense with the sliding interferometer, and there is no real difficulty in determining P without it. For this purpose we use a modified form of (4), viz.:

$$\frac{P'}{P} = \frac{\lambda}{\lambda'}\left(1 + \frac{x^2}{8} - \frac{x'^2}{8}\right), \quad \text{..........................(5)}$$

expressing P'/P as a function of λ/λ', regarded as known, and of the diameters. To test a proposed (integral) value of P, we calculate P' from (5). If the result deviates from an integer by more than a small amount (depending upon the accuracy of the observations), the proposed value of P is to be rejected. In this way, by a process of exclusion the true value is ultimately arrived at.

The details of the best course will depend somewhat upon circumstances. It will usually be convenient to take first a ratio of wave-lengths not differing much from unity. Thus in my actual operations the mechanical measure of the distance between the plates was 4·766 mm., and the first optical observations calculated related to the two yellow lines of mercury. The ratio of wave-lengths, according to the measurements of Fabry and Perot, is 1·003650; giving after correction for the measured diameters 1·003641 as the ratio P'/P. From the mechanical measure we find as a rough value of P, $P = 16460$. Calculating from this, we get $P' = 16519\cdot92$, not sufficiently close to an integer. Adding 22 to P we find as corresponding values

$$P = 16482, \qquad P' = 16542\cdot00,$$

giving P as closely as it can be found from these observations. This makes the value of P for the cadmium-red ring observed at the same time about 14824, and this should not be in error by more than ± 30.

Having obtained an approximate value of P for the cadmium red, we may now conveniently form a table, of which the first column contains all the so far admissible (say 60) integral values of P. The other columns contain the results by calculation from (5) of comparisons between other radiations and the cadmium red. The second and third columns, for example, may relate to cadmium green and cadmium blue. These almost suffice to fix the value of P, but any lingering doubt will be removed by additional columns relating to mercury green and mercury yellow (more refrangible). An extract from the table (p. 317) may make the matter clearer.

Inasmuch as the ratio of cadmium red to cadmium green is 1·2659650, very nearly 5 : 4, only every fourth number for red is admissible on this

* *Loc. cit.* p. 112.

ground alone. If we consider a number such as 14803 not excluded by the comparison with cadmium green, we see that while it would pass the mercury green test, it is condemned by the cadmium blue and still more by the mercury yellow test. The only possible value of P is found to be 14814.

The criticism may probably suggest itself that, although other values of P may be excluded, the agreement of the row containing 14814 with integers is none too good. It is to be remembered that these observations were of a preliminary character, and were taken without the full precautions with regard to temperature afterwards found to be necessary. The formula at the basis of the calculation assumes that e, the thickness of the plate, is constant,

Cd red	Cd green	Cd blue	Hg green	Hg yellow
14788	18721·03	19836·04	17435·24	
9				
14790				
1	....	19840·07		
2	18726·09			
3				
4	....	19844·09		
5	18729·89			
6	18731·15			
7	....	19848·12		
8				
9	18734·95	19850·80		
14800	....	19852·14		
1				
2				
3	18740·01	19856·16	17459·04	16518·68
4				
5				
6	....	19860·19		
7	18745·08			
8	....	19862·87		
9				
14810	18748·88			
1	18750·14	19866·90	17462·36	
2				
3				
4	18753·94	19870·92	17465·90	16530·96
5				

but in fact it changes with temperature. On this account alone erroneous results will be obtained unless the observations are well alternated, so as to eliminate such effects. The numbers finally arrived at, in substitution for the row in the table, are

14814, 18753·95, 19870·95, 17465·97, 16531·00.

The deviations from integers still outstanding have their origin in a complication which must be admitted to be a drawback to the method and might at first sight be estimated even more seriously. The optical thickness

e of the plate, on which everything depends, is not really constant, as has been assumed, when we pass from one part of the spectrum to another somewhat distant from the first. The question is discussed by Fabry and Perot. If, to take account of this factor, we denote the thicknesses for the two wavelengths by e_λ, $e_{\lambda'}$, we have

$$\frac{p'}{p} = \frac{e_{\lambda'}}{e_\lambda} \frac{\lambda}{\lambda'},$$

and accordingly in place of (5)

$$\frac{P'}{P} = \frac{\lambda e_{\lambda'}}{\lambda' e_\lambda} \left(1 + \frac{x^2}{8} - \frac{x'^2}{8}\right). \quad \ldots\ldots\ldots\ldots\ldots\ldots(6)$$

But although I was prepared to find the calculated values of P' differing somewhat from integers, I was disturbed by the amount and at first by the direction of the difference. For in their paper of 1899* Fabry and Perot remark:—"La surface optique du métal pour la radiation rouge est, par suite, située un peu plus profondément dans le métal que celle de la lumière verte, et à une distance de 4 $\mu\mu$." At this rate e_λ (red) would exceed $e_{\lambda'}$ (green), and the introduction of the new factor in (6) would increase, and not remove, the discrepancy. It would seem, however, that the passage above quoted is in error and inconsistent with the discussion given in the later paper†, itself indeed embarrassed by several misprints‡.

The amount of the correction required to bring the number for cadmium green up to an integer—about $2\frac{1}{2}$ parts in a million—is $2\frac{1}{2}$ times as great as one would expect from Fabry and Perot's indications§. As to this, it may be observed that the wave-lengths employed in the calculation of the cadmium radiations are those of Michelson, and were obtained by a method free of the complication now under discussion. If these are correct, as there is no reason to doubt, and if there is no mistake in the identification of the ring—and there can be none here—it follows that the change of optical thickness in passing from red to green is determined by the numbers given and may be used to correct ratios of wave-lengths not previously known with precision.

If we wish to make the results of the present method entirely independent, we must obtain material from observation sufficient to allow the variation of thickness with wave-length to be eliminated, that is, we must use the same silvered plates at two different distances. In Fabry and Perot's

* _Ann. de Chimie_, xvi. p. 311.

† _Loc. cit._ pp. 120—124.

‡ Of these it may be worth while to note that the sign of 6·6 $\mu\mu$ on p. 123, line 5 should apparently be − instead of +.

§ It is known that the effect depends upon the thickness of the silver films; perhaps also upon the process used in silvering and upon the condition of the surfaces in other respects. Surfaces that have stood some time in air are almost certain to be contaminated with layers of volatile greasy matter.

work the sliding interferometer was employed; the silvered surfaces were brought to very small distances, and the coincidences of two band systems, *e.g.* cadmium red and cadmium green, were observed, the telescope being focussed upon the plate, and not as before for infinity. It appears that excellent results were obtained in this way, affording material for eliminating the complication due to change of optical thickness.

It is rather simpler in principle, and has the incidental advantage of allowing the sliding interferometer to be dispensed with, if we follow the same method for the small as for the greater distance. If the calculation be conducted on the same lines as before by means of (5), we ought to obtain the same *fractional* part again in the value of P', *e.g.* ·95 for cadmium green referred to cadmium red. For, as we see from (6), the proportional error in P'/P as calculated from (5) is $(e_{\lambda'} - e_\lambda)/e_\lambda$. In the second set of operations, writing η for e, we find as the proportional error $(\eta_{\lambda'} - \eta_\lambda)/\eta_\lambda$, in which $\eta_{\lambda'} - \eta_\lambda = e_{\lambda'} - e_\lambda$; so that the proportional errors are as $\eta_\lambda : e_\lambda$, or inversely as P or P'. Thus the absolute error in P', as calculated from (5), is unaffected by the change of e to η. If the fractional part is not recovered, within the limits of error, it is a proof that the assumed ratio of wave-lengths calls for correction, and the discrepancy gives the means for effecting such correction.

The above procedure is the natural one, when it is a question of identifying a ring or of confirming ratios of wave-lengths already presumably determined with full accuracy; but when the object is to find more accurately wave-lengths only roughly known, it has an air of indirectness. Otherwise, we have as before,

$$2e_\lambda = p\lambda, \qquad 2e_{\lambda'} = p'\lambda';$$

and again for a smaller interval between the surfaces,

$$2\eta_\lambda = \pi\lambda, \qquad 2\eta_\lambda = \pi'\lambda'.$$

Hence

$$2\,(e_\lambda - \eta_\lambda) = (p - \pi)\,\lambda, \qquad 2\,(e_{\lambda'} - \eta_{\lambda'}) = (p' - \pi')\,\lambda';$$

and $e_\lambda - \eta_\lambda = e_{\lambda'} - \eta_{\lambda'}$, so that

$$\frac{\lambda}{\lambda'} = \frac{p' - \pi'}{p - \pi}. \quad \dots\dots\dots\dots\dots\dots\dots(7)$$

Hence p, π, p', π' are the ordinal numbers at the centre. They are to be deduced, as before, from the integral numbers proper to the rings actually observed and from the measured angular diameters of these rings.

It is obvious that p and π must not be nearly equal. If p be the larger number corresponding to the greater interval, π should not exceed $\frac{1}{4}p$. On the other hand, too great a reduction of π would lead to difficulties on account of the increased angular diameter of the rings. Perhaps it was for this reason that Fabry and Perot adopted an altered course. In my experiments the longer interval was, as already mentioned, about 5 mm., and the shorter interval was about 1 mm., so that the angular diameter of the rings was rather more than doubled in the latter case.

The facility with which angular diameters larger than usual could be observed is due, in part at any rate, to the special construction of my apparatus. MM. Fabry and Perot employed a fixed interference-gauge and a fixed telescope, measuring the diameters of the rings by an eyepiece micrometer. There are, I think, some advantages in a modified arrangement, whereby it becomes possible to refer the rings to a wire fixed in the optic axis of the telescope. To this end the wire is made vertical, and the rings are brought to coincidence with it by a rotation of the gauge, which is mounted upon a turntable giving movement round a vertical axis. The middle plane of the gauge is vertical and adjusted so as to include the axis of rotation. In this way of working the reference wire is backed always by the same light, whether opposite sides of one ring or of different rings are under observation. It is perhaps a more important advantage that the same part of the object-glass is always in use, and to a better approximation the same parts of the plates of the gauge. The diaphragm which limits the latter should be as close to the plates as possible (or to their image near the eye), but when the multiple reflexions are taken into account it is impossible to secure that exactly the same part should always be in action.

The revolving turntable carried with it a thick strip of plate-glass upon which was scratched a radial line. The point observed described a circle of 10 inches radius, and the rotation was measured by means of a travelling microscope reading to ·001 inch. The angles involved are sufficiently small to allow the diameter of a ring to be taken as proportional to the difference of readings at the microscope.

As regards the gauge itself, the plates are by Brashear. For the mounting of the 5 mm. gauge, which is of brass, I am indebted to my son Mr R. J. Strutt. The 1 mm. gauge is of iron and was made by my assistant Mr Enock. They are much after the design of Fabry and Perot. For the final adjustment to parallelism the eye is moved in various directions across the line of vision so as to bring different parts of the plates into action, and for this purpose it may be desirable to increase the aperture. A dilatation of the rings means that the corresponding parts of the plates need approximation by additional pressure. The aperture employed in the actual measurements was of about 9 mm. diameter.

The (achromatic) object-glass of the telescope is of 15 inches focus. In rigid connexion with it is the vertical reference wire accurately adjusted to focus, and close to the wire a small frame suitable for carrying the horizontal slits (cut out of thin sheet zinc) necessary for the isolation of the various colours*. The eyepiece is a single lens of 5 inches focus, *mounted independently*, so that it can be re-adjusted without fear of disturbing the object-glass and reference wire. The change of position required for the best seeing in passing from red to blue or even from red to green is so great as to occasion surprise that good results can be attained in the absence of such a provision†.

The separation of the colours was usually effected by direct-vision prisms held between the eyepiece and the eye. Of these two were available. The larger containing (in all) three prisms was usually the more convenient, but sometimes a smaller and more dispersive combination containing five prisms was preferred. It is better to use more dispersion than unduly to narrow the slit. The refracting edges of the prisms are, of course, horizontal. In order to secure that the proper parts of the ring systems should be visible, the axis of the telescope was adjusted in the vertical plane with substitution for the slit of a horizontal wire coincident with the middle line of the former.

The advantage of this arrangement is that the ring systems (or at least so much of them as is necessary) of the various radiations emitted by one source of light are all in view at the same time.

In some cases, direct-vision prisms held between the 5-inch eye-lens and the eye do not suffice. The soda lines, for example, require a high dispersion. Even the yellow lines of mercury, which are about three times as far apart as the soda lines, could not be fully separated by the prisms already spoken of. Here a good deal depends upon chance. If the rings of one mercury system happen to bisect approximately those of the other system, both can be measured in the interferometer-gauge, and the only question which remains open is the distinction of the two systems. For this purpose a prism of moderate power, by which one system is lifted a little relatively to the other, suffices. If, however, the two ring systems chance to be nearly in coincidence, a much more powerful dispersion is required in order to measure them separately.

In such cases recourse was had to a special direct-vision prism of glass and bisulphide of carbon through which a selected ray of the spectrum passes without refraction at all at any of the surfaces‡. In this instrument the upper edge of the beam traverses 20 inches of glass and the lower edge 20 inches of bisulphide of carbon. This prism cannot be inserted between the eyepiece already described and the eye, which latter must be placed at the image of the object-glass. Additional lenses are therefore required.

* Fabry and Perot, *C. R.* March 27, 1904.
† Especially in using the method of coincidences. I ought perhaps to mention that my eyes have now very little power of accommodation.
‡ *Nature*, LX. p. 64 (1899); *Scientific Papers*, Vol. IV. p. 394.

R. V. 21

These are merely ordinary spectacle-lenses and constitute a telescope of unit magnifying power. A more precise description is postponed, as I am not sure that I have as yet hit upon the best arrangement. It may suffice to say that with this instrument rings formed of spectral rays even closer than the soda lines could be readily separated, and that without too great a contraction of the slit limiting the visible portion of the rings.

The source of light, sometimes very small, was focussed upon the diaphragm at the gauge, and it is necessary that the aperture be completely filled with light. This gives the ratio between the distances of the lens from the source (u) and from the gauge (v). Again, the angular diameter of the field of light, which must not be too small, fixes the ratio of the aperture of the lens to v; so that only the absolute scale of the three quantities is left open. It is desirable that the lens be achromatic. I have used a one-inch lens from a small opera-glass, and this worked well with $u = 2\frac{1}{2}$ inches and $v = 4$ feet.

As sources of light in experiments involving high interference, vacuum-tubes are by far the most convenient, and their introduction is one of the many services which Optics owes to Prof. Michelson. At the head of the list stands the helium tube, both on account of its not requiring to be heated and also of the brilliancy of the yellow radiation. Hitherto, however, the wave-lengths have not been measured with the highest accuracy. The tube that I have employed was made some years ago by my son and had already seen a good deal of service in experiments designed to answer the question: "Is Rotatory Polarization influenced by the Earth's Motion?*" From the overpowering brilliancy of the yellow line, it may be inferred that the pressure is not very low. Mercury too, for which the principal wave-lengths have been determined with great accuracy by Fabry and Perot, is convenient as requiring only a very moderate heating; and cadmium, in spite of the higher temperature demanded, is indispensable. Not only is the cadmium red by general consent the ultimate standard, but a comparison of the red and green ring systems, even without a prism, gives rapid information as to the condition of the gauge, slightly variable from time to time on account of temperature and of necessary readjustments. Thus, in most of my observations, the red ring under measurement was in very approximate coincidence with a green ring. If, owing to rise of temperature, this ring had so far expanded as to make it advisable to substitute the next interior one, there could still be no uncertainty as to the order (one higher) of the ring actually under observation.

As cadmium tubes appear to have been found troublesome, it may be well to describe a simple construction specially adapted to private workers whose skill in glass-blowing is limited. It was thought that alloying and

* *Phil. Mag.* Vol. IV. p. 215 (1902) [*Scientific Papers*, Vol. v. p. 58].

consequent expansion of platinum sealings was a likely source of difficulty, and these were accordingly dispensed with. The diagram exhibits half the complete tube. The working capillary A, the enlargement BD, and the lateral tube C for attachment to the pump are much as usual. But the enlargement is continued by a second capillary DE, perhaps $1\frac{1}{2}$ mm. in diameter and 15 cm. long, through which passes with approximate fit a straight aluminium wire, serving as electrode. The air-tight joint at E between the wire and the glass is made with sealing-wax. The length DE

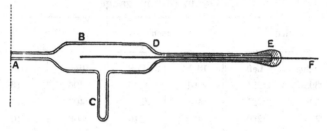

must be sufficient to allow E to remain cool, although D, enclosed in a copper case, is hot enough to keep the cadmium vapour uncondensed. The lateral tube C projects from the case, and the cadmium condensed in it may need to be driven back from time to time by temporary application of the flame of a spirit-lamp or bunsen-burner.

This construction, used with cadmium, mercury, and thallium, has so far answered my expectations. Cadmium tubes, apart from failures by cracking, are said often to deteriorate rapidly. My experience did not contradict this; for after four or five evenings' work the *red* radiation, which at first had been very brilliant, was no longer serviceable, although the green did not seem to have suffered much. At this stage the tube was re-exhausted and then appeared to behave differently, the red radiation being much better maintained. One must suppose that something deleterious had been emitted and been pumped away. There is much in the behaviour of vacuum-tubes which at present defies explanation.

To excite the electric discharge a large Ruhmkorff, actuated by five small storage-cells, was usually employed. Sometimes, especially in the comparison of the cadmium radiations, an alternate current was substituted; but there was no perceptible difference in the measurements. In this case a transformer of home construction was fed from a De Meritens magneto machine.

The radiations from zinc (and occasionally from cadmium) were obtained by an arrangement similar to Fabry and Perot's "trembler"*. The behaviour was very capricious. Sometimes, even when actuated by five secondary cells only, the zinc rings were magnificent; but the deterioration was usually rapid as the zinc points lost their metallic surfaces. This change appears to

* *C. R.* 130. p. 406 (1900).

be independent of oxidation. When the current was from a dynamo giving about 80 volts, the apparatus was less troublesome, but even then required careful management. The fineness of the points needs to be accommodated to the current employed.

As an example of the observations and calculations therefrom, I will take a series of Dec. 20, 1905, relative to the three radiations from the cadmium vacuum-tube. In this series the temperature conditions were more favourable than usual.

Cadmium (5 mm. gauge).

Red		Green		Blue	
Right	Left	Right	Left	Right	Left
·398	·221	·401	·217	·406	·213
·399	·221	·402	·217	·404	·211
·399	·220	·402	·216	·406	·210
·3987	·2207	·4017	·2167	·4053	·2113
Diff. = ·1780		Diff. = ·1850		Diff. = ·1940	

The numbers entered are the actual readings of the microscope in inches for settings on the right and left sides of the rings. Each horizontal row constitutes really a complete set. In order to eliminate temperature effects as far as possible, the readings are taken in a certain sequence. Thus in the first row the sequence was Red (R), Green (R), Blue (R), Blue (L), Green (L), Red (L). The differences, representing the diameters of the rings, are thus appropriate to the *middle* of the time occupied. If, as happened here and usually, the temperature was rising, so that the rings dilated, the first reading (·398) on the red is too small, but the error is compensated in the last reading (·221), which is equally too small. As a matter of convenience the next row would be taken in the reverse order, beginning with a repetition of Red (L), and so on.

Since the radius of the circle described by the point of observation is 10 inches, the angular diameters (x) of the rings are as follows:—

	Red	Green	Blue
x	·01780	·01850	·01940
x^2	$10^{-4} \times 3{\cdot}168$	$10^{-4} \times 3{\cdot}422$	$10^{-4} \times 3{\cdot}764$
$\frac{1}{8}x^2$	$10^{-4} \times {\cdot}3960$	$10^{-4} \times {\cdot}4277$	$10^{-4} \times {\cdot}4705$
Diff.	....	$10^{-4} \times {\cdot}0317$	$10^{-4} \times {\cdot}0745$

The calculation now proceeds by means of (5). If P refer to cadmium red and P' to green, we have with Michelson's values of the wave-lengths:

$$\frac{P'}{P} = 1\cdot2659650\,(1 - \cdot00000317) = 1\cdot2659610,$$

which with $P = 14814$ gives

$$P' = 14814 + 3939\cdot945 = 18753\cdot945.$$

In like manner for the blue referred to the red,

$$\frac{P'}{P} = 1\cdot3413733\,(1 - \cdot00000745) = 1\cdot3413633,$$

whence　　　　　　$P' = 14814 + 5056\cdot955 = 19870\cdot955.$

The wave-lengths of the various radiations from a single source can thus be compared with great ease, and but little fear of temperature error. A set of observations from which this error is practically eliminated can be made in a short time and a few repetitions give all the security necessary. But the situation is not so favourable when we compare radiations from different sources. More time is occupied and there is corresponding opportunity for temperature change. It is necessary to alternate the observations, taking the first source twice and the second once, or preferably the first three times and the second twice. Even with this precaution I believe that temperature change was the principal source of error in the results of a single evening's work.

In the observations with an interval of one millimetre between the silvered surfaces, the influence of temperature is of course much less perceptible. For a similar reason the identification of the rings is a much easier matter. I will give as a specimen a series of operations (Feb. 9) in which helium was compared with cadmium. The first and third sets, each containing a repetition, related to cadmium; the second set (twice repeated) related to helium. Only the mean diameter for each set is here recorded:—

Cadmium.

	Red	Green	Blue
I........	·749	·636	·641
III........	·752	·643	·643
Mean......	·750	·639	·642

Helium.

	Red 7065	Red 6678	Yellow 5876	Green 5016	Green 4922	Blue 4713	Violet 4472
II.	·704	·673	·721	·630	·684	·665	·637

The first question is as to the ratio P'/P derived by (5) from these numbers for cadmium. The integral value of P for the cadmium red ring was 3328. From this we find

$$P' \text{ cad. green} = 4213{\cdot}946,$$
$$P' \text{ cad. blue } = 4464{\cdot}935,$$

on the basis of Michelson's wave-lengths. For the green the fractional part is practically identical with that deduced above from one set of observations with the 5 mm. gauge. In the case of the blue the fractional part is now distinctly lower.

The above are the results of work on single evenings. On the mean of all the comparisons with the two intervals there resulted :—

Cadmium.

	5 mm.	1 mm.
Red	14814	3328
Green	18753·95	4213·95
Blue..............	19870·95	4464·94

As already explained the agreement of the fractional parts constitutes a complete verification of Michelson's ratios of wave-lengths, accurate to one part in 2 millions in the case of red and blue and to a still closer accuracy in the case of red and green. And it appears further that the phase-changes, upon which depend the deviations from integers, are decidedly greater than in the examples recorded by Fabry and Perot.

The above results for cadmium suffice to indicate what deviations from integral values are to be expected when any radiation is compared with cadmium red assumed integral. In so far as the expected fractional parts appear in the results, so far are the ratios of wave-lengths assumed in the calculation verified. The following are the wave-lengths, reckoned in air at

15° C. and 760 mm. pressure, whose ratios to cadmium red have been verified by my observations to about one part in a million :—

$$
\begin{array}{ll}
\text{Cadmium} \left\{ \begin{array}{l} 6438\cdot4722 \\ 5085\cdot8240 \\ 4799\cdot911 \end{array} \right\} & \text{Michelson.} \\[2em]
\text{Mercury} \left\{ \begin{array}{l} 5790\cdot659 \\ 5769\cdot598 \\ 5460\cdot742 \\ 4358\cdot343 \end{array} \right\} & \text{Fabry and Perot.} \\[2.5em]
\text{Zinc} \left\{ \begin{array}{l} 6362\cdot345 \\ 4810\cdot535 \\ 4722\cdot164 \\ 4680\cdot138 \end{array} \right\} & \text{Fabry and Perot.} \\[2.5em]
\text{Soda} \left\{ \begin{array}{l} 5895\cdot932 \\ 5889\cdot965 \end{array} \right\} & \text{Fabry and Perot.}
\end{array}
$$

I have spoken of an agreement to about 1 part in a million. In several cases the confirmation was decidedly closer. In one only, that of zinc red, did there appear an indication of a disagreement rather outside the limits of error. My observations would point to a wave-length about 1 millionth part greater than that of Fabry and Perot; but in view of the difficulty of observations with the trembler, I am not disposed to insist upon it. The soda observations were on light from a cadmium vacuum-tube in which soda accidentally presented itself. The numbers quoted from Fabry and Perot relate to a soda-flame.

As an example in which the ratios of wave-lengths were less accurately known beforehand, I will give some details relative to helium, beginning with observations of Feb. 9 by the 1 mm. gauge, already referred to. The Table I. annexed gives, in the second column the wave-lengths of the various helium lines recorded by Runge*, in the fifth the same reduced to Michelson's scale as employed by Fabry and Perot. The third column gives the corrections for obliquity as calculated from the observations with the 1 mm. apparatus already recorded, the fourth the differences from the corresponding quantity for cadmium red. Taking, for example, the helium ray of longest wave-length in comparison with cadmium red, we get by (5)

$$P' = 3328 \frac{6438\cdot472}{7065\cdot22}(1 + \cdot000084) = 3033\cdot03.$$

These numbers should be integers, were the wave-lengths accurate, and were there no phase change. On account of the phase change as determined from the cadmium observations, the fractional parts should be those entered in the 7th column. The differences are trifling, except in the cases of 5016

* *Astrophysical Journal*, January 1896.

TABLE I. (1 mm.)

	λ (Range)	$10^4 \times \frac{1}{2}x^2$	Diff.	λ (reduced)	P		Correction in millionths	λ (corrected)
Cd red	......	7·04	......	......	3328	......	......	......
He	7065·48	6·20	·84	7065·22	3033·03	·02	+3	7065·24
"	6678·37	5·67	1·37	6678·12	3200·01	·01	+0	6678·12
"	5875·870	6·49	·54	5875·653	3646·98	·98	+0	5875·653
"	5015·732	4·96	2·08	5015·547	4273·06	·95	+26	5015·677
"	4922·096	5·84	1·20	4921·912	4353·96	·95	+2	4921·922
"	4713·252	5·53	1·51	4713·079	4547·03	·94	+20	4713·173
"	4471·646	5·08	1·98	4471·480	4792·93	·93	0	4471·480

TABLE II. (5 mm.)

	λ	P		Correction in millionths	λ (corrected)
Cd green	5086	18753·95	·95	......	......
He	7065	13498·97	·02	−4	7065·192
"	6678	14282·07	·01	+4	6678·147
"	5876	16232·89	·98	−6	5875·618
"	5016	19016·46	·95	+27	5015·682
"	4922	19378·00	·95	+3	4921·927
"	4713	......	......	......	(4713·173)
"	4471	......	......	......	(4471·480)

and 4713. The proportional corrections by which the λ's of column 5 are to be increased are set out in column 8 expressed in millionths; but of course an accuracy of 4 or 5 millionths is hardly to be expected in results from a single set of observations with the 1 mm. gauge.

In the observations (Table II.) with the 5 mm. gauge the comparisons were with the cadmium *green*, for which P is assumed to be 18753·95, corresponding to 14814·00 for cadmium red. The numbers given embody the results of three days' observations, but they do not include the wave-lengths 4713, 4472. The procedure is the same as for Table I. If the observations with the 5 mm. gauge stood alone, we should be in doubt whether P for 5016 should be 19016·95, or 19015·95. The results with the 1 mm. gauge show that the latter alternative must be chosen. Except in this respect, the 5 mm. results are independent; and they are of course to be preferred as presumably more accurate. The final numbers for helium are therefore those given in column 6 of Table II.

The only further remark that I will make is that the observations on the helium yellow (5876) are not improbably somewhat embarrassed by a companion of feeble luminosity which could not be separated. In the 5 mm. apparatus the two components would be nearly but not quite in coincidence.

[1911. See further *Phil. Mag.* Vol. xv. p. 548, 1908; *This Collection*, Vol. v. Art. 327.]

317.

ON THE EXPERIMENTAL DETERMINATION OF THE RATIO OF THE ELECTRICAL UNITS.

[*Philosophical Magazine*, Vol. XII. pp. 97—108, 1906.]

An able discussion of the principal determinations of the above quantity, usually denoted by v, has been given in the Reports of the Paris Physical Congress (1900) by H. Abraham—himself a contributor to the series. This ground it is unnecessary to retraverse, but I desire to place on record one or two suggestions which have occurred to me but which I may probably have no opportunity of myself putting into practice.

The most approved methods involve the construction either of a condenser or of an electrometer, of which in the first case the capacity, and in the second the potential, can be calculated in electrostatic measure. The first method, on the whole, offers the greatest advantages, and I preferred it when (about 1882, and with the advice of Prof. Stuart) the Cambridge condenser was designed*. In this method two currents are compared by a galvanometer. The first is that due to a given electromotive force in a resistance whose value is known in electromagnetic measure. The second is the intermittent current due to the same electromotive force charging n times per second a condenser whose capacity is known from the data of construction in electrostatic measure. The comparison may be conducted by the aid of Wheatstone's bridge.

There are, however, one or two matters as to which doubts may arise. Thus it is essential that the commutator by whose action the condenser is periodically charged and discharged, should introduce no electromotive force on its own account. A more serious doubt hangs over the behaviour of the galvanometer. It is assumed that this instrument indicates exactly the *mean* current, whether the current be steady or intermittent. The principal error to be feared, arising from a somewhat oblique position of the needle

---

* For description see J. J. Thomson, *Phil. Trans.* 1883, p. 711; Thomson and Searle, *Phil. Trans.* 1890, p. 586.

and its temporary magnetization under the condensed charging currents, would be eliminated by reversing the battery. But is it certain that the axial magnetization remains constant, even when this axis is strictly perpendicular to the magnetic forces due to the currents*?

Another question relates to the leads connecting the condenser with the remainder of the apparatus. These must themselves have capacity, and the effect is easily allowed for† if the capacity is definite. It is here that a doubt arises. Consider for example the coaxial cylinders of the Cambridge condenser. When the condenser is to be in action, a leading wire is brought into contact with the inner surface of the inner cylinder. A rupture of this contact throws the condenser out of action; but whatever be done with the end of the lead, its electrical situation is not the same as before. It is only in very special cases, if at all, that capacities can be *added* by simply making contacts.

Passing on to the condenser itself, we may notice that in almost all cases it has been necessary to provide a guard-ring, on the principle first introduced by Lord Kelvin. This leads to complications, though perhaps not very serious ones. Thomson and Searle have shown how to allow for the guard-ring in the calculation of electrostatic capacity, and how to connect it with the bridge in the electromagnetic measurement. It is a further slight complication that the potential is not quite the same for the guard-ring and for the main part of the condenser.

It has occurred to me that a condenser, not very different from the Cambridge one, may be so arranged as not only to dispense with the guard-ring, but also to eliminate all questions connected with the capacity of the leads. The principle is that of the variable condenser described in Maxwell's *Electricity*, § 127, and further considered below. There are three outer A, G, D and two inner cylinders B, F, the components of two pairs being of the same length; and the outer surfaces of the inner cylinders and the inner surfaces of the outer cylinders being accurately worked to the same diameters. One pair A and B are mounted coaxially upon an insulating base and remain undisturbed. The other parts are movable and allow of the formation of two condensers. In the first of these the third outer cylinder D is mounted upon A so that the inner surfaces correspond. Upon the accurately worked top of B is placed a disk C of the same diameter, and D is also closed above by a plate E. The leads make contact with the cylinders A, B at their bases. Of this condenser and its leads the capacity is unknown.

---

* It is possible that the difficulty arising from the uncertain behaviour of steel magnets might be obviated by the use of a galvanometer of the so-called d'Arsonval type. The string galvanometer of Einthoven (Drude, *Ann.* XII. p. 1059, 1903) would appear to be specially suitable.

† Thomson and Searle, *loc. cit.*

In the second arrangement the long pair of cylinders F, G are interpolated, G resting directly upon A and F upon B, while C is removed so as to close F in place of B. The third outer cylinder D with its cover E now rests upon G instead of A. In this way we obtain a second condenser. Although its capacity is unknown, the *increase* of capacity is accurately that of the intermediate cylinders F, G considered as forming parts of infinitely prolonged wholes. That is, if l be the length, b the larger and a the smaller radius, the increase of capacity is $\frac{1}{2}l \div \log(b/a)$.

The circumstance that in this method the smaller capacity is much greater than that of the leads alone is scarcely an objection. In the approximate formula the electromagnetic capacity is proportional to the resistance

Fig. 1.

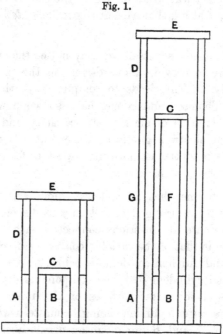

of the opposite member of the Wheatstone quadrilateral, so that it is merely with the *difference* of resistances needed in this branch that we are mainly concerned. The resistance that must be *added* as we pass from one condenser to the other can be determined with full accuracy.

The length l and the smaller diameter $2a$ are readily measured. The inner diameter $2b$ of the outer cylinder is less easily dealt with; and even if the error were no greater than for $2a$, it would be seriously multiplied in $\log(b/a)$, which is approximately proportional to $(b-a)$. In the Cambridge apparatus the interval between the cylinders was intended to be found by gauging the space with water, and the process is described by Thomson and

Searle (p. 600). If this plan be adopted, there is no need to measure b otherwise. If v be the included volume,

$$v = \pi l \, (b^2 - a^2),$$

and

$$C = \frac{\tfrac{1}{2} l}{\log (b/a)} = \frac{l}{\log \left(1 + \dfrac{v}{\pi l a^2}\right)},$$

or approximately

$$C = \frac{\pi l^2 a^2}{v}. \, {}^{*}$$

It is to be remarked that by this method we determine what we really require, *i.e.* the *mean* value of $b - a$.

In carrying out the necessary measurements there should be no difficulty over l or a. The evaluation of v is more troublesome, and the principal uncertainty would seem to arise out of the possible presence of air-bubbles. Thomson and Searle used a vacuum towards the later stages of the filling. Perhaps it would be an improvement to have a vacuum (? from carbonic acid) from the first, and to introduce the previously boiled water from *below*. It would be possible, though probably more elaborate, to determine v without water by the behaviour of included air.

The investigation of the formula for the electromagnetic measure of the capacity as derived from observations with Wheatstone's bridge is given in Maxwell's *Electricity*, §§ 775, 776, but so succinctly that the full bearing of it may easily be misunderstood. Thus Thomson[†] speaks of it as "only an approximation," and substitutes a fuller treatment. After pointing out that in simple circuit the combination of commutator (period T) and condenser (capacity C) is equivalent to a resistance R, where $R = T/C$[‡], Maxwell proceeds to consider the bridge arrangement. "Let us suppose that...a zero deflexion of the galvanometer has been obtained, first with the condenser and commutator, and then with a coil of resistance R_1 in its place, then the quantity $T \div [2] \, C$ will be measured by the resistance of the circuit of which the coil R_1 forms part, and which is completed by the remainder of the conducting system including the battery. Hence the resistance R, which we have to calculate, is equal to R_1, that of the resistance-coil, together with R_2, the resistance of the remainder of the system (including the battery), the extremities of the resistance-coil being taken as the electrodes of the system."

---

* In the Cambridge condenser $l = 61$ cm., $2a = 23\tfrac{1}{2}$ cm., and $2b - 2a = 1\cdot1$ cm. I do not know that these dimensions are susceptible of much improvement.

† *Phil. Trans.* 1883, p. 708.

‡ Maxwell has $2C$ in place of C, inasmuch as he supposes the charge of the condenser to be *reversed* instead of merely annulled.

"Using the notation of Art. 347 [see figure], and supposing the condenser and commutator substituted for the conductor AC in Wheatstone's Bridge,

Fig. 2.

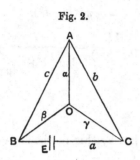

and the galvanometer inserted in OA, and that the deflexion of the galvanometer is zero, then we know that the resistance of a coil, which placed in AC would give a zero deflexion, is

$$b = \frac{c\gamma}{\beta} = R_1. \quad \dots\dots\dots\dots\dots\dots\dots\dots\dots\dots\dots(3)$$

The other part of the resistance, R_2, is that of the system of conductors AO, OC, AB, BC, and OB, the points A and C being considered as the electrodes. Hence

$$R_2 = \frac{\beta(c+a)(\gamma+\alpha) + ca(\gamma+\alpha) + \gamma\alpha(c+a)}{(c+a)(\gamma+a) + \beta(c+a+\gamma+a)}. \quad \dots\dots(4)^*$$

"In this expression a denotes the internal resistance of the battery and its connexions, the value of which cannot be determined with certainty; but by making it small compared with the other resistances, this uncertainty will only slightly affect the value of R_2.

"The value of the capacity of the condenser in electromagnetic measure is

$$C = \frac{T}{[2](R_1 + R_2)}. \quad \dots\dots\dots\dots\dots\dots\dots(5)"$$

Apart from the difference of notation, (5) is the same as the formula arrived at by Prof. Thomson. Maxwell's idea would appear to have been that it makes no difference to the galvanometer in OA whether in AC we have the resistance R_1, which gives the ordinary balance, or the commutator and condenser, provided that the condition be satisfied that the same integral current passes from A to C in both cases†. In considering the fulfilment of this condition we must remember that the difference of potential $(A - C)$ at A and C under the steady current is not the same as that $(A' - C')$ to which

---

* In Maxwell's statement a and α are interchanged in the first term of the denominator.

† $E.g.$ in the case of steady currents the introduction of an electromotive force into AC has no effect, provided the resistance of that branch be so altered as to satisfy the above condition.

the condenser is charged. The latter corresponds to the rupture of AC, so that no current there passes. The condition may be expressed:

$$\frac{\text{Capacity}}{T}(A' - C') = \frac{A - C}{R_1};$$

and what we have further to consider is the relation between $A' - C'$ and $A - C$.

Let E' be the electromotive force which must act in R_1 in order to stop the current in it. Then $E' = A' - C'$. From another point of view the zero current in AC may be regarded as the resultant effect of two independent electromotive forces E, E' acting in the system composed of R_1 and the other resistances. Thus

$$\frac{E'}{R_1 + R_2} = \frac{A - C}{R_1},$$

so that

$$\frac{A' - C'}{R_1 + R_2} = \frac{A - C}{R_1}.$$

And

$$\frac{\text{Capacity}}{T} = \frac{1}{R_1 + R_2}, \text{ simply.}$$

But although the condenser method may be the best, it is not so perfect but that a desire remains to see results so obtained confirmed otherwise. The construction of an absolute electrometer is beset with difficulties, some of which have been remarked upon by M. Abraham. In point of theory the

Fig. 5.

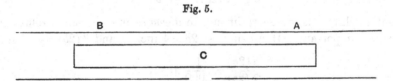

best arrangement is that described by Maxwell (*Electricity*, § 127, 1873) in which (fig. 5) an inner cylinder C moves coaxially in the interior of fixed coaxial cylinders A, B. It will suffice to suppose that C and A are at potential zero while B is at electrostatic potential B.

"The capacities of the parts of the cylinders near the [gap] and near the ends of the inner cylinder will not be affected by the [motion] provided a considerable length of the inner cylinder enters each of the hollow cylinders. Near the ends of the hollow cylinders, and near the ends of the inner cylinder[*], there will be distributions of electricity which we are not yet able to calculate, but the distribution near the [gap] will not be altered by the motion of the inner cylinder provided neither of its ends comes near the [gap], and the distributions at the ends of the inner cylinder will move with

* A solution of this problem for the case of *two dimensions* has been given by Prof. J. J. Thomson (*Recent Researches*, § 237, 1893).

it, so that the only effect of the motion will be to increase or diminish the length of those parts of the inner cylinder where the distribution is similar to that on an infinite cylinder."

Thus if a be the radius of the inner cylinder and b of the outer, the force with which the former is drawn into B is

$$F = \frac{\frac{1}{4}B^2}{\log(b/a)}.$$

It appears that F depends only on the *ratio* of the diameters of the cylinders B, C. Suppose for example that $2a = 2$, $2b = 4$ (perhaps in inches), then $\log(b/a) = \log 2 = {\cdot}69$. If the potential B correspond to 2000 volts,

$$B = \frac{2 \times 10^{11}}{v} = \frac{2 \times 10^{11}}{3 \times 10^{10}} = \frac{20}{3},$$

and $F = 16$ dynes, or mgs. weight. This is rather small; but since $F \propto B^2$, we get 64 mgs. for 4000 volts, and 144 mgs. for 6000 volts.

As regards the effect of errors in the fundamental measurements b, a, we have if $y = \log(b/a)$,

$$\frac{dy}{y} = \frac{db}{by} + \frac{da}{ay};$$

or with the above proportions

$$\frac{dy}{y} = \frac{d(2b)}{{\cdot}7(2b)} - \frac{d(2a)}{{\cdot}7(2a)}.$$

The outer cylinder is the more difficult to measure, but a given absolute error in it is less important. If we suppose $2b = 4$ inches, and $d(2b) = \frac{1}{1000}$ inch,

$$\frac{d(2b)}{{\cdot}7(2b)} = \frac{1}{3000} \text{ about,}$$

and the proportional error in y is halved when we pass to that of B.

It should be borne in mind that what we have to do with here is *not* the . *mean* diameters of the cylinders, if such diameters vary.

This form of absolute electrometer was employed in the researches of Hurmuzescu[*], who mounted the cylinders on a torsion balance, so that the motion was horizontal and not strictly axial. Some advantages are attained in this arrangement, especially perhaps that of being able to reverse the force and so to double the rather inadequate value of the subject of measurement; but upon the whole it appears to me preferable to suspend the moving cylinder C vertically in an ordinary balance, C and the upper fixed cylinder A

---

[*] *Ann. d. Chim.* x. p. 433, 1897. This author erroneously attributes Maxwell's reasoning above, by which the unknown parts of the electrical distribution are eliminated, to Bichat and Blondlot (1886).

being in connexion with earth by wires which may be very fine*. The force is then evaluated in gravitation measure; and it may of course in effect be doubled by duplicating the cylinders on the other side of the balance.

When we come to actual design the question at once obtrudes itself as to how long the cylinders really need to be. In theory it is easy to treat them as infinite, but in practice some concession must be made. In particular the weight of C must not be increased unnecessarily.

The penetration of potential arising at the gap between B and A into the annulus between C and A is easily investigated. For practical purposes it suffices to treat the problem as in two dimensions. If r be the radius and z (the axial coordinate) be measured from the end of A, the potential V in the annulus may be taken to be

$$V = \Sigma H e^{-m\pi z/(b-a)} \sin \frac{m\pi(r-a)}{b-a},$$

where m is an integer and H an arbitrary constant variable with m. At the surfaces of the cylinders where $r = a$ or $r = b$, V vanishes. The term whose influence extends furthest is that for which $m = 1$. Limiting ourselves to this, we take, since $e^\pi = 23\cdot2$,

$$V = H_1 (23\cdot2)^{-z/(b-a)} \sin \frac{\pi(r-a)}{b-a},$$

showing that when $z = b - a$ the value of V is already reduced to one twenty-third part of that at $z = 0$. When $z = 4(b-a)$, that is at a distance from the end equal to four times the thickness of the annulus, this term is attenuated 290,000 times, and it is safe to conclude that the whole disturbance of potential may be neglected. A similar argument applies to the annulus between B and C; so that a total length of 8 or 9 times the thickness of the annulus—8 or 9 inches in the example spoken of above—should amply suffice.

There is less objection to increasing unnecessarily the length of the fixed cylinders, but even here there must be some limit. It is easy to see that the prolongation of the upper cylinder A at zero potential above C, also at zero potential, is of little importance. But the case of the charged cylinder B requires further consideration.

For the sake of definiteness it is convenient to suppose B provided with a metallic bottom as in the figure annexed, and the question is as to the

---

* Further screens in connexion with earth may be introduced to protect the moving parts of the balance more effectually from the influence of the electricity upon B.

effect of the finite distance l of this bottom. If we regard the bottom as a moveable piston at the same potential (taken as zero) as the cylinder B which it closes, we may regard it as mechanically attached to the suspended cylinder C, in spite of electrical attachment to B. In this case, i.e. when C and the bottom of B move together as a rigid body, Maxwell's argument applies exactly as if the cylinder B were infinite. The correction of which we are in search is therefore equal to the electrical force of attraction which tends to draw the piston towards C.

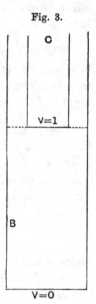

Fig. 3.

The potential V in the interior of B is expressible by means of Bessel's functions in terms of the values of V over the plane which includes the bottom of C. Thus, if r be the radius vector and z the vertical axial coordinate measured downwards from the above plane, we may write

$$V = \Sigma A J_0 (kr) \sinh k (l - z),$$

where k has in succession a series of values such that

$$J_0 (kb) = 0,$$

b being the radius of B. Each term in the above satisfies Laplace's general equation and reduces to zero on the walls and on the bottom of the cylinder. If σ denote the density of electricity on the bottom ($z = l$), we have

$$4\pi\sigma = dV/dz \,(= l),$$

and for the force of attraction upon the bottom as a whole

$$\text{Force} = \frac{1}{8\pi} \iint \left(\frac{dV}{dz}\right)^2_l dS,$$

dS representing an element of area. Now

$$\frac{dV}{dz \,(= l)} = - \Sigma k A J_0 (kr);$$

and thus, since the products of the various terms must vanish when integrated over the circle $r = b$,

$$\text{Force} = \frac{1}{4} \int_0^b r\,dr\, \Sigma k^2 A^2 [J_0 (kr)]^2 = \tfrac{1}{8} \Sigma k^2 b^2 A^2 J_0'^2 (kb),$$

by a known theorem. The values of kb are 2·404, 5·520, 8·654, &c.

When $z = 0$, $V = \Sigma A J_0 (kr) \sinh kl;$

and the values of A can be found if we know that of V over the whole circle $r = b$. As usual we have

$$\int_0^b V J_0 (kr)\, r\, dr = A \sinh kl \int_0^b J_0{}^2 (kr)\, r\, dr$$

$$= \tfrac{1}{2} b^2 A \sinh kl\, J_0'^2 (kb);$$

so that
$$\text{Force} = \Sigma \frac{\tfrac{1}{2}k^2 \left[\int_0^b V J_0 (kr)\, r\, dr \right]^2}{b^2 . \sinh^2 kl . J_0'^2 (kb)}.$$

This would determine the force if we knew the value of V over the whole circle $r = b$. We know that $V = 1$ from $r = 0$ to $r = a$, and that from $r = a$ to $r = b$ it falls from 1 to 0, but we do not know the precise law of this fall. A pretty good estimate of the integral would be made by taking $V = 1$ up to a radius $\tfrac{1}{2}(a + b)$ and afterwards $V = 0$. In this case we get*

$$\int_0^b V J_0 (kr)\, r\, dr = - \frac{\tfrac{1}{2}(a+b)}{k} J_0' [\tfrac{1}{2} k (a+b)];$$

and
$$\text{Force} = \Sigma \frac{(a+b)^2 J_0'^2 [\tfrac{1}{2} k (a+b)]}{8b^2 . \sinh^2 kl . J_0'^2 (kb)}.$$

If the object be merely to find an upper limit to the force of attraction, we may suppose the value $V = 1$ to extend up to $r = b$, and this in practical cases will not really alter the result very much. Writing $a = b$ in the above formula, we get the simple expression

$$\text{Force} = \tfrac{1}{2} \Sigma [\sinh kl]^{-2}.$$

This is the force with which the piston is attracted towards C when the potential-difference is unity, and it expresses the excess of the force by which C is drawn in over that which would act if $l = \infty$. In the case above considered of $b = 2a$, the latter is equal to about $\tfrac{1}{3}$.

As has been stated, the principal value of kl is $2\cdot404\, l/b$. If we suppose that the distance between C and the bottom of B is equal to the diameter of the latter, $l = 2b$, and $\sinh (4\cdot808)$ may be identified with $\tfrac{1}{2}e^{4\cdot808}$, while the other terms corresponding to higher values of k may be neglected. Hence

$$\text{Correction to force} = 2e^{-9\cdot6} = (6500)^{-1}.$$

This compares with $\tfrac{1}{3}$: so that if the correction be neglected altogether the error would be less than 1 in 2200. This error, although finally halved, is too great. It would be necessary either to increase the value of l a little, or to calculate the correction and allow for it.

Perhaps the weakest point in the use of an absolute electrometer on these lines is the rather high potential required to attain the necessary sensitiveness. There should be no difficulty over $\tfrac{1}{20}$ mg., especially if the necessary

* See for example, *Theory of Sound*, § 204, equation (8).

changes could be made without taking the moving parts of the balance off their knife edges.   But even then 2000 volts would scarcely suffice, and it is likely that 3000 or 4000 would prove necessary*.

The objection to a high potential is not so much the difficulty of obtaining it with steadiness as the risk of a brush-like discharge through the air, the occurrence of which would probably be fatal to the success of the measurements.   In order to diminish the risk, the edges of the cylinders A, B, C should be rounded off, as can be done without theoretical objection.

I scarcely know whether the necessity of measuring a high potential in electromagnetic measure (say in volts) is to be regarded as a disadvantage in this method of determining v.   It would seem that such measurements are needed in any case and that they constitute a separate problem.

Upon the whole, while still disposed to give the preference to the condenser, I am of opinion that the electrometer method is worthy of further trial.

* Hurmuzescu employed about 2000 volts.

318.

ON THE INTERFERENCE-RINGS, DESCRIBED BY HAIDINGER, OBSERVABLE BY MEANS OF PLATES WHOSE SURFACES ARE ABSOLUTELY PARALLEL.

[*Philosophical Magazine*, Vol. XII. pp. 489—493, 1906.]

THE importance which these rings have acquired in recent years, owing to the researches of [Lummer], Michelson, and of Fabry and Perot, lends interest to the circumstances of their discovery. It seems to be usually supposed that Haidinger merely observed the rings, without a full appreciation of the mode of formation. Thus Mascart* writes: "C'est par ce procédé, que Haidinger les a observées le premier avec une lame de mica, mais sans en donner la véritable explication." A reference to the original papers will, I think, show that Haidinger, in spite of one or two slips, understood the character of the rings very well, and especially the distinction between them and the rings usually named after Newton and dependent upon a variable thickness in the thin plate.

In the first memoir (Pogg. *Ann.* LXXVII. p. 219, 1849) the bands formed by *reflexion* are especially discussed. A spirit-flame with salted wick, seen by reflexion at considerable obliquity in a mica plate, is traversed by approximately straight bands running perpendicularly to the plane of incidence. Talbot had observed phenomena in many respects similar.... But the yellow and black lines, observed by Talbot in thin blown glass, differ in character from the lines from mica, though both are dependent upon the interference of light. In the case of the glass the interference is due to the fact that the thickness of the glass is variable, and the lines are localised at the plate. The lines from the mica behave differently. However the plate may be turned round in its own plane, the yellow and dark lines remain perpendicular to the plane of incidence. The two surfaces of the mica are absolutely

* *Ann. de Chim.* t. XXIII. p. 128 (1871).

parallel to one another, and accordingly the phenomenon is the same in all azimuths. The lines appear sharper and more distinct, the nearer the mica be held to the eye, in contradistinction to the lines from glass which then become more and more indistinct and finally disappear.

The bands are due to interference of light reflected at the two surfaces. The difference of path for the rays reflected at the front and back surfaces amounts for the bright bands to a whole number of wave-lengths *plus* a half wave-length, for the dark bands to a whole number of wave-lengths simply.

The dark parallel lines are seen with the greatest distinctness in the reflected light. There is then a striking contrast between the reflected bright light and the black due to its absence, the plate being backed by a dark ground. If the plate is held in an oblique position between the eye and the flame, the parallel lines are seen directly, but there is a much less striking contrast with the bright parts.

At this time Haidinger had not succeeded in seeing the complete rings, but in a later memoir (Pogg. *Ann.* XCVI. p. 453, 1854) he returns to the subject and shows how the obstacle to the incident light caused by the head of the observer may be overcome with the aid of a glass plate inclined at 45°. The incident light on its way to the mica is reflected at the glass plate, while on its return it traverses the plate and so reaches the eye.

The observation of the transmitted rings is of the simplest possible character. It is sufficient to look through the plate of mica at a sheet of white paper illuminated either from in front or from behind by the homogeneous light of the spirit-flame. The rings are complementary to those seen by reflexion. They are, however, much less intense, being due to the interference of the powerful directly transmitted light with the much feebler light twice reflected in the interior of the plate.

The distinction between "Berührungs-ringe" and "Plattenringe" is again emphasised, the former depending upon a variable thickness, the latter upon a variable obliquity. We may well agree with Haidinger when he concludes: "Die *Plattenringe* am Glimmer bilden also eine Classe von Interferenz-Erscheinungen für sich, die einfachste, die es geben kann, wie ich diess in der vorhergehenden Zeilen mit hinreichender Evidenz nachgewiesen zu haben glaube*."

It is interesting to remark that Haidinger's rings, rather than Newton's, are those directly explained by the usual calculation due to Young, Poisson,

---

* It should perhaps be noted that Haidinger omits the factor μ (refractive index) in the expression for the retardation on which the interference depends, viz. $2\mu e \cos \theta$, where e denotes the thickness of the plate and θ the angle of refraction. Also that, probably by a slip of the pen, he speaks of the retardation as increasing with the obliquity.

and Airy, where plane waves of light are supposed to be incident upon a parallel plate. The application to a plate of variable thickness cannot be more than approximately correct. That the indirect rather than the direct application should have been (until lately) the more familiar may be attributed to the great difficulty of preparing artificial surfaces of the necessary accuracy. The demand for equality of thickness is satisfied naturally in plates of mica obtained by cleavage, and again when a layer of water rests upon mercury*.

There is no difficulty in repeating Haidinger's observation. The transmitted rings are best seen by holding the mica close to the eye (focused for infinity) and immediately in front of a piece of finely ground glass behind which is placed a salted Bunsen flame†. If the mica be very thin, of the kind sold by photographic dealers—perhaps ·05 mm. thick, the rings are on too large a scale. But if the plate be inclined to the line of vision, the circular arcs are well seen and, owing to the enhanced reflexions, exhibit more contrast than is attainable at perpendicular incidence. When it is desired to examine the complete rings, the plate should be much thicker. I have experimented especially with two plates, ·185 mm. and ·213 mm. thick, and have observed some novel effects, evidently dependent upon the double refraction of the mica, hitherto it would seem not taken into account.

Very cursory observation on these plates, held squarely, showed that with the thinner ·185 mm. plate the inner rings were well seen, while with the thicker one they were not. Familiarity with Fabry and Perot's apparatus at once suggested that the complication might be due to a double system of rings, corresponding to the two D lines, accidentally coincident in the first case but interfering with one another in the second. It soon appeared, however, that the duplicity of sodium light was not the cause. The substitution of a *helium* vacuum-tube for the salted Bunsen made no material difference. And further, calculation showed that the two soda systems would be practically in coincidence in both cases. Thus, if we take as the mean thickness ·20 mm. and a refractive index of 1·5, the relative retardation is $2 \times 1·5 \times ·20$ or ·60 mm. The wave-length for soda light is $5·9 \times 10^{-4}$ mm., so that the order of the rings under observation is about $·60 \div 5·9 \times 10^{-4}$, or very near 1000. Now the wave-lengths of the two soda lines differ by about one-thousandth part, and thus the two ring-systems are almost in coincidence. As the thickness increases from ·20 mm., the concordance would be lost, but complete discordance would not occur until a thickness of ·30 mm. was reached. Practically in both cases the ring-systems may be considered to be in coincidence. But

* *Nature*, XLVIII. p. 212 (1893); *Scientific Papers*, IV. p. 54.

† According to Prof. Wood's recommendation the salting is best effected with the aid of a piece of asbestos, previously soaked in brine, wrapped round the tube of the Bunsen and forming a prolongation of it.

although the duplicity of the soda line is not the cause, there is in fact a second ring-system, owing to the double refraction of the mica. That this is the case is easily proved with the aid of a nicol capable of rotation about its axis. When the ring-system of the thicker plate is examined with the nicol, there are four positions at right angles to one another at which the inner rings become distinct. But in adjacent positions, *i.e.* positions distant 90°, of the nicol the ring-systems seen are different. If one has a bright centre, the other has a dark centre. When no nicol is used, or when the nicol occupies positions at 45° to those above mentioned, the ring-systems interfere, and little or nothing is visible, at any rate near the centre. When the thinner plate is employed, the ring-system is really double, but does not appear to be so, since the components are approximately in coincidence. In this case the appearance is but little altered by the use of a nicol however held.

It is only near the centre of the system that the rings are obscured when the thicker plate is used without a nicol. Further out, the rings become distinct enough. A closer examination shows, however, that this statement needs qualification. Along four directions, apparently at right angles, radiating from the centre, there are regions of no definition. These regions are narrow, so narrow that they might at first escape observation, and they constitute, as it were, *spokes* of the ring-system. It was natural to suppose that these spokes represented places where the rings of each system bisected the intervals between the rings of the other system—a conjecture supported by the fact that the spokes disappeared when a nicol was introduced in the positions suitable for rendering the inner rings distinct. The effect was to make distinct the outer rings all round the circumference. Further confirmation was afforded by the introduction behind the mica of cross-wires and a collimator-lens, serving to indicate a fixed direction. This was pointed at a spoke, so that without a nicol no bands were visible in the neighbourhood of the cross-wires. The nicol was then introduced in such a position as to give maximum distinctness, and the cross-wires adjusted to coincide with the centre of a bright band. A rotation of the nicol through 90° showed that in the band-system then visible the cross-wires marked the centre of a dark in place of a bright band.

The disappearance of the rings at the places where the brightest places in one system bisect the intervals between the brightest places of the other system depends of course upon the width of the bright rings being not much less than half of the complete period. If, as in the Fabry-Perot apparatus, the bright rings are much narrower, both systems should become visible. I thought therefore that it would be of interest to silver lightly on both sides a portion of the thicker plate, the more as, apart altogether from the spokes, the whole effect would be improved owing to the enhanced reflexions. By the chemical method, as ordinarily used for silvering glass surfaces, I did not succeed; but there was no difficulty in getting the required deposits by the method of electrical discharge *in vacuo* using a silver cathode*. With the silvered plate the nature of the whole phenomenon, including the character of the spokes, was evident on simple inspection. At the spokes both ring-systems could now be seen, forming a compound system of half the original period. If we neglect the circular character of the rings, the effect may be imitated with straight bands, as shown [below]. From a piece of striped

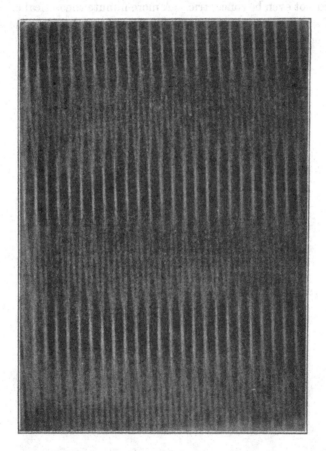

stuff, in which the bright bands are of width equal to about a quarter of the complete period, a photographic negative was taken. The prints from this negative are made with *two* exposures, between which the paper is slightly rotated. In this way two systems of bands are impressed, crossing one another at a small angle.

The fact that the two concentric ring-systems cross one another proves of course that they cannot both be strictly circular. Complete circularity in all

* Wright, *Amer. Journ. Sci.* xiv. p. 169 (1877).

cases would require a wave-surface in the form of two concentric spheres, and such is not the character of optical double refraction. An exception would occur in the case of a uniaxal crystal cut perpendicularly to the axis, which would then be an axis of symmetry for the whole phenomenon. Another comparatively simple case would arise if the surfaces of the plate were parallel to the axis of a uniaxal crystal. In general one at least of the ring-systems would be elliptical as in the observations; and it would seem that the systems need not even be concentric. A more minute theoretical examination might be of some interest, especially in comparison with observations upon a plate whose optical constants were known.

319.

ON OUR PERCEPTION OF SOUND DIRECTION*.

[*Philosophical Magazine*, Vol. XIII. pp. 214—232, 1907.]

IT is some thirty years ago since I executed a rather extensive series of experiments in order to ascertain more precisely what are the capabilities of the ears in estimating the direction of sounds†. It appeared that when the alternative was between right and left, the discrimination could be made with certainty and without moving the head, even although the sounds were pure tones. Nor was any difficulty introduced by the requirement that the ears should be stopped at the moment when the sounds commenced.

On the other hand, if the question was whether a sound were situated in front or behind the observer, no pronouncement could be made in the case of pure tones. The impossibility of distinguishing front and back carries with it further confusions relating to cases where the sound may be obliquely situated. But with sounds of other character and notably with the speaking voice, front and back could often be distinguished. It is understood, of course, that the head was kept still. A slight rotation, bringing a pure tone (originally situated exactly in front or exactly behind) to the right or the left, gives the information that was previously lacking.

The discrimination between right and left is usually supposed to be explicable by the greater intensity of sensation experienced by the ear which lies nearer to the sound. When the pitch is pretty high, there is no doubt that this explanation is adequate. A whistle of pitch f^{IV}, preferably blown from a gas-bag, is much better heard with the nearer than with the further ear. "A hiss is also heard very badly with the averted ear. This observation

---

* This paper formed the substance of the Sidgwick lecture given at Cambridge on November 10, 1906, and (except the last two or three pages) was written before the delivery of the lecture. I have learned since from Dr L. More that three years ago at Cincinnati he made experiments which led him to similar conclusions. It is to be hoped that Dr More will publish an account of his work, the more as it was conducted on lines different from mine.

† *Nature*, XIV. p. 32 (1876); *Phil. Mag.* III. p. 546 (1877); *Phil. Mag.* XIII. p. 340 (1882). *Scientific Papers*, I. pp. 277, 314; II. p. 98.

may be made by first listening with both ears to a steady hiss on the right or left, and then closing one ear. It makes but little difference when the further ear is closed, but a great difference when the nearer ear is closed. A similar observation may be made upon the sound of running water." In a modified form of the experiment the ear, say the right, nearest to the falling water is stopped with the right hand. The comparatively feeble sound then heard may be much increased if the left hand be so held at a little distance out as to reflect the sound into the left ear. The effect remains conspicuous even when the hand is held out at full arm's length. Of course a reflector larger than the hand is still more effective.

The discrimination between the right and left situations of high sounds is thus easily explained upon the intensity theory; but this theory becomes less and less adequate as the pitch falls. At a frequency of 256 (middle $c = c'$) the difference of intensities at the two ears is far from conspicuous. At 128 it is barely perceptible. But although the difference of intensities is so small, the discrimination of right and left is as easy as before.

There is nothing surprising in the observation that sounds of low pitch are nearly as well heard with the further as with the nearer end. When the wave-length amounts to several feet, it is not to be expected that the sound (originating at a distance) could be limited to one side of the head. The question is well illustrated by calculations relating to the incidence of plane waves upon a rigid spherical obstacle, and the results may conveniently be repeated here from *Theory of Sound*, § 328. [See also p. 151.]

$\dfrac{2\pi c}{\lambda}$	μ	$F + iG$	$F^2 + G^2$
	1	$\cdot 522 + \cdot 139\,i$	$\cdot 294$
$\frac{1}{2}$	-1	$\cdot 159 - \cdot 484\,i$	$\cdot 260$
	0	$\cdot 430 - \cdot 217\,i$	$\cdot 232$
	1	$\cdot 668 + \cdot 238\,i$	$\cdot 503$
1	-1	$- \cdot 440 - \cdot 303\,i$	$\cdot 285$
	0	$+ \cdot 322 - \cdot 365\,i$	$\cdot 237$
	1	$\cdot 797 + \cdot 234\,i$	$\cdot 690$
2	-1	$\cdot 250 + \cdot 506\,i$	$\cdot 318$
	0	$- \cdot 154 - \cdot 577\,i$	$\cdot 365$

In this table $2\pi c$ is the circumference of the sphere, and λ is the wavelength of the sound. The symbol μ denotes the cosine of the angle at the centre of the sphere between the direction of the sound ($\mu = 1$) and the point upon the sphere at which the intensity is to be reckoned. $F + iG$ denotes the (complex) condensation, and $F^2 + G^2$ the intensity. In the present question of a sound situated say to the right of the observer, the intensity at the right ear corresponds to $\mu = +1$, and at the left ear to $\mu = -1$. In the case of the head the circumference $(2\pi c)$ may be taken at about 2 feet or a little less, so that $2\pi c/\lambda = \frac{1}{2}$ corresponds about to middle c, or frequency 256. It will be seen that the difference of intensities for $\mu = \pm 1$ is only about 10 per cent. of the whole intensity.

For still smaller values of $2\pi c/\lambda$, i.e. in the present application for still graver notes, the difference of intensities may be adequately expressed by a very simple formula. It appears that

$$(F^2 + G^2)_{\mu=1} - (F^2 + G^2)_{\mu=-1} = \frac{3}{4}\left(\frac{2\pi c}{\lambda}\right)^4,$$

while at the same time the total value of $F^2 + G^2$ approximates to ·25. A fall in pitch of an octave thus reduces the difference of intensities 16 times. At frequency 128 the difference would be decidedly less than one per cent. of the whole; and from this point onwards it is difficult to see how the difference could play any important part.

So far as I am aware no explanation of the above difficulty, emphasised in 1876, has been arrived at. A few months since I decided to repeat, and if possible to extend, the observations, commencing with frequency 128. Two forks of this pitch were mounted in the open air at a considerable distance apart, and were electrically maintained, one driving the other. In connexion with each was a resonator which could be put out of action by interposing (without contact) the blade of a knife or a piece of card. An observer with eyes closed, placed between the two forks and so turned as to have them upon his right and his left, could tell with certainty which resonator was in action. The ears may be open all the time; or, what is in some respects better, they may be closed while the changes at the resonators are being made, and afterwards opened simultaneously. If one ear be opened, the sound will appear to be on that side; but when the second ear is also opened, the sound assumes its correct position, whether or not this involves a reversal of the earlier judgement. When the sounds were in front and behind, instead of to the right and the left, several observers agreed that a discrimination between front and back could not be made.

In another method of experimenting a single resonator and fork (which need not be electrically maintained) suffice, but more than one assistant may be required. The observer, either on his feet, or more conveniently seated upon a rotating stool, is turned round until he loses his bearings. There is no difficulty in this*, but some precautions are needed to prevent the bearing

---

* The process is aided by the illusion of a reverse rotation when the real rotation has stopped.

being afterwards recovered. The wind may act as a tell-tale. It is often necessary to cover the eyes with the hands as well as to close the eye-lids in order sufficiently to exclude the light. Until all is ready for a judgement, the ears are kept closed by pressure with the thumbs, and it is usually advisable to keep the thumbs in motion and thus to cause miscellaneous noises loud enough to drown any residue of the sound under observation. Pure tones of pitch 128 and 256 yielded by this method results in agreement with those already described.

The turn-table facilitates observation upon the question as to the relative loudness with which a sound is heard according as the source is on the same, or the opposite, side as the ear in use. In my own case I thought I could detect an advantage when the source was on the same side as the open ear, while others could detect no difference. This relates to pitch 128. At 256 the advantage is quite marked.

In considering whether the discrimination between right and left at pitch 128 can really be attributable to the small intensity-difference, it occurred to me that, if so, the judgement might perhaps be disturbed by the introduction of an obstacle, such as a piece of board, near the head of the observer and on the same side as the sound. But it was found that no mistakes could thus be induced, although in each trial the observer did not know whether the board were in position or not. Another circumstance, unfavourable to the intensity theory, may also be mentioned. It was found that the observer on the turn-table could sometimes decide between the right and the left before un-stopping his ears.

The next step in the investigation appeared to be the examination of pure tones of still graver pitch. A globe, such as are sold to demonstrate the combustion of phosphorus in oxygen gas, was sounded with the aid of a hydrogen flame*. Careful observation revealed little or no trace of overtones. The frequency was about 96 vibrations per second, thus by the interval of the Fourth graver than the 128 forks. At the temperature of the observations this would correspond to a wave-length of about 12 feet. If we use this value in the formula already given, we find for the proportional difference of intensities on the two sides of a sphere of 2 feet circumference only about 2 parts per thousand.

Observation in the open air showed that there was no difficulty whatever in deciding whether this low sound was on the right or the left. Several observers agreed that the discrimination was quite as easy as in the case of forks of pitch 128. On the other hand, as was to be expected, the front and back situations could not be discriminated.

At this stage a reconsideration of theoretical possibilities seemed called for. There could be no doubt but that relative intensities at the two ears

* *Phil. Mag.* VII. p. 149 (1879); *Scientific Papers*, I. p. 407.

play an important part in the localization of sound. Thus if a fork of what-ever pitch be held close to one ear, it is heard much the louder by that ear, and is at once referred instinctively to that side of the head. It is impossible to doubt that this is a question of relative intensities. On the other hand, as we have seen, there are cases where this explanation breaks down. When a pure tone of low pitch is recognized as being on the right or the left, the only alternative to the intensity theory is to suppose that the judgement is founded upon the difference of *phases* at the two ears. But even if we admit, as for many years* I have been rather reluctant to do, that this difference of phase can be taken into account, we must, I think, limit our explanation upon these lines to the cases of not very high pitch. For what is the differ-ence of phase at the two ears when a sound reaches the observer—say from the right? It is easy to see that the retardation of distance at the left ear is of the order of the semi-circumference of the head, say one foot. At this rate the retardation for middle c ($c' = 256$) is nearly one quarter of a period; for c'' (512) nearly half a period; for c''' (1024) nearly a whole period, and so on. Now it is certain that a phase-retardation of half a period affords no material for a decision that the source is on the right rather than on the left, seeing that there is no difference between a retardation and an acceleration of half a period. It is even more evident that a retardation of a whole period, or of any number of whole periods, would be of no avail. In the region of some-what high pitch a judgement dependent upon phase would seem to be hardly possible, especially when we reflect that the phase-differences enter by degrees, rising from zero when a sound is directly in front or behind to a maximum or minimum in the extreme right and left positions.

As to whether there is any difficulty in localizing to the right or left a tone of pitch 512, an early observation, conducted with the aid of two forks and resonators of this pitch, gave an answer in the negative. The localization was as easy and as distinct at pitch 512 as at pitch 256. But it is quite possible that 512 is not sufficiently near the particular pitch for which the retardation would have the value of precisely half a period.

The calculations for a spherical obstacle already quoted give some informa-tion upon this point. From the values of $F + iG$ corresponding to $\mu = \pm 1$ for the case of $2\pi c/\lambda = 1$, it appears that there is approximate opposition of phase. A closer examination and comparison of the three cases shows that exact opposition will occur at a somewhat higher value, say $2\pi c/\lambda = 1\cdot1$. Calculating from this and taking $2\pi c$ at 2 feet, I find at the temperature of the observations a pitch about a minor third *above* 512, as that corresponding to a phase-difference of half a period.

Naturally, in its application to the head, the calculation is not very trustworthy, and I thought it important to make sure by actual experiment

* Conf. *Theory of Sound*, § 385.

that there is no pitch in this region for which the discrimination of right and left is at all uncertain. Tuning-forks not being available, I fell back upon "singing-flames," *i.e.* tubes, usually of metal and about 1 inch in diameter, maintained in vibration by hydrogen flames. In order to eliminate overtones, the tubes were provided near their centres with loosely-fitting rectangular blocks about two diameters long, held in position by the friction of attached springs. In this way and with the precaution of not sounding them more loudly than was necessary, the tones were, it is believed, sufficiently pure. Trials were made in the open air on many occasions, the pitch ranging in all from d' to g''', and there was never the smallest suspicion of a difficulty in discriminating right and left. In the region from c'' to g'' of special interest, the pitch was varied by half semitones with the aid of sliding prolongations of the tube. During most of the observations the listener was placed upon the rotating stool and was in ignorance of the real position of the source. This precaution is of course desirable; but after a good deal of practice I found that I was able to trust the direct sensual impression. In the case of right or left the impression is always distinct and always correct; but in trying to discriminate front and back there is usually no distinct impression, and when there is it often turns out to be wrong.

We may fairly conclude that in this region of pitch (above $c'' = 512$) the discrimination of right and left is not made upon the evidence of phase-differences, or at any rate not upon this evidence alone. And this conclusion leads to no difficulty; for, as has already been explained, the difference of intensities at the two ears gives adequate foundation for a judgement. It would seem that at high pitch, above c'', the judgement is based upon intensities; but that at low pitch, at any rate below c (128), phase-differences must be appealed to.

It remained to confirm, if possible, the suggestion that not only are we capable of appreciating the phase-differences with which sounds of equal intensity may reach the two ears, but that such appreciation is the foundation of judgements as to the direction of sounds—in particular of the right and left effects. The obvious method is to conduct to the two ears separately two pure tones, nearly but not quite in unison. During the cycle, or beat, the phase-differences assume all possible values; and the mere recognition of the cycle is evidence of some appreciation of phase-differences. Experiments on these lines are not new. In 1877 Prof. S. P. Thompson* demonstrated "the existence of an interference in the perception of sounds by leading separately to the ears with india-rubber pipes the sounds of two tuning-forks struck in separate apartments, and tuned so as to 'beat' with one another—the 'beats' being very distinctly marked in the resulting sensation, although the two sources had had no opportunity of mingling externally, or of acting

* *Phil. Mag.* Nov. 1878.

jointly, on any portion of the air-columns along which the sound travelled. The experiment succeeded even with vibrations of so little intensity as to be singly inaudible." And in an observation of my own*, where tones supposed to be moderately pure were led to the ears with use of telephones, a nearly identical conclusion was reached. But although the cycle was recognized, in neither case apparently was there any suggestion of a right and left effect.

In repeating the experiment recently I was desirous of avoiding the use of telephones or tubes in contact with the ears, under which artificial conditions an instinctive judgment would perhaps be disturbed. It seemed that it might suffice to lead the sounds through tubes whose open ends were merely in close proximity one to each ear, an arrangement which has the advantage of allowing the relative intensities to be controlled by a slight lateral displacement of the head towards one or other source. Two forks of frequency 128, independently electrically maintained, were placed in different rooms. Associated with each was a suitably tuned resonator from the interior of which a composition (gas) pipe led the sound through a hole in a thick wall to the observer in a third room. With the aid of closed doors and various other precautions of an obvious character, the sounds were fairly well isolated. But each resonator emits a rather loud sound into its neighbourhood, a little of which might eventually reach the other. To eliminate this cause of disturbance more completely, a second resonator † of like pitch was employed in association with each fork, so situated that the phases of vibration in the two resonators were opposed. By a little adjustment it was possible to provide that but little sound radiated externally from the combination, though the internal vibrations might be as vigorous or more vigorous than when one resonator was employed alone. These arrangements were so successful that when one fork singly was in action, the sound was imperceptible from the tube belonging to the other, even though the open end were pressed firmly into the ear, by which the effect is enormously increased. The open ends of the two pipes may thus be regarded as sources of sound of constant intensity.

In the greater number of experiments the observer, leaning over a table for the sake of steadiness, placed his head between the pipes, which were at such a distance apart that one or two inches separated the open ends from the adjacent ears. At the very first trial on July 31, the period of the cycle being 5 seconds, Lady Rayleigh and I at once experienced a distinct right and left effect, the sound appearing to transfer itself alternately from the one side to the other. When the effect was at its best, the sound seemed to lie entirely on the one side or on the other.

---

* *Phil. Mag.* II. p. 280 (1901); *Scientific Papers*, IV. p. 553.

† Three out of the four resonators consisted of "Winchester" bottles from which the necks had been removed.

The beat may be slowed down until it occupies 40 or even 70 seconds, thus giving opportunity for more leisurely observation. The position of the head should be so chosen that the right and left effects are equally distinct. Under these circumstances it is found that the sound seems to be predominantly on the right or on the left for almost the whole of the cycle, the transitions occupying only small fractions of the whole time. The observations may be made with the ears continuously open, or, as in some of the outdoor experiments, the ears may be opened and closed simultaneously at short intervals. It is perhaps better still, keeping the ears open, to close periodically with the thumbs the open ends of the tubes from which the sounds issue. When the tubes are closed, no sound is audible. It should be said that although the best results require the position of the head to be carefully chosen, right and left effects are perceived through a considerable range. It is only necessary that the intensities be approximately equal.

These results are quite decisive, if we can assume that the sounds were sufficiently isolated—that nothing appreciable could pass from the open end of a pipe to the wrong ear. It was easy to verify that when one pipe and the opposite ear were closed, next to nothing could be heard; but it may perhaps be argued that the test is not delicate enough. The risk of error from this cause is diminished by approximating the open ends to their respective ears. Many experiments of this kind were made, but without influencing the results. Finally, short lengths of rubber tubing were provided, by means of which the ears could be connected almost air-tight with the pipes. In this case, to avoid being deafened, it was necessary to reduce the sounds by withdrawing the resonators from their respective forks. The right and left effect remained fully marked. Another argument to show that the effects cannot be explained by sounds passing round the head will be mentioned presently.

A question of great importance still remains to be considered. Before the laboratory experiments can be accepted as explanatory of the discrimination of right and left when a single sound is given in the open, it is necessary to show that in the former the sensation of right (say) is associated with a phase-difference such that the vibration reaching the right ear *leads*. There was no difficulty in obtaining a decision. While one observer listens as described for right and left effects, a second observes the maxima and minima of the beat as heard by one ear situated symmetrically with respect to the two sources. In the case of sounds of higher pitch to be considered later some precaution is required here; but for the present sounds, corresponding to a wave-length of nearly 9 feet, there is no difficulty. Under good conditions the minimum, represented by a *silence*, is extremely well marked, and can often be signalled to within half a second. This signal, corresponding to *opposition* of phases, gives the required information to the first observer. If a signal for the maximum, representing phase-agreement, is desired, it is best made by halving the intervals between the silences.

The results can be stated without the slightest ambiguity. The *transitions* between right and left effects correspond to agreement and opposition of phase, not usually recognized by the first observer as maxima and minima of sound. When the vibration on the right is the quicker, the sensation of right follows agreement of phase, and (what is better observed) the sensation of left follows opposition of phase. And similarly when the vibration on the left is the quicker, the left sensation follows agreement, and the right follows opposition of phase. The question which fork is vibrating the quicker is determined in the usual way by observing the effect upon the period of the beat of the addition or removal of a small load of wax. If for example the beat is *slowed* by loading the right fork, we may be sure that that fork was originally the quicker. A large number of comparisons of this kind have been made at various times, and in no case (at this pitch) has the rule been violated. It is not a little remarkable that by merely listening to right and left effects aided by signals giving the moment of phase-opposition, it is possible conversely to pronounce which fork is the quicker, although the difference of frequencies may not exceed ·02 vibration per second.

It should not pass unnoticed that the laboratory experiments cover a wider field than the observations in the open. In the latter case, if the single sound of pitch 128 is in front or behind, there is agreement of phases at the two ears. As the position becomes more and more oblique, the phase-difference increases; but it can never exceed a moderate amount, about one-eighth of a period, which is attained when the position is precisely to the left or right. Phase-differences in the neighbourhood of half a period do not occur. From the laboratory experiments it appears that the right and left effects are not subject to this limitation; but that, for example, a right effect is experienced when the vibration reaching the right ear leads, whether the amount of the lead be small or whether it approaches the half period.

The right and left or, as I shall sometimes say for brevity, the *lateral*, sensations observable in this way are so conspicuous that I was curious to inquire how I had contrived to miss them in the earlier experiments with telephones already alluded to. The apparatus was the same as before. In the neighbourhood of the electro-magnet driving each fork (128) was placed a small coil of insulated wire whose circuit was completed through a telephone. The double wires connected to the telephones were passed through the perforated wall. In order to weaken the higher overtones, thick sheets of copper intervened between the electro-magnets and the coils. But when the telephones were held to the ears, the sounds were perceived to be of a more mixed character than I had expected; and I am forced to the conclusion that I must formerly have overestimated their approximation to the character of pure tones. Although the cycle could be recognized, a distinct lateral effect was not perceived, and the failure was evidently connected with the composite character of the sounds. By loading the disks of the telephones with penny-pieces (attached at the centres with wax) the higher components could be better eliminated. It was then possible to fix the attention upon the fundamental tone and to recognize its transference from left to right during the cycle. But the effect was by no means so conspicuous as with the tubes, and might perhaps be missed by an unprepared observer*.

The subject now under consideration is illustrated by a curious observation accidentally made in the course of another inquiry. A large tuning-fork of frequency about 100, mounted upon a resonance-box, was under examination with a Quincke tube. This consisted of a piece of lead pipe more than one-quarter of a wave-length long, one end of which was inserted into the resonance-box. At a distance of one-eighth of a wave-length from the outer end a lateral tube was attached which communicated with one ear by means of an india-rubber prolongation. When the second ear was closed, it appeared to make no difference to the sound whether or not the outer end of the Quincke tube was closed with the thumb. But when the second ear was open, marked changes in the sound accompanied the opening and closing of the Quincke tube. On the view hitherto held, it would appear very paradoxical that a change not affecting the sound heard in either ear separately should be able to manifest itself so conspicuously. It was easily recognized that the alterations observed were of the nature of right and left effects, and that they could be explained by the local *reversal of phase* which accompanied the closing of the Quincke tube.

The conclusion, no longer to be resisted, that when a sound of low pitch reaches the two ears with approximately equal intensities but with a phase-difference of one-quarter of a period, we are able so easily to distinguish at which ear the phase is in advance, must have far reaching consequences in the theory of audition. It seems no longer possible to hold that the vibratory character of sound terminates at the outer ends of the nerves along which the communication with the brain is established. On the contrary, the processes in the nerve must themselves be vibratory, not of course in the gross mechanical sense, but with preservation of the period and retaining the characteristic of phase—a view advocated by Rutherford, in opposition to Helmholtz, as long ago as 1886. And when we admit that phase-differences

---

* Subsequently by a much heavier loading (53 gms.) the telephone-plates were tuned approximately to pitch 128, as could be verified by tapping them with the finger. To find room for these extra loads, the ear-pieces of the telephones had to be modified. The sounds now heard were very approximately pure tones, and the lateral effects were as distinct as those observed when the sounds were conveyed through tubes. It is easy to understand that considerable complication must attend an accompaniment of octave and higher harmonics, which would transfer themselves from right to left more rapidly than the fundamental tone.

at the two ears of tones in unison are easily recognized, we may be inclined to go further and find less difficulty in supposing that phase-relations between a tone and its harmonics, presented to the *same* ear, are also recognizable.

The discrimination of right and left in the case of sounds of frequency 128 and lower, so difficult to understand on the intensity theory, is now satisfactorily attributed to the phase-differences at the two ears. The next observations relate to pure tones of pitch 256. Two large forks of this pitch were used, such as are commonly to be found in collections of acoustical apparatus. Tuned with wax so as to give beats of 3 or 4 seconds period, they may be held (after excitation) by their stalks one to each ear, preferably by an assistant. The sensation of transference from right to left was fully marked, and when the conditions were good, especially in respect of equality of intensities at the two ears, the whole of the sound seemed to come first from one side and then from the other. The method of holding in the fingers is satisfactory as regards the isolation of the sounds. Practically nothing of either fork can be heard by the further ear. But there is a little difficulty in maintaining quite constant the relative positions of forks and ears.

In another arrangement, which has certain advantages, the forks are mounted, stalks upwards, on a sort of crown, in such a fashion that the free ends of the forks are about opposite the ears. A sketch by Mr Enock is reproduced in the figure (p. 358). If the crown be sufficiently large, it can be adapted to various sized heads with the aid of pads. At the back, attached to the crown, is a forked tube of brass, symmetrically shaped, whose open ends abut upon the faces of the forks. The short limb, forming the stalk, is prolonged by india-rubber tubing and so connected with the ear of the assistant observer. If the forks are vibrating equally, a well-defined silence marks the moment of phase-opposition. To excite and maintain the vibrations, a violin-bow is employed in the usual manner.

Very good observations may be made in this way if the vibrations of the forks are equally and sufficiently maintained. The assistant, listening through the forked tube, is able to give a sharp signal at the moment of phase-opposition. A transition in respect of right and left effect occurs at this moment, and the sequence rule already stated defining whether the transition is from right to left or from left to right is found to be obeyed.

There are some advantages, of course, in an experimental arrangement allowing the sounds to be uniformly maintained. As in Helmholtz's vowel investigations, the 256 forks can be driven by the 128 interrupter-forks already employed, and in each case the frequency of the driven fork is the exact double of the frequency of the driving fork. The observations may be made in two ways. Either the 256 forks may themselves be brought close to the ears, leading wires being conveyed through the wall; or, what is on the whole preferable, the method employed for the 128 per second tones may be followed. In this case each 256 fork is associated with two resonators, vibrating in opposite phases, with one of which the pipe leading to the observation room is connected. The isolation was good, each sound being inaudible through the tube provided for the other.

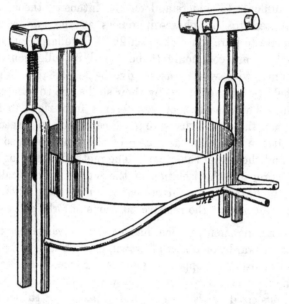

Excellent results were obtained in this way. With good adjustment the transitions between right and left were sharply marked and the sequence rule already formulated was obeyed. It is again to be noted that right and left effects are observable in the neighbourhood of phase-opposition, a situation which does not arise when one sound in the open influences both ears.

Unless the open ends are pretty close to the ears, the sound has more tendency to travel round the head to the wrong ear than was observed at pitch 128. This may raise a question whether, after all, the right and left effects may not be due to small differences of intensities at the two ears varying periodically. The best answer to this objection is to consider what would be the consequence of such invasions. Suppose that at one moment the vibration on the right is in advance by one quarter period, so that a full right sensation is being experienced. The retardation in travelling round the head will at this pitch be about one quarter period, so that the sound starting from the right in advance will on arrival at the left be in approximate phase-agreement with the principal sound there. On the other hand, the sound starting from the left, already a quarter period in arrear, will on arrival at the right be in approximate phase-opposition with the principal sound on the right. The effect of travel round the head is therefore to augment the sound on the left and to diminish the sound on the right. This would evidently tend to cause a sensation of sound on the *left*, and cannot therefore be the explanation of the observed sensation of a sound on the right. The same considerations will apply, if in less degree, to sounds of pitch 128, and to sounds somewhat higher in pitch than 256.

The next sounds to be experimented upon in order of pitch were from forks giving *e′* of 320 vibrations per second. These could not be driven from the 128 per second interrupters, and were merely held to the ears in the fingers of an assistant. The right and left effect was very marked, but there was a little difficulty at first in fixing the moment of phase-opposition. After a few trials it became sufficiently clear that the rule was the same as at the lower pitch, viz. that the quicker fork asserts itself after the maximum of the beat, corresponding to phase-agreement.

From this point, as the pitch rises, the observations become more difficult, partly no doubt on account of purely experimental complications, but also, I believe, because the effects are themselves less well marked. From two forks of pitch *g′*, electrically driven from the 128 interrupters and provided with resonators in connexion with pipes, fairly distinct right and left effects were obtained, but at first there were discrepancies as to which effect followed phase-opposition. These appear to have been due to faulty observation of the phase of opposition. As might have been anticipated, the moment of silence representing phase-opposition varies with the position in the room of the ear of the assistant observer. To secure a satisfactory signal at the higher pitches, this position requires to be carefully chosen. In the later experiments a resonator was always employed, whose mouth was symmetrically situated with respect to the open ends of the pipes, which are the proximate sources of sound, connexion with the ear of the assistant observer being through a suitable rubber tube. After this there was no ambiguity, the rule of the lower pitch being uniformly followed. But when the open ends of the pipes were not very close to the ears, perhaps 2 inches distant, the right and left effects seemed to two observers (including myself) not only to be rather obscured, but to be concentrated into the neighbourhood of phase-opposition. A third observer, however, heard the right and left effects more strongly, and with less apparent concentration towards phase-opposition.

On the theory that passage of sound round the head had something to do with these complications, the open ends of the pipes were brought much closer to the ears, but without fitting air-tight, the resonators being re-adjusted so as to diminish the loudness. In this condition of things the two observers experienced the right and left effect more normally and without special concentration in the region of phase-opposition. But the observation is certainly more difficult than at lower pitches, and I believe that the effects are really less pronounced.

Experiments similarly conducted with forks of pitch c'' (512) gave results of the same character. When the open ends of the pipes were quite close to the ears, the right and left effect was pretty good and fairly distributed. In this three observers concurred. But a slight withdrawal of the pipes introduced confusion, the extent of which, however, appeared to vary with the observer. In all cases the right and left effects, when sufficiently marked to be observed, obeyed the sequence rule.

At pitch e'' (640) the results were not very different. The open ends of the pipes being close to the ears, but not fitted air-tight, only pretty good right and left effects could be observed, and these appeared to be crowded towards phase-opposition. The sequence rule was obeyed.

Finally, trials by the same method were made with forks of pitch g'' ($6 \times 128 = 768$). No particular difficulty was encountered in satisfying the necessary experimental conditions, but the results were of a nondescript character. Even when the open ends of the pipes were close to the ears, I could not satisfy myself that I experienced any right and left effect. Another observer thought he heard a little. It seems clear that at any rate the limit was being approached.

It will be understood that some of these observations were not made without difficulty. Probably an experimenter new to the work would feel misgivings with respect to some even of the easier decisions. But all the more important results have the concurrence of at least three observers*. I regard it as established that up to pitch g' phase-differences are attended with marked lateral effects. They are probably the principal basis on which discriminations of right and left are founded—at any rate below c' (256).

As has already been suggested, it was reasonable to anticipate that phase-difference would cease to avail as an indicator at high pitch. Up to about e' the conditions are favourable. At this pitch the phase-difference at the ears affected by a distant sound increases from zero when the source is in front or behind to a maximum of a quarter period (in one or other direction) when the source is on the right or the left in the line of the ears. This is the phase-difference for which one would expect the lateral sensation to be most intense, so that up to this pitch the lateral sensation would keep step with the true lateralness of the source. At a point somewhat higher in pitch it would seem that complications must enter. The maximum sensation (corresponding to a phase-difference of a quarter period) would occur while the source was still in an oblique position, and the sensation of lateralness would diminish while the true lateralness was still increasing. At a pitch in the neighbourhood of e'' (640) the maximum phase-difference would rise to half a period, a phase-difference which could not give rise to lateral sensation at all.

---

* Lady Rayleigh, Mr Enock, and myself.

Thus, although there might be right and left sensations from sources obliquely situated, these sensations would fail when most needed, that is when the source is really in the line of the ears. In this case a perception of phase-differences would seem to do more harm than good. At a pitch a little higher, ambiguities of a misleading and dangerous kind would necessarily enter. For example, the same sensations might arise from a sound a little on the left and from another fully on the right.

On the whole it appears that the sensation of lateralness due to phase-difference disappears in the region of pitch where there would be danger of its becoming a misleading guide. It is not suggested that there is any precise numerical coincidence. If it were a question of calculating a pitch precisely, it might be necessary to look beyond the size of modern adult heads to those of our ancestors, perhaps in a very distant past. It is fortunate that when difference of phase fails, difference of intensity comes to our aid. Perhaps it is not to be expected that we should recognize intuitively the very different foundations upon which our judgment rests in the two cases.

A rather difficult question arises as to whether in the laboratory experiments it is possible to distinguish the phases of agreement and of opposition. Not unnaturally perhaps, the apparent movements of the sound from right to left and back are liable to be interpreted as parts of a general movement of revolution, so that, for example, phase-agreement may correspond to the front and phase-opposition to the back position. In a particular case the question is as to the direction of the revolution, whether clockwise or counter-clockwise. With respect to this, my observers frequently disagreed, from which I am disposed to conclude that in these experiments phase-agreement and opposition are not definitely connected with front and back sensations.

At this point there seems to be some discrepancy with the observations of Prof. S. P. Thompson, who found* that "when two simple tones in unison reach the ears in opposite phases, the sensation of the sound is localized at the back of the head." In Prof. Thompson's most striking experiment a microphone is connected in series with a battery and two similar Bell telephones, one of the telephones being provided with a commutator by which the direction of the current through it can be reversed. When the current flows similarly through the telephones, a light tap near the microphone is heard in the ears; but when the current is reversed in one of them a sensation is experienced "only to be described as of some one tapping with a hammer *on the back of the skull from the inside.*" In some (rather inadequate) experiments I have not succeeded in repeating this observation.

The other branch of the subject, which I had hoped to treat in this paper, is the discrimination between the front and back position when a sound is

* *Phil. Mag.* November 1878, p. 391.

observed in the open; but various obstacles have intervened to cause delay. Among these is the fact that (at 64 years of age) my own hearing has deteriorated. Thirty years ago it was only pure tones, or at any rate musical notes free from accompanying noises, that gave difficulty. Now, as I find to my surprise, I fail to discriminate, even in the case of human speech. It is to be presumed that this failure is connected with obtuseness to sounds of high pitch, such as occur especially in the sibilants. For some years I have been aware that I could no longer hear as before many of the high notes from bird-calls, such as I employ with sensitive flames for imitating optical phenomena. If, as seems the only possible explanation, the discrimination of front and back depends upon an alteration of *quality* due to the external ears, it was to be expected that it would be concerned with the higher elements of the sound. In this matter it would not be surprising if individual differences manifested themselves, apart from deafness. A "paddle-box" formation of the external ear, if not ornamental, may have practical advantages.

My assistant, Mr Enock, is able to make discriminations between front and back, though I think not so well as I used to be able to do. Experiments of this kind are easily tried on a lawn in the open, the observer closing his eyes and ears, with if necessary a movement of the thumbs over the latter to drown residual external sounds. At the moment of observation the ears are of course opened. In observing sounds from sources not conveniently move-able, such as the ticking of a clock, the rotating stool is useful.

As had been expected, Mr Enock's judgment was liable to be upset by the operation of little reflecting flaps situated just outside the ears. The arrangement was that of Prof. Thompson's "*pseudophone*"*, whereby the reflectors, whose planes were at an angle of 45° with the line of the ears, could be rotated in a manner unknown to the observer about that line as axis. In my use of it the two reflectors were always adjusted symmetrically. Thus, if the reflectors were so turned as to send into the ears sounds from the front, no mistakes were made, as if the action were co-operative with the natural action of the external ears. On the other hand, if the collars carrying the reflecting flaps were turned through 180° so as to reflect into the ears sounds from *behind*, frequent mistakes ensued. I hope before long to be able to confirm and extend these observations.

In conclusion, I will remark that the facts now established have a possible practical application. In observing fog-signals at sea it is of course of great importance to be able to estimate the bearing. If a sound is of sufficiently long duration (5 or 6 seconds), it is best by turning the body or head to bring it apparently to the right and to the left, and to settle down into the position facing it, where no lateral effect remains. If, as for most fog-signals,

---

* *Phil. Mag.* November 1879.

the duration be decidedly less than this, it may be preferable to keep still; but we are then liable to serious errors, should the signal happen to come from nearly in front or nearly behind. A judgment that the signal is to the right or left may usually be trusted, but a judgment that it comes from in front or behind is emphatically to be distrusted. If, for example, the sound seems to come from a position 45° in *front* of full right, we must be prepared for the possibility that it is really situated 45° *behind* full right. A combination of 3 or 4 observers facing different ways offers advantages. A comparison of their judgments, attending only to what they think as to right and left and disregarding impressions as to front and back, should lead to a safe and fairly close estimate of direction.

[1911. See further on the subject of phase-perception Myers and Wilson, *Proc. Roy. Soc.* Vol. 80 A, p. 260, 1908; Rayleigh, *ibid.* 83 A, p. 61, 1909.]

320.

ACOUSTICAL NOTES.

[*Philosophical Magazine*, Vol. XIII. pp. 316—333, 1907.]

Sensations of Right and Left from a revolving Magnet and Telephones.—Multiple Harmonic Resonator.—Tuning-Forks with slight Mutual Influence.—Mutual Reaction of Singing Flames.—Longitudinal Balance of Tuning-Forks.—A Tuning-Fork Siren and its Maintenance.—Stroboscopic Speed Regulation.—Phonic Wheel and Commutator.

Sensations of Right and Left from a revolving Magnet and Telephones.

AMONG the methods available for the production of a pure tone in a telephone circuit is that where the electromotive force has its origin in the revolution of a small magnet about an axis perpendicular to its length, the magnet acting inductively upon a neighbouring coil which forms part of the telephone circuit. It was by experiments made partly in this manner that I formerly* determined the minimum of current necessary for audibility in the telephone. In connexion with recent work upon the origin of the lateral sensation in binaural audition† I have again employed this method, and I now propose to give a brief account of the results, which were not available in time for incorporation in the paper just cited.

The object of the experimental arrangements is the separate presentation to the two ears of pure tones, practically in unison, in such a manner as to allow the effect of a variation in the phase-relationship to be appreciated. When the sounds proceed from tuning-forks vibrating independently, the phase-difference passes cyclically through all degrees, and if the beat be slow enough, there is good opportunity for observation. But it is not possible to stop anywhere, nor in some uses of the method to bring into juxtaposition phase-relationships which differ finitely. I thought that it would be of interest to observe under conditions which would allow any particular phase-relation to be maintained at pleasure, and to this the revolving magnet method naturally lends itself.

---

* *Phil. Mag.* Vol. XXXVIII. p. 285 (1894); *Scientific Papers*, Vol. IV. p. 109.
† *Phil. Mag.* [6] Vol. XIII. p. 214 (1907). [See preceding paper.]

The propulsion is by means of wind (under about two inches water pressure) from a well regulated bellows. The blade forming the magnet may be bent wind-mill fashion and receive the wind directly, but in the present experiments it was combined with a diminutive turbine, the whole revolving about a vertical axis. The speed was about 190 per second, giving in the telephones a note of pitch g. Two inductor-coils* were used, the circuit of each being completed through a telephone. The planes of the coils were vertical, their centres being at the same level as the magnet. One was fixed and the other was so mounted that it could revolve about an axis coincident with that of the magnet and turbine. The angle between the planes represents of course the phase-difference of the periodic electromotive forces, subject it may be to an ambiguity of half a period, dependent on the way the connexions are made. If the circuits are similar, as is believed, the phase-difference of the currents and of the electromotive forces is the same. The telephone-discs were loaded, but not so heavily as to bring them into tune with the sounds employed. The circuit of one telephone included a commutator by which the current through the instrument could be reversed, corresponding to a phase-change of 180°.

In commencing observations the first step is to adjust to equality the sounds heard from the two telephones. This can be effected by varying the distances between the magnet and the inductor-coils. The telephones are then brought into simultaneous action at the two ears, and the effect is observed. A rotation of the movable coil may then be made, or the current in one telephone may be reversed by means of the commutator. The results were for the most part in harmony with what had been expected from the experiments with forks. But one anomaly must be noted, relating to the neutral condition where no pronouncement can be made in favour of either right or left. This should occur when the phases of vibration at the ears are either the same or precisely opposed; and it had been expected that the condition would be realized when the planes of the inductor-coils were strictly parallel. There is no difficulty in determining the neutral position, where neither right nor left has the advantage in either state of the commutator; but I was surprised to find that according to my own judgment the neutral position deviated very appreciably, perhaps 10° and on one occasion even more, from that of parallelism. At first I supposed that the explanation of the anomaly was to be sought in the behaviour of the telephone plates, whose vibrations may not have the same relation in the two cases to the electric currents actuating them. It is possible that this cause of disturbance may have been operative to some extent; but that it was not a complete account of the matter became evident later when it was found that in Mr Enock's

---

* They were also employed in the 1894 experiments and are there spoken of as " wooden coils " from the fact that the wire is wound upon wood.

judgment the neutral position did coincide sensibly with parallelism of the coils. There is no doubt at all but that the judgments of the two observers really differed; each repudiated the setting of the coil satisfactory to the other.

In the judgment of the individual observer, the neutral position can be determined with considerable precision by the observed absence of lateral effect, in conformity with the results of the tuning-fork experiments. In using the commutator in order to ascertain that no change in respect of lateral effect accompanies reversal, a complication arises from the fact that during reversal one circuit is momentarily broken. The telephones may be removed from the ears during commutation. If this be not done, care must be taken that the judgment made relates to the permanent effect, or errors may ensue due to the momentary action of the sound on one ear only.

When the neutral position of the coil is departed from, a lateral sensation—say to the right—is experienced, and this increases until the displacement reaches 90°. A reversal at the commutator changes the *right* into a *left* sensation, having the same effect as a rotation of the coil through 180°.

When the adjustment is such that the combined lateral sensation (to the right) is a maximum, it is interesting to observe the effect of applying the telephones to the ears consecutively. If the right telephone be the first applied the sensation of course is of a sound to the right. When the left telephone follows, the sound remains on the right and appears louder. If on the other hand the left telephone be the first applied, the sound appearing originally to be on the left transfers itself to the right as the second telephone comes into action. Under the best conditions there seems to be nothing remaining over on the left.

The results are thus confirmatory of those obtained from tuning-forks. Unquestionably we are able to take account of the phase-difference at the two ears, and this in the case of low pitch is the foundation of the secure judgment as to direction that we are able to form when a single sound is heard from the right or from the left. With respect to the convenience of the two methods of experimenting, much of course depends upon what appliances are available. In most laboratories, I suppose, the tuning-forks would be preferred.

Multiple Harmonic Resonator.

The use of Helmholtz resonators to demonstrate the compound character of a musical note is now familiar. The harmonic component tone which has the pitch of the resonator is specially reinforced and so rendered conspicuous even to untrained ears. By changing the resonator, the fundamental tone or any of the harmonics may be intensified in succession.

Such effects are rendered far more striking if the necessary changes of pitch can be brought about in a single resonator, which then remains continuously connected with the ear. We may do a little in this direction with a resonator of the usual König pattern. Choosing one of somewhat high pitch and listening to a harmonium note two or three octaves down, we find that various harmonics swell out in turn as we pass the finger over the aperture, thereby gradually lowering the pitch to which the resonator responds.

The idea is carried out more completely if the resonator is provided with a number of separate apertures, any or all of which can be completely closed with the fingers. According to the simple approximate theory* the natural frequency (N) of the resonator is given by

$$N = \frac{a}{2\pi} \sqrt{\left(\frac{c}{S}\right)}, \quad \dots\dots\dots\dots\dots\dots(1)$$

where a is the velocity of sound in air, S the volume included in the resonator, and c the electrical conductivity between the interior and exterior calculated upon the supposition that air is a conductor of unit specific conducting power and that the walls behave as insulators. For a circular aperture in a thin wall c is equal to the diameter of the aperture. If there are several apertures, well separated from one another, c is equal to the *sum* of the diameters, or as we may write it,

$$c = d_1 + d_2 + d_3 + \dots \quad \dots\dots\dots\dots\dots(2)$$

Hence, if the first aperture acting alone give the fundamental tone, the first and second together the octave, the first three together the twelfth and so on, we have so far as relative magnitudes are concerned,

$$d_1 = 1, \quad d_1 + d_2 = 4, \quad d_1 + d_2 + d_3 = 9, \text{ &c.};$$

or $\qquad d_1 : d_2 : d_3 \dots\dots = 1 : 3 : 5 \dots \quad \dots\dots\dots\dots(3)$

The ratios (3) may give some idea of the proportions, but for many reasons—among them the neglect of the thickness of the walls—they are only roughly applicable. There is no reason for insisting on a circular form of aperture; indeed, when the aperture is large, an elongated form lends itself better to closure by a finger. Extreme cases excluded, the effectiveness of an aperture depends mainly upon its *area*.

When, as would be especially likely to happen in the case of the fundamental tone, a simple aperture would be very small, it may be well to replace it by a channel of finite length. If R be the radius of a tube of circular section and L its length,

$$c = \frac{\pi R^2}{L + \frac{1}{2}\pi R}, \quad \dots\dots\dots\dots\dots\dots(4)$$

---

* *Theory of Sound*, §§ 304, 305, 306.

from which it will be seen that the area of aperture may be much increased as compared with what would be admissible if $L = 0$.

Two compound resonators on this principle have been constructed. The first was made from the upper part of a glass bottle which had been cut off square near the neck. Over the wide opening a rather stout zinc plate was cemented through which the various apertures were bored. Through a cork, fitted into the neck, passed a short piece of brass tubing, by means of which and a suitable india-rubber prolongation connexion between the ear and the interior of the resonator was established. In tuning the instrument it is necessary to begin with the lowest tone and care must be taken to complete the adjustment of each aperture, or channel, before the next is attempted. Further details are hardly required. If the resonance is improved by shading an aperture with the finger, it is a sign that the aperture is already too large. If on the other hand the resonance improves up to complete withdrawal of the finger, the aperture may still be too small. Before finally enlarging the aperture it may be well to ascertain that the resonance is improved by a partial use of the one next in order.

The second resonator was constructed entirely of metal. It consists of an elliptical box of sheet zinc, with top and bottom also of zinc and slightly dished for the sake of enhanced rigidity (fig. 1). The capacity (S) is about

Fig. 1.

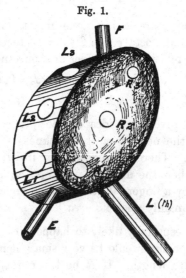

140 c.c. The apertures for the fundamental tone (F) and for the octave L(th) are formed of brass tubing soldered into position. The other passages are simple perforations in the wall and in the top of the box. E represents a short length of brass tubing over which is slipped an india-rubber attachment passing to the ear. To sound the fundamental tone of about 128

vibrations per second (B of my harmonium), F, about 8 mm. in diameter and 20 mm. long, is alone open. For the octave tone L(th), 11 mm. in diameter and 53 mm. long, is opened *in addition*. R_3 then gives the twelfth, R_2 the double octave, R_1 the higher third, L_3 the sixth component (octave + twelfth), L_2 the harmonic seventh, and finally L_1 the triple octave. The diameter of R_3 is about 5 mm. and that of L_1 (the largest aperture) say 13 mm. The letters are intended to indicate the fingering. Thus L_1, L_2, L_3 denote the first, second, and third fingers of the left hand; R_1, R_2, R_3 the corresponding fingers of the right hand. The octave tube L(th) is closed with the thumb of the left hand.

The performance of this instrument is very satisfactory. The seventh and eighth components are a little weak, but the others, and especially the twelfth and higher third, are loudly heard. The experimenter should bear in mind that when working in-doors much depends upon the precise position. The room is intersected with nodes and loops of approximately stationary vibrations, whose position varies from one tone to another. If a particular harmonic is ill heard, it may only be that the situation of the resonator is unfavourable. A motion of a few inches will often make a great difference. Usually the effects are best when the resonator is held pretty close to the reed in action.

In general a harmonium note is the most convenient for experiment and demonstration, but other instruments are of course available. A man's voice singing the proper note (B as above) gives excellent results.

Tuning-Forks with slight Mutual Influence.

Two forks giving 128 vibrations per second are independently maintained, each making and breaking its own contacts at a mercury cup. If mutual influence be altogether excluded, the "beat" may be made as slow as we please. But although the electric circuits may be entirely distinct, if the forks stand on the same table there may be enough mutual influence to bring about absolute unison. The best method of observation is by Lissajous' figures. The permanence of the ellipse is a sign that there is mutual control and that absolute unison is established; otherwise the ellipse undergoes more or less slowly the usual transformations.

A series of observations on this subject were made in 1901. Mutual influence may arise from both forks being connected with the same battery. If the electric circuits are in a series which includes two Grove cells, the forks keep together indefinitely; but this arrangement is rather akin to the familiar one in which a single interrupter-fork drives another, the second having no break of its own. Even when the fork circuits are in parallel and are fed from two Grove cells*, there is, or may be, sufficient reaction to

---

* The internal resistance of the cells comes into play here.

maintain absolute unison. The feebler the reaction, the more nearly must the natural frequencies approach to identity. When the reaction is just insufficient for control, it is interesting to watch the cycle of the beat, as revealed by Lissajous' figure. At one part of the cycle the changes are very slow and at the opposite part relatively very quick.

In another set of experiments the electric circuits were quite distinct, each fork being driven by a separate Grove cell. A sufficient mutual reaction could be obtained through the air. To this end a large resonator was constructed by cementing a wooden plate over the opening of a bell-glass. In the plate were two similar apertures, to which the free ends of the forks were presented, the pitch of the resonator being equal to that of the forks (128). In this way an adequate control was secured, but the margin was narrow.

Fig. 2.

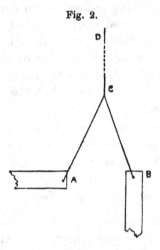

A more powerful controlling reaction accompanies a connexion between the two forks by means of slender cotton threads. The arrangement employed is indicated in the figure. A and B are the free ends of the upper prongs seen from above. To them is attached a Y-shaped thread $ABCD$, the tension of which can be adjusted at D. When the control is established, the Lissajous' ellipse is stationary and usually open. An ellipse closed in upon its major axis would indicate that the natural frequencies were identical, independently of the control. By touching a fork judiciously with the rubber tip of the exciting hammer, the phase may be disturbed without stopping the electric maintenance. If one fork be touched, the ellipse closes in, while a similar operation upon the other fork opens it out. In a short time the ellipse settles back, showing that the original phase-relationship is recovered.

Mutual Reaction of Singing Flames.

In a former paper* I discussed the mutual influence of organ-pipes nearly in unison, showing that the disturbances depend upon the approximation of the open ends, and not sensibly upon the circumstance that they may take their wind from the same source. When the reaction suffices, only one note is sounded, and that is usually higher in pitch than the notes of either pipe separately. It is proposed to record the results of some observations of a similar character recently made upon so-called singing flames, *i.e.* tubes caused to speak by means of hydrogen flames.

The tubes were of glass from the same length, each 30 cm. long and 16 mm. internal diameter. The hydrogen bottles were also similar and were provided with burners formed from 14 cm. lengths of glass drawn down at the upper ends. Very small flames suffice. The tubes were held vertically and so that their upper (and lower) ends were at the same level.

Fig. 3.

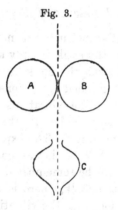

When the distance between the tubes is considerable, say 30 cm., and draughts are avoided, fairly slow beats may be obtained by suitable tuning, as by approach of the finger to one end of that tube which vibrates the quicker. But when the distance is reduced to perhaps 8 or 10 cm., a difficulty begins to be experienced in producing slow beats. Either they are rather quick or else, when the tuning does not allow of that, they disappear altogether, the vibrations as it were engaging. On the margin where beats still occur, their character is peculiar. They appear unsymmetrical, the swell being protracted and the fall hurried. The phenomenon is the same as that observed optically in the case of forks (p. 370). When the tubes are as close as possible—they may conveniently be tied together with string,— even moderately slow beats are excluded. In this situation the sound is much attenuated, indicating that the phases of vibration are opposite, at any rate in the ideal case. The ideal case is, however, rather difficult to attain.

* *Phil. Mag.* Vol. VII. p. 149 (1879) ; *Scientific Papers,* Vol. I. p. 409.

There should be complete cessation of the principal tone in a resonator (C) whose mouth is held near the upper ends (A, B) symmetrically in the median plane (fig. 3). Frequently a better silence may be reached by moving round a little, and even then it is not absolute. By use of the finger to give a finishing touch to the tuning, the most silent position may be driven to the median plane, but even so the residual tone may not be quite extinguished. It is evident that the ideal condition is easily disturbed a little by slight failures of symmetry, probably connected with the flames.

As so far described, the disappearance of the principal tone, sometimes very nearly realized, leaves a considerable amount of octave outstanding. A remedy may be applied by the insertion, at the middles of the tubes, of rectangular blocks of wood, about two diameters long and forming a loose fit. They are held in their places by springs. In this way the outstanding octave may be very much reduced.

Longitudinal Balance of Tuning-Forks.

The vibrations of a well-constructed tuning-fork are approximately isolated and are conveyed to the stalk in only a limited degree. When, as in the ordinary use of small forks, the stalk is pressed against a sounding-board, the principal tone is attended by a considerable accompaniment of octave, especially at first when the vibrations are vigorous. The substitution of a suitably tuned resonance-box for the sounding-board may easily render the octave sound preponderant*. The experiments now to be recorded were an attempt to ascertain how far it was possible to carry the isolation of the principal tone. It should be remembered that however complete may be the isolation as regards the stalk, there is necessarily a certain small amount of direct communication from the vibrating prongs to the surrounding air. For our present purpose this is to be disregarded.

At first sight it may appear that the desired state of things must be very approximately attained in the usual construction where the prongs are parallel. Something will depend upon the manner in which the transition takes place between prongs and stalk. In what follows I have more particularly in view a construction in which the prongs form a U of tolerably uniform thickness, to which a cylindrical stalk is attached without much excess of metal at the junction. As a rough approximation we may suppose that the inertia of the fork is concentrated at the ends of the prongs. Then if the fork be free in space, these ends can move only backwards and forwards along the line joining them. The question we have to ask is—

---

* *Phil. Mag.* Vol. III. p. 456 (1877); *Scientific Papers*, Vol. I. p. 318. Even when the box is tuned to the fundamental note, the octave and twelfth are often easily audible. I have observed this effect with three different 256-forks when mounted upon a particular resonance-box. There was no suggestion of looseness or chatter.

does the stalk remain at rest? A little consideration makes it fairly clear that in the case of parallel prongs the answer is in the negative. As the prongs approach one another the curvature of the bend is increased and the stalk moves along its length *outwards, i.e.* away from the prongs*. Similarly half a period later the opening of the prongs is accompanied by an approach of the stalk. Under these conditions if the stalk be brought into contact with a sounding-board, a motion of the first order is communicated and the principal tone is heard.

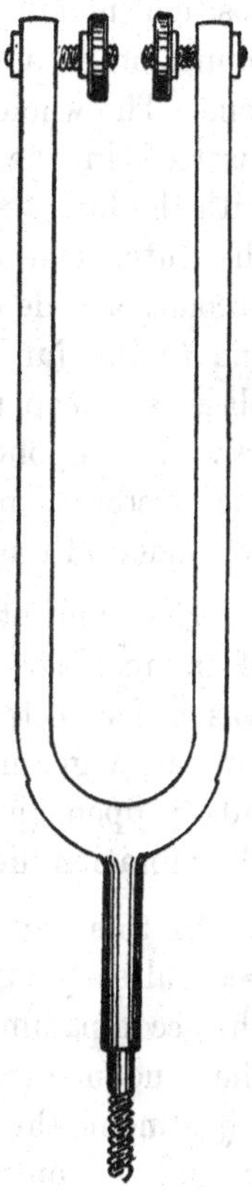

It is evident that the effect to be expected when the prongs are parallel may be compensated by a suitable permanent bending of the prongs *inwards*, or what comes to the same by a suitable *loading* on the inner sides. The motion of the stalk during the vibration is then composed of two parts which have opposite signs—the one already considered depending on the variable curvature at the bend, the other on the obliquity of the prongs to the line of motion at the ends. It would appear then that by this adjustment it should be possible to secure that the stalk remains at rest, so far as motion of the first order is concerned. It is assumed that everything is symmetrical, so that the stalk, if it moves at all, does so along its length and (in view of its dimensions relatively to the wave-length of vibration in steel) practically as a rigid body.

Fig. 4.

For the purposes of the experiments a large fork was constructed by Mr Enock. The U was from a single length of steel 60 cm. long and of section 1·275 cm. square. The prongs were parallel, 5·35 cm. apart (inside measurement), and the stalk was attached by brazing (fig. 4).

In its unloaded condition it gave 128 vibrations per second and could be screwed to a resonance-box appertaining to a large fork by König of the same pitch. When excited by bowing it emitted a very powerful sound and, largely in consequence, came somewhat rapidly to rest. The isolation of the vibrations was thus far from complete.

The principal loads, of 40 gms. each, were in the form of nuts and travelled along screws passing through the prongs near their ends and parallel to the direction of vibration. Suitable lock-nuts kept all tight. In consequence of the loading, the pitch fell about a major third, and the tuning of the resonance-box had to be readjusted by a piece of board obstructing the mouth. It soon appeared that, as had been expected, when the loads

---

* Somewhat as if by a violent local bending at the middle of the U the prongs were brought into contact throughout their whole length.

were outside the prongs the sound diminished as they were moved inwards as far as possible. To obtain a minimum, the loads must be *inside* the prongs; and a great falling off was readily achieved by adjustment of their position.

At this stage considerable difficulty was experienced in appreciating the quality of the residual sound, but it was suspected that most of it was octave, in spite of the fact that the resonance-box was tuned to the fundamental tone. The device appropriate to stop tones of a particular pitch from gaining access to the ear is a Quincke tube. A straight length of composition metal tubing, open at both ends, was provided with a lateral connexion at a distance from the outer end amounting to $\frac{1}{4}\lambda$ of the *octave* tone. The whole length was nearly the double of this, and the other end was inserted in the resonance-box. At the same time the ear was connected with the lateral branch with the aid of an india-rubber prolongation. When the outer end of the straight tube is closed with the thumb, that end becomes a node of the stationary vibrations of octave pitch executed therein, and as the junction with the lateral tube is distant $\frac{1}{4}\lambda$ from the end that place is a loop, and consequently no (octave) vibration is propagated to the ear. The application of the thumb accordingly has the effect of freeing the possibly compound sound from its octave component, while it leaves the fundamental tone in full vigour.

The application of this test proved at once that by far the greater part of the residual sound heard when the loads were in approximate adjustment was in fact octave. Immediately after bowing, when the vibrations of the fork are vigorous, a loud sound is heard when the outer end of the Quincke tube is open, but comparative silence ensues when the thumb is applied. As the vibration dies down, closing the end has less effect.

In this way it appeared that in reality a great measure of success had been already attained in isolating the fundamental tone, only obscured by the accompaniment of octave in unexpected amount. A sensible revival of the fundamental tone ensued when the loads were rotated from their best adjustment through a quarter turn each, corresponding to a lateral shift inwards or outwards of $\frac{1}{80}$ inch. Since the test is of such delicacy, we may perhaps consider the isolation of the fundamental tone to be practically complete.

The fact remains, and must not be slurred over, that it was not possible by any adjustment of the loads to eliminate the fundamental tone *entirely*. The residual sound did not come directly through the air from the prongs, but was propagated through the stalk to the resonance-box. It is a little difficult to trace the nature of this residue. Upon the supposition that the vibrations of the various parts of the fork are all in one phase, and of complete geometrical and mechanical symmetry in the construction of the fork, it would appear that some adjustment of the loads *must* eliminate the fundamental tone. There was, indeed, evidence of actual lack of symmetry, which could not in any case be mathematically perfect. When with the aid of a handle the fork was held horizontally so that its stalk rested upon a wooden edge supported in its turn upon the top of the resonance-box, sound was heard from the box, which varied as the fork rotated round its stalk as axis and in fact nearly vanished in two asymmetrical positions. It would seem that the residual fundamental tone heard in the more normal use may be connected with a *lateral* movement of the stalk, dependent upon some failure of symmetry.

As a variation upon the above arrangement the prongs were now bent inwards so that at the outer ends the distance from metal to metal was reduced from 53·5 mm. to 38 mm., a bending intended nearly to represent the effect of the loads. With the 40 gm. loads it was no longer possible to reduce the fundamental tone to silence and, as soon appeared, for this reason that the proper position for the loads was unattainable, being that occupied by the metal of the prongs themselves. When smaller (20 gm.) loads were substituted, interior positions could be found allowing the elimination of the fundamental tone to about the same degree of perfection as before. In listening with the Quincke tube to the dying sound with alternate application and removal of the thumb at the outer end, it was recognized that the low tone was practically gone (thumb on) while the octave (thumb off) was still fairly audible. About the same displacement of the loads as before ($\frac{1}{80}$ inch) was sufficient to cause a perceptible augmentation of the residual fundamental tone.

As to whether these results can be turned to practical account in the construction of forks, we must remember that if a fork is to be used in conjunction with a sounding-board or resonance-box a too complete isolation of the fundamental tone would defeat the intention. On the other hand if, as in Helmholtz's vowel experiments, a fork is to be employed to excite an air resonator placed near the ends of its prongs, a suitable turning inwards of these prongs and consequent quiescence of the stalk would be of advantage.

In conclusion attention may be drawn to the circumstance that a symmetrical *bell* with stalk attached would not need any particular adjustment in order to ensure the isolation of the vibrations of the first order. If the stalk tend to move outwards when contraction occurs along one particular diameter of the circumference, the same tendency must repeat itself half a period later when the contraction is transferred to the diameter at right angles to the first. A similar remark would apply to a symmetrical compound fork, such as we may imagine to be produced by cutting away all the material of the bell, except in the neighbourhood of two perpendicular meridians.

A Tuning-Fork Siren and its Maintenance.

When in 1901 I was experimenting upon the work absorbed in various cases of the production of sound* I had at my disposal a Trinity House "Manual" Fog-horn. In this instrument the wind is generated by cylinders and pistons, and a much higher pressure is available than is usual in laboratory apparatus. Among the experiments then tried was the substitution of what I called a tuning-fork siren for the natural reed and conical horn of the instrument. Fitted to a wind-chest was a metal plate, carefully faced internally and carrying a rectangular aperture about 10 mm. broad. This aperture could be nearly closed by a plate 3 mm. wider, which vibrated laterally in front of it. The vibrating plate was mounted upon the side of one prong of a fork making 128 vibrations per second. When the fork was at rest, the aperture was obstructed and the fit of the moving and fixed plates was so good that the leakage of wind was not serious. But as the fork vibrated the aperture was in part uncovered, and that twice during each complete vibration of the fork, so that the pitch of the instrument, considered as a siren, was 256. The fork was driven electrically from another interrupter-fork of the same pitch situated outside. Fitted to the aperture externally was a resonating tube whose length could be adjusted to give the maximum effect. One of the objects was to be able to vary the resonance without disturbing the maintenance or the pitch of the siren. The pressure employed was sometimes as low as 2·5 cm. of mercury, and the consumption of wind about $3\frac{1}{2}$ litres per second, corresponding to ·015 horse-power. At this rate of working the pumps could be kept going by hand, or rather by legs, for a moderate length of time.

One unexpected effect presented itself, which seems worthy of record. As has been mentioned, the intention had been to keep the fork in motion electrically. But it was found that, at any rate after being once started, it remained in vigorous vibration under the action of the *wind alone*, although the electric connexion was cut off. It will be observed that this case is altogether different from that of a *reed*, where the tongue approaches and recedes from the aperture normally. It is more analogous to the æolian harp, where, as I have formerly shown†, the vibration is executed in a plane perpendicular to the direction of the wind. So far as I am aware, no adequate mechanical explanation of this singular behaviour has been given.

---

* *Phil. Mag.* Vol. VI. p. 292 (1903). [Vol. V. p. 129.]

† *Phil. Mag.* Vol. VII. p. 149 (1879); *Scientific Papers*, Vol. I. p. 413.

Fig. 5.

Stroboscopic Speed Regulation.

The stroboscopic method* has often been employed for testing and regulating the speed of revolving shafts. I used it extensively in my determinations of absolute electrical units†, referring the speed of revolving coils to the frequency of vibration of an electro-magnetically maintained fork. But I doubt whether even now the convenience of this method for general purposes is appreciated. A few years ago the late Mr Gordon drew for me upon card rows of alternate black and white "teeth" from 20 to 40. Photographs from this upon flexible paper could be mounted upon a revolving shaft so as to form reentrant circles of teeth for observation by intermittent view. In my use of it an electrically maintained fork of large dimensions and of home construction was employed. The fork was provided with solid (platinum) contacts and made 64 vibrations per second. At the ends of the prongs were blackened plates of thin metal perforated with slits, so disposed as to be opposite to one another in the equilibrium position. When the vibrations were excited by one or two cells, there were 128 views per second through the slits.

Viewed past the fork, some of the circles of revolving teeth appear nearly stationary. Usually two neighbouring circles can be picked out, which appear to revolve slowly in opposite directions. From these data the necessary information is obtained in a way that need not be further explained. It is thought that it may be of service to give a reproduction of Gordon's drawing (see Plate, fig. 5). Photographic copies can easily be made upon any desired scale adapted to the shafts round which it is intended to mount them. Care should be taken to effect the junction properly, so that the circles of teeth are continued through it without irregularity.

Phonic Wheel and Commutator.

By the use of the phonic wheel, invented independently by La Cour and myself‡, the speed of revolving shafts may be not merely compared with a fork but automatically governed thereby. I have used this method for driving a commutator of the kind required for passing a regular succession of condenser charges through a galvanometer, as for example in determining the ratio of the electrical units (velocity of light). The contacts required are such that a piece A in connexion with the insulated pole of the condenser shall make contacts alternately with a piece B representing the insulated

---

* Plateau (1836); Töpler, *Phil. Mag.* Jan. 1867.

† See for example *Proc. Roy. Soc.* Vol. xxxii. p. 104 (1881); *Scientific Papers*, Vol. ii. p. 8.

‡ *Phil. Trans.* 1883, p. 295. *Scientific Papers*, Vol. i. p. 355; Vol. ii. p. 179.

pole of the battery and another C connected to earth and to the other pole of the battery and condenser. Of course the contacts must be good, and it is essential that both be not made at the same time.

As in Thomson and Searle's work*, the commutator was of the usual type adapted to a revolving motion, except that the cycle of contacts was repeated so as to occur twice in each revolution. The developed form is shown in the accompanying diagram (fig. 6) where the shaded parts represent brass pieces, separated by ebonite insulation. Springs lightly bearing against the exterior continuous portions of metal correspond to B and C, while A corresponds to a brush bearing near the centre and making contact alternately with the two metal pieces. Provision was made for varying the pressures at these contacts during the running and without disturbing the insulation. The problem is to secure a uniform rotation of this commutator, whose diameter was 28 mm.

Fig. 6.

The phonic wheel, mounted on the same shaft as the commutator, takes its time from a vibrating fork (44 per second†) acting as interrupter of an electric current. The current (about 4 amperes) is from three secondary cells and excites not only the electro-magnet by which the vibrations of the fork are maintained but also the electro-magnet of the phonic wheel. Four soft iron armatures are mounted round the circumference of the drum and in their passage complete approximately the magnetic circuit. The holes through which the fork is viewed are also four in number.

The most advantageous action of the regulating current occurs when one armature passes for each complete vibration of the fork. Under these circumstances the prong, or rather a projecting wire attached for the purpose, is seen stationary and *single*. There are then $11 (= \frac{1}{4} \times 44)$ revolutions of the wheel per second and 22 charges and discharges of the condenser. But the wheel may also be run at double or triple this speed, and then the projecting wire is in general seen doubled or tripled. The regulating current from the fork is of itself capable of maintaining the rotation at single or double speed when once the necessary *engagement* has been secured. For this purpose the speed must be raised to the required point by means of a string passed round the shaft and worked with the fingers, and even then it may be only after many trials that engagement ensues. For the triple

---

* *Phil. Trans.* 1890, p. 607.

† For the design of steel vibrators and for rough determinations of frequency, especially when below the limit of hearing, the theoretical formula is often convenient. We may take

$$\text{frequency} = 84600t/l^2,$$

where l is the total length of a prong and t the thickness in the plane of vibration, both being reckoned in centimetres. (*Theory of Sound*, § 177.) At any rate the *octave* is never uncertain.

speed the power of the fork current is insufficient, and recourse must be had to independent driving by an electric motor or water-power engine. If the driving power be in excess, the fork currents are equally capable of holding the wheel back. The whole behaviour is evident on observation of the fork through the revolving apertures, and so long as the engagement lasts the wheel can never gain nor lose a complete cycle relatively to the fork.

I have used the commutator thus driven to observe the charges of a condenser in their passage through a galvanometer. The only inconvenience was the necessity of a considerable separation between the galvanometer and the rest of the apparatus to obviate magnetic disturbance. The galvanometer deflexion was steady and apparently independent of the force exercised at the springs of the commutator. I believe that the arrangement might be used with advantage in such work as determining the ratio of the electrical units.

321.

ON THE PASSAGE OF SOUND THROUGH NARROW SLITS.

[*Philosophical Magazine*, Vol. XIV. pp. 153—161, 1907.]

THEORY leads to the curious conclusion that plane waves of sound incident upon a parallel infinitely thin reflecting screen in which is perforated a narrow slit, are transmitted in a degree depending but little upon the width of the slit. If the plane of the screen be at $x = 0$, and if the waves incident on the negative side be denoted by

$$\phi = \cos(nt - kx), \quad \dots\dots\dots\dots\dots\dots\dots\dots\dots(1)$$

then the waves diverging from the slit upon the positive side have the expression

$$\psi_p = -\left(\frac{\pi}{2kr}\right)^{\frac{1}{2}} \frac{\cos(nt - kr - \frac{1}{4}\pi)}{\gamma + \log(\frac{1}{4}kb)}, \quad \dots\dots\dots\dots\dots(2)$$

in which $2b$ is the width of the slit, r the perpendicular distance of any point from it, γ = Euler's constant ($\cdot 5772$), and as usual $k = 2\pi/\lambda$, the wave-length [*]. These equations apply also to the case of light incident upon a perfectly reflecting screen, provided that the electric vector is perpendicular to the length of the slit.

I thought that it would be of interest to examine the question experimentally, but the difficulties in the way have turned out to be more considerable than had been expected. In dealing with sound-vibrations we have a large choice of wave-lengths, down to 1 inch, or less if a sensitive flame be employed as percipient. But, even so, there are formidable obstacles in the way of realizing the theoretical conditions. In practice the screen must be limited; and then if the source be situated far behind, sound readily passes the boundary, and with the aid of reflexions reaches the ear more effectively by this course than through the slit; while if the source be near, the waves incident upon the slit are not sufficiently plane. Even with the shortest waves an adequate approximation to the infinitely wide and at the same time infinitely thin screen of theory seems unrealizable. An attempt was next made to obviate the difficulty by boxing up the source of sound so

---

[*] *Phil. Mag.* Vol. XLIII. p. 259 (1897); *Scientific Papers*, Vol. IV. p. 291.

as to ensure that nothing could reach the ear otherwise than through the slit. Even this demand is not met without much care, and when it is met there is no security that the amplitude and phase of the vibration at different parts of the length of the slit shall be the same. As to this length itself, it is evident that it ought to amount to a considerable multiple of the length of a wave.

An effective test on these lines of the escape of sound through a narrow slit seeming hopeless, attention was turned to the opposite extreme where the length of the slit is regarded as a small, in place of a large, multiple of the wave-length. The expression replacing (2) is now*

$$\psi_p = M \frac{\cos (nt - kr)}{r}, \quad \dots\dots\dots\dots\dots\dots\dots(3)$$

where M denotes the electrical *capacity* of a plate having the size and shape of the aperture, and situated at a distance from all other electrified bodies. So far as I am aware, M has not been calculated for a rectangular aperture; but for an ellipse of semi-major axis a and eccentricity e

$$M = \frac{a}{F(e)}, \quad \dots\dots\dots\dots\dots\dots\dots\dots\dots(4)$$

F being the symbol of the complete elliptic function of the first kind. When $e = 0$, $F(e) = \frac{1}{2}\pi$; so that for a circular aperture $M = 2a/\pi$.

If the ellipse be very elongated,

$$F(e) = \log \frac{4}{\sqrt{(1 - e^2)}} = \log \frac{4a}{b}, \quad \dots\dots\dots\dots\dots(5)$$

if b be the semi-axis minor, so that in this case

$$M = \frac{a}{\log (4a/b)}. \quad \dots\dots\dots\dots\dots\dots\dots(6)$$

The introduction of this value into (3) shows the same comparative independence of the magnitude of the width of the small aperture as was manifested in (2). It is understood that the longer dimension $2a$ of the ellipse, as well as the shorter, is to be a small fraction of λ.

In the earlier experiments the latter condition was but imperfectly fulfilled. The source of sound was a bird-call† giving a wave-length of $1\frac{1}{5}$ inch or 30 mm. The wave-length is ascertained in the usual way by placing a high-pressure sensitive flame in the stationary system compounded of the direct waves and of those reflected perpendicularly from a movable reflector. The displacement of the reflector required to pass from a maximum to a maximum or from a minimum to a minimum of excitation, measures the *half* wave-length. When the sensitiveness of the flame is suitably adjusted,

---

* See equations (14), (15) of memoir cited, or *Theory of Sound*, § 292.
† *Theory of Sound*, § 371, 2nd edition.

the observation of the minimum, characterized by recovery of the flame, admits of great precision. The bird-call was mounted in a glazed earthen-ware drain-pipe. This was closed at the hinder end with a wooden disk perforated by two tubes, one serving as an inlet and the other as an outlet for the wind, and in front by a second disk upon which the adjustable slit was mounted. As has already been hinted, the isolation of the sound requires much precaution. Although the jaws of the slit fitted well, the sound could not be prevented from escaping without the aid of grease. The wind was from a bag of heavy rubber cloth suitably weighted and controlled by obser-vation of a manometer, and the outlet tube was continued to a point outside the window by a rubber prolongation. It was necessary to cement up every joint until the whole was air-tight, otherwise sound could be heard through a rubber tube connected with the ear and presented at the outer end to the place under test. I should say that my own ears are not now effective for observations at this pitch, but Mr Enock's are very sensitive to it.

The slit was of tin-plate with jaws carefully filed to a knife-edge. The length was half an inch. The width could be reduced to about 2 thousandths of an inch without too great an uncertainty of measurement. With these arrangements it had been expected that the desired observations could be made without difficulty. An intensity observed whether by ear or flame at a certain short distance in front and with a certain width of slit might be recovered at a greater distance with a wider slit. On the basis of the usual law of attenuation of sound with distance a measure would be attained of the effect of widening the slit.

But it soon appeared that nothing of value could be obtained on these lines. In listening with the ear through a rubber tube whose open end (usually provided with a conical termination of glass) was moved to and from the slit, the intensity on receding was found not to fall in a continuous manner, but to be subject to alternate risings and fallings, almost as if it were due to a system of stationary waves. The effect was not equally apparent with the flame, but it is difficult to make good observations with a flame unless it can be maintained in a fixed position.

The earlier experiments were made in the laboratory where the floor, ceiling, and walls might act as reflectors, but I was surprised at the vigour of the alternations in view of the proximity of the place of observation to the source. A transfer of the apparatus to the outside proved indeed to be no remedy. With the slit facing upwards, observations along the vertical line passing through it indicated alternations nearly as marked as before. Although it is impossible to avoid obstacles altogether—the observer himself and the wind-bag would be the principal ones—the result seemed unfavour-able to the reflexion theory.

At this stage I thought that the limitation of the disk upon which the slit was mounted—its diameter was about 9 inches—might be a source of complication, and the apparatus was modified so as to permit the disk to form part of the general floor of the laboratory. For this purpose a smaller containing jar was necessary and the opportunity was taken to replace the bird-call by a small whistle or open organ-pipe whose pitch lay more within the capacity of my own ears. Preliminary experiments with the flame and movable reflector showed that (as had been expected from the pitch f^V and dimensions of the pipe) the complete wave-length was $2\frac{1}{2}$ inches, giving $1\frac{1}{4}$ inch periods between nodes or between loops.

Experiments with the apparatus thus mounted exonerated the disk, for alternations of a marked kind were still recognized commencing at a few inches' distance from the slit, which might now be regarded as situated in an infinite impermeable plane. The character of these alternations was peculiar. Sometimes they occurred in a period of $1\frac{1}{4}$ inch, as if due to a reflector in front which might perhaps be the ceiling of the room. But as a rule the more conspicuous feature was a period of about $2\frac{1}{4}$ inches, and this sometimes manifested itself in a condition of great purity.

Many times the thought occurred that I had misestimated the octave, and that the fundamental wave-length was really double what had been supposed, but this suggestion could not be maintained. After all, if reflexions are admitted, the period of the alternations is not limited to the half wave-length. Even when stationary waves are formed truly in one dimension, i.e. in parallel planes, it is possible by crossing them obliquely to encounter periods which may exceed the half wave-length in any proportion. The whole wave-length would in fact be a reasonable average value. In particular, the whole wave-length would be the period due to the interference of the direct sound with one arriving from a distant reflector in a perpendicular direction.

An origin of this kind would afford an explanation of a several times suspected difference of behaviour between the ear and the sensitive flame. As regards the former, it makes but little difference in which direction a sound arrives at the end of the hearing-tube; but it is otherwise with the latter, whose excitation is due, not to variable pressure, but to variable motion. If, as usual, the flame burns vertically, it is insensitive to sounds arriving in a vertical direction, and even in the horizontal plane the sensitiveness is of a semi-circular character, vanishing in two opposite azimuths*. It may thus easily happen in the above experiment that the flame may be insensitive to reflected sounds, which are nevertheless capable of influencing the ear.

* *Nature*, Vol. xxxviii. p. 208 (1888); *Scientific Papers*, Vol. iii. p. 24.

The difficulty of accepting the explanation by reflected waves is in great degree that of understanding how they can be powerful enough in comparison with the direct sound at such short distances from the source. In forming a judgment we must bear in mind that it is amplitudes that are compounded and not intensities, so that as in the theory of Newton's rings seen by *transmission*, a sound which would be inaudible by itself may be competent to cause a very perceptible variation in the loudness of another.

Fig. 1.

In order to obtain further evidence a modification was introduced in the terminal of the hearing-tube. Thus if this be a T-piece (fig. 1) of such dimensions that the head of the T measures half a wave-length, discrimination will be made between different directions somewhat as in the case of the flame. If the head be parallel to a wave-front, the two openings co-operate; but on the other hand if the head be perpendicular to the wave-front, the phases at the ends are opposed and nothing is propagated to the ear. This is the reciprocal of the effect noticed on a former occasion[*] that open organ-pipes emit little sound in the direction of their length. The T-piece may be used to eliminate a reflected sound travelling at right angles to the one which it is desired to isolate.

Fig. 2.

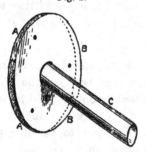

For my purpose a more symmetrical arrangement was preferable which should treat similarly sounds arriving in *all* directions perpendicular to the length of the stalk. In the sketch (fig. 2) AA and BB are circular disks of tin-plate of which A is complete while B is perforated to receive the stalk C. The former is held in position by three distance-pieces extending across, and

---

[*] *Phil. Mag.* Vol. VI. p. 304 (1903). [Vol. v. p. 139.]

the diameter AA measures $1\frac{7}{8}$ inch (47 mm.). The distance between the plates is $\frac{1}{4}$ inch. When the hearing-tube is provided with this appliance as a terminal, sounds are heard with full intensity when they arrive in a direction parallel to the stalk inasmuch as phase-agreement then obtains all round the circumference, but if the pitch be that of the whistle a sound arriving in any direction perpendicular to the stalk fails to penetrate C on account of interference. The theory is given in the appendix to this paper.

I may note in passing that a similar apparatus has been constructed upon a larger scale with disks of thick mill-board 10 inches in diameter and about half an inch apart. This is adapted to pitch c''' of 1024 vibrations per second, and it works well. The note may be given upon a harmonium at a few feet distance. When the stalk points towards the reed the sound is very loud, but it falls off in oblique positions and becomes faint when the reed lies in the plane of the disks. In the open air the apparatus may be used to find the direction in which a sound arrives.

With these aids I had hoped to be able to eliminate reflexions sufficiently to realize the continuous diminution of intensity which should attend recession from a single source whether situated in the open or on the bounding wall of a semi-infinite space; but these hopes have been disappointed. In the laboratory an adjustment of the disks to parallelism with the floor (in which the slit was situated) should eliminate reflexions from the walls. There remains the ceiling; but so far as this was flat it should give rise to alternations with a period of the half wave-length. As a fact the whole wave-length was often observed, but the ceiling was somewhat curved and there were of course other obstacles in the room. Not much better success was attained out of doors, the apparatus being placed upon a lawn and the slit facing upwards. Here again the horizontal position of the terminal disks should have eliminated reflexions from obvious obstacles, but alternations in the period of the full wave-length were usually apparent. The question would sometimes suggest itself whether visible obstacles were necessary at all to cause reflexions. In fog-signalling, echos, sometimes up to 30 seconds' duration, have been observed when the sea was smooth and there was no visible cause of reflexion. But these are usually attributed to a streaky condition of the air, a cause which could scarcely have operated in evening experiments over a lawn. That reflected waves, arriving more or less horizontally, were still concerned is suggested by the observation that the anomalous alternations were intensified by holding the terminal disks nearly vertical so as to attenuate the direct sound. Possibly there was diffuse reflexion from the grass. Whatever the cause of disturbance may have been, it rendered hopeless any attempt to compensate width of slit by alteration of distance, as had been the original intention. Indeed it would hardly have been worth while to describe at so much length the difficulties encountered, were it not that they may probably embarrass other observers unprepared for them, and that the terminal disk-apparatus has an independent interest.

The best that I have been able to do is by altering the *length* of the slit so as to compensate variations of width. For this purpose a sliding plate was provided cutting off equal lengths from the two ends. Observations have been made both in the laboratory and outside upon the lawn. In both cases the slit faced upwards, and the sound (f^V) was observed by ear through a hearing-tube of rubber, provided at the further end with the disk-apparatus already described and held in a clip with stalk vertical. In the first arrangement of the slit the width might be ·002 inch and the length one-half an inch. The sound reaching the ear would be observed and as far as possible retained in the memory. An assistant would then alter the width, say to ·010 inch, and the length say to one-quarter inch, and the observer would endeavour to decide which of the two sounds was the louder. It was found as the mean result of two observers that compensation ensued when the length was reduced from ·5 inch to ·28 inch. A similar change of length compensated an alteration of width from ·004 to ·020 inch.

It will be seen that at any rate the sound is much less sensitive to an alteration of width than to one of length. If we apply the formula (6) applicable to an ellipse, identifying the length of the slit with $2a$ and its width with $2b$, we get using logarithms to base 10

$$\text{(i)} \quad M' = \frac{\cdot 25}{\log_{10}(4 \times \cdot 25 \div \cdot 001)} = \cdot 083,$$

$$\text{(ii)} \quad M' = \frac{\cdot 14}{\log_{10}(4 \times \cdot 14 \div \cdot 005)} = \cdot 068 ;$$

so that the formula in question gives results not very wide of the mark. Other observations also were in fair accordance.

APPENDIX.

Mean potential over the circumference of a circle whose plane is parallel to the direction of propagation of plane waves.

P is any point on the circle. $OP = r$, $OM = x$. Potential at P

$$= e^{ikx} = e^{ikr \cos \theta}.$$

Hence for the mean we have

$$\frac{1}{\pi} \int_0^\pi e^{ikr \cos \theta} \, d\theta = J_0(kr).$$

$J_0(kr)$ vanishes when $kr = 2\pi r/\lambda = 2\cdot 404$, or $2r = \cdot 77\,\lambda$.

If the plane of the circle be inclined at an angle α to the direction of propagation, $J_0(kr)$ is replaced by $J_0(kr \cos \alpha)$. If $\alpha = \frac{1}{2}\pi$, so that the plane is parallel to the waves, we find unity for the mean value, as was to be expected.

When the potential to be averaged varies in two dimensions only, even though the waves may not be plane, we may proceed by the method of *Theory of Sound*, § 339. Using polar coordinates and omitting as before the time-factor, we have

$$\psi = A_0 J_0(kr) + \ldots + (A_n \cos n\theta + B_n \sin n\theta) J_n(kr) + \ldots,$$

n being integral.

Thus mean ψ over circle of radius $r = A_0 J_0(kr)$. At the centre of the circle, $\psi_{r=0} = A_0$; so that mean ψ over circle $r = \psi_{r=0} \times J_0(kr)$, and vanishes as before when $J_0(kr) = 0$.

Fig. 3.

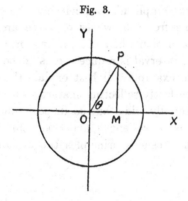

A similar method applies in three dimensions. Thus (*Theory of Sound*, § 330) we have in general

$$\psi = -2ikS_0 \frac{\sin kr}{kr} + \text{terms in spherical harmonics of orders 1, 2, &c.,}$$

vanishing when integrated over a sphere and also vanishing when $r = 0$. Hence

$$\text{Mean } \psi \text{ over sphere } r = \psi_{r=0} \times \frac{\sin kr}{kr}.$$

Accordingly mean ψ vanishes if $kr = n\pi$, or $r = \frac{1}{2}n\lambda$. If $kr = (n + \frac{1}{2})\pi$,

$$\text{Mean } \psi \text{ over } r = \psi_{r=0} \times \frac{\pm 1}{(n + \frac{1}{2})\pi}.$$

For an example, reference may be made to the case of plane waves treated in § 334.

322.

ON THE DYNAMICAL THEORY OF GRATINGS.

[*Proceedings of the Royal Society*, A, Vol. LXXIX. pp. 399—416, 1907.]

IN the usual theory of gratings, upon the lines laid down by Fresnel, the various parts of the primary wave-front after undergoing influences, whether affecting the phase or the amplitude, are conceived to pursue their course as if they still formed the fronts of waves of large area. This supposition, justifiable as an approximation when the grating interval is large, tends to fail altogether when the interval is reduced so as to be comparable with the wave-length. A simple example will best explain the nature of the failure. Consider a grating of perfectly reflecting material whose alternate parts are flat and parallel and equally wide, but so disposed as to form a groove of depth equal to a quarter wave-length, and upon this let light be incident perpendicularly. Upon Fresnel's principles the central regularly reflected

Fig. 1.

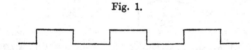

image must vanish, being constituted by the combination of equal and opposite vibrations. If the grating interval be large enough, this conclusion is approximately correct and could be verified by experiment. But now suppose that the grating interval is reduced until it is less than the wave-length of the light. The conclusion is now entirely wide of the mark. Under the circumstances supposed there are no lateral spectra and the *whole* of the incident energy is necessarily thrown into the regular reflection, which is accordingly total instead of evanescent. A closer consideration shows that the recesses in this case act as resonators in a manner not covered by Fresnel's investigations, and illustrates the need of a theory more strictly dynamical.

The present investigation, of which the interest is mainly optical, may be regarded as an extension of that given in *Theory of Sound* [*], where plane

* Second edition, § 272 a, 1896.

waves were supposed to be incident perpendicularly upon a regularly corrugated surface, whose form was limited by a certain condition of symmetry. Moreover, attention was there principally fixed upon the case where the wave-length of the corrugations was long in comparison with that of the waves themselves, so that in the optical application there would be a large number of spectra. It is proposed now to dispense with these restrictions. On the other hand, it will be supposed that the *depth* of the corrugations is small in comparison with the length (λ) of the waves.

The equation of the reflecting surface may be taken to be $z = \zeta$, where ζ is a periodic function of x, whose mean value is zero, and which is independent of y. By Fourier's theorem we may write

$$\zeta = c_1 \cos px + c_2 \cos 2px + s_2 \sin 2px + \ldots + c_n \cos npx + s_n \sin npx + \ldots$$
$$= \tfrac{1}{2}c_1 (e^{ipx} + e^{-ipx}) + \tfrac{1}{2}(c_n - is_n) e^{inpx} + \tfrac{1}{2}(c_n + is_n) e^{-inpx} + \ldots, \ldots\ldots\ldots(1)$$

the wave-length (ϵ) of the corrugation being $2\pi/p$. Formerly the s terms

Fig. 2.

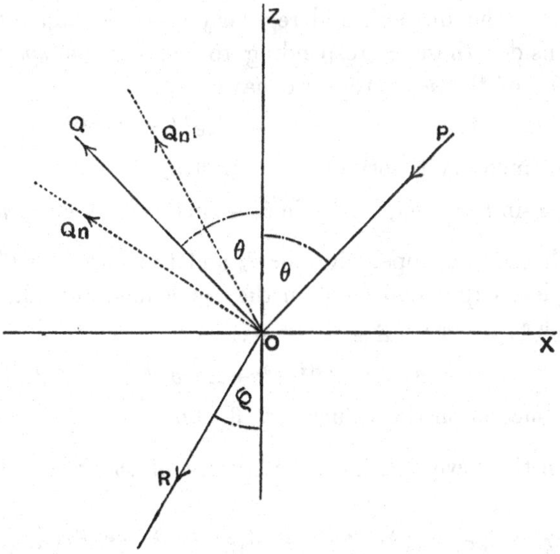

were omitted and attention was concentrated upon the case where c_1 was alone sensible. The omission of the s terms makes the grating symmetrical, so that at perpendicular incidence the spectra on the two sides are similar. It is known that this condition is often, and indeed advantageously, departed from in practice.

The vibrations incident at obliquity θ, POZ, fig. 2, are represented by

$$\psi = e^{ik(Vt + z\cos\theta + x\sin\theta)}, \ldots\ldots\ldots\ldots\ldots\ldots\ldots\ldots(2)$$

where $k = 2\pi/\lambda$, and V is the velocity of propagation in the upper medium. Here ψ satisfies in all cases the same general differential equation, but its significance must depend upon the character of the waves. In the acoustical application, to which for the present we may confine our attention, ψ is the velocity-potential. In optics it is convenient to change the precise interpretation according to circumstances, as we shall see later.

The waves regularly reflected along OQ are represented by

$$\psi = A_0 e^{ik(Vt - z\cos\theta + x\sin\theta)}, \quad\dots\dots\dots\dots\dots(3)$$

in which A_0 is a (possibly complex) coefficient to be determined. In all the expressions with which we have to deal the time occurs only in the factor e^{ikVt}, running through. For brevity this factor may be omitted.

In like manner the waves regularly refracted along OR into the lower medium have the expression

$$\psi_1 = B_0 e^{ik_1(z\cos\phi + x\sin\phi)}, \quad\dots\dots\dots\dots\dots(4)$$

ϕ being the angle of refraction; and, by the law of refraction,

$$k_1 : k = V : V_1 = \sin\theta : \sin\phi. \quad\dots\dots\dots\dots\dots(5)$$

In addition to the incident and regularly reflected and refracted waves, we have to consider those corresponding to the various *spectra*. For the reflected spectra of the nth order we have

$$\psi = A_n e^{ik(-z\cos\theta_n + x\sin\theta_n)} + A'_n e^{ik(-z\cos\theta'_n + x\sin\theta'_n)}, \quad\dots\dots\dots(5')$$

where, by the elementary theory of these spectra,

$$\epsilon\sin\theta_n - \epsilon\sin\theta = \pm n\lambda, \quad \text{or} \quad \sin\theta_n - \sin\theta = \pm n\lambda/\epsilon = \pm np/k. \quad\dots(6)$$

We shall choose the upper sign for θ_n and the lower for θ'_n. In virtue of (6) the complete expression for ψ in the upper medium takes the form

$$\psi \cdot e^{-ikx\sin\theta} = e^{ikz\cos\theta} + A_0 e^{-ikz\cos\theta} + \dots$$
$$+ A_n e^{inpx} e^{-ikz\cos\theta_n} + A'_n e^{-inpx} e^{-ikz\cos\theta'_n} + \dots, \quad\dots(7)$$

where n has in succession the values 1, 2, 3, etc.

Similarly, in the lower medium the spectra of the nth order are represented by

$$\psi_1 = B_n e^{ik_1(z\cos\phi_n + x\sin\phi_n)} + B'_n e^{ik_1(z\cos\phi'_n + x\sin\phi'_n)}, \quad\dots\dots\dots(8)$$

where

$$\sin\phi_n - \sin\phi = \pm np/k_1. \quad\dots\dots\dots\dots\dots(9)$$

Accordingly, for the complete expression of ψ_1, we have with use of (5),

$$\psi_1 \cdot e^{-ikx\sin\theta} = B_0 e^{ik_1 z\cos\phi} + \dots + B_n e^{inpx} e^{ik_1 z\cos\phi_n} + B'_n e^{-inpx} e^{ik_1 z\cos\phi'_n}. \quad(10)$$

We must now introduce boundary conditions to be satisfied at the transition between the two media when $z = \zeta$. It may be convenient to commence with a very simple case determined by the condition that $\psi = 0$. The whole of the incident energy is then thrown back, and is distributed between the regularly reflected waves and the various reflected spectra.

We proceed by approximation depending on the smallness of ζ. Expanding the exponentials on the right side of (7), we get

$$(1 + A_0)(1 - \tfrac{1}{2}k^2\zeta^2\cos^2\theta + \ldots) + (1 - A_0)(ik\zeta\cos\theta + \ldots)$$
$$+ A_n e^{inpx}(1 - ik\zeta\cos\theta_n + \ldots) + A'_n e^{-inpx}(1 - ik\zeta\cos\theta'_n + \ldots) = 0. \quad (11)$$

In this equation the value of ζ is to be substituted from (1), and then in accordance with Fourier's theorem the coefficients of the various exponential terms, such as e^{inpx}, e^{-inpx}, are to be separately equated to zero. As the first approximation, we get from the constant term (independent of x)

$$1 + A_0 = 0, \quad \ldots\ldots\ldots\ldots\ldots\ldots\ldots\ldots (12)$$

and from the terms in e^{inpx}, e^{-inpx},

$$A_n = -ik\cos\theta(c_n - is_n), \quad A'_n = -ik\cos\theta(c_n + is_n). \quad \ldots\ldots(13)$$

Thus, as was to be expected, A_n, A'_n are of the first order in ζ, and if we stop at the second order inclusive, (11) may be written

$$1 + A_0 + 2ik\zeta\cos\theta + A_n e^{inpx}(1 - ik\zeta\cos\theta_n) + A'_n e^{-inpx}(1 - ik\zeta\cos\theta'_n) = 0.$$
$$\ldots\ldots(14)$$

For the second approximation to A_0 we get

$$1 + A_0 - \tfrac{1}{2}k^2\cos\theta\,\Sigma\,(c_n^2 + s_n^2)(\cos\theta_n + \cos\theta'_n) = 0. \quad \ldots\ldots(15)$$

By means of (13) and (15) we may verify the principle that the energies of the incident, and of all the reflected vibrations taken together, are equal. The energy corresponding to unit of wave-front of the incident waves may be supposed to be unity, and for the other waves $\mathrm{mod}^2 A_0$, $\mathrm{mod}^2 A_1$, $\mathrm{mod}^2 A'_1$, etc. But what we have to consider are not equal areas of wave-front, but areas corresponding to the same extent of reflecting surface, i.e., areas of wave-front proportional to $\cos\theta$, $\cos\theta_1$, $\cos\theta'_1$, etc. Hence,

$$\cos\theta\,.\,\mathrm{mod}^2 A_0 + \Sigma\cos\theta_n\,.\,\mathrm{mod}^2 A_n + \Sigma\cos\theta'_n\,.\,\mathrm{mod}^2 A'_n = \cos\theta, \ldots(16)$$

with which the special approximate values already given are in harmony. In the formation of (16) only real values of $\cos\theta_n$, $\cos\theta'_n$ are to be included. If $p > k$, no real values exist, i.e., there are no lateral spectra. The regular reflection is then total, and this without limitation upon the magnitude of the c's. The question is further considered in *Theory of Sound*, § 272 a.

In pursuing a second approximation for the coefficients of the lateral spectra, we will suppose for the sake of brevity that the s terms in (1) are omitted. From the term involving e^{inpx} in (14), we get with use of (13),

$$A_n = -ik\cos\theta\,c_n + \tfrac{1}{2}k^2\cos\theta\cos\theta'_n\,.\,c_n c_{2n}$$
$$+ \tfrac{1}{2}k^2\cos\theta\,\{(c_1\cos\theta_{n-1} + c_{2n-1}\cos\theta'_{n-1})\,c_{n-1}$$
$$+ (c_2\cos\theta_{n-2} + c_{2n-2}\cos\theta'_{n-2})\,c_{n-2} + \ldots$$
$$+ (c_1\cos\theta_{n+1} + c_{2n+1}\cos\theta'_{n+1})\,c_{n+1}$$
$$+ (c_2\cos\theta_{n+2} + c_{2n+2}\cos\theta'_{n+2})\,c_{n+2} + \ldots\}, \ldots\ldots\ldots(17)$$

in which the first (descending) series is to terminate when the suffix in $\cos \theta_{n-r}$ is equal to unity.

The value of A'_n may be derived from (17) by interchange of θ' and θ in $\cos \theta_{n-r}$, $\cos \theta'_{n-r}$, $\cos \theta_{n+r}$, $\cos \theta'_{n+r}$, $\cos \theta$ remaining unchanged. As a particular case of (17), we have, for the spectra of the first order,

$$A_1 = - ikc_1 \cos \theta + \tfrac{1}{2} k^2 c_1 c_2 \cos \theta \cos \theta'_1$$
$$+ \tfrac{1}{2} k^2 \cos \theta \{c_2 (c_1 \cos \theta_2 + c_3 \cos \theta'_2)$$
$$+ c_3 (c_2 \cos \theta_3 + c_4 \cos \theta'_3) + \ldots\}, \quad \ldots\ldots\ldots\ldots(18)$$
$$A'_1 = - ikc_1 \cos \theta + \tfrac{1}{2} k^2 c_1 c_2 \cos \theta \cos \theta_1$$
$$+ \tfrac{1}{2} k^2 \cos \theta \{c_2 (c_1 \cos \theta'_2 + c_3 \cos \theta_2)$$
$$+ c_3 (c_2 \cos \theta'_3 + c_4 \cos \theta_3) + \ldots\}, \quad \ldots\ldots\ldots\ldots(19)$$

the descending series in (17) disappearing altogether.

If the incidence is normal, $\cos \theta = 1$, $\cos \theta'_n = \cos \theta_n$, and thus A_n, A'_n become identical and assume specially simple forms. Referring to (7), we see that in this case

$$\psi = e^{ikz} + A_0 e^{-ikz} + 2A_1 e^{-ikz \cos \theta_1} \cos px + \ldots + 2A_n e^{-ikz \cos \theta_n} \cos npx + \ldots, (20)$$

in which, to the second order,

$$A_0 = - 1 + k^2 \Sigma c_n^2 \cos \theta_n. \quad \ldots\ldots\ldots\ldots\ldots(21)$$
$$A_n = - ikc_n + \tfrac{1}{2} k^2 \cos \theta_n . c_n c_{2n}$$
$$+ \tfrac{1}{2} k^2 \{(c_1 + c_{2n-1}) c_{n-1} \cos \theta_{n-1} + (c_2 + c_{2n-2}) c_{n-2} \cos \theta_{n-2} + \ldots$$
$$+ (c_1 + c_{2n+1}) c_{n+1} \cos \theta_{n+1} + (c_2 + c_{2n+2}) c_{n+2} \cos \theta_{n+2} + \ldots\}. \quad \ldots\ldots(22)$$

If we suppose that in (1) only c_1 and c_2 are sensible, we have

$$A_0 = - 1 + k^2 c_1^2 \cos \theta_1 + k^2 c_2^2 \cos \theta_2, \quad \ldots\ldots\ldots\ldots(23)$$
$$A_1 = - ikc_1 + \tfrac{1}{2} k^2 c_1 c_2 (\cos \theta_1 + \cos \theta_2), \quad \ldots\ldots\ldots(24)$$
$$A_2 = - ikc_2 + \tfrac{1}{2} k^2 c_1^2 \cos \theta_1, \quad \ldots\ldots\ldots\ldots\ldots\ldots(25)$$
$$A_3 = \tfrac{1}{2} k^2 c_1 c_2 (\cos \theta_1 + \cos \theta_2), \quad \ldots\ldots\ldots\ldots\ldots(26)$$

while A_4, A_5, etc. vanish to the second order of small quantities inclusive.

There is no especial difficulty in carrying the approximations further. As an example, we may suppose that c_1 is alone sensible in (1), so that we may write

$$\zeta = c \cos px, \quad \ldots\ldots\ldots\ldots\ldots\ldots\ldots(27)$$

and also that the incidence is perpendicular. For brevity we will denote $k \cos \theta_n$ or $k \cos \theta'_n$ by μ_n. The boundary condition ($\psi = 0$) becomes by (7) in this case,

$$e^{ik\zeta} - e^{-ik\zeta} + (A_0 + 1) e^{-ik\zeta} + 2A_1 e^{-i\mu_1\zeta} \cos px + \ldots\ldots$$
$$+ 2A_n e^{-i\mu_n\zeta} \cos npx + \ldots\ldots = 0, \ldots\ldots(28)$$

in which

$$e^{ik\zeta} - e^{-ik\zeta} = 4i \{J_1 (kc) \cos px - J_3 (kc) \cos 3px + J_5 (kc) \cos 5px - \ldots\}, \ldots(29)$$
$$e^{-ik\zeta} = J_0 (kc) - 2J_2 (kc) \cos 2px + \ldots$$
$$- i \{2J_1 (kc) \cos px - 2J_3 (kc) \cos 3px + \ldots\}, \quad \ldots\ldots(30)$$

with similar expressions for $e^{-i\mu_1\zeta}$, $e^{-i\mu_2\zeta}$, etc.   By Fourier's theorem the terms independent of x, in $\cos px$, $\cos 2px$, etc., must vanish separately.   The first gives

$$(A_0+1)J_0(kc) - 2iA_1J_1(\mu_1c) - 2A_2J_2(\mu_2c) + 2iA_3J_3(\mu_3c) + 2A_4J_4(\mu_4c) + \dots = 0.$$
$$\dots(31)$$

The term in $\cos px$ gives

$$2iJ_1(kc) - i(A_0+1)J_1(kc) + A_1\{J_0(\mu_1c) - J_2(\mu_1c)\}$$
$$- iA_2\{J_1(\mu_2c) - J_3(\mu_2c)\} - A_3\{J_2(\mu_3c) - J_4(\mu_3c)\} + \dots = 0. \quad \dots(32)$$

The term in $\cos 2px$ gives

$$-(A_0+1)J_2(kc) - iA_1\{J_1(\mu_1c) - J_3(\mu_1c)\}$$
$$+ A_2\{J_0(\mu_2c) + J_4(\mu_2c)\} + \dots = 0. \quad \dots\dots\dots\dots(33)$$

The term in $\cos 3px$ gives

$$-2iJ_3(kc) + i(A_0+1)J_3(kc)$$
$$- A_1\{J_2(\mu_1c) - J_4(\mu_2c)\} - iA_2\{J_1(\mu_2c) + J_5(\mu_2c)\}$$
$$+ A_3\{J_0(\mu_3c) - J_6(\mu_3c)\} + \dots = 0. \quad \dots\dots\dots\dots(34)$$

We see from these that A_0+1 is of the second order in kc, that A_1 is of the first order, A_2 of the second order, A_3 of the third order, and so on. Expanding the Bessel's functions, we find, to the second order inclusive, as in (23), (24), (25), (26),

$$\left.\begin{array}{ll} A_0 = -1 + k\mu_1c^2, & A_1 = -ikc, \\ A_2 = \tfrac{1}{2}k\mu_1c^2, & A_3 = 0, \end{array}\right\} \quad \dots\dots\dots\dots(35)$$

A_4, etc., vanishing.   To the third order inclusive (34) now gives

$$A_3 = \tfrac{1}{24}ikc^3(k^2 - 3\mu_1^2 + 6\mu_1\mu_2). \quad \dots\dots\dots\dots(36)$$

From (33) to the same order we have still for A_2,

$$A_2 = \tfrac{1}{2}k\mu_1c^2, \quad \dots\dots\dots\dots\dots\dots\dots\dots(37)$$

and from (32)

$$A_1 = -ikc + \tfrac{1}{8}ikc^3(k^2 + 4k\mu_1 + 2\mu_1\mu_2 - 3\mu_1^2). \quad \dots\dots\dots(38)$$

These are complete to the third order of kc inclusive.   To this order A_4, A_5, etc., vanish.

So far as the third order of kc inclusive, A_0 remains as in (35); but it is worth while here to retain the terms of the fourth order.   We find from (31)

$$A_0 = -1 + k\mu_1c^2 + \tfrac{1}{8}kc^4(k^2\mu_1 - 4k\mu_1^2 + 2\mu_1^3 - 2\mu_1^2\mu_2 + \mu_1\mu_2^2). \quad \dots(39)$$

It is to be noted that k, μ_1, μ_2 are not independent.   By (6), with $\theta = 0$,

$$\mu_n^2 = k^2\cos^2\theta_n = k^2 - n^2p^2, \quad \dots\dots\dots\dots\dots\dots(40)$$

so that

$$\mu_1^2 = k^2 - p^2, \qquad \mu_2^2 = k^2 - 4p^2,$$

and

$$3k^2 - 4\mu_1^2 + \mu_2^2 = 0. \quad \dots\dots\dots\dots\dots\dots(41)$$

By use of (41) it may be verified to the fourth order that when μ_1, μ_2 are real, so that the spectra of the second order are actually formed,

$$\text{mod}^2 A_0 + \frac{2\mu_1}{k} \text{mod}^2 A_1 + \frac{2\mu_2}{k} \text{mod}^2 A_2 = 1, \quad \ldots\ldots\ldots(42)$$

expressing the conservation of vibratory energy.

When μ_1 is real, but not μ_2, we may write $\mu_2 = -i\nu_2$, where ν_2 is positive. In this case

$$A_0 = -1 + k\mu_1 c^2 + \tfrac{1}{8}kc^4 (k^2\mu_1 - 4k\mu_1^2 + 2\mu_1^3 - \mu_1\nu_2^2) + \tfrac{1}{4}ikc^4\mu_1^2\nu_2,$$

$$A_1 = -ikc + \tfrac{1}{8}ikc^3 (k^2 + 4k\mu_1 - 3\mu_1^2) + \tfrac{1}{4}kc^3\mu_1\nu_2;$$

and in virtue of (41) to the fourth order,

$$\text{mod}^2 A_0 + \frac{2\mu_1}{k} \text{mod}^2 A_1 = 1. \quad \ldots\ldots\ldots\ldots(43)$$

Again, if μ_1, μ_2 are both imaginary, equal, say, to $-i\nu_1$, $-i\nu_2$, we have from (39) with separation of real and imaginary parts,

$$A_0 = -1 + \tfrac{1}{2}k^2\nu_1^2 c^4 - i(k\nu_1 c^2 + \text{terms in } c^4),$$

so that, to the fourth order,

$$\text{mod}^2 A_0 = 1, \quad \ldots\ldots\ldots\ldots\ldots(44)$$

expressing that the regular reflection is now total.

In the acoustical interpretation for a gaseous medium ψ represents the velocity-potential, and the boundary condition $(\psi = 0)$ is that of constant pressure. In the electrical and optical interpretation the waves are incident from air, or other dielectric medium, upon a perfectly conducting and, therefore, perfectly reflecting corrugated substance. Here ψ represents the electromotive intensity Q parallel to y, that is parallel to the lines of the grating, the boundary condition being the evanescence of Q.

We now pass on to the boundary condition next in order of simplicity, which ordains that $d\psi/dn = 0$, where dn is drawn normally at the surface of separation. Since the surfaces $z - \zeta = 0$, $\psi = \text{constant}$, are to be perpendicular, the condition expressed in rectangular co-ordinates is

$$\frac{d\psi}{dz} - \frac{d\psi}{dx}\frac{d\zeta}{dx} = 0, \quad \ldots\ldots\ldots\ldots(45)$$

ψ being given by (7) and ζ by (1).

For the purposes of the first approximation, we require in $d\psi/dx$ only the part independent of the c's and s's, since $d\zeta/dx$ is already of the first order. Thus at the surface

$$\frac{d\psi}{dx} = ik \sin\theta \, e^{ikx\sin\theta} (1 + A_0).$$

Also, correct to the first order,

$$\frac{d\psi}{dz} = ik\, e^{ikx\sin\theta} [\cos\theta\,\{1 - A_0 + (1 + A_0)\,ik\zeta\cos\theta\} - \ldots\ldots$$

$$- \cos\theta_n A_n\, e^{inpx} - \cos\theta'_n A'_n e^{-inpx}].$$

Thus (45) gives

$$\cos\theta\,(1 - A_0) + \cos^2\theta\,(1 + A_0)\,ik\zeta - \cos\theta_n A_n\, e^{inpx}$$

$$- \cos\theta'_n A'_n e^{-inpx} - \ldots\ldots - \sin\theta\,(1 + A_0)\frac{d\zeta}{dx} = 0. \quad\ldots\ldots\ldots(46)$$

From the term independent of x we see that, as was to be expected,

$$A_0 = 1. \quad\ldots\ldots\ldots\ldots\ldots\ldots\ldots\ldots\ldots(47)$$

Also
$$A_n\cos\theta_n = i\,(c_n - is_n)\,\{k\cos^2\theta - np\sin\theta\}, \quad\ldots\ldots\ldots(48)$$

$$A'_n\cos\theta'_n = i\,(c_n + is_n)\,\{k\cos^2\theta + np\sin\theta\}. \quad\ldots\ldots\ldots(49)$$

When $n = 1$ in (48), (49), we may put $s_1 = 0$. These equations constitute the complete solution to a first approximation.

For the second approximation we must retain the terms of the first order in $d\psi/dx$. Thus from (5), (7)

$$e^{-ikx\sin\theta}\frac{d\psi}{dx} = ik\,[\sin\theta\,\{1 + A_0 + (1 - A_0)ik\cos\theta\}$$

$$+ \sin\theta_n A_n e^{inpx} + \sin\theta'_n A'_n e^{-inpx}]$$

$$= ik\,\{2\sin\theta + \sin\theta_n A_n\, e^{inpx} + \sin\theta'_n A'_n\, e^{-inpx}\},\ldots(50)$$

since to the first order inclusive $A_0 = 1$. Also

$$e^{-ikx\sin\theta}\frac{d\psi}{dz} = ik\cos\theta\,\{1 - A_0 + 2ik\zeta\cos\theta\}$$

$$- ik\cos\theta_n A_n e^{inpx}\,(1 - ik\zeta\cos\theta_n) - ik\cos\theta'_n A'_n e^{-inpx}\,(1 - ik\zeta\cos\theta'_n).\ldots(51)$$

Thus by (45) the boundary condition is

$$\cos\theta\,(1 - A_0) + 2ik\zeta\cos^2\theta - 2\sin\theta\frac{d\zeta}{dx}$$

$$- A_n e^{inpx}\left\{\cos\theta_n - ik\cos^2\theta_n\zeta + \sin\theta_n\frac{d\zeta}{dx}\right\}$$

$$- A'_n e^{-inpx}\left\{\cos\theta'_n - ik\cos^2\theta'_n\zeta + \sin\theta'_n\frac{d\zeta}{dx}\right\} = 0. \quad\ldots\ldots(52)$$

In the small terms we may substitute for A_n, A'_n their approximate values from (48), (49).

In (52) the coefficients of the various terms in e^{inpx}, e^{-inpx} must vanish separately. In pursuing the approximation we will write for brevity

$$\zeta = \zeta_1 e^{ipx} + \zeta_{-1} e^{-ipx} + \ldots + \zeta_n e^{inpx} + \zeta_{-n} e^{-inpx}, \quad\ldots\ldots\ldots(53)$$

where
$$\zeta_1 = \zeta_{-1} = \tfrac{1}{2}c_1,$$

and
$$\zeta_n = \tfrac{1}{2}\,(c_n - is_n), \quad \zeta_{-n} = \tfrac{1}{2}\,(c_n + is_n). \quad\ldots\ldots\ldots(54)$$

The term independent of x gives A_0 to the second approximation. Thus

$$\cos \theta (1 - A_0) + iA_n (k \cos^2 \theta_n + np \sin \theta_n) \zeta_{-n}$$
$$+ iA'_n (k \cos^2 \theta'_n - np \sin \theta'_n) \zeta_n = 0. \quad \ldots(55)$$

In (55), as follows from (6),

$$k \cos^2 \theta_n + np \sin \theta_n = k \cos^2 \theta - np \sin \theta,$$
and $\qquad k \cos^2 \theta'_n - np \sin \theta'_n = k \cos^2 \theta + np \sin \theta.$

Hence with introduction of the values of A_n, A'_n from (48), (49),

$$\cos \theta (1 - A_0) = \ldots\ldots + \frac{c_n^2 + s_n^2}{2 \cos \theta_n} (k \cos^2 \theta - np \sin \theta)^2$$
$$+ \frac{c_n^2 + s_n^2}{2 \cos \theta'_n} (k \cos^2 \theta + np \sin \theta)^2 + \ldots\ldots, \quad \ldots\ldots\ldots(56)$$

as might also be inferred from (48), (49) alone, with the aid of the energy equation—

$$\cos \theta = \cos \theta \bmod^2 A_0 + \ldots + \cos \theta_n \bmod^2 A_n + \cos \theta'_n \bmod^2 A'_n. \quad \ldots(57)$$

From the term in e^{inpx} in (52) we get

$$\cos \theta_n A_n = 2i (k \cos^2 \theta - np \sin \theta) \zeta_n$$
$$+ iA'_n (k \cos^2 \theta'_n - 2np \sin \theta'_n) \zeta_{2n} + \ldots\ldots$$
$$+ iA_{n-r} (k \cos^2 \theta_{n-r} - rp \sin \theta_{n-r}) \zeta_r$$
$$+ iA_{n+r} (k \cos^2 \theta_{n+r} + rp \sin \theta_{n+r}) \zeta_{-r}$$
$$+ iA'_{n-r} \{k \cos^2 \theta'_{n-r} - (2n - r) p \sin \theta'_{n-r}\} \zeta_{2n-r}$$
$$+ iA'_{n+r} \{k \cos^2 \theta'_{n+r} - (2n + r) p \sin \theta'_{n+r}\} \zeta_{2n+r}. \quad \ldots\ldots\ldots(58)$$

In (58) r is to assume the values 1, 2, 3, etc., the descending series terminating when $n - r = 1$.

The corresponding equation for A'_n may be derived from (58) by changing the sign of n, with the understanding that

$$A_{-m} = A'_m, \qquad A'_{-m} = A_m; \qquad \theta_{-m} = \theta'_m, \qquad \theta'_{-m} = \theta_m. \quad \ldots(59)$$

If the incidence be perpendicular, so that $\theta'_m = - \theta_m$, and if $\zeta_{-m} = \zeta_m$, which requires that $s_m = 0$, the values of A'_n and A_n become identical.

If $n = 1$, the descending series in (58) make no contribution. We have

$$\cos \theta_1 A_1 = 2i (k \cos^2 \theta - p \sin \theta) \zeta_1 + iA'_1 (k \cos^2 \theta'_1 - 2p \sin \theta'_1) \zeta_2$$
$$+ iA_2 (k \cos^2 \theta_2 + p \sin \theta_2) \zeta_{-1} + iA_3 (k \cos^2 \theta_2 + 2p \sin \theta_2) \zeta_{-2} + \ldots$$
$$+ iA'_2 (k \cos^2 \theta'_2 - 3p \sin \theta'_2) \zeta_3 + iA'_3 (k \cos^2 \theta'_3 - 4p \sin \theta'_3) \zeta_4 + \ldots. \quad (60)$$

We will now introduce the simplifying suppositions that $\theta = 0$, $s_m = 0$, making $A'_n = A_n$, and also that only c_1 and c_2 are sensible, so that

$\zeta_3 = \zeta_4 = \ldots\ldots = 0$. We will also, as before, denote $k \cos \theta_n$ or $k \cos \theta'_n$ by μ_n. Accordingly (60) gives, with use of (6), (48), (49),

$$A_1 = \frac{ik^2 c_1}{\mu_1} - \frac{k^2 c_1 c_2}{2\mu_1^2}(\mu_1^2 + 2p^2) - \frac{k^2 c_1 c_2}{2\mu_1 \mu_2}(\mu_2^2 + 2p^2). \quad\ldots\ldots(61)$$

In like manner, we get from (58)

$$A_2 = \frac{ik^2 c_2}{\mu_2} - \frac{k^2 c_1^2}{2\mu_1 \mu_2}(\mu_1^2 - p^2), \quad\ldots\ldots\ldots\ldots(62)$$

$$A_3 = -\frac{k^2 c_1 c_2 (\mu_1 + \mu_2)}{2\mu_3}\left\{1 - \frac{2p^2}{\mu_1 \mu_2}\right\}, \quad\ldots\ldots\ldots(63)$$

$$A_4 = -\frac{k^2 c_2^2}{2\mu_2 \mu_4}(\mu_2^2 - 4p^2), \quad\ldots\ldots\ldots\ldots(64)$$

after which A_5, A_6, etc., vanish to this order of approximation. In any of these equations we may replace μ_n^2 by its value from (6), that is $k^2 - n^2 p^2$.

The boundary condition of this case, i.e., $d\psi/dn = 0$, is realised acoustically when aerial waves are incident upon an immovable corrugated surface. In the interpretation for electrical and luminous waves, ψ represents the magnetic induction (b) parallel to y, so that the electric vector is perpendicular to the lines of the grating, the boundary condition at the surface of a perfect reflector being $db/dn = 0$.

We have thus obtained the solutions for the two principal cases of the incidence of polarised light upon a perfect corrugated reflector. In comparing the results for the first order of approximation as given in (13) for the first case and in (48), (49) for the second, we are at once struck with the fact that in the second case, though not in the first, the intensity of a spectrum may become infinite through the evanescence of $\cos \theta_n$ or $\cos \theta'_n$, which occurs when the spectrum is just disappearing from the field of view. But the effect is not limited to the particular spectrum which is on the point of disappearing. Thus in (61) A_1, giving the spectrum of the *first* order, becomes infinite as the spectrum of the *second* order disappears ($\mu_2 = 0$). Regarded from a mathematical point of view, the method of approximation breaks down. The problem has no definite solution, so long as we maintain the suppositions of perfect reflection, of an infinite train of simple waves, and of a grating infinitely extended in the direction perpendicular to its ruling. But under the conditions of experiment, we may at least infer the probability of abnormalities in the brightness of any spectrum at the moment when one of higher order is just disappearing, abnormalities limited, however, to the case where the electric displacement is perpendicular to the ruling*. It may be remarked

---

* See a "Note on the Remarkable Case of Diffraction Spectra described by Professor Wood," recently communicated to the *Philosophical Magazine*, Vol. XIV. p. 60, 1907. [Art. 323.]

that when the incident light is unpolarised, the spectrum about to disappear is polarised in a plane parallel to the ruling.

In both the cases of boundary conditions hitherto treated, the circumstances are especially simple in that the aggregate reflection is perfect, the whole of the incident energy being returned into the upper medium. We now pass on to more complicated conditions, which we may interpret as those of two gaseous media of densities σ and σ_1. Equality of pressures at the interface requires that

$$\sigma\psi = \sigma_1\psi_1, \quad\dots\dots\dots\dots\dots\dots\dots\dots(65)$$

and we have also to satisfy the continuity of normal velocity expressed by

$$d\psi/dn = d\psi_1/dn, \quad\dots\dots\dots\dots\dots\dots(66)$$

or, as in (45),

$$\frac{d(\psi-\psi_1)}{dz} - \frac{d(\psi-\psi_1)}{dx}\frac{d\zeta}{dx} = 0, \quad\dots\dots\dots\dots(67)$$

ψ and ψ_1 being given by (7), (10). We must content ourselves with a solution to the first approximation, at least for general incidence.

From (65),

$$\frac{\sigma}{\sigma_1}\{1 + A_0 + (1 - A_0)\,ik\zeta\cos\theta + A_n\,e^{inpx} + A'_n e^{-inpx}\}$$

$$= B_0(1 + ik_1\zeta\cos\phi) + B_n\,e^{inpx} + B'_n e^{-inpx}. \quad\dots(68)$$

Distinguishing the various components in ζ as in (53), we find

$$\frac{\sigma}{\sigma_1}(1 + A_0) = B_0, \quad\dots\dots\dots\dots\dots\dots\dots(69)$$

$$\frac{\sigma}{\sigma_1}A_n - B_n = i\zeta_n\left\{k_1\cos\phi B_0 - \frac{\sigma}{\sigma_1}(1 - A_0)\,k\cos\theta\right\}, \dots\dots\dots(70)$$

$$\frac{\sigma}{\sigma_1}A'_n - B'_n = i\zeta_{-n}\left\{k_1\cos\phi B_0 - \frac{\sigma}{\sigma_1}(1 - A_0)\,k\cos\theta\right\}. \quad\dots\dots(71)$$

In forming the second boundary condition (67) we require in $d(\psi-\psi_1)/dx$ only the part independent of ζ. Thus

$$\frac{d(\psi-\psi_1)}{dx} = ik\sin\theta\,e^{ikx\sin\theta}\{1 + A_0 - B_0\}.$$

Also

$$e^{-ikx\sin\theta}\frac{d\psi}{dz} = ik\cos\theta\,\{1 - A_0 + (1 + A_0)\,ik\zeta\cos\theta\}$$

$$- ik\cos\theta_n A_n e^{inpx} - ik\cos\theta'_n A'_n e^{-inpx},$$

$$e^{-ikx\sin\theta}\frac{d\psi_1}{dz} = ik_1\cos\phi B_0(1 + ik_1\zeta\cos\phi)$$

$$+ ik_1\cos\phi_n B_n\,e^{inpx} + ik_1\cos\phi'_n B'_n e^{-inpx}.$$

Thus (67) takes the form

$$ik \cos \theta \,(1 - A_0) - ik_1 \cos \phi B_0$$
$$- k^2 \cos^2 \theta \,(1 + A_0)\, \zeta + k_1^2 \cos^2 \phi B_0 \zeta$$
$$- e^{inpx} \{ik \cos \theta_n A_n + ik_1 \cos \phi_n B_n\}$$
$$- e^{-inpx} \{ik \cos \theta'_n A'_n + ik_1 \cos \phi'_n B'_n\}$$
$$= ik \sin \theta \,(1 + A_0 - B_0)\, d\zeta/dx. \quad\quad\quad (72)$$

From the part independent of x we get

$$k \cos \theta \,(1 - A_0) - k_1 \cos \phi B_0 = 0, \quad\quad\quad (73)$$

and from the parts in e^{inpx}, e^{-inpx}

$$k \cos \theta_n A_n + k_1 \cos \phi_n B_n$$
$$= i\zeta_n \{k^2 \cos^2 \theta \,(1 + A_0) - k_1^2 \cos^2 \phi B_0 - npk \sin \theta \,(1 + A_0 - B_0)\}, \quad (74)$$

and a similar equation involving A'_n, B'_n.

From (69), (73) we find

$$A_0 = \frac{\dfrac{\sigma_1}{\sigma} - \dfrac{\cot \phi}{\cot \theta}}{\dfrac{\sigma_1}{\sigma} + \dfrac{\cot \phi}{\cot \theta}}, \qquad B_0 = \frac{2}{\dfrac{\sigma_1}{\sigma} + \dfrac{\cot \phi}{\cot \theta}}. \quad\quad\quad (75)$$

Again, eliminating B_n between (70), (74), we get, with the use of (5),

$$A_n \{k \cos \theta_n + k_1 \cos \phi_n . \sigma/\sigma_1\} = \frac{2ik^2 \zeta_n}{D} \left[\frac{\sigma_1}{\sigma} \cos^2 \theta - \frac{\sigma_1 - \sigma}{\sigma} \frac{np}{k} \sin \theta \right.$$
$$\left. - \frac{\sin^2 \theta \cos \phi}{\sin^2 \phi} \left\{ \cos \phi - \cos \phi_n + \frac{\sigma}{\sigma_1} \cos \phi_n \right\} \right], \quad (76)$$

D denoting the denominators in (75).

The equations (75) for the waves regularly reflected and refracted are those given (after Green) in *Theory of Sound*, § 270. They are sufficiently general to cover the case where the two gaseous media have different constants of compressibility (m, m_1) as well as of density (σ, σ_1). The velocities of wave propagation are connected with these quantities by the relation, see (5),

$$k_1^2 : k^2 = \sin^2 \theta : \sin^2 \phi = V^2 : V_1^2 = m/\sigma : m_1/\sigma_1. \quad\quad\quad (77)$$

In ideal gases the compressibilities are the same, so that

$$\sigma_1 : \sigma = \sin^2 \theta : \sin^2 \phi. \quad\quad\quad (78)$$

In this case (75) gives

$$A_0 = \frac{\sin 2\theta - \sin 2\phi}{\sin 2\theta + \sin 2\phi} = \frac{\tan (\theta - \phi)}{\tan (\theta + \phi)}, \quad\quad\quad (79)$$

Fresnel's expression for the reflection of light polarised in a plane perpendicular to that of incidence. In accordance with Brewster's law the reflection vanishes at the angle of incidence whose tangent is V/V_1.

In like manner the introduction of (78) into (76) gives, after reduction,

$$A_n \{k \cos\theta_n + k_1 \cos\phi_n . \sigma/\sigma_1\}$$
$$= 2ik^2 \zeta_n \cot\theta \tan(\theta-\phi) \{\cos(\theta+\phi)\cos(\theta-\phi)$$
$$- \cos\phi(\cos\phi - \cos\phi_n) - np/k . \sin\theta\}. \quad\ldots\ldots\ldots\ldots(80)$$

If the wave-length of the corrugations be very long, $p = 0$, $\cos\phi_n$ becomes identical with $\cos\phi$, and thus A_n vanishes when $\cos(\theta+\phi) = 0$, that is at the same (Brewsterian) angle of incidence for which $A_0 = 0$, as was to be expected.   In general $A_n = 0$, when

$$\cos(\theta+\phi)\cos(\theta-\phi) = \cos\phi(\cos\phi - \cos\phi_n) + np/k . \sin\theta. \ldots(81)$$

If we suppose that np/k is somewhat small, we may obtain a second approximation to the value of $\cos(\theta+\phi)$.   Thus, setting in the small terms $\theta + \phi = \tfrac{1}{2}\pi$, we get

$$\cos(\theta+\phi) = \tfrac{1}{2}\sec\theta\{\cos\phi - \cos\phi_n + np/k\}.$$

Here $\cos\phi_n = \cos\phi - np/k_1 . \tan\phi = \cos\phi - np/k . \cot^2\theta,$

so that $$\cos(\theta+\phi) = \frac{np}{2k\sin^2\theta\cos\theta}. \quad\ldots\ldots\ldots\ldots\ldots(82)$$

This determines the angle of incidence at which to a second approximation (in np/k) the reflected vibration vanishes in the nth spectrum.

Since, according to (82) with n positive, $\theta + \phi < \tfrac{1}{2}\pi$, and $\theta_n > \theta$, it seemed not impossible that (82) might be equivalent to $\cos(\theta_n + \phi) = 0$, forming a kind of extension of Brewster's law.   It appears, however, from (6) that

$$\cos(\theta_n+\phi) = \frac{np}{k\cos\theta}\left(\frac{1}{2\sin^2\theta} - 1\right), \quad\ldots\ldots\ldots\ldots(83)$$

so that the suggested law is not observed, although the departure from it would be somewhat small in the case of moderately refractive media.

For the other spectrum of the nth order we have only to change the sign of n in (82), (83).

When np/k is not small, we must revert to the original equation (81). Even this, it must be remembered, depends upon a first approximation, including only the first powers of the ζ's.

Another special case of interest occurs when $\sigma_1 = \sigma$, so that in the acoustical application the difference between the two media is one of compressibility only.   The introduction of this condition into (75) gives

$$A_0 = \frac{\tan\phi - \tan\theta}{\tan\phi + \tan\theta} = \frac{\sin(\phi-\theta)}{\sin(\phi+\theta)}, \quad\ldots\ldots\ldots\ldots(84)$$

the other Fresnel's expression.

Again, from (76),

$$A_n \{k \cos \theta_n + k_1 \cos \phi_n\} = \frac{2ik^2\zeta_n}{D} \left\{\cos^2\theta - \frac{\sin^2\theta \cos^2\phi}{\sin^2\phi}\right\},$$

whence
$$A_n = \frac{2ik\zeta_n \cos\theta \sin(\phi - \theta)}{\sin\phi \cos\theta_n + \sin\theta \cos\phi_n}. \quad\quad\quad\quad\quad (85)$$

In this case the vibration in the nth spectrum does not vanish at any angle of incidence.

We have now to consider the application of our solutions to electro-magnetic vibrations, such as constitute light, the polarisation being in one or other principal plane. In the usual electrical notation,

$$V^2 = 1/K\mu, \quad\quad\quad\quad V_1^2 = 1/K_1\mu_1,$$

K, K_1 being the specific inductive capacities, and μ, μ_1 the magnetic permeabilities; while in the acoustical problem,

$$V^2 = m/\sigma, \quad\quad\quad\quad V_1^2 = m_1/\sigma_1.$$

The boundary conditions are also of the same general form. For instance, the acoustical conditions

$$\sigma\psi = \sigma_1\psi_1, \quad\quad\quad\quad d\psi/dn = d\psi_1/dn,$$

may be written

$$(\sigma\psi) = (\sigma_1\psi_1), \quad\quad\quad\quad \sigma^{-1}d(\sigma\psi)/dn = \sigma_1^{-1}d(\sigma_1\psi_1)/dn;$$

and in the upper medium where σ is constant it makes no difference whether we deal with ψ or $\sigma\psi$. Thus if in the case of light we identify ψ with β, the component of magnetic force parallel to y, the conditions to be satisfied at the surface are the continuity of β and of $K^{-1}d\beta/dn$*.

Comparing with the acoustical conditions, we see that K replaces σ, and consequently (by the value of V^2) μ replaces $1/m$. Hence, in the general solution (75), (76), it is only necessary to write K in place of σ. For optical purposes we may usually treat μ as constant. This corresponds to the special supposition (78), so that (79), (80) apply to light for which the magnetic force is parallel to the lines of the grating, or the electric force perpendicular to the lines, i.e., in the plane of incidence.

From (76) we may fall back upon (48) by making $K_1 = \infty$, $\mu_1 = 0$, in such a way that V_1, and therefore ϕ, remains finite.

The other optical application depends upon identifying ψ with Q, the electromotive intensity parallel to y, i.e., parallel to the lines of the grating. The conditions at the surface are now the continuity of Q and of $\mu^{-1}dQ/dn$. Equations (75), (76) become applicable if we replace σ by μ. If μ be invariable, this is the special case of (84), (85); so that these equations are

---

* See *Phil. Mag.* Vol. xii. p. 81 (1881); *Scientific Papers*, Vol. i. p. 520.

applicable to light when the electric vibration is parallel to the lines of the grating, or perpendicular to the plane of incidence. The associated Fresnel's expression (79) or (84) suffices in each case to remind us of the optical circumstances.

In order to pass back from (76) to (13), we are to suppose $K_1 = \infty$, μ_1 (or σ_1) $= 0$, so that ϕ remains finite. Thus $D = \cot \phi / \cot \theta$, and the only terms to be retained in (76) are those which include the factor σ / σ_1.

The polarisation of the spectra reflected from glass gratings was noticed by Fraunhofer:—"Sehr merkwürdig ist es, dass unter einem gewissen Einfallswinkel ein Theil eines durch Reflexion entstandenen Spectrums aus *vollständig polarisirtem Lichte* besteht. Dieser Einfallswinkel ist für die verschiedenen Spectra sehr verschieden, und selbst noch sehr merklich für die verschiedenen Farben ein und desselben Spectrums. Mit dem Glasgitter $\epsilon = 0.0001223$ ist polarisirt: $E(+I)$, d.i., der *grüne* Theil dieses ersten Spectrums, wenn $\sigma = 49°$ ist; $E(+II)$, oder der grüne Theil in dem zweiten auf derselben Seite der Axe liegenden Spectrum, wenn $\sigma = 40°$ beträgt; endlich $E(-I)$, oder der grüne Theil des ersten auf der entgegengesetzten Seite der Axe liegenden Spectrums, wenn $\sigma = 69°$. Wenn $E(+I)$ vollständig polarisirt ist, sind es die übrigen Farben dieses Spectrums noch unvollständig*."

In Fraunhofer's notation σ is the angle of incidence, here denoted by θ, and λ $(E) = 0.00001945$ in the same measure (the Paris inch) as that employed for ϵ, so that $p/k = \lambda/\epsilon = 0.159$. If we suppose that the refractive index of the glass was 1·5, we get

Order	θ	$\sin \theta$	$\sin \phi$	ϕ	$\theta + \phi$
$E(+I)$	49°	0·755	0·503	30° 11′	79° 11′
$E(+II)$	40°	0·643	0·429	25° 25′	65° 25′
$E(-I)$	69°	0·934	0·623	107° 33′	107° 33′

On the other hand, from (82) we get for $E(+I)$ $\theta + \phi = 77° 44′$, for $E(+II)$ 59° 48′, and $E(-I)$ 104° 45′, a fair agreement between the two values of $\theta + \phi$, except in the case of $E(+II)$.

It appears, however, that the neglect of p^2 upon which (82) is founded is too rough a procedure. By trial and error I calculate from (81) for $E(+I)$ $\theta = 48° 52′$; for $E(+II)$ $\theta = 42° 17′$; for $E(-I)$ $\theta = 65° 46′$. These agree perhaps as closely as could be expected with the observed values, considering that they are deduced from a theory which neglects the square of the depth of the ruling. The ordinary polarising angle for this index (1·5) is 56° 19′.

* Gilbert's *Ann. d. Physik*, Vol. LXXIV. p. 337 (1823); *Collected Writings*, Munich, 1888, p. 134.

It would be of interest to extend Fraunhofer's observations; but the work should be in the hands of one who is in a position to rule gratings himself. On old and deteriorated glass surfaces polarisation phenomena are liable to irregularities.

In the hope of throwing light upon the remarkable observation of Professor Wood*, that a frilled collodion surface shows an enhanced reflection, I have pursued the calculation of the regularly reflected light to the second order in ζ, the depth of the groove, limiting myself, however, to the case of perpendicular incidence and to the supposition that ζ_1 and its equal ζ_{-1} are alone sensible. Although the results are not what I had hoped, it may be worth while to record the principal steps.

Retaining only the terms independent of x, we get from the first condition (65),

$$\sigma/\sigma_1 \cdot \{(1+A_0)(1-k^2\zeta_1^2) - 2ik\zeta_1\cos\theta_1 A_1\} = B_0(1-k_1^2\zeta_1^2) + 2ik_1\zeta_1\cos\phi_1 B_1, \quad\ldots(86)$$

and from the second condition (67),

$$k(1-A_0)(1-k^2\zeta_1^2) + 2ik^2\zeta_1 A_1\cos^2\theta_1 - k_1 B_0(1-k_1^2\zeta_1^2) - 2ik_1^2\zeta_1 B_1\cos^2\phi_1$$
$$= -2ip^2\zeta_1(A_1-B_1). \quad\ldots\ldots\ldots\ldots\ldots\ldots\ldots(87)$$

Eliminating $B_0(1-k_1^2\zeta_1^2)$, and remembering that

$$k^2\cos^2\theta_1 + p^2 = k^2, \qquad k_1^2\cos^2\phi_1 + p^2 = k_1^2,$$

we get $\quad k(1-A_0) - \sigma/\sigma_1 \cdot k_1(1+A_0) + 2ik^2\zeta_1 A_1$

$$+ \sigma/\sigma_1 \cdot 2ikk_1\zeta_1 A_1\cos\theta_1 + 2ik_1^2\zeta_1 B_1\cos\phi_1 - 2ik_1^2\zeta_1 B_1 = 0, \quad\ldots(88)$$

in which we are to substitute the values of A_1, B_1 from (70), (74). From this point it is, perhaps, more convenient to take the principal suppositions separately.

Let, as in (78), $\qquad \sigma_1 : \sigma = \sin^2\theta : \sin^2\phi = k_1^2 : k^2;$

we have $\qquad\qquad A_0 = \dfrac{k_1-k}{k_1+k}, \qquad B_0 = \dfrac{2k^2}{k_1(k_1+k)},$

and accordingly, from (70), (74),

$$k^2 A_1 - k_1^2 B_1 = 2ik^2\zeta_1(k_1-k), \qquad k\cos\theta_1 A_1 + k_1\cos\phi_1 B_1 = 0;$$

so that $\qquad\qquad A_1\{k\cos\phi_1 + k_1\cos\theta_1\} = 2ik\zeta_1(k_1-k)\cos\phi_1.$

Hence, from (88),

$$\frac{1-A_0}{1+A_0} = \frac{k}{k_1} + \frac{2k\zeta_1^2(k_1^2-k^2)}{k_1}\left\{1 - \frac{(k_1^2-k^2)\cos\theta_1\cos\phi_1}{k_1(k\cos\phi_1 + k_1\cos\theta_1)}\right\}. \quad\ldots(89)$$

Again, when $\sigma_1 = \sigma$,

$$A_0 = \frac{k-k_1}{k+k_1}, \qquad B_0 = \frac{2k}{k+k_1},$$

and from (70), (74),

$$A_1 = B_1 = \frac{2ik\zeta_1(k-k_1)}{k\cos\theta_1 + k_1\cos\phi_1}. \quad\ldots\ldots\ldots\ldots\ldots(90)$$

* *Physical Optics*, p. 145.

The introduction of these into (88) gives

$$\frac{1-A_0}{1+A_0} = \frac{k_1}{k} - \frac{2\,(k_1{}^2 - k^2)\,\zeta_1{}^2}{k}\left\{k_1 - \frac{k_1{}^2 - k^2}{k\cos\theta_1 + k_1\cos\phi_1}\right\} . \quad \ldots\ldots(91)$$

The question is whether the numerical value of A_0 is increased or diminished by the term in $\zeta_1{}^2$. In (89) it is easy to recognise that in the standard case of k_1 greater than k (air to glass in optics) the term in $\zeta_1{}^2$ is positive, θ_1 and ϕ_1 being supposed real. The effect of the second term is thus to bring the right-hand member nearer to unity than it would otherwise be, and thus to *diminish* the reflection. Again, in (91), the second term is negative, even when $\cos\theta_1 = 0$, as we may see by introducing the appropriate value of $\cos\phi_1$, viz., $\sqrt{(1 - k^2/k_1{}^2)}$. The effect is therefore to subtract something from k_1/k, which is greater than unity, and thus again to diminish the reflection.

If in (89), (91) we neglect the terms in $(k_1{}^2 - k^2)^2\,\zeta_1{}^2$, which will be specially small when the two media do not differ much, the formulæ become independent of the angles θ_1 and ϕ_1. In both cases the effect is the same as if the refractive index, supposed greater than unity, were diminished in the ratio $1 - 2\,(k_1{}^2 - k^2)\,\zeta_1{}^2 : 1$. It appears then that the present investigation gives no hint of the enhanced reflection observed in certain cases by Professor Wood.

[1911. For some extensions of this theory see Voigt, *Proceedings of the Göttingen Society*, Jan. 1911.]

323.

NOTE ON THE REMARKABLE CASE OF DIFFRACTION SPECTRA DESCRIBED BY PROF. WOOD.

[*Philosophical Magazine*, Vol. XIV. pp. 60—65, 1907.]

IN the *Philosophical Magazine* for Sept. 1902 Prof. Wood describes the extraordinary behaviour of a certain grating ruled upon speculum metal, which exhibits what may almost be called discontinuities in the distribution of the brightness of its spectra. Thus at a certain angle of incidence this grating will show one of the *D*-lines of sodium, *and not the other*. In fig. 1, p. 398, Prof. Wood gives ten diagrams fixing the positions (in terms of wave-length) of bright and dark bands in the spectrum at various angles of incidence ranging from 4° 12′ on the same side of the normal as the spectrum to 5° 45′ on the other side. In general there may be said to be two bands which approach one another as the angle of incidence diminishes, coincide when the incidence is normal, and open out again as the angle increases upon the other side. In the tenth diagram there is a third band whose behaviour is different and still more peculiar. In the movement of the two bands the rate of progress along the normal spectrum is the same for each. The above represents the cycle when the grating is in air. "If a piece of plane-parallel glass is cemented to the front of the grating with cedar-oil the cycle is quite different. In this case we have a pair of unsymmetrical shaded bands which move in the same direction as the angle of incidence is changed."

An important observation relates to polarization. "It was found that *the singular anomalies were exhibited only when the direction of vibration (electric vector) was at right angles to the ruling.* On turning the nicol through a right angle all trace of bright and dark bands disappeared. The bands are naturally much more conspicuous when polarized light is employed."

The production of effects changing so suddenly with the wave-length would appear to require the cooperation of a large number of grating-lines.

But, as the result of an experiment in which all but about 200 lines were blocked out, Prof. Wood was compelled to refer the matter to the form of the groove. To this cause one would naturally look for an explanation of the difference between this grating and others ruled with the same interval, but it does not appear how the discontinuity itself can have its origin in the form of the groove.

The first step towards an explanation would be the establishment of a relation between the wave-lengths of the bands and the corresponding angles of incidence; and at the time of reading the original paper I was inclined to think that the determining circumstance might perhaps be found in the passing off of a spectrum of higher order. Thus in the spectrum under observation of the first order, an abnormality might be expected at a particular wave-length if in the third order light of this wave-length were just passing out of the field of view, *i.e.* were emerging tangentially to the grating surface. The verification or otherwise of this conjecture requires a knowledge of the grating interval (ϵ). This is not given in the published paper; but on hearing from Prof. Wood that there were 14,438 lines to the inch, I made at once the necessary calculation.

If θ be the angle of incidence for which light of wave-length λ is just passing off in the nth spectrum,

$$\epsilon(1 \pm \sin\theta) = n\lambda. \quad\dots\dots\dots\dots\dots\dots(1)$$

In the first diagram the angle of incidence is $4° 12'$ and the wave-lengths of the bands are given as 609 and 517, or in centimetres 6.09×10^{-5} and 5.17×10^{-5}. Also $\epsilon = 2.540/14438$ cm., and $\sin\theta = .0732$. Using these data in (1), we find for the larger wave-length $n = 3.10$, or $n = 2.68$, according as the upper or the lower sign is taken. Again, for the smaller wave-length we find with the upper sign $n = 3.65$, and with the lower $n = 3.15$. To reconcile these numbers with the suggested relation it is necessary to suppose that 609 is passing off in the third spectrum on the same side as that on which the light is incident, and 517 in the third spectrum upon the other side. But the agreement of 3.10 and 3.15 with the integer 3.00 seemed hardly good enough, and so the matter was put aside until recently, when my attention was recalled to it in reading an article by Prof. Ames* on Rowland's ruling-machines, from which it appeared that gratings have been ruled with three different spaces, viz. 14,438, 15,020, and 20,000 lines to the inch. If we permit ourselves to suppose that the number of lines in the special grating is really 15,020 to the inch in place of 14,438, the alteration would be in the right direction, 3.10 becoming 2.98 and 3.15 becoming 3.03, so that the mean would be about correct.

In view of this improved agreement it seems worth while to consider how

* John Hopkins University Circular. Notes from the Physical Laboratory, Ap. 1906.

far the position of the bands recorded in the other diagrams would accord with the formula

$$\lambda = \tfrac{1}{3}\epsilon\,(1 \pm \sin \theta), \quad\dots\dots\dots\dots\dots\dots\dots\dots(2)$$

taking ϵ to correspond with ruling at the rate of 15,020 to the inch. In one respect there is a conspicuous agreement with Prof. Wood's observations. For if λ_1, λ_2 are the two values of λ in (2), we have at once

$$\lambda_1 + \lambda_2 = \tfrac{2}{3}\epsilon, \quad\dots\dots\dots\dots\dots\dots\dots\dots(3)$$

so that the two bands move equally in opposite directions as θ changes.

The results calculated from (2) for comparison with diagrams (2)...(10) (fig. 1) are given below.

	Calculation		Observation		
θ	λ_1	λ_2	λ_1	λ_2	No.
2° 37′	590	538	589	537	2
0° 15′	566	561	566	559	3
0° 5′	564	563	562	561	4
0° 0′	564	564	561	561	5
0° 5′	564	563	......	......	6
1° 15′	576	551	575	549	7
1° 53′	582	545	581	542	8
2° 38′	590	538	589	538	9
5° 45′	620	507	619	506	10

The numbers headed "observation" are measured from Prof. Wood's diagrams; but owing to the width and unsymmetrical form of some of the bands they are liable to considerable uncertainty. It would appear that (with the exception of the third band in diagram (10)) all the positions are pretty well represented by (2).

As regards the observations when the face of the grating was cemented to glass with cedar-oil, we have in place of (1)

$$\epsilon\,(1 \pm \sin \theta') = n\lambda',$$

where λ' is the wave-length and θ' the angle of incidence in the oil. Now if μ be the refractive index of the oil,

$$\lambda' = \lambda/\mu, \qquad \sin \theta' = \sin \theta\,.\,/\mu,$$

so that

$$\epsilon\,(\mu \pm \sin \theta) = n\lambda, \quad\dots\dots\dots\dots\dots\dots\dots\dots(4)$$

if as usual θ and λ are measured in air.

In the diagrams of Prof. Wood's fig. (2) there are four angles of incidence. The bands are markedly unsymmetrical and the numbers entered in the following table are those corresponding to the sharp edge. The values for n are calculated from (4) on the supposition that $\mu = 1\cdot5$, the lower sign being chosen if the angles on the first side are regarded as positive. The wave-lengths observed correspond pretty well with the passing off of the fourth and fifth spectra on the opposite side to that upon which the light is incident. There seems to be nothing corresponding to the passing

θ	λ	n
12° 8′	541	4·03
7° 8′	590, 469	3·94, 4·96
3° 53′	610, 489	3·97, 4·95
−2° 29′	655, 529	3·99, 4·93

off of spectra on the same side. Upon the whole there appears to be confirmation of the idea that the abnormalities are connected with the passing off of higher spectra, especially if the suggested value of ϵ can be admitted.

The argument which led me to think that something peculiar was to be looked for when spectra are passing off may be illustrated from the case of plane waves of sound, incident upon a parallel infinitely thin screen in which are cut apertures small in comparison with λ. The problem for a single aperture was considered in *Phil. Mag.* Vol. XLIII. p. 259, 1897[*], from which it appears that corresponding to an incident wave of amplitude unity the wave diverging from the aperture on the further side has the expression

$$\psi = M \frac{e^{-ikr}}{r}, \quad \dots\dots\dots\dots\dots\dots\dots\dots\dots(5)$$

where $k = 2\pi/\lambda$, r is the distance from the aperture of the point where the velocity-potential ψ is reckoned, and M represents the electrical *capacity* of a conducting disk having the size and shape of the aperture, and situated at a distance from all other electrical bodies. In the case of a circular aperture of radius a,

$$M = 2a/\pi. \quad \dots\dots\dots\dots\dots\dots\dots\dots(6)$$

The expression (5) applies in general only when the aperture is so small that the distance between any two points of it is but a small fraction of λ. It may, however, be extended to a series of equal small apertures disposed at equal intervals along a straight line, provided that the distance between consecutive members of the series is a multiple of λ. The condition is then satisfied that any two points, whether on the same or on different apertures, are separated by a distance which is very nearly a precise multiple of λ. The expression for the velocity-potential may be written

$$\psi = M' \frac{e^{-ikr_1}}{r_1} + M' \frac{e^{-ikr_2}}{r_2} + \dots, \quad \dots\dots\dots\dots\dots(7)$$

[*] Or *Scientific Papers*, Vol. IV. p. 283.

where r_1, r_2, &c., are the distances of any point on the further side of the screen from the various apertures, and M' is the electrical capacity of each aperture, now no longer isolated, but subject to the influence of the others similarly charged.

It is not difficult to see that if the series of apertures is infinitely extended, M' approaches zero. For, if ϵ be the distance between immediate neighbours, and we consider the condition of the system when charged to potential unity, we see that the potential at any member due to the charges on the other members has the value

$$\frac{2M'}{\epsilon}(1 + \tfrac{1}{2} + \tfrac{1}{3} + \dots) = M' \times \infty .$$

Accordingly $M' = 0$, indicating that the efficiency of each aperture in allowing waves to pass to the further side of the screen is destroyed by the cooperative reaction of the series of neighbours. The condition of things now under contemplation is that in which one of the lateral spectra formed by the series of holes (considered as a grating) is in the act of passing off, and it is evident that the peculiar interference is due to this circumstance*. The argument applies even more strongly, if less simply, to an actual grating formed by a series of narrow parallel and equidistant slits cut in an infinite screen.

The case of a reflecting grating differs in some important respects from that above considered. An investigation applicable to light is now nearly completed†. It confirms the general conclusion that peculiarities are to be looked for at such angles of incidence that spectra of higher order are just passing off, but (it is especially to be noted) only when the polarization is such that the electric vector is perpendicular to the grating.

P.S. June 5.—In anwer to further inquiry Prof. Wood tells me that he thinks the ruling may perhaps be 15,020 to the inch, but (the grating being for the time out of his hands) he is not able to speak with certainty.

[1911. In a letter of Nov. 2, 1911 Prof. Wood writes "I think that there is no doubt about the grating constant. It must have been 15,020, as the 14,000 machine was not in use at the time it was made. I am practically sure of this." And in a subsequent letter referring to more recent observations "The thing is evidently a little more complicated than your first theory, but I think you are on the right track."]

* If ϵ be not a precise multiple of λ, the series $1 + \tfrac{1}{2} + \tfrac{1}{3} + \dots$ ad infin. would be replaced by

$$e^{-ik\epsilon} + \tfrac{1}{2}e^{-2ik\epsilon} + \tfrac{1}{3}e^{-3ik\epsilon} + \dots,$$

which is equivalent to

$$-\log\{2\sin(\tfrac{1}{2}k\epsilon)\} + \tfrac{1}{2}i(k\epsilon - \pi).$$

† [1911. See preceding paper.]

324.

ON THE LIGHT DISPERSED FROM FINE LINES RULED UPON REFLECTING SURFACES OR TRANSMITTED BY VERY NARROW SLITS.

[*Philosophical Magazine*, Vol. XIV. pp. 350—359, 1907.]

THE problem of the incidence of plane waves upon a cylindrical obstacle, whose radius is small in comparison with the length of the waves and whose axis is parallel to their plane, is considered in *Theory of Sound*, § 343, also *Scientific Papers*, Vol. IV. p. 314; but it is now desired to carry the approximation further and also to make some applications. On the other hand, we shall confine ourselves to the cases of perfect reflexion where the boundary conditions are simplest.

The primary waves, travelling in the negative direction, are represented by $\phi = e^{ik(at+x)}$, where a is the velocity of propagation and $k = 2\pi/\lambda$, λ being the wave-length. Dropping the time-factor for brevity, we shall write

$$\phi = e^{ikx} = e^{ikr\cos\theta} = J_0(kr) + 2iJ_1(kr)\cos\theta + \dots + 2i^n J_n(kr)\cos n\theta + \dots, \quad \dots(1)$$

J_n being the Bessel's function usually so denoted, so that

$$J_n(z) = \frac{z^n}{2^n \Gamma(n+1)} \left\{ 1 - \frac{z^2}{2(2n+2)} + \frac{z^4}{2 \cdot 4(2n+2)(2n+4)} - \dots \right\}. \quad \dots(2)$$

In (1) x and r are measured from the centre of the section of the cylinder, whose length is supposed parallel to the axis of z.

The secondary waves diverging from the obstacle are represented by

$$\psi = B_0 D_0(kr) + B_1 D_1(kr)\cos\theta + B_2 D_2(kr)\cos 2\theta + \dots, \quad \dots\dots(3)$$

where

$$D_0(z) = -\left(\frac{\pi}{2iz}\right)^{\frac{1}{2}} e^{-iz} \left\{ 1 - \frac{1^2}{1 \cdot 8iz} + \frac{1^2 \cdot 3^2}{1 \cdot 2 \cdot (8iz)^2} - \dots \right\}$$

$$= \left(\gamma + \log\frac{iz}{2}\right) \left\{ 1 - \frac{z^2}{2^2} + \frac{z^4}{2^2 \cdot 4^2} - \dots \right\}$$

$$+ \frac{z^2}{2^2} - \frac{z^4}{2^2 \cdot 4^2}(1 + \tfrac{1}{2}) + \frac{z^6}{2^2 \cdot 4^2 \cdot 6^2}(1 + \tfrac{1}{2} + \tfrac{1}{3}) - \dots, \quad \dots(4)$$

γ being Euler's constant ($\cdot$5772...), and the other D's are related to D_0 according to

$$D_n(z) = (-2z)^n \left(\frac{d}{d \cdot z^2}\right)^n D_0(z). \quad \dots\dots\dots\dots(5)$$

The first expression in (3) is available when z is large and the second when z is small. It should be remarked that the notation is not quite the same as in the papers referred to.

The leading term in D_0 when z is large is

$$D_0(z) = -\left(\frac{\pi}{2iz}\right)^{\frac{1}{2}} e^{-iz}, \quad \dots\dots\dots\dots(6)$$

and in finding the leading term in $D_n(z)$ by (5) it suffices to differentiate only the factor e^{-iz}. Thus when z is great

$$D_n(z) = -i^n \left(\frac{\pi}{2iz}\right)^{\frac{1}{2}} e^{-iz}. \quad \dots\dots\dots\dots(7)$$

Accordingly when in (3) ψ is required at distances from the cylinder very great in comparison with λ, we may take

$$\psi = -\left(\frac{\pi}{2ikr}\right)^{\frac{1}{2}} e^{-ikr}\{B_0 + iB_1 \cos\theta - B_2 \cos 2\theta + \dots\}. \quad \dots\dots(8)$$

We have now to consider the boundary conditions to be satisfied at the surface of the cylinder $r = c$, and we will take first the condition that

$$\phi + \psi = 0 \quad \dots\dots\dots\dots(9)$$

at this surface. We have at once from (1) and (3) in virtue of Fourier's theorem

$$B_0 = -J_0(kc) \div D_0(kc), \qquad B_1 = -2iJ_1(kc) \div D_1(kc),$$

and generally

$$B_n = -2i^n J_n(kc) \div D_n(kc). \quad \dots\dots\dots\dots(10)$$

In like manner if the condition to be satisfied at the surface of the cylinder is

$$\frac{d\phi}{dr} + \frac{d\psi}{dr} = 0, \quad \dots\dots\dots\dots(11)$$

we get, using C_0, C_1 &c. in place of B_0, B_1 &c.,

$$C_0 = -J_0{}'(kc) \div D_0{}'(kc),$$
$$C_n = -2i^n J_n{}'(kc) \div D_n{}'(kc), \quad \dots\dots\dots\dots(12)$$

the dashes denoting differentiation.

The next step is to introduce approximations depending upon the smallness of kc. In addition to (4) we have

$$D_1(z) = -\frac{dD_0}{dz} = -\frac{1}{z}\left\{1 - \frac{z^2}{2^2} + \dots\right\}$$

$$+ \left(\gamma + \log\frac{iz}{2}\right)\left\{\frac{z}{2} - \dots\right\} - \frac{z}{2} + \dots, \quad \dots\dots(13)$$

$$D_2(z) = -\frac{2}{z^2} - \frac{1}{2} + \dots, \quad \dots\dots\dots\dots(14)$$

and so on. Also

$$D_0'(z) = \frac{1}{z} + \frac{z}{4} - \frac{z}{2}\left(\gamma + \log\frac{iz}{2}\right), \quad \dots\dots\dots\dots(15)$$

$$D_1'(z) = \frac{1}{z^2} + \frac{1}{2}\left(\gamma + \frac{1}{2} + \log\frac{iz}{2}\right), \quad \dots\dots\dots\dots(16)$$

$$D_2'(z) = \frac{4}{z^3}. \quad \dots\dots\dots\dots\dots\dots\dots\dots\dots\dots\dots(17)$$

Using these, we find

$$-B_0^{-1} = \gamma + \log\left(\tfrac{1}{2}ikc\right) + \tfrac{1}{4}k^2c^2\left(1 + \tfrac{5}{32}k^2c^2\right), \quad \dots\dots\dots(18)$$

$$-iB_1 = k^2c^2\left[1 + \tfrac{1}{2}k^2c^2\left\{\gamma - \tfrac{3}{4} + \log\left(\tfrac{1}{2}ikc\right)\right\}\right], \quad \dots\dots\dots(19)$$

$$B_2 = -\tfrac{1}{8}k^4c^4. \quad \dots\dots\dots\dots\dots\dots\dots\dots\dots\dots\dots(20)$$

Referring to (8) we see that when kc is small the predominant term is the symmetrical one dependent on B_0. Retaining only this term, we have as the expression for the secondary waves

$$\psi = \frac{e^{-ikr}}{\gamma + \log\left(\tfrac{1}{2}ikc\right)}\left(\frac{\pi}{2ikr}\right)^{\frac{1}{2}}, \quad \dots\dots\dots\dots\dots\dots(21)$$

as in the papers referred to. Relatively to this, the term in $\cos\theta$ is of order k^2c^2, and that in $\cos 2\theta$ of order k^4c^4.

Passing on to the second boundary condition (11), we have

$$C_0 = \frac{k^2c^2}{2}\left\{1 + \frac{k^2c^2}{2}\left(\gamma - \frac{3}{4} + \log\frac{ikc}{2}\right)\right\}, \quad \dots\dots\dots\dots(22)$$

$$iC_1 = k^2c^2\left\{1 - \frac{k^2c^2}{2}\left(\gamma + \frac{5}{4} + \log\frac{ikc}{2}\right)\right\}, \quad \dots\dots\dots\dots(23)$$

$$-C_2 = -\frac{k^4c^4}{8}. \quad \dots\dots\dots\dots\dots\dots\dots\dots\dots\dots\dots(24)$$

When these values are introduced into (8), we see that the terms in C_0 and C_1 are of equal importance. Limiting ourselves to these, we have

$$\psi = -k^2c^2e^{-ikr}\left(\frac{\pi}{2ikr}\right)^{\frac{1}{2}}\left(\tfrac{1}{2} + \cos\theta\right), \quad \dots\dots\dots\dots(25)$$

the symbolical expression which gives the effect of the incidence of aerial waves upon a rigid and fixed obstacle (*Theory of Sound*, § 343). Fully to interpret it we must restore the time-factor and finally reject the imaginary part, thus obtaining

$$\psi = -\frac{2\pi \cdot \pi c^2}{r^{\frac{1}{2}}\lambda^{\frac{3}{2}}}\left(\tfrac{1}{2} + \cos\theta\right)\cos\frac{2\pi}{\lambda}\left(at - r - \tfrac{1}{8}\lambda\right), \quad \dots\dots\dots(26)$$

corresponding to the primary waves

$$\phi = \cos\frac{2\pi}{\lambda}\left(at + x\right). \quad \dots\dots\dots\dots\dots\dots(27)$$

In the application to electric or luminous vibrations the present solution is available for the case of primary waves

$$c^* = e^{ikx}, \quad \dots\dots\dots\dots\dots\dots\dots\dots\dots\dots(28)$$

incident upon a perfectly conducting, *i.e.* reflecting, cylinder, c^* denoting the magnetic induction [parallel to z], for which the condition to be satisfied at the surface is $dc^*/dr = 0$. Accordingly the secondary waves are given by (25) with c^* written for ψ. This is the case of incident light polarized in a plane parallel to the length of the cylinder.

For incident light polarized in the plane perpendicular to the length of the cylinder the primary waves have the expression

$$R = e^{ikx}, \quad \dots\dots\dots\dots\dots\dots\dots\dots\dots\dots(29)$$

where R denotes the electromotive intensity parallel to z. In this case the secondary waves are given by (21) with R in place of ψ. It appears that if the incident waves in the two cases are of equal intensity, the secondary

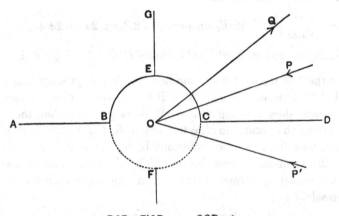

$$POD = P'OD = \alpha; \quad QOD = \phi.$$

waves are of different orders of magnitude, R preponderating. Thus if unpolarized light be incident, the scattered light is polarized in the plane perpendicular to the length of the cylinder, and the polarization is complete if the cylinder be small enough.

It is now proposed to make application of these solutions to meet the problem in two dimensions of the incidence of plane waves upon a *perfectly* reflecting plane surface from which rises an excrescence, also composed of perfectly reflecting material, and having the form of a semi-cylinder whose axis lies in the plane. We shall show that it is legitimate to substitute the complete cylinder, provided that we suppose incident upon it *two* sets of plane waves adjusted to one another in a special manner.

In the figure $ABECD$ represents the actual reflector, upon which are incident waves advancing along PO, a direction making an angle α with the surface OD. The secondary disturbance is required at a great distance (r) along OQ inclined to OD at an angle ϕ. But for the present we suppose the cylinder to be complete and the plane parts of the reflector AB, CD to be abolished; and in addition to the waves advancing along PO we consider others of the same quality advancing along a line $P'O$ equally inclined to the surface upon the other side. The angle θ of previous formulæ is represented now by POQ, $P'OQ$ whose values are $\phi - \alpha$ and $\phi + \alpha$. Thus we may write

$$\cos \theta = \cos (\phi - \alpha), \qquad \cos \theta' = \cos (\phi + \alpha),$$

so that

$$\cos \theta + \cos \theta' = 2 \cos \alpha \cos \phi, \qquad \cos \theta - \cos \theta' = 2 \sin \alpha \sin \phi.$$

The two sets of waves advancing along PO, $P'O$ will be supposed to be of equal amplitude; but we shall require to consider two distinct suppositions as to their phases. In dealing with R the supposition is that the phases are precisely opposed. In this case we obtain from (8) as the complete expression of the secondary waves

$$R = -\left(\frac{\pi}{2ikr}\right)^{\frac{1}{2}} e^{-ikr} \{2iB_1 \sin \alpha \sin \phi - 2B_2 \sin 2\alpha \sin 2\phi + \ldots\}, \quad \ldots(30)$$

the term in B_0 disappearing, while the values of B_1, B_2 are given by (19), (20).

Each of the two separate solutions here combined, primary and secondary terms included, satisfies the condition $R = 0$ at the surface of the cylinder and so of course does the aggregate. It is easy to see that the aggregate further satisfies the condition $R = 0$ along AB, CD where θ, θ' are equal, the contributions from the two solutions being equal and opposite. Hence (30) gives the secondary waves due to the incidence of primary waves along PO upon the reflecting surface $ABECD$; and the expression for the primary waves themselves is

$$R = e^{ik (x \cos \alpha + y \sin \alpha)} - e^{ik (x \cos \alpha - y \sin \alpha)}, \quad \ldots(31)$$

x being parallel to OD and y parallel to OG, so that

$$x = r \cos \phi, \qquad y = r \sin \phi.$$

In like manner the c^* solution may be built up. In this case we have to give the same phase to the two component primaries. Corresponding to the incident

$$c^* = e^{ik (x \cos \alpha + y \sin \alpha)} + e^{ik (x \cos \alpha - y \sin \alpha)}, \quad \ldots(32)$$

we have for the secondary disturbance

$$c^* = -\left(\frac{\pi}{2ikr}\right)^{\frac{1}{2}} e^{-ikr} \{2C_0 + 2iC_1 \cos \alpha \cos \phi - 2C_2 \cos 2\alpha \cos 2\phi + \ldots\}. \quad \ldots(33)$$

Each solution, consisting of primary and associated secondary, satisfies over the surface of the cylinder $dc^*/dn = 0$, dn being an element of the normal. And over the plane part AB, CD the two solutions contribute equal and opposite components to dc^*/dn. Hence all the conditions are satisfied for the incidence of waves along PO upon the compound reflecting surface $ABECD$.

The problem is now solved for the two principal cases of polarized incident light. If the incident light be unpolarized, the condition as regards polarization of the scattered light turns upon the value of

$$\Pi = \frac{iB_1 \sin \alpha \sin \phi - B_2 \sin 2\alpha \sin 2\phi + \dots}{C_0 + iC_1 \cos \alpha \cos \phi - C_2 \cos 2\alpha \cos 2\phi + \dots}, \quad \dots\dots\dots\dots(34)$$

in which the values of $B_1, B_2, \dots, C_0, C_1, \dots$ are to be substituted from (18), (19), (20), and from (22), (23), (24).

If we stop at the first approximation, neglecting B_2, C_2, &c., we have

$$\Pi = -\frac{2 \sin \alpha \sin \phi}{1 + 2 \cos \alpha \cos \phi}. \quad \dots\dots\dots\dots\dots(35)$$

From (34) or (35) we see that the value of Π is symmetrical as between α and ϕ, an example of the general law of reciprocity (*Theory of Sound*, § 108 &c.).

If $\alpha = 0$, or if $\phi = 0$, Π vanishes without appeal to approximations. This means that c^* preponderates, or that the scattered light is polarized in a plane parallel to the length of the cylinder. The conclusion follows approximately although α be not very small, provided ϕ be also small.

According to (35) Π becomes infinite when

$$1 + 2 \cos \alpha \cos \phi = 0, \quad \dots\dots\dots\dots\dots\dots(36)$$

for example when $\alpha = 45°$, $\phi = 135°$.

If we take $\alpha = 40°$, $\phi = 130°$ so as to avoid the directly reflected rays, we have $\Pi = -67$, so that there is nearly complete polarization in the plane perpendicular to the length of the cylinder.

If we suppose $\phi = 180° - \alpha$, so that observation is made nearly in the direction of the regularly reflected rays, (35) becomes

$$\Pi = -\frac{2 \sin^2 \alpha}{1 - 2 \cos^2 \alpha}. \quad \dots\dots\dots\dots\dots(37)$$

The scattered light is unpolarized when $\Pi = \pm 1$. If we make this supposition in (37) we find $\alpha = 30°$. This angle separates the two kinds of polarization. Thus when α is small, $\Pi = 2 \sin^2 \alpha$; when $\alpha = 30°$, $\Pi = 1$; when $\alpha = 45°$, $\Pi = \infty$; when $\alpha = 90°$, $\Pi = -2$.

By use of (34) the approximation may be carried further. As an example we may take the case of perpendicular incidence and observation, so that $\alpha = 90°$, $\phi = 90°$. Thus by (34)

$$\Pi = \frac{iB_1}{C_0 - C_2} = -2\left(1 + \tfrac{1}{4} k^2 c^2\right). \quad \dots\dots\dots\dots(38)$$

It may be well to recall that in the results which we have obtained the angles α, ϕ are measured from the *surface* and not, as is usual in Optics, from the normal. Again, if it be desired to attach significance to the *sign* of II, we must remember that in one case we were dealing with c^* and in the other with R.

The above given theoretical investigation was undertaken in order to see how far an explanation could be arrived at of some remarkable observations by Fizeau*, relating to the light dispersed at various angles from fine lines or scratches traced upon silver and other reflecting surfaces. In every case the incident and dispersed ray is supposed to be perpendicular to the lines, so that the problem is in two dimensions. The most striking effects are observed when the incident and dispersed rays are both highly oblique and upon the same side of the normal to the surface. The dispersed light is then strongly, sometimes almost completely, polarized, and the plane of polarization is *parallel* to the direction of the lines, *i.e.* perpendicular to the plane of incidence. A silver surface, polished by rubbing with ordinary rouge in one direction, shows these effects well, and even a piece of tin-plate, treated similarly with cotton-wool, suffices. The plate is to be held obliquely and the incident rays should come from a window or sky-light behind the observer. It is of importance to avoid stray light and especially any that could reach the eye by specular reflexion. Observation with a nicol shows at once that the light is strongly polarized and in the opposite way to that regularly reflected from a glass plate similarly held.

Under the microscope a single line may be well observed, especially when strongly lighted by sunlight. Fizeau found that when the incidence is oblique and the observation normal ($\alpha =$ small, $\phi = 90°$), or equally when the incidence is normal and the observation oblique ($\alpha = 90°$, $\phi =$ small), the above specified polarization obtains, provided the line be very fine; otherwise the polarization may be reversed. When the incidence is oblique and the light nearly retraces its course (α and ϕ both small and of the same sign), the polarization is more complete and also less dependent upon extreme fineness. When α and ϕ are both in the neighbourhood of $90°$, the polarization becomes insensible. If the incidence is oblique and the angle of observation in the neighbourhood of the regularly reflected light, traces of reversed polarization are to be detected.

My own observations are in essential agreement with Fizeau. At first accidental scratches upon silver surfaces which had been worked in one direction were employed. Afterwards I had the opportunity of observing specially fine lines ruled with a diamond by a dividing-engine, for which I am indebted to Lord Blythswood. In the latter case the plate was of speculum-metal.

* *Annales de Chimie*, Vol. LXIII. p. 385 (1861); Mascart's *Traité d'Optique*, § 645.

It will be seen that the theory agrees with observation well in some respects, but fails in others. When α and ϕ are both less than 90° and of the same sign, the polarization expressed by (35) sufficiently represents the facts. But there is little in the observations to confirm the strongly reversed polarization which should occur when the denominator in (35) becomes small. One defect of correspondence in the conditions of theory and experiment is obvious. The former relates to semi-cylindrical *excrescences*, while the observations are made upon light dispersed from scratches which are mainly *depressions*. In order to examine the question thus arising, a glass plate provided with suitable scratches was coated chemically with silver upon which copper was afterwards deposited by electrolysis. When the coating thus obtained was stripped from the glass, a highly reflecting surface was obtained in which the original scratches are represented by precisely fitting protuberances. But even with this I was unable to find the strongly reversed polarization to be expected according to (35) when (36) is nearly satisfied.

If we trace back the denominator in (35), we find that it is derived from the factor $(\frac{1}{2} + \cos\theta)$ in (26), and that its evanescence depends upon the antagonistic effects of the terms which are symmetrical and proportional to $\cos\theta$. The precise form of this factor is doubtless connected with the assumption of a circular cross-section, but the discrepancy from observation seems almost too complete to be attributed to such departures from the theoretical shape. As other possible sources of discrepancy we may note the assumption of reflecting power which is absolutely complete, and again that the dimensions of the line are small in comparison with the wavelength. It may be that lines sufficiently fine to justify (35) in its integrity would not reflect enough light to be visible. At the same time the evanescence of Π with α or ϕ does not demand such a high degree of fineness.

In the memoir already cited Fizeau treats also the polarization impressed upon light which traverses fine slits. Thus (p. 401): "Une lame d'argent très-mince, déposé chimiquement sur le verre, a été rayée en ligne droite, avec de l'émeri très-fin; c'était un fragment de la lame désignée précédemment par la lettre (A), et dont l'épaisseur a été trouvée de 1/3400 de millimètre. Un grand nombre de stries avaient traversé la couche d'argent de manière à donner naissance à autant de fentes d'une ténuité extrême. Ces lignes lumineuses étant observées, à l'aide d'un analyseur, au microscope éclairé par la lumière solaire, ont présenté les phénomènes de polarisation déjà décrits, c'est-à-dire qu'un grand nombre d'entre elles étaient polarisée dans un plan perpendiculaire à leur longueur.

"Mais en observant avec plus d'attention les moins lumineuses de toutes ces lignes, c'est-à-dire celles qui devaient être les plus fines, on en a trouvé un certain nombre qui présentaient un phénomène de sens opposé, c'est-à-dire qu'elles étaient polarisées dans un plan parallèle à leur longueur, les unes totalement, les autres partiellement; cet effet étant accompagné de phéno-mènes de coloration semblables à ceux qui ont été signalés dans les lignes qui donnent la polarisation perpendiculaire."

The passage of electric or luminous waves through a fine slit in a thin perfectly conducting screen was considered by me in a memoir published ten years ago*. If the electric vector is parallel to the length of the slit, the amplitude of the transmitted vibration is proportional to the square of the width of the slit; but if the electric vector is perpendicular to the length of the slit, the transmitted vibration involves the width only as a logarithm—see equation (46)—much as in equation (21) of the present paper. If the incident vibration be unpolarized and the slit be very fine, the latter component preponderates in the transmitted waves, viz. the direction of polarization is parallel to the length of the slit, in accordance with Fizeau's observations upon light transmitted by apertures of minimum width.

* *Phil. Mag.* Vol. xliii. p. 259 (1897); *Scientific Papers*, Vol. iv. p. 283. See also *Phil. Mag.* July 1907. [*Scientific Papers*, Vol. v. p. 405.]

325.

ON THE RELATION OF THE SENSITIVENESS OF THE EAR TO PITCH, INVESTIGATED BY A NEW METHOD.

[*Philosophical Magazine*, Vol. xiv. pp. 596—604, 1907.]

IN a former research* I examined the sensitiveness of the ear to sounds of different pitch with results which were thus summarized:—

$$c', \text{ frequency} = 256, \quad s = 6\cdot0 \times 10^{-9},$$
$$g', \qquad „ \qquad = 384, \quad s = 4\cdot6 \times 10^{-9},$$
$$c'', \qquad „ \qquad = 512, \quad s = 4\cdot6 \times 10^{-9},$$

no reliable distinction appearing between the two last numbers. "Even the distinction between 6·0 and 4·6 should be accepted with reserve; so that the comparison must not be taken to prove much more than that the condensation necessary for audibility varies but slowly in the singly dashed octave." Here s denotes the condensation (or rarefaction) which in one respect is a maximum and in another a minimum. It is the maximum condensation which occurs during the course of the vibration, but the vibration (and s with it) is the minimum capable of impressing the ear in a progressive wave. The method employed depended upon a knowledge of the rate at which energy was emitted from a resonator under excitation by a freely vibrating tuning-fork. The amplitude of vibration of the prongs of the fork was under continuous observation with the aid of a microscope. From this could be inferred the energy in the fork at any time and the rate at which it was lost. The loss was greatest when the resonator was in action, and the excess was taken to represent the energy converted into sound. From this again the condensation in the progressive waves at a given distance could be calculated. It was remarked that the numbers thus obtained were "somewhat of the nature of upper limits, for they depend upon the assumption that all the dissipation *due to the resonator* represents production of sound. This may not be strictly the case even with the moderate amplitudes here in question, but the uncertainty is far less than in the case of resonators or organ-pipes caused to speak by wind."

* *Phil. Mag.* Vol. xxxviii. p. 365 (1894); *Scientific Papers*, Vol. iv. p. 125.

In a careful re-examination of this question, M. Wien*, working with the telephone, finds not only a still higher degree of sensitiveness but a much more rapid variation with pitch. In the following extract from his Table xiv., N represents the frequency and Δ the proportional excess of pressure, equal to γs, where γ is the ratio of specific heats of air (1·41). The higher degree of sensitiveness may be partly explained by the greater precautions taken to ensure silence and by the sounds under observation being rendered intermittent; or, on the other hand, my estimates of sensitiveness may have been too low in consequence of the already named assumption that all the excess of damping due to the resonator represented production of sound. With respect to the dependence on pitch, Wien remarks that my own observations† on the minimum current in the telephone necessary for audibility at various frequencies, support his view of the question. Certainly this is their tendency. At the time when these observations were made the whole *modus operandi* of the telephone was still involved in doubt, and my object in these observations was rather to elucidate

N	Δ	N	Δ
50	$1·6 \times 10^{-7}$	1600	$1·4 \times 10^{-11}$
100	$1·1 \times 10^{-8}$	3200	$1·4 \times 10^{-11}$
200	$1·0 \times 10^{-9}$	6400	$2·3 \times 10^{-11}$
400	$1·2 \times 10^{-10}$	12800	$8·0 \times 10^{-11}$
800	$2·3 \times 10^{-11}$		

the action of this instrument. Even now there are points which remain obscure, for example the easy audibility of sounds when the iron disk is replaced by one of copper or aluminium. It is to be presumed that the movements of the disk then depend upon electric currents induced in it. If so, they would follow different laws from those governing the behaviour of a simply magnetized disk; and in the case of iron complications would ensue from the cooperation of both causes.

Again, though this is partly a matter of definition, I am of opinion that the sensitiveness of the ear is best investigated with the ear free. When a telephone, pressed closely up, is employed, the situation is materially altered. For example, the natural resonances of the ear-passage must be seriously disturbed.

The above objections do not apply to some of Wien's determinations, where the ear was placed at a distance from the telephone and the vibrations

---

* Pflüger's *Arch.* Vol. xcvii. p. 1 (1903).

† *Phil. Mag.* Vol. xxxviii. p. 285 (1894); *Scientific Papers*, Vol. iv. p. 109.

of the plate were directly measured; and his conclusions must necessarily carry great weight. But I could not forget that my own experiments in 1894 had been carefully made, and I was desirous, if possible, of checking the results by some new method different from those previously employed either by Wien or myself. The difficulties of the problem are considerable; but it occurred to me that, so far as the important question of the dependence of sensitiveness upon pitch is concerned, they might be turned by calling to our aid the general principle of dynamical similarity*. Thus if vibrations are communicated to the air from the prongs of a tuning-fork, we are unable to calculate the theoretical connexion between the invisible aerial vibrations and the visible amplitude of the prongs. But if there are two precisely similar forks of different dimensions, and each communicating vibrations, a *comparison* may be effected. In the first place, if the material is the same, the times of vibration of the forks (regarded as uninfluenced by the air) are as the linear dimensions. And further, what is more important for our purpose, the condensations at *corresponding* points in the surrounding air will be the same, provided the amplitudes of vibrations at the prongs be themselves in the proportion of the linear dimensions. Corresponding points are, of course, such as are similarly situated with respect to the vibrating forks, the distances from corresponding points of the forks being in proportion to the linear dimensions of the latter. Since times and distances are altered in the same proportion, velocities are unchanged. In conformity with this, the velocity-potentials in the two systems are as the linear dimensions.

It appears then that by means of the principle of similarity we can obtain aerial condensations which may be recognized to be equal in spite of a change of pitch. As has been said, equality occurs when the amplitudes of the solid vibrators are as the linear dimensions. In virtue of the principle of superposition as applicable to the small vibrations of either system, we are not limited to the case of equal condensations. The ratio of condensations can be inferred from the ratio of amplitudes by introduction of the factor expressing the linear magnification.

My first intention had been to use forks for the actual experiments. But apart from the difficulty of obtaining them of the necessary geometrical similarity over a sufficient range, it appeared that the communication of vibration to the air was inadequate. At a suitable distance there was danger that the sounds might prove inaudible. In connexion with this the difficulty of supporting the forks has to be considered. It is essential that no sound capable of influencing the results shall reach the ear by way of the supports, to which the principle of similarity can hardly be extended;

---

* The application of this principle to acoustical problems is discussed in *Theory of Sound*, 2nd ed. § 381.

and the danger of disturbance from this source increases if the direct communication of vibration to the air is too much enfeebled. On the other hand the arrangements must, if possible, be such as will render adequate the optically observed amplitude up to the point at which the ear is beginning to fail. We have in fact to steer as best we may between difficulties on opposite sides.

The requirements of the case seem to be best met by using thin open metal cans vibrating after the manner of bells. They were constructed by Mr Enock from tin-plate or ferrotype plate and were maintained in vibration electrically. The wall is a simple cylinder and there is a flat bottom of similar material. In the case of the largest can, giving 85 complete vibrations per second, the height was 8 inches and the diameter 4·5 inches. During the vibration the circular bottom bends, and thus the can must not be held fast round the lower circumference. Support was usually given at the centre only, by means of a short length of tube attached with solder. It is possible, however, as will be explained more fully later, to give support at four points of the lower circumference, or rather along two diameters of the base perpendicular to one another. These diameters are at 45° to the line of the electromagnet by which the vibrations are maintained. In order to avoid communicating vibration, the metal handle of the can was attached to a large bung, resting upon a leaden slab, supported in its turn from the floor by a tall retort-stand. Before or after observations the bung could be lifted with the fingers and security taken that all the sound heard came direct from the vibrating can.

The bar electromagnet, by which the iron substance of the can is attracted, lies just below a diameter of the upper rim and is supported from the centre of the base. Since the electromagnet acts as an obstacle to the aerial vibrations in the region where they are strongest, care must be taken that in passing from one can to another geometrical similarity extends to the external form of the magnet and its accessories. The current was supplied by an interrupter-fork and usually both cans under comparison were driven (alternately) from the same fork, so that no question could arise with respect to the accuracy of the musical intervals. In constructing the cans the thickness of the plate employed was taken proportional to the other linear dimensions, but this alone would not suffice to secure an accurate tuning. The final adjustment for the greatest possible response to the intermittent electric current was effected with wax, required only in small quantities.

Trouble was sometimes experienced from the intrusion of undesired tones. A shunt across the mercury break of the interrupter-fork, as employed by Helmholtz, was useful, and an improvement would often follow a readjustment of the position of the electromagnet. No observations were taken until false sounds had been rendered entirely subordinate, if not inaudible. In every case the ear was placed in the plane of the rim opposite the *loops* L (fig. 1), where the radial motion is greatest, two of which face the poles of the electromagnet. All four positions were utilized, and either the intensity of vibration or the common distance was varied until the average intensity was judged suitable. The intensity aimed at was such that the sound was *just easily* audible, but sometimes a little more was allowed. The amplitude of vibration at a loop L was then measured by means of a microscope provided with a micrometer-scale and focused upon starch grains carried by the rim. In passing to the can under comparison the distance of observation is no longer variable at pleasure but must be taken in proportion to the linear dimension. Thus if the second can be on half the scale of the first and sound the octave above, the distance must be halved. If, when listened to in the four positions, the sound is judged too strong or

Fig. 1.

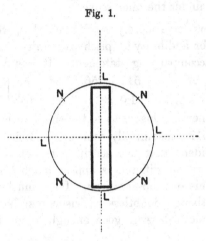

too faint, the vibration must be modified (by varying interposed resistance) until the former audibility is reproduced. The amplitude of vibration at a loop is then measured with a second microscope similar to the first. The distances of the ear, measured from the rim, varied from 8 inches to 24 inches. The sounds under comparison were usually estimated independently by Mr Enock and myself. A slight tendency on my part to estimate the graver sounds as the louder was suspected, but the difference was of no importance. Other observers also have taken part occasionally, and there was sufficient repetition on different days to eliminate chance errors. It will suffice to record the mean results.

In the comparison of cans of dimensions in the ratio of 2:1 and making 128 and 256 vibrations per second, it was found that for equal audibility at distances in the ratio of 2:1 the radial amplitude of the larger can required to be 4·0 times that of the smaller. Equal aerial condensations at the points of observation require amplitudes in the ratio of $2:1$, from which we infer that for equal audibilities the condensation needed at pitch 128 is the *double* of that needed at pitch 256. In like manner observations with another pair of cans showed that the condensation needed at 256 was 1·6 of that needed at 512 vibrations per second. It did not appear feasible by this method to go to higher pitch, but the range could be extended at the other end. For this purpose the largest can already spoken of was constructed, whose dimensions relatively to the 128 can were as $3:2$. In this case the interval was a Fifth, and the comparison showed that the condensation necessary for audibility at 85 per second was almost precisely the double of that needed at pitch 128. So far no interval had been attempted exceeding the octave; but subsequently confirmation was obtained by a direct comparison between the cans vibrating 256 and 85 per second. With large intervals the difficulties are increased, as the amplitude of the smaller can is too minute for satisfactory measurement under the microscope.

Since the numbers have merely a relative value, we may call the condensation necessary for audibility at pitch 512 unity. The results are then summarized in the accompanying statement. It was rather to my surprise

N	512	256	128	85
s	1·0	1·6	3·2	6·4

that I found my former conclusion as to the small variation of sensitiveness in the octave 256—512 substantially confirmed. Below 256 and especially below 128, it is evident that the sensitiveness of the ear falls off more rapidly; but even here the differences appear much less than those calculated by Wien from his own observations. I am much at a loss to explain the discrepancy. Although doubtless criticisms may be made, I should have supposed that both methods were good enough to yield fairly approximate results.

To give a general idea of the trend, a plot of the values of s is given in fig. 2, the logarithm of the periodic time being taken as abscissa. It would appear that the minimum s, corresponding to maximum sensitiveness of the ear, would not be reached under 1024 vibrations per second, and perhaps not until an octave higher, in accordance with Wien's conclusions.

I take this opportunity of recording a few observations on the mode of vibration of these cans, although the results have no immediate connexion with the main subject of this note. The theory of the vibration of thin cylindrical shells vibrating without extension of the middle surface* indicates two distinct types, of which one is excluded by the action of the plane disk forming the bottom of a can. The remaining type is defined by the equations

$$\delta r = sz \sin s\phi, \qquad a\delta\phi = z \cos s\phi, \qquad \delta z = -s^{-1}a \sin s\phi, \quad ......(1)$$

---

* *Proceedings Royal Society*, Vol. XLV. p. 105 (1888); *Scientific Papers*, Vol. III. p. 217.

in which z is measured upwards from the bottom, and the angle ϕ is measured round the circumference, the radius being a. δr, $a\delta\phi$, and δz are the radial, circumferential, and axial displacements of the point whose equilibrium position is defined by a, ϕ, z. When $z = 0$, *i.e.* at the junction of the cylindrical and plane parts, δr and $a\delta\phi$ vanish, but δz remains finite except when $\sin s\phi = 0$. Thus if δz be constrained to remain zero all round the circumference of the bottom, no vibration of this kind is possible; but the bottom may be supported at the places defined by $\sin s\phi = 0$, which are situated under the nodes N (fig. 1) of the radial motion at the upper rim.

Fig. 2.

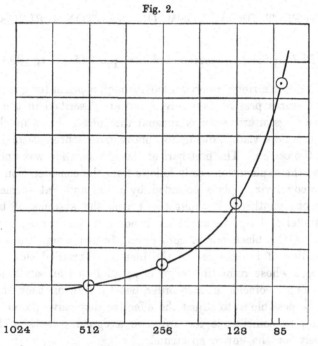

1024   512   256   128   85

If, as in the above experiments, we limit ourselves to the principal vibration for which $s = 2$, and if the height be l, we have for the maximum amplitudes

$$\delta r_{z=l} = 2l, \qquad a\delta\phi_{z=l} = l, \qquad \delta z = \tfrac{1}{2}a. \quad\ldots\ldots\ldots\ldots\ldots(2)$$

These relations were verified by observation under the microscope. The ratio of the maximum radial to the maximum circumferential motion at the upper rim was found to be almost exactly $2:1$. The accurate observation of δz, as found at the bottom, was more difficult on account of its relative smallness. In the actual case (85 per second) $l = 8$ in., $2a = 4\tfrac{1}{2}$ in., so that $8l/2a = 14$. This is the theoretical ratio of the maximum radial motion at the rim to the maximum axial motion. By observation the mean number was 15, as close to theory as could be expected.

326.

EFFECT OF A PRISM ON NEWTON'S RINGS.

[*Philosophical Magazine*, Vol. xv. pp. 345—351, 1908.]

WHEN Newton's rings are regarded through a prism (or grating) several interesting features present themselves, and are described in the "Opticks." Not only are rings or arcs seen at unusual thicknesses, but a much enhanced number of them are visible, owing to approximate achromatism—at least on one side of the centre. The first part of the phenomenon was understood by Newton, and the explanation easily follows from the consideration of the case of a true wedge, viz. a plate bounded by plane and flat surfaces slightly inclined to one another. Without the prism, the systems of bands, each straight parallel and equidistant, corresponding to the various wave-lengths (λ) coincide at the black bar of zero order, formed where the thickness is zero at the line of intersection of the planes. Regarded through a prism of small angle whose refracting edge is parallel to the bands, the various systems no longer coincide at zero order, but by drawing back the prism, it will always be possible so to adjust the effective dispersive power as to bring the nth bars to coincidence for any two assigned colours, and therefore approximately for the entire spectrum.

"In this example the formation of visible rings at unusual thicknesses is easily understood; but it gives no explanation of the increased numbers observed by Newton. The width of the bands for any colour is proportional to λ, as well after the displacement by the prism as before. The manner of overlapping of two systems whose nth bars have been brought to coincidence is unaltered; so that the succession of colours in white light, and the number of perceptible* bands, is much as usual.

"In order that there may be an achromatic *system* of bands, it is necessary that the width of the bands near the centre be the same for the various colours. As we have seen, this condition cannot be satisfied when the plate

---

* Strictly speaking the number of visible bands is doubled, inasmuch as they are now formed on *both sides* of the achromatic band.

is a true wedge; for then the width for each colour is proportional to λ. If, however, the surfaces bounding the plate be *curved*, the width for each colour varies at different parts of the plate, and it is possible that the blue bands from one part, when seen through the prism, may fit the red bands from another part of the plate. Of course, when no prism is used, the sequence of colours is the same whether the boundaries of the plate be straight or curved."

In the paper* from which the above extracts are taken, the question was further discussed, and it appeared that the bands formed by cylindrical or spherical surfaces could be made achromatic, so far as small variations of λ are concerned, but only under the condition that there be a finite separation of the surfaces at the place of nearest approach. If a denote the smallest distance, the region of the nth band may form an achromatic system if

$$a = \tfrac{1}{4} n \lambda. \quad\dots\dots\dots\dots\dots\dots\dots\dots\dots\dots\dots\dots\dots(1)$$

At the time pressure of other work prevented my examining the question experimentally. Recently I have returned to it and I propose now to record some observations and also to put the theory into a slightly different form more convenient for comparison with observation.

For the present purpose it suffices to treat the surfaces as *cylindrical*, so that the thickness is a function of but one coordinate x, measured along the surfaces in the direction of the refraction. The investigation applies also to spherical surfaces if we limit ourselves to points lying upon that diameter of the circular rings which is parallel to the refraction†. If we choose the point of nearest approach as the origin of x, the thickness may be taken to be

$$t = a + bx^2, \quad\dots\dots\dots\dots\dots\dots\dots\dots\dots\dots\dots(2)$$

where b depends upon the curvatures. The black of the nth order for wavelength λ occurs when

$$\tfrac{1}{2} n \lambda = a + bx^2, \quad\dots\dots\dots\dots\dots\dots\dots\dots\dots(3)$$

or

$$x = \surd\{(\tfrac{1}{2} n \lambda - a)/b\}, \quad\dots\dots\dots\dots\dots\dots\dots(4)$$

so that

$$\frac{dx}{d\lambda} = \frac{\tfrac{1}{4} n}{\surd b \cdot \surd(\tfrac{1}{2} n \lambda - a)}. \quad\dots\dots\dots\dots\dots\dots(5)$$

The nth band, formed actually at x, is seen displaced under the action of the prism. The amount of the linear displacement (ξ) is proportional to the distance D at which the prism is held, so that we may take approximately

$$\frac{d\xi}{d\lambda} = -\beta \cdot D, \quad\dots\dots\dots\dots\dots\dots\dots\dots\dots(6)$$

---

* "On Achromatic Interference Bands," *Phil. Mag.* Vol. xxvii. pp. 77, 189 (1889); *Scientific Papers*, Vol. iii. p. 313.

† In the paper referred to the general theory of curved achromatic bands is considered at length.

β representing the dispersive power of the prism, or grating. The condition that the nth band may be achromatic (for small variations of λ) is accordingly

$$\frac{d(x+\xi)}{d\lambda} = 0, \dots\dots\dots\dots\dots\dots\dots\dots\dots(7)$$

or

$$\frac{1}{16}\frac{n^2}{\beta^2 D^2 b} = \tfrac{1}{2}n\lambda - a, \dots\dots\dots\dots\dots(8)$$

a quadratic in n. The roots of the quadratic are real, if

$$\beta^2 D^2 b > a/\lambda^2. \dots\dots\dots\dots\dots\dots\dots(9)$$

If a be zero, the condition (9) is satisfied for all values of D, so that at whatever distance the prism be held there is always an achromatic band. And if a be finite, the condition can still always be satisfied if the prism be drawn back far enough.

From (8) if n_1, n_2 be the roots,

$$\frac{1}{n_1} + \frac{1}{n_2} = \frac{\lambda}{2a}. \dots\dots\dots\dots\dots\dots(10)$$

Again, if $a = 0$, that is if the plates be in contact, $n_1 = 0$, and

$$n_2 = 8\lambda\beta^2 D^2 b. \dots\dots\dots\dots\dots\dots(11)$$

The order of the achromatic band increases with the dispersive power of the prism and with the distance at which it is held. The corresponding value of x from (4) is

$$x = 2\lambda\beta D. \dots\dots\dots\dots\dots\dots\dots(12)$$

If a be finite, there is no achromatic band so long as D is less than the value given in (9). When D acquires this value, the roots of the quadratic are equal, and

$$\frac{1}{n_1} = \frac{1}{n_2} = \frac{\lambda}{4a},$$

or

$$n_1 = n_2 = 4a/\lambda. \dots\dots\dots\dots\dots\dots(13)$$

This is the condition formerly found for an achromatic *system* of bands. If D be appreciably greater than this, two values of n satisfy the condition, viz. there are two separated achromatic bands, though no achromatic *system*. From (8)

$$n_1 n_2 = 16ab\beta^2 D^2. \dots\dots\dots\dots\dots(14)$$

Thus if D be great, one of the roots, say n_2, becomes great, while the other, see (10), approximates to $2a/\lambda$, that is to half the value appropriate to the achromatic system (13).

There is no particular difficulty in following these phenomena experimentally, though perhaps they are not quite so sharply defined as might be expected from the theoretical discussion, probably for a reason which will be alluded to presently. It is desirable to work with rather large and but very slightly curved surfaces. In my experiments the lower plate was an optical "flat" by Dr Common, about six inches in diameter and blackened behind. The upper plate was wedge-shaped with surfaces which had been intended to be flat but were in fact markedly convex. In order to see the bands well, it is necessary that the luminous background, whether from daylight or lamp-light, be uniform through a certain angle, and yet this angle must not be too large. Otherwise it is impossible to eliminate the light reflected from the upper surface of the upper plate, which to a great extent spoils the effects. In my case it sufficed to use gas-light diffused through a ground-glass plate whose angular area was not so great but that the false light could be thrown to one side in virtue of the angle between the upper and lower surfaces of the wedge*. It will be understood that these precautions are needed only in order to see the effects at their best. The most ordinary observation and appliances suffice to exhibit the main features.

Another question which I was desirous of taking the opportunity to examine was one often propounded to me by my lamented friend Lord Kelvin, viz. the nature of the obstruction usually encountered in trying to bring two surfaces nearly enough together to exhibit the rings of low order. In favour of the view that the obstacle is merely dust and fibres, I remember instancing the ease with which a photographic print, *enamelled* by being allowed to dry in contact with a suitably prepared glass plate, could be brought back into optical contact after partial separation therefrom. My recent observations with the glass plates point entirely in the same direction. However carefully the surfaces are cleaned by washing and wiping—finally with a dry hand, the rings of low order can usually be attained only at certain parts of the surface†. If we attempt to shift them to another place chosen at random, they usually pass into rings of higher order or disappear altogether. On the other hand, when rings of low order have once been seen at a particular place, it is usually possible to lift the upper glass carefully and to replace it without losing the rings at the place in question. I have repeatedly lifted the glass when the centre of the system was showing the white of the first order or even the darkening (I do not say black) corresponding to a still closer approximation, and found the colour recovered under no greater force than the weight of the glass. Some *time* is required, doubtless in order that the air may escape, for the complete recovery of the original closeness; but in the absence of foreign matter it appears that there is no other obstacle to an approximation of say $\frac{1}{8}\lambda$.

In making the observations it is convenient to introduce a not too small magnifying lens of perhaps 8 inches focus and to throw an image of the

---

* Compare "Interference Bands and their Applications," *Scientific Papers*, Vol. IV. p. 54.

† The plates are here supposed to be brought together without sliding. By a careful sliding together of two surfaces, the foreign matter may be extruded, as in Hilger's echelon gratings, where optical contact is attained over considerable areas.

source of light upon the pupil of the eye. With the glasses in contact it is easy to trace the rise in the order of the achromatic band as the eye and prism are drawn back. As regards the latter a direct-vision instrument of moderate power (three prisms in all) is the most suitable. An interval between the glasses may be introduced by stages. When the approximation is such as to show colours of the 3rd or 4th orders at the centre, it becomes apparent that the best achromatic effects are attained when the prism is at a certain distance, and that when this distance is exceeded the more achromatic places are separated by a region where the bands are fringed with colour. This feature becomes more distinct as the interval is still further increased, so that without the prism only faint rings or none at all can be perceived. For the greater intervals the interposition of a piece of mica at one edge is convenient. In judging of the degree of achromatism, I found that narrow coloured borders could be recognized as such much more easily by one of my eyes than by the other, and the difference did not seem to depend on any matter of focusing.

In observing bands of rather high order, the question obtruded itself as to whether the achromatism was *anywhere* complete. It will have been remarked that the theoretical discussion, as hitherto given, relates only to a small range of wave-length and that no account is taken of what in the telescope is called *secondary* colour. So long as this limitation is observed, the character of the dispersive instrument does not come into play. It appeared, however, not at all unlikely that even with gas-light the range of wave-length included might be too great to allow of this treatment being adequate; and with daylight, of course, the case would be aggravated. It is thus of interest to examine what law of dispersion is best adapted to secure compensation and in particular to compare the operation of a prism and a grating.

As to the law of dispersion to be aimed at, we have from (4), if $\lambda = \lambda_0 + \delta\lambda$,

$$x = \left\{ \frac{\frac{1}{2}n\lambda_0 - a}{b} \right\}^{\frac{1}{2}} \left\{ 1 + \frac{\frac{1}{2}n\delta\lambda}{\frac{1}{2}n\lambda_0 - a} - \frac{1}{8} \left(\frac{\frac{1}{2}n\delta\lambda}{\frac{1}{2}n\lambda_0 - a} \right)^2 + \dots \right\}. \quad \dots\dots(15)$$

If ξ be the displacement due to the instrument, ξ should be a similar function of $\delta\lambda$. In this matter the constant terms (independent of $\delta\lambda$) are of no account, and the terms in $\delta\lambda$ may be adjusted to one another, as already explained, by suitably choosing the distance D. In pursuing the approximation, what we are concerned with is the ratio of the term in $(\delta\lambda)^2$ to that in $\delta\lambda$. And in (15) this ratio is

$$-\frac{1}{8} \frac{n\delta\lambda}{\frac{1}{2}n\lambda_0 - a}; \quad \dots\dots\dots\dots\dots\dots(16)$$

thus in the particular cases

$$a = 0, \qquad -\frac{1}{4}\frac{\delta\lambda}{\lambda_0}; \qquad \dots\dots\dots\dots\dots(17)$$

$$a = \tfrac{1}{4}n\lambda_0, \qquad -\frac{1}{2}\frac{\delta\lambda}{\lambda_0}. \qquad \dots\dots\dots\dots\dots(18)$$

Corresponding expressions are required for the dispersive instruments. In any particular case they could of course be determined; but no very simple rules are available in general. If the intrinsic dispersion be small—the necessary effect being arrived at by increasing D, we may make the comparison more easily. Thus in the case of the grating the variable part of ξ is proportional to $\delta\lambda$ simply, so that the ratio of the second and third terms, corresponding to (16), is zero. And in the case of the prism if we assume Cauchy's law of dispersion, viz. $\mu = A + B\lambda^{-2}$, we get in correspondence with (16)

$$-\frac{3}{2}\frac{\delta\lambda}{\lambda_0}. \qquad \dots\dots\dots\dots\dots\dots\dots(19)$$

So far as these expressions apply, it appears that the dispersion required is between that of a grating and of a prism, and that especially when $a = 0$ the grating gives the better approximation. It would be possible to combine a grating and a prism in such a way as to secure an intermediate law, the dispersions cooperating although the deviations (in the case of a simple prism) would be in opposite directions.

I have made observations with a grating, using for the purpose a photographic reproduction upon bitumen*. This contains lines at the rate of 6000 to the inch and gives very brilliant spectra of the first order. I thought that I could observe the superior achromatism of the most nearly achromatic bands as compared with those given by the prism, but the conditions were not very favourable. The dispersive power was so high that the grating had to be held very close, and the multiplicity of spectra was an embarrassment. If it were possible to prepare a grating with not more than 3000 lines to the inch, and yet of such a character that most of the light was thrown into one of the spectra of the first order, it might be worth while to resume the experiment and, as suggested, to try for a more complete achromatism by combining with the grating a suitable prism.

* *Nature*, Vol. LIV. p. 332 (1896); *Scientific Papers*, Vol. IV. p. 226.

327.

FURTHER MEASUREMENTS OF WAVE-LENGTHS, AND MISCELLANEOUS NOTES ON FABRY AND PEROT'S APPARATUS.

[*Philosophical Magazine*, Vol. xv. pp. 548—558, 1908.]

IN a former paper* I described a modified form of apparatus and gave the results of some measurements of wave-lengths, partly in confirmation of numbers already put forward by Fabry and Perot and partly novel, relating to helium. I propose now to record briefly some further measures by the same method, together with certain observations and calculations relating thereto of general optical interest.

The apparatus was arranged as before, the only change being in the interference-gauge itself. The distance-pieces, by which the glasses are kept apart, were now of *invar*, with the object of diminishing the dependence upon temperature. The use of invar for this purpose was suggested by Fabry and Perot, but I do not know whether it has actually been employed before. The alloy was in the form of nearly spherical balls, 5 mm. in diameter, provided with projecting tongues by which they were firmly fitted to the iron frame. The springs, holding the glasses up to the distance-pieces, were of the usual pattern. The whole mounting was constructed by Mr Enock, and it answered its purpose satisfactorily. There is no doubt, I think, as to the advantage accruing from the use of invar.

The measurements were conducted as explained in the earlier paper. The first set related to *zinc* which was compared with cadmium. Both metals were used in vacuum-tubes, of the pattern already described, with electrodes merely cemented in. It was rather to my surprise that I found ordinary soft glass available in the case of zinc, but no difficulty was experienced. The former observations with the "trembler" suggested a wave-length for zinc red about one-millionth part greater than that (6362·345) given by Fabry and Perot. This correction has been confirmed, and I would

---

* *Phil. Mag.* [6] Vol. XI. p. 685 (May 1906). [This Collection, Vol. v. p. 313.]

propose 6362·350, as referred to Michelson's value of the cadmium red, viz. 6438·4722. No difficulty was experienced in identifying the order of the rings by the method formerly described and dependent upon observations with the gauge alone.

The results of the measurements upon helium were not in quite such close accord with the earlier ones as had been expected. Both sets are given below for comparison.

Wave-lengths of Helium.

II.	III.
7065·192	7065·200
6678·147	6678·150
5875·618	5875·625
5015·682	5015·680
4921·927	4921·930
(4713·173)	4713·144
(4471·480)	4471·482

The two last entries under II., enclosed in parentheses, were obtained with the 1 mm. apparatus, and could not be expected to be very accurate. Preference may be given to III. throughout.

These measurements of wave-lengths were not further pursued, partly because it was understood that other observers were in the field and partly because my own vision, though not bad, is less good than it was. In particular at the blue end of the spectrum I found difficulty. It is evident that work of this sort should be undertaken only under the best conditions.

One of the less agreeable features of the method is the complication which arises from the optical distance between the surfaces being slightly variable with the colour. In the earlier observations with a 5 mm. apparatus I was surprised to find the change amounting to $2\frac{1}{2}$ parts per million between cadmium red and cadmium green. In the light of subsequent experience I am disposed to think that the silver surfaces must have been slightly tarnished. At any rate in the later measurements I found the difference much less, indeed scarcely measurable. It will be understood that no final uncertainty in the ratio of wave-lengths arises from this cause. Whatever the change may prove to be, it can be allowed for.

Thirty Millimetre Apparatus.

In this instrument the object was to construct a gauge with a much greater distance than usual between the plates, but otherwise on the same general plan as that of Fabry and Perot. The distance-piece $A A$, fig. 1, consisted of a 30 millimetre length of glass tubing, each end being provided with three protuberances, equally spaced round the circumference, at which the actual contacts took place. The removal of the intervening material and the shaping of the protuberances were effected with a file moistened with turpentine.

Fig. 1.

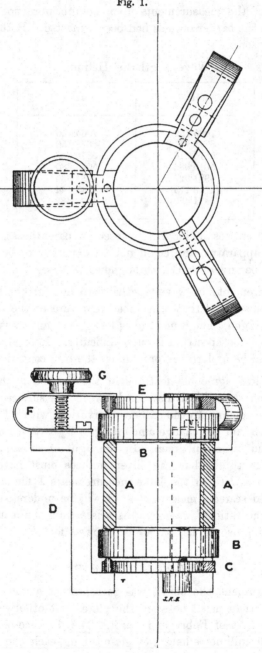

Against this distance-piece the glass plates BB are held by the arrangement shown in fig. 1. The lower plate B rests upon a brass ring C to which the brass castings D are rigidly attached. The upper ring E is connected with the castings only through the steel springs F. Both rings are provided with protuberances in line with those on the glass cylinder, and the pressure is regulated by the screws G. The whole was constructed by Mr Enock. Some little care is required in putting the parts together to avoid scratching the half-silvered faces; but when once the apparatus is set up its manipulation is as easy as that of the ordinary type.

In all interference-gauges it is desirable that the distance-pieces be adjusted as accurately as possible. For although a considerable deficiency in this respect may be compensated by regulating the pressures (see below), the adjustment thus arrived at is less durable, at least in my experience. Even when the distance-pieces are themselves well adjusted, it is advisable to employ only moderate pressures.

Observations with the 30 mm. gauge have been made upon helium, thallium, cadmium, and mercury. In the first case the (yellow) rings are faint, the retardation being not far from the limit. Indeed when at first it was attempted to adjust the plates with helium, the rings could not be found. With thallium also the rings were rather faint, but with mercury and cadmium there was no difficulty.

Magnifying Power.

At a distance of 30 mm. the rings are rather small, and one is tempted to increase the magnifying power of the observing telescope. As to this there should be no difficulty if the aperture could be correspondingly increased. But although the plates themselves may be large enough, an excessive strain may thus be thrown upon the accuracy of the figuring and upon the adjustment to parallelism. If, on the other hand, the aperture be not increased, the illumination of the image fails and the extra magnifying may do more harm than good.

A means of escape from this dilemma is to effect the additional magnification in one direction only, which in the present case answers all purposes. When straight interference-bands, or spectrum lines, are under observation, there is no objection to astigmatism, and we may merely replace the ordinary eyepiece of the telescope by a cylindrical lens or by a combination of spherical and cylindrical lenses. This arrangement can be employed in the present instance, but the result is not satisfactory. A complete focusing, leading to a point-to-point correspondence between image and object, may however be attained by suitably sloping the object-lens of the telescope. In this way excellent observations upon interference-rings are possible under a magnifying power which otherwise would be inadmissible, as entailing too great a loss of light. The subject will be more fully treated in a special paper*.

Adjustment for Parallelism.

If the surfaces are flat, and well-adjusted, Haidinger's rings depend entirely upon obliquity. A slight departure from parallelism shows itself by an expansion or contraction of the rings as the eye is moved about so as to bring different parts of the surfaces into play. In making this observation the eye must be adjusted to infinity, if necessary with the aid of spectacle-glasses, and it may be held close to the plates; but a telescope is not needed or even desirable. If the departure from parallelism be considerable, no rings at all are visible; but there is an intermediate state of things where circular arcs may be seen by an eye drawn back somewhat and focused upon the plates.

The character of these bands is intermediate between those of Newton's and Haidinger's rings, the retardation depending *both* upon the varying direction in which the light passes the plates and reaches the eye and also upon the varying local thickness. If we take, as origin of rectangular coordinates in the plane of the plates, the place corresponding to normal passage of the light, the retardation due to obliquity is as $-(x^2 + y^2)$. The retardation due to local thickness is represented by a linear function of x and y, so that the variable part of it may be considered to be proportional to x. Hence the equation of the bands is

$$\alpha x - x^2 - y^2 = \text{constant},$$

α being positive if x is considered positive in the direction of increasing thickness. Accordingly the bands are in the form of concentric circles and the coordinates of the centre are

$$x = \tfrac{1}{2}\alpha, \qquad y = 0.$$

When curved arcs are seen by an eye looking at the plates perpendicularly, the greatest thickness lies upon the *concave* side of the arcs. The perpendicular direction of vision may be tested by observing the reflexion of the eye itself in the silvered surface.

Behaviour of Vacuum-Tubes.

The form of vacuum-tube described in the first paper, and depending on sealing-wax for air-tightness, continues to give satisfaction. As already mentioned, though made of soft glass, they are available for zinc, and the cadmium tubes have lasted well with occasional re-exhaustion. It is advisable to submit them to this operation when the *red* light begins to fall

* [See *Proc. Roy. Soc.* Vol. LXXXI. p. 26 (1908); This Collection, Vol. V. Art. 328, Part II.]

off. After one or two re-exhaustions the condition seems to be more durable.

With thallium my experience has been rather remarkable. The green light is very brilliant and offers a further advantage as being comparatively free from admixture with other colours*. But Fabry and Perot found thallium tubes to be very short-lived, sometimes lasting only a few minutes. I have used but one thallium tube, of the same construction as the others, and charged with a little thallium chloride. This tube has been used without special care on many occasions—I cannot say how many, but probably seven or eight times,—and it does not appear to have deteriorated at all. It looks as though the chloride had decomposed and metal had deposited upon the aluminium electrodes. But what the circumstances can be that render my experience so much more favourable I am at a loss to conjecture.

The same form of tube answers well for mercury, but with this metal there is usually no difficulty.

Control of the figure of the glasses by bending.

Very good plates can now be procured from the best makers, but on careful testing they usually show some deficiency, mostly of the nature of a slight general curvature. Thus when in Fabry and Perot's apparatus the adjustment for parallelism is made as perfect as possible, the rings may be observed to dilate a little as the eye moves outwards in any direction from the centre towards the circumference of the plates. This indicates a general convexity.

It occurred to me that an error of this kind might be approximately corrected by the application of bending forces to one of the plates—it does not matter which. The easiest way to carry out the idea is to modify the apparatus in such a way that the points of application of external pressure are not exactly opposite the contacts with the distance-pieces, but are displaced somewhat inwards or outwards in the radial direction (fig. 2). If the plates are too convex, the points of pressure must be displaced outwards. In this form I have tried the experiment with a certain degree of success, but the displacements that I could command (1 mm. only) were too small in relation to the thickness of the plate. If it were intended to give this plan a proper trial, which I think it would be worth in order to render a larger aperture than usual available, the plates, or at least one of them, should be prepared of extra diameter, so that the bending forces could act with a longer leverage and at

Fig. 2.

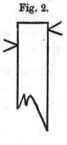

a greater distance from the parts to be employed optically. Such a construction need not involve a much enhanced cost, inasmuch as the outer parts would not need to be optically accurate.

It may be worth while to consider the question here raised more generally. The problem is so to deform one surface, by forces and couples applied at the boundary, as to compensate the joint errors of the two surfaces and render the distance between them constant. If we take rectangular coordinates x, y in the plane of the surface with origin at the centre, the deformation obtainable in this way is expressed by terms in the value of ζ (the other coordinate of points on the surface) proportional to x, y, x^2, xy, y^2, x^3, x^2y, xy^2, y^3. For such terms are arbitrary in the solution of the general equation of equilibrium of a plate, viz.

$$\left(\frac{d^2}{dx^2} + \frac{d^2}{dy^2}\right)^2 \zeta = 0.$$

Of these terms those in x and y correspond of course merely to the adjustment for parallelism, and those of the second degree to curvature at the centre. The conclusion is that we may always, by suitable forces applied to the edge, render the distance between the plates constant, so far as terms of third order inclusive.

Another inference from the same argument is that, in any optical apparatus, approximately plane waves of light may be freed from curvature and from unsymmetrical aberration (expressed by terms of the third order) by means of reflexion at a plate to the boundary of which suitable forces are applied. And the surface of the plate need not itself be more than approximately flat.

Figuring by Hydrofluoric Acid.

It would be poor economy to employ any but the best surfaces in measuring work needing high accuracy; but there are occasions when all that is needed can be attained by more ordinary means. Common plate-glass is rarely good enough*; but I have found that it can be re-figured with hydrofluoric acid so as to serve fairly well, and the process is one of some interest. From what has been said already it will be understood that it is not necessary to make both surfaces plane, but merely to fit them together, which can be effected by operations conducted upon one only.

Pieces of selected plate-glass, about ¼ inch thick and of a size suited to the interference-gauges, were roughly shaped by chipping. The best surfaces were superposed and the character of the *fit* examined by soda-light. One

---

* If the surfaces are so shaped that the interference-bands presented on superposition are hyperbolic, much may be gained by limiting the aperture to a narrow slit corresponding to one of the asymptotes, especially if the magnification used is in one direction only.

glass being rotated upon the other, the most favourable relative azimuth was chosen; and by means of suitable marks upon the edges the plates were always brought back to the chosen position.

The principles upon which the testing is conducted have been fully explained in a former paper*. In the present case the surfaces are so close to one another that no special precautions are required. With a little management the contact is so arranged that a moderate number of bands are visible. If the fit were perfect, or rather if the surfaces were *capable* of being brought into contact throughout, these bands would be straight, parallel and equi-distant. Any departure from this condition is an error which it is proposed to correct. The sign of the error can be determined without moving the glasses by observing the effect of *diminishing* the obliquity of reflexion, which *increases* the retardation. Thus if a band is curved, and the change in question causes the band to move with convexity forwards, it is a sign that material needs to be removed from the parts of the glass occupied by the ends of the band. Such an operation will tend to straighten the band. If, however, the movement take place with concavity forward, then material needs to be removed from the middle parts. In every case the rule is that by removal of glass the bands, or any parts of them, can be caused to move in the same direction as that in which they move when the obliquity of reflexion is diminished.

In carrying out the correction, the plate on which it is intended to operate is placed below, and it is convenient if it be held in some form of steady mounting so that the upper plate can be removed and replaced in the required position without trouble. The acid, two or three times diluted, is applied with a camel's hair brush and after being worked about for a few seconds is removed suddenly with a soft cloth. Endeavour should be made to keep the margin of the wetted region moving in order to obviate the formation of hard lines. Success depends of course upon judgment and practice, and the only general advice that can be given is to make a great many bites at the cherry, and to keep a record of what is done each time by marking suitably on one of a system of circles drawn upon paper and representing the surface operated on. After each application of acid the plates are re-examined by soda-light and the effect estimated. The difficulty is that in most cases the bands are not reproduced in the same form. In one presentation the error may reveal itself as a curvature of the bands and in another as an inequality in the spacing of bands fairly straight. Often by a little humouring the original form may be approximately recovered, and in any case the general rule indicates what needs to be done.

---

* "Interference Bands and their Applications," *Nature*, Vol. XLVIII. p. 212 (1893); *Scientific Papers*, Vol. IV. p. 54.

By this method I have prepared two pairs of plates which perform very fairly well, but of course only when placed in the proper relative position. The operations, though prolonged, are not tedious, and I doubt not that with perseverance better results than mine might be achieved. The surface of the glass under treatment suffers a little from the development of previously invisible scratches in the manner formerly explained, but the defect hardly shows itself in actual use. I have not ventured to apply the method to surfaces already very good such as those supplied by the best makers for use in Fabry and Perot's apparatus; but I should be tempted to do so if I came across a pair suffering from slight general *concavity*. The application of acid would then be at the outer parts. In the best glasses that I possess the error is one of convexity.

Effect of Pressure in Fabry and Perot's Apparatus.

The observation that the rings were more sensitive than had been expected to the pressure by which the plates are kept up to the distance-pieces, led to a calculation on Hertz's theory of the relation between the change of interval and the pressure. If two spheres of radii r_1 and r_2 and of material for which the elastic constants in Lamé's notation are λ_1, μ_1, λ_2, μ_2, are pressed together with a force P, the relation between P and the distance (α) through which the centres approach one another, as the result of the deformation in the neighbourhood of the contact, is

$$P = \frac{4}{3\pi} \left(\frac{r_1 r_2}{r_1 + r_2} \right)^{\frac{1}{2}} \frac{\alpha^{\frac{3}{2}}}{\theta_1 + \theta_2} *,$$

where

$$\theta_1 = \frac{\lambda_1 + 2\mu_1}{4\pi\mu_1 (\lambda_1 + \mu_1)}, \qquad \theta_2 = \frac{\lambda_2 + 2\mu_2}{4\pi\mu_2 (\lambda_2 + \mu_2)}.$$

In the case of materials which satisfy Poisson's condition, $\lambda = \mu$, and we may take as sufficiently approximate

$$\theta_1 = \frac{3}{8\pi\mu_1}, \qquad \theta_2 = \frac{3}{8\pi\mu_2};$$

so that

$$P = \frac{32}{9} \left(\frac{r_1 r_2}{r_1 + r_2} \right)^{\frac{1}{2}} \frac{\mu_1 \mu_2 \alpha^{\frac{3}{2}}}{\mu_1 + \mu_2}.$$

In the application that we have to make, one of the spheres is of steel (invar) and of radius $r_1 = \cdot 25$ cm., while the other is of glass and of radius $r_2 = \infty$. Further, for the steel we may take $\mu_1 = 8 \cdot 2 \times 10^{11}$, and for glass $\mu_2 = 2 \cdot 4 \times 10^{11}$, and thus

$$P = 3 \cdot 30 \times 10^{11} . \alpha^{\frac{3}{2}},$$

α being in cm. and P in dynes. It will be convenient for our purpose to

* See Love's *Math. Theory of Elasticity*, § 139.

reckon α in wave-lengths (equal say to 6×10^{-5} cm.) and P in kilograms, taking the dyne as equal to a milligram weight. On this understanding

$$P = \cdot 15\,\alpha^{\frac{3}{2}},$$

signifying that to cause an approach of one wave-length the force required is ·15 kilogram. If P and α undergo small variations,

$$dP = \tfrac{3}{2}(\cdot 15)\,\alpha^{\frac{1}{2}}\,d\alpha = \tfrac{3}{2}(\cdot 15)^{\frac{2}{3}}\,P^{\frac{1}{3}}\,d\alpha,$$

$dP/d\alpha$ being somewhat dependent upon the total pressure P.

For the purposes of experiment a spring-balance was mounted upon the frame of the apparatus (carrying the distance-pieces) so as to diminish the pressure exerted over one of the distance-pieces, that is to diminish the pressure by which *one* of the plates was held up to *one* of the distance-pieces. Starting from perfect parallelism of the plates and keeping the eye carefully fixed so as to receive the light from the centre of the plates, it was observed that to cause a shift of one band (helium yellow) the spring-balance needed to exert a pull of ·78 kilo. At this stage the plates were of course no longer parallel and a moderate shift of the eye would cause a displacement independently of any change in the spring-balance. At the same time the rings lost their sharpness. On this account it is hardly practicable to use a shift of more than one band, and indeed a smaller shift of half a band was usually preferred. The total force required to compensate the spring of the apparatus, and so to relieve the compression of the distance-piece on this side, was 2·4 kilos. This is what is represented by P in the above formula, while $dP = -\cdot 78$.

In order now to compare theory and observation we must remember that the one band (corresponding to half a wave-length) observed at the centre implies three times as great a shift at the particular distance-piece where alone the force was varied. Thus the observed dP corresponds to $d\alpha = -\tfrac{3}{2}$. For this $d\alpha$, the calculated dP is

$$dP = -\tfrac{9}{4}(\cdot 15)^{\frac{2}{3}}(2\cdot 4)^{\frac{1}{3}} = -\cdot 85 \text{ kilo.}$$

The agreement with the observed $-\cdot 78$ is certainly as good as could have been expected.

In considering what differences of distance are to be expected when the plates are adjusted to parallelism under different pressures, we must remember that the above calculation and observation relates to the compression which may occur at the contact of a single distance-piece with a single plate. There are in all six contacts of this kind, and we may conclude that when no special pains are taken to regulate the absolute pressures employed, a shift of six bands or more on remounting need not cause surprise.

328.

ON THE ABERRATION OF SLOPED LENSES AND ON THEIR ADAPTATION TO TELESCOPES OF UNEQUAL MAGNIFYING POWER IN PERPENDICULAR DIRECTIONS.

[*Proceedings of the Royal Society*, A, Vol. LXXXI. pp. 26—40, 1908.]

THE present paper consists of two parts to a large extent independent. The reader who does not care to follow the details of calculation may take the results relative to unsymmetrical aberration for granted. The subject of the second part is somewhat larger than the title. It treats of the advantage which often attends a magnification unequal in different directions and of the methods available for attaining it. Among these is the method of the sloped object-lens. Such sloping introduces in general unsymmetrical aberration. The intention of the first part is largely to show how this may be minimised so as to become unimportant.

PART I.

Before proceeding to actual calculations of the aberrations of a sloped lens, it may be well to consider briefly the general character of a pencil of rays affected with unsymmetrical aberration.

The axis of the pencil being taken as axis of z, let the equation of the wave-surface, to which all rays are normal, be

$$z = \frac{x^2}{2\rho} + \frac{y^2}{2\rho'} + \alpha x^3 + \beta x^2 y + \gamma x y^2 + \delta y^3 + \dots \dots \dots \dots (1)$$

The principal focal lengths, measured from $z = 0$, are ρ and ρ. In the case of symmetry about the axis, ρ and ρ' are equal, and the coefficients of the terms of the third order vanish. The aberration then depends upon terms of the fourth order in x and y, and even these are made to vanish in the formulæ for the object-glasses of telescopes by the selection of suitable curvatures. In the theory of imperfectly constructed spectroscopes and of sloped lenses it is necessary to retain the terms of the third order, but we may assume a plane of symmetry $y = 0$, which is then spoken of as the *primary* plane. The equation of the wave-surface thus reduces to

$$z = \frac{x^2}{2\rho} + \frac{y^2}{2\rho'} + \alpha x^3 + \gamma x y^2, \quad \dots\dots\dots\dots\dots(2)$$

terms of higher order being omitted. In (2) ρ is the primary and ρ' the secondary focal length.

The equation of the normal at the point x, y, z is

$$z - \zeta = \frac{\xi - x}{x/\rho + 3\alpha x^2 + \gamma y^2} = \frac{\eta - y}{y/\rho' + 2\gamma x y}, \quad \dots\dots\dots\dots(5)$$

and its intersection with the plane $\zeta = \rho$ occurs at the point determined approximately by

$$\xi = -\rho\,(3\alpha x^2 + \gamma y^2), \qquad \eta = \frac{\rho' - \rho}{\rho'}\,y - 2\rho\gamma x y, \dots\dots\dots(6)$$

terms of the third order being omitted.

According to geometrical optics, the thickness of the image of a luminous line (parallel to y) at the primary focus is determined by the extreme value of ξ, and for good definition it is necessary to reduce this thickness as much as possible. To this end it is necessary in general that both α and γ be small.

We will now examine more closely the character of the image at the primary focus in the case of a pencil originally of circular section. Unless $\rho' = \rho$, the second term in the value of η in (6) may be neglected. The rays proceeding from the circle $x^2 + y^2 = r^2$ intersect the plane $\zeta = \rho$ in the parabola

$$\frac{\rho\rho'^2\,(3\alpha - \gamma)}{(\rho' - \rho)^2}\,\eta^2 - \xi = 3\alpha\rho r^2; \quad \dots\dots\dots\dots\dots(7)$$

and the various parabolas corresponding to different values of r differ from one another only in being shifted along the axis of ξ. To find out how much of the parabolic arcs is included, we observe that for any given value of r the value of η is greatest when $x = 0$. Hence the rays starting in the secondary plane give the remainder of the boundary of the image. Its equation, formed from (6) after putting $x = 0$, is

$$\eta^2 = -\frac{(\rho' - \rho)^2}{\rho\rho'^2\gamma}\,\xi, \quad \dots\dots\dots\dots\dots\dots(8)$$

and represents a parabola touching the axis of η at the origin. The whole of the image is included between this parabola and the parabola of form (7) corresponding to the maximum value of r.

The width of the image when $\eta = 0$ is $3\alpha\rho r^2$, and vanishes when $\alpha = 0$, *i.e.*, when there is no aberration for rays in the primary plane. In this case the two parabolic boundaries coincide, and the image is reduced to a linear arc. If, further, $\gamma = 0$, this arc becomes *straight*, and then the image of a short luminous line (parallel to y) is perfect to this order of approximation at the primary focus. In general, if $\gamma = 0$, the parabola (8) reduces to the straight line $\xi = 0$; that is to say, the rays which start in the secondary plane remain in that plane.

We will now consider the image formed at the secondary focus. Putting $\zeta = \rho'$ in (5), we obtain

$$\xi = \frac{\rho - \rho'}{\rho}\, x, \qquad \eta = - 2\gamma\rho'xy. \quad\quad\quad\quad\quad (9)$$

If $\gamma = 0$, the secondary focal line is formed without aberration, but not otherwise. In general, the curve traced out by the rays for which $x^2 + y^2 = r^2$, is

$$\left(\frac{\rho}{\rho - \rho'}\right)^2 \xi^2 + \frac{(\rho - \rho')^2\, \eta}{4\gamma^2\rho^2\rho'^2 \xi^2} = r^2 \quad\quad\quad\quad\quad (10)$$

in the form of a figure of 8 symmetrical with respect to both axes. The rays starting either in the primary or in the secondary plane pass through the axis of ξ, the thickness of the image being due to the rays for which $x = y = r/\sqrt{2}$.[*]

Or if in order to find the intersection of the ray with the primary plane we put $\eta = 0$ in (5), we have approximately

$$\xi = \frac{(\rho - \rho')\,x}{\rho}, \qquad \zeta = \frac{1}{1/\rho' + 2\gamma x},$$

showing that ζ is constant only when $\gamma = 0$.

The calculation of aberration for rays in the primary plane is carried out in the paper cited for the case of a thin lens sloped through a finite angle. If the curvature of the first surface be $1/r$ and of the second $1/s$, and if μ be the refractive index, the focal length f_1 in the primary plane is given by

$$\frac{1}{f_1} = \frac{\mu c' - c}{c^2}\left(\frac{1}{r} - \frac{1}{s}\right), \quad\quad\quad\quad\quad (11)$$

and the condition that there shall be no aberration is

$$\frac{(2\mu' + 1)\,c}{u} + \frac{\mu'^2}{s} + \frac{\mu' - \mu'^2 + 1}{r} = 0. \quad\quad\quad\quad\quad (12)$$

Here u is the distance of the radiant point from the lens, ϕ the obliquity of the incident ray, ϕ' of the refracted ray, $c = \cos\phi$, $c' = \cos\phi'$, and $\mu' = \mu \cos\phi'/\cos\phi$.

---

[*] The above is taken from my "Investigations in Optics," *Phil. Mag.* 1879; *Scientific Papers*, Vol. I. p. 441, and following. Some errata may be noted:—p. 441, line 9, insert y as factor in the first term of η; p. 443, line 9, for (7) read (8), line 10, for η read ξ.

A result, accordant with (12), but applicable only when ϕ is small, was given in another form by Mr Dennis Taylor in *Astron. Soc. Monthly Notices*, Ap., 1893.

If the incident rays be parallel, $u = \infty$, and the condition of freedom from aberration is

$$-\frac{r}{s} = \frac{1 + \mu' - \mu'^2}{\mu'^2}. \qquad\qquad (13)$$

As appears from (11), opposite signs for r and s indicate that both surfaces are convex.

If $\phi = 0$, $\mu' = \mu$, so that (13) gives, in this case,

$$-\frac{r}{s} = \frac{1 + \mu - \mu^2}{\mu^2} \ldots \qquad\qquad (14)$$

Thus, if $\mu = 1\cdot5$, the aberration vanishes for small obliquities when $s = -9r$. This means a double convex lens, the curvature of the hind surface being one-ninth of that of the front surface. If $s = \infty$, that is, if the lens be plano-convex with curvature turned towards the parallel rays,

$$1 + \mu - \mu^2 = 0, \qquad\qquad (15)$$

or $$\mu = \tfrac{1}{2}(1 + \sqrt5) = 1\cdot618.$$

Returning to finite obliquity, we see from (13) that whatever may be the index and obliquity of the lens, it is possible so to choose its form that the aberration shall vanish. If the form be plano-convex, the condition of no aberration is

$$1 + \mu' - \mu'^2 = 0, \qquad\qquad (16)$$

or $$\mu' = \mu \cos \phi'/\cos \phi = 1\cdot618.$$

Here $\cos \phi' > \cos \phi$, and the ratio of the two cosines increases with obliquity from unity to infinity. Hence if $\mu > 1\cdot618$, there can be no freedom from aberration at any angle. When $\mu = 1\cdot618$, the aberration vanishes, as we have seen, when $\phi = 0$. If μ be less than $1\cdot618$, the aberration vanishes at some finite angle. For example, if $\mu = 1\cdot5$, this occurs when $\phi = 29°$.

In many cases the aberration of rays in the secondary plane is quite as important as that in the primary plane. In my former paper I gave a result applicable to a plano-convex lens, on the curved face of which parallel light falls. It was found that the secondary aberration vanished when the relation between obliquity and refractive index was such that

$$\sin^2 \phi = \frac{3\mu^2 - \mu^4 - 1}{3 - \mu^2}. \qquad\qquad (17)$$

For small values of ϕ this gives the same index as before (15), inasmuch as

$$\mu^4 - 3\mu^2 + 1 = (\mu^2 - 1)^2 - \mu^2 = (\mu^2 - \mu - 1)(\mu^2 + \mu - 1).$$

I inferred that for a plano-convex lens of index 1·618 neither kind of aberration is important at moderate slopes.

Having no note or recollection of the method by which (17) was obtained, and wishing to confirm and extend it, I have lately undertaken a fresh investigation, still limiting myself, however, to *parallel* incident rays. For simplicity, the lens may be supposed to come to a sharp circular edge, the plane containing this edge being that of XY. The centre of the circle is the origin, and the axis of Z is the axis of the lens. The incident rays are parallel to the plane ZX, and make an angle Φ with OZ; so that Φ is the angle of incidence for the ray which meets the first surface of the lens at its central point. Everything is symmetrical with respect to the *primary* plane $y = 0$. It will suffice to consider the course of the rays which meet the lens close to its edge, of which the equation is $x^2 + y^2 = R^2$, if $2R$ be the diameter.

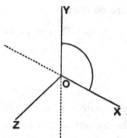

In order to carry out the calculation conveniently, we require general formulæ connecting the direction-cosines of the refracted ray with those of the incident ray and of the normal to the surface. If we take lengths AP, AQ along the incident and refracted rays proportional to μ, μ', the indices of the medium in which the rays travel, and drop perpendiculars PM, QN upon the normal MAN, then by the law of refraction the lines PM, NQ are equal and parallel; and the projection of $PA + AM$ on any axis is equal to the projection of $NA + AQ$ on the same axis. Thus if l, m, n are the direction-cosines of the incident ray, l', m', n' of the refracted ray, p, q, r of the normal taken in the direction from the medium in which the light is incident, ϕ, ϕ' the angles of incidence and refraction,

$$\mu l - \mu \cos \phi . p = - \mu' \cos \phi' . p + \mu' l'$$

and two similar equations. Hence

$$(\mu'l' - \mu l)/p = (\mu'm' - \mu m)/q = (\mu'n' - \mu n)/r = \mu' \cos \phi' - \mu \cos \phi. \quad ...(18)^*$$

Also
$$\cos \phi = lp + mq + nr, \quad ....(19)$$

and ϕ' is given by
$$\mu' \sin \phi' = \mu \sin \phi. \quad ....(20)$$

For our purpose there is no need to retain the two refractive indices, and for brevity we will suppose that the index outside the lens is unity and inside it equal to μ; so that in the above formulæ we are to write $\mu = 1$, $\mu' = \mu$. Hence

$$(\mu l' - l)/p = (\mu m' - m)/q = (\mu n' - n)/r = \mu \cos \phi' - \cos \phi. \quad ...(21)$$

Equation (19) remains as before, while (20) becomes

$$\mu \sin \phi' = \sin \phi. \quad ....(22)$$

* See Herman's *Geometrical Optics*, Cambridge, 1900, p. 22.

For the first refraction at the point x, y, we have

$$l = \sin \Phi, \quad m = 0, \quad n = \cos \Phi;$$

and if χ_1 be the angle which the normal to the first surface at the edge of the lens makes with the axis,

$$p = \sin \chi_1 . x/R, \quad q = \sin \chi_1 . y/R, \quad r = \cos \chi_1;$$

so that

$$\frac{\mu l' - \sin \Phi}{x/R . \sin \chi_1} = \frac{\mu m'}{y/R . \sin \chi_1} = \frac{\mu n' - \cos \Phi}{\cos \chi_1} = \mu \cos \phi' - \cos \phi = C_1, \quad (23)$$

and

$$\cos \phi = \sin \Phi \sin \chi_1 . x/R + \cos \Phi \cos \chi_1. \ldots \ldots \ldots (24)$$

In like manner if l'', m'', n'' be the direction-cosines of the twice refracted ray, p', q', r' those of the second normal, we may take

$$\frac{l'' - \mu l'}{x/R . \sin \chi_2} = \frac{m'' - \mu m'}{y/R . \sin \chi_2} = \frac{n'' - \mu n'}{\cos \chi_2} = \cos \psi' - \mu \cos \psi = C_2, \ldots (25)$$

if ψ, ψ' be respectively the angles of incidence and refraction at the second surface.

Here

$$\cos \psi = l'p' + m'q' + n'r'. \ldots \ldots \ldots \ldots (26)$$

Eliminating l', m', n' between (23) and (25), we get

$$l'' = \sin \Phi + (C_1 \sin \chi_1 + C_2 \sin \chi_2) x/R,$$

$$m'' = (C_1 \sin \chi_1 + C_2 \sin \chi_2) y/R,$$

$$n'' = \cos \Phi + C_1 \cos \chi_1 + C_2 \cos \chi_2.$$

The equation of the ray after passage through the lens is

$$\frac{\xi - x}{l''} = \frac{\eta - y}{m''} = \frac{\zeta}{n''}. \ldots \ldots \ldots \ldots (27)$$

The aberration in the secondary plane (depending on γ) is most simply investigated by inquiring where the ray (27) meets the primary plane $\eta = 0$. For the co-ordinates of the point of intersection,

$$\xi = x - \frac{l''y}{m''} = -\frac{R \sin \Phi}{C_1 \sin \chi_1 + C_2 \sin \chi_2}, \ldots \ldots \ldots \ldots (28)$$

$$\zeta = -\frac{n''y}{m''} = -R \frac{\cos \Phi + C_1 \cos \chi_1 + C_2 \cos \chi_2}{C_1 \sin \chi_1 + C_2 \sin \chi_2}. \ldots \ldots \ldots (29)$$

In interpreting (28), (29) we must remember that ζ is now measured parallel to the axis of the lens and not, as in the preliminary discussion, along the principal ray. Freedom from aberration requires that the line determined by varying x and y in (28), (29) should be perpendicular to the principal ray, or that $\zeta \cos \Phi + \xi \sin \Phi$ should be constant. And

$$-\frac{\zeta \cos \Phi + \xi \sin \Phi}{R} = \frac{1 + (C_1 \cos \chi_1 + C_2 \cos \chi_2) \cos \Phi}{C_1 \sin \chi_1 + C_2 \sin \chi_2}. \ldots \ldots (30)$$

Before proceeding further it may be well to compare (30) with known results when the aberration is neglected. For a first approximation we may identify ϕ and ϕ' with Φ and Φ', and also ψ and ψ' with Φ' and Φ respectively. Thus

$$C_1 = -C_2 = \mu \cos \Phi' - \cos \Phi. \quad\dots\dots\dots\dots\dots(31)$$

Again, if r, s be the radii of the surfaces, we have, neglecting χ^2,

$$\chi_1 - \chi_2 = R/r - R/s; \quad\dots\dots\dots\dots\dots(32)$$

and thence, from (30),

$$-\frac{1}{\zeta \cos \Phi + \xi \sin \Phi} = (\mu \cos \Phi' - \cos \Phi)\left(\frac{1}{r} - \frac{1}{s}\right), \quad\dots\dots\dots(33)$$

the usual formula for the secondary focal length. The reckoning is such that the signs of r and s are opposite in the case of a doubly convex lens. We have now to proceed to a second approximation and inquire under what conditions (30) is independent of the particular ray chosen. In the numerator it is sufficient to retain the first power of χ_1, χ_2, so that we may take $\cos \chi_1$, $\cos \chi_2$ equal to unity; but in the denominator, which is already a small quantity of the first order, we must retain the terms of the second order in χ_1, χ_2. It is not necessary, however, to distinguish between the sines of χ_1, χ_2 and the angles themselves. The first step is to determine corrections to the approximate values of C_1 and C_2 expressed in (31).

For $\cos \phi$ itself we have, from (24),

$$\cos \phi = \cos \Phi + \chi_1 x/R . \sin \Phi;$$

and again

$$\mu \cos \phi' = \sqrt{\{\mu^2 - 1 + \cos^2 \phi\}} = \mu \cos \Phi' + \frac{\sin \Phi \cos \Phi}{\mu \cos \Phi'} \frac{\chi_1 x}{R},$$

so that

$$C_1 = (\mu \cos \Phi' - \cos \Phi)\left\{1 - \frac{\chi_1 x}{R} \frac{\sin \Phi}{\mu \cos \Phi'}\right\}. \quad\dots\dots\dots\dots(34)$$

In like manner, for C_2 in (26),

$$p' = \chi_2 x/R, \qquad q' = \chi_2 y/R, \qquad r' = 1,$$

so that

$$\mu \cos \psi = C_1 + \cos \Phi + \sin \Phi . \chi_2 x/R = \mu \cos \Phi' + \frac{x \sin \Phi}{R}\left(\frac{\chi_1 \cos \Phi}{\mu \cos \Phi'} - \chi_1 + \chi_2\right);$$

and

$$\cos \psi' = \sqrt{\{1 - \mu^2 + \mu^2 \cos^2 \psi\}}$$

$$= \cos \Phi + \frac{\mu \cos \Phi' \sin \Phi . x/R}{\cos \Phi}\left(\frac{\chi_1 \cos \Phi}{\mu \cos \Phi'} - \chi_1 + \chi_2\right);$$

whence

$$C_2 = (\cos \Phi - \mu \cos \Phi')\left\{1 - \frac{x \tan \Phi}{R}\left(\frac{\chi_1 \cos \Phi}{\mu \cos \Phi'} - \chi_1 + \chi_2\right)\right\}. \quad\dots\dots(35)$$

Thus, if we write $\mu' = \mu \cos \Phi'/\cos \Phi$,

$$C_1 + C_2 = \frac{x \sin \Phi}{R}(\mu' - 1)(\chi_2 - \chi_1); \quad\dots\dots\dots\dots(36)$$

and $C_1\chi_1 + C_2\chi_2$

$$= \cos\Phi\,(\mu'-1)\,(\chi_1-\chi_2) + \frac{(\mu'-1)\,x\sin\Phi}{\mu'R}\,\{-\chi_1{}^2 - (\mu'-1)\,\chi_1\chi_2 + \mu'\chi_2{}^2\},$$

$$\qquad\qquad\qquad\qquad\qquad\qquad\qquad\qquad\qquad\qquad\qquad ...(37)$$

in which $\mu'\chi_2{}^2 - (\mu'-1)\,\chi_1\chi_2 - \chi_1{}^2 = (\chi_2-\chi_1)\,(\mu'\chi_2 + \chi_1).$

Accordingly, $\zeta\cos\Phi + \xi\sin\Phi$

$$= -\frac{R}{(\mu'-1)\cos\Phi\,(\chi_1-\chi_2)}\left\{1 + \frac{x\tan\Phi}{\mu'R}\,[(\chi_2-\chi_1)\,(\mu'^2-\mu')\cos^2\Phi + \mu'\chi_2 + \chi_1]\right\}$$

$$\qquad\qquad\qquad\qquad\qquad\qquad\qquad\qquad\qquad\qquad\qquad ...(38)$$

and the condition of no aberration is

$$(\chi_2-\chi_1)\,(\mu'^2-\mu')\cos^2\Phi + \mu'\chi_2 + \chi_1 = 0. \qquad\qquad ...(39)$$

Since $\chi_1,\ \chi_2$ are inversely proportional to r and s, we may write (39) in the form

$$\frac{1-(\mu'^2-\mu')\cos^2\Phi}{r} + \frac{\mu'+(\mu'^2-\mu')\cos^2\Phi}{s} = 0, \qquad ...(40)$$

where $$\mu' = \mu\cos\Phi'/\cos\Phi. \qquad\qquad\qquad\qquad ...(41)$$

If $s = \infty$, so that the second surface is flat, we have as the special form of (40)

$$1-(\mu'^2-\mu')\cos^2\Phi = 0; \qquad\qquad\qquad ...(42)$$

or in the case where $\Phi = 0$,

$$1+\mu-\mu^2 = 0, \qquad\qquad\qquad\qquad ...(43)$$

the same condition as that (15) required to give zero aberration in the *primary* plane for small obliquities. In the case of finite obliquities we may write (42) in terms of μ,

$$\mu\cos\Phi\cos\Phi' = \mu^2\cos^2\Phi' - 1, \qquad\qquad ...(44)$$

or if we take the square of both sides of the equation,

$$\mu^2\,(1-\sin^2\Phi)\,(1-\sin^2\Phi') = (\mu^2\cos\Phi' - 1)^2.$$

Of this the left-hand side may be equated to

$$(1-\sin^2\Phi)\,(\mu^2-\sin^2\Phi) = \mu^2 - (\mu^2+1)\sin^2\Phi + \sin^4\Phi,$$

while on the right we have

$$(\mu^2-1-\sin^2\Phi)^2 = (\mu^2-1)^2 - 2\,(\mu^2-1)\sin^2\Phi + \sin^4\Phi;$$

so that $$\sin^2\Phi = \frac{3\mu^2-\mu^4-1}{3-\mu^2}, \qquad\qquad\qquad ...(45)$$

as formerly found (see (17)).

In interpreting (45), which we may also write in the form

$$\sin^2\Phi = \frac{(\mu^2-\mu-1)\,(\mu^2+\mu-1)}{\mu^2-3}, \qquad\qquad ...(46)$$

we must bear in mind that it covers not only the necessary equation (44), but also the equation derived from (44) by changing the sign of one of the members.   For instance, if we put $\mu = 1$ in (46), we derive $\sin^2 \Phi = \frac{1}{2}$, or $\Phi = 45°$; but on referring back we see that these values satisfy, not (44), but

$$- \mu \cos \Phi \cos \Phi' = \mu^2 \cos^2 \Phi' - 1.$$

The transition occurs when $\cos \Phi = 0$, or $\Phi = 90°$, when (45) gives $\mu^2 = 2$, or $\mu = 1\cdot4142$. For smaller values of μ there is no solution of (44). Onwards from this point, as μ increases, Φ diminishes. For example, when $\mu = 1\cdot5$, $\sin^2 \Phi = \frac{11}{12}$, whence $\Phi = 73°$. The diminution of Φ continues until $\mu^2 - \mu - 1 = 0$, or $\mu = 1\cdot618$, when $\Phi = 0$, so that this is the value suitable for a plano-convex lens at small obliquities. After this value of μ is exceeded, $\sin^2 \Phi$ in (46) is negative until $\mu^2 = 3$, or $\mu = 1\cdot732$. When this point is passed, $\sin^2 \Phi$ becomes positive, but a real value of Φ is not again reached. We infer that in the case of a plano-convex lens (curved face presented to parallel rays) there can be no freedom from secondary aberration unless μ lies between the rather narrow limits $1\cdot414$ and $1\cdot618$.

If the plano-convex lens be so turned as to present its plane face to the parallel rays, $r = \infty$; and (40) requires that

$$\mu' + (\mu'^2 - \mu') \cos^2 \Phi = 0,$$

which cannot be satisfied, since $\mu' > 1$.

Leaving now the particular case of the plano-convex lens, let us suppose in the general formula (40) that $\Phi = 0$. We have

$$\frac{1 + \mu - \mu^2}{r} + \frac{\mu^2}{s} = 0, \quad\dots\dots\dots\dots\dots\dots(47)$$

from which we see that, whatever may be the value of μ, compensation may be attained by a suitable choice of the ratio $r : s$. If $\mu < 1\cdot618$, r and s have opposite signs, that is, the lens is double convex; while if $\mu > 1\cdot618$, r and s have the same sign, or the lens is of the meniscus form. For example, if $\mu = 1\cdot5$, (47) gives $s = -9r$, so that the lens is double convex, the hind surface having one-ninth the curvature of the front surface.

We have seen that the aberrations in both the primary and the secondary planes are eliminated for small obliquities in the case of a plano-convex lens if $\mu = 1\cdot618$. The question arises whether this double elimination is possible at finite obliquities if we leave both the form of the lens and the refractive index arbitrary. It appears that this can *not* be done. The necessary condition is by (13), (40)

$$- \frac{s}{r} = \frac{\mu'^2}{1 + \mu' - \mu'^2} = \frac{\mu' + (\mu'^2 - \mu') \cos^2 \Phi}{1 - (\mu'^2 - \mu') \cos^2 \Phi},$$

or

$$\frac{\mu'}{1 + \mu' - \mu'^2} = \frac{\mu' - (\mu' - 1) \sin^2 \Phi}{1 + \mu' - \mu'^2 + (\mu'^2 - \mu') \sin^2 \Phi},$$

whence

$$(\mu'^2 - 1) \sin^2 \Phi = 0, \quad\dots\dots\dots\dots\dots\dots(48)$$

which can be satisfied only by $\Phi = 0$, since $\mu' > 1$.

Since it is not possible to destroy both the primary and secondary aberrations when the angle of incidence is finite, it only remains to consider a little further in detail one or two special cases.

We have already spoken of the plano-convex lens; but for a more detailed calculation it may be well to form the equation for absence of primary aberration analogous to (45). From (16),

$$\mu \cos \phi' \cos \phi = \mu^2 - 1, \quad \dots\dots\dots\dots\dots\dots(49)$$

whence, if we square both sides,

$$\sin^4 \phi - (\mu^2 + 1) \sin^2 \phi + 3\mu^2 - \mu^4 - 1 = 0,$$

giving

$$\sin^2 \phi = \frac{\mu^2 + 1 \pm \sqrt{5}(\mu^2 - 1)}{2}, \quad \dots\dots\dots\dots(50)$$

so that

$$\sin^2 \phi = 1 \cdot 618034 - 0 \cdot 618034 \, \mu^2, \quad \dots\dots\dots\dots(51)$$

the other root being excluded if $\mu > 1$. It may be remarked that there is no distinction between ϕ here and Φ in (45).

The following table will give an idea of the values of ϕ from (51) and (45) for which the plano-convex lens of variable index is free from aberration in the primary and secondary planes respectively.

μ	ϕ Primary plane	ϕ Secondary plane
1·0000	90° 0′	—
1·4142	38° 11′	90° 0′
1·5000	28° 29′	73° 13′
1·5500	21° 24′	58° 37′
1·5900	13° 38′	39° 45′
1·6000	10° 55′	32° 25′
1·6100	7° 16′	22° 1′
1·6180	0° 0′	0° 0′

In the above the curved face is supposed to be presented to the parallel rays. If the lens be turned the other way, $r = \infty$, and (13) gives $\mu' = 0$, an equation which cannot be satisfied. In this case neither the primary nor the secondary aberration can be destroyed at any angle.

Next suppose that the lens is equi-convex, so that $s = -r$. In this case (13) gives

$$\mu'^2 - \tfrac{1}{2}\mu' - \tfrac{1}{2} = 0, \quad \dots\dots\dots\dots\dots\dots(52)$$

whence $\mu' = 1$, or $-\tfrac{1}{2}$, of which the latter has no significance. Also from (40) we get $\mu' = 1$. It appears that neither aberration can vanish for an equi-convex lens, unless in the extreme case $\mu = 1$, $\phi = 0$, when the lens produces no effect at all.

PART II.

It is a common experience in optical work to find the illumination deficient when an otherwise desirable magnification is introduced. Sometimes there is no remedy except to augment the intensity of the original source of light, if this be possible. But in other cases the defect may largely depend upon the manner in which the magnification is effected. With the usual arrangements magnifying takes place equally in the two perpendicular directions, though perhaps it may only be required in one direction. For example, in observations upon the spectrum, or upon interference bands, there is often no need to magnify much, or perhaps at all, in the direction parallel to the lines or bands. If, nevertheless, we magnify equally in both directions, there may be an unnecessary and often very serious loss of light.

In discussing this matter there is another distinction to be borne in mind. Sometimes it is not necessary or advantageous that there should exist a point-to-point correspondence between the object and the image. It suffices that a point in the object be represented in the image by a narrow line. This happens, for example, in the use of Rowland's concave gratings. A conspicuous instance occurs in the refractometer which I described in connection with observations upon argon and helium*. Here while the object-glass of the telescope was as usual, a very high magnification in one direction was secured by the use, as sole eye-piece, of a cylindrical lens taking the form of a glass rod 4 mm. in diameter. An equal magnification in both directions, such as would have been afforded by the usual spherical eye-pieces, would have so reduced the light as to make the observations impossible.

Whenever the field of view varies only in one dimension, there is usually no loss, and there may even be gain in the presence of astigmatism. In other cases a point-to-point correspondence between image and object is desirable or necessary, and the question arises how it may best be attained otherwise than by the use of a common telescope, which limits the magnification in the two directions to equality. I had occasion to consider this problem in connection with observations upon Haidinger's rings as observed with a Fabry and Perot apparatus. Here the field is symmetrical about an axis, and all the advantage that magnification can give is secured though it take place only in one direction. At the same time light is usually saved by abstaining from magnifying in the second direction also. In this way the circular rings assume an elongated elliptical form—a transformation which in no way prejudices observation by simple inspection. The question

---

* *Roy. Soc. Proc.* Vol. LIX. p. 198 (1896); *Scientific Papers*, Vol. IV. pp. 218, 364.

whether light is saved, as compared with symmetrical magnification, depends of course upon the aperture available in the two directions. In a Fabry and Perot apparatus this is usually somewhat restricted.

One simple solution of the problem, available when the light is homogeneous, may be found in the use of a *magnifying prism*, that is a prism so held that the emergence is more nearly grazing than the incidence. In this way we may obtain a moderate magnification in one direction combined with none at all in the second direction. A magnification equal in both directions may then be superposed with the aid of a common telescope. This method would probably answer well in certain cases, but it has its limitations. Moreover, the accompanying deviation of the rays through a large angle would often be inconvenient.

If we are allowed the use of cylindrical lenses, or of lenses whose curvature though finite is different in the two planes, we may attain our object with a construction analogous to that of a common telescope. Suppose that the eye-piece is constituted of a spherical and a contiguous cylindrical convex lens. In one plane the power of the eye-piece is greater than in the other perpendicular plane. Thus, if the object-glass be composed of spherical lenses only, there cannot be complete focusing. With the spherical lens or lenses of the object-glass, mounted as usual, it is necessary to combine a cylindrical lens of comparatively feeble power, which may be either convex or concave. All that is necessary to constitute a telescope in the full sense of the word, that is an apparatus capable of converting incident parallel rays into emergent parallel rays, is that the usual condition connecting the focal lengths of object-glass and eye-piece should be satisfied for the two principal planes taken separately. The magnifying powers in the two planes may thus be chosen at pleasure; and since there is symmetry with respect to both planes the apparatus is free from the unsymmetrical aberration expressed in (1).

When the magnifying desired is considerable in both planes, there is but little for the cylindrical component of the object-glass to do, and it occurred to me that it might be dispensed with, provided a moderate slope were given to the single (spherical) lens. In the earlier experiments the object-glass was a nearly equi-convex lens of 14 inches focus. The eye-piece was a combination of a spherical lens of 6 inches focus with a cylindrical lens of $2\frac{1}{2}$ inches focus, so that the focal lengths of the combination were about 2 inches and 6 inches in the principal planes, giving a *ratio* of magnifications as three to one. With the above object-lens the actual magnifications would be about 2 and 6. During the observations the axis of the telescope was horizontal and that of the cylindrical lens vertical, so that the higher magnification was in the horizontal direction of the field. During the adjustments it is convenient to examine a cross formed by horizontal and vertical lines, ruled upon paper well illuminated and placed at a sufficient distance.

When the object-lens stands square, there is, of course, no position of the compound eye-piece which allows both constituents of the cross to be seen in focus together. If we wish to pass from the focus for the horizontal to that necessary for the vertical line, we must push the eye-piece in. In order to focus both at once we must slope the object-lens. And since while both the primary and secondary focal lengths are diminished by obliquity the former is the *more* diminished, it follows that the sloping required is in the vertical plane, the lens being rotated about its horizontal diameter. If we introduce obliquity by stages, we find that the displacement of the eye-piece required to pass from one focus to the other gradually diminishes until an obliquity is reached which allows both lines of the cross to be in focus simultaneously. At a still higher obliquity the relative situation of the two foci is reversed. In the actual experiment with the 14-inch object-lens, the critical obliquity was roughly estimated at about 30°.

The above apparatus worked fairly well when tried upon interference rings from a thallium vacuum tube. But it was evident that the image suffered somewhat from aberration. A better result ensued when the magnification in both directions was increased by the substitution of an object-lens of 24 inches focus, although this also was equi-convex.

Being desirous of testing the method of the sloped lens under more favourable conditions, I procured from Messrs Watson a lens of baryta crown glass of index for mean rays 1·59, and of *plano*-convex form. The aperture was about $1\frac{3}{4}$ inch, and the focal length 24 inches. When this was combined with the compound eye-piece already described, the performance was very good, if, in accordance with the indications of theory, the curved face of the object-lens was presented to the incident light. The test may be made either upon a cross or upon a system of concentric circles drawn upon paper. The angle of slope giving the best effect was now very sharply defined. When, however, the object-lens was reversed, so as to present its plane face to the incident rays, no good result could be attained, evidently in consequence of aberration. The change in the character of the image was now very apparent when the eye was moved up and down, the rings appearing more elliptical as the eye moved in the direction of the nearest part of the edge of the sloped lens. Next to nothing of this effect could be observed when the object-lens was used in the proper position. It is scarcely necessary to say that care must be taken to ensure that the axis, about which the lens is turned, is truly perpendicular to the axis of the cylindrical component of the eye-piece.

Altogether it appears that the combination of sloped object-lens with compound cylindrical eye-piece constitutes a satisfactory solution of the problem. I believe that it may be applied with advantage in the many cases which arise in the laboratory where an unsymmetrical magnifying best meets the conditions. The question as to the precise index to be chosen for the plano-convex lens remains to some extent open. Possibly a somewhat higher index, *e.g.* 1·60, or even 1·61, might be preferred to that which I have used.

With the view to the design of future instruments, it may be convenient to set out the formula giving the distance between the primary and secondary foci of the object-lens as dependent upon the obliquity ϕ. If f_1, f_2 be the primary and secondary focal lengths, it is known (compare (33)) that

$$\frac{f_0}{f_1} = \frac{\mu \cos \phi' - \cos \phi}{(\mu - 1) \cos^2 \phi}, \quad \dots\dots\dots\dots\dots\dots(53)$$

$$\frac{f_0}{f_2} = \frac{\mu \cos \phi' - \cos \phi}{\mu - 1}, \quad \dots\dots\dots\dots\dots\dots(54)$$

f_0 being the focal length corresponding to $\phi = 0$; so that

$$\frac{f_2 - f_1}{f_0} = \frac{(\mu - 1) \sin^2 \phi}{\mu \cos \phi' - \cos \phi}. \quad \dots\dots\dots\dots\dots\dots(55)$$

In this

$$\mu \cos \phi' - \cos \phi = \sqrt{(\mu^2 - \sin^2 \phi)} - \sqrt{(1 - \sin^2 \phi)} = (\mu - 1)\left\{1 + \frac{\sin^2 \phi}{2\mu}\right\}$$

approximately. Hence

$$\frac{f_2 - f_1}{f_0} = \sin^2 \phi \left\{1 - \frac{\sin^2 \phi}{2\mu}\right\}, \quad \dots\dots\dots\dots\dots\dots(56)$$

from which the required obliquity is readily calculated when the nature of the eye-piece and the focal length of the object-lens are given.

P.S., June 6.—From von Rohr's excellent *Theorie und Geschichte des Photographischen Objectivs*, Berlin, 1899, I learn that Rudolf and, at a still earlier date (1884), Lippich had proposed a different method of obtaining a diverse magnification, and one that I had overlooked. This consists in the use of an eye-piece formed by crossing two cylindrical lenses of different powers. The two lenses are mounted, not close together, but at such distances from the image as to render parallel the rays diverging from it in the two planes separately. In this method the object-lens remains square to the axis of the instrument. Lippich had the same object in view as that which guided me. I have tried his method with success, obtaining an image as good, or nearly as good, as that afforded by the sloped lens. I understand that Professor S. P. Thompson also has used a similar device.

HAMILTON'S PRINCIPLE AND THE FIVE ABERRATIONS OF VON SEIDEL.

[*Philosophical Magazine*, Vol. xv. pp. 677—687, 1908.]

LARGELY owing to the fact that the work of Hamilton, and it may be added of Coddington, remained unknown in Germany and that of v. Seidel in England, it has scarcely been recognized until recently how easily v. Seidel's general theorems relating to optical systems of revolution may be deduced from Hamilton's principle. The omission has been supplied in an able discussion by Schwarzschild, who expresses Hamilton's function in terms of the variables employed by Seidel, thus arriving at a form to which he gives the name of Seidel's Eikonal*. It is not probable that Schwarzschild's investigation can be improved upon when the object is to calculate complete formulae applicable to specified combinations of lenses; but I have thought that it might be worth while to show how the number and nature of the five constants of aberration can be deduced almost instantaneously from Hamilton's principle, at any rate if employed in a somewhat modified form.

When we speak, as I think we may conveniently do, of five constants of aberration, there are two things which we should remember. The first is that the five constants do not stand upon the same level. By this I mean, not merely that some of them are more important in one instrument and some in another, but rather that the nature of the errors is different. In earlier writings the term aberration was, I think, limited to imperfect focusing of rays which, issuing from one point, converge upon another. Three of the five aberrations are of this character; but the remaining two relate, not to imperfections of focusing, but to the position of the focus. It is, in truth, something of an accident that, *e.g.* in photography, we desire to focus distant objects upon a *plane*. The second thing to which I wish to refer is that, although Seidel did much, four out of the five aberrations were pretty fully discussed by Airy and Coddington before his time. To these

---

* The word Eikonal was introduced by Bruns.

authors is due the rule relating to the curvature of images, generally named after Petzval, so far, at any rate, as it refers to combinations of *thin* lenses.

Some remarks are appended having reference to systems of less highly developed symmetry.

According to Hamilton's original definition of the characteristic function V, it represents the time taken by light to pass from an initial point (x', y', z') to a final point (x, y, z), and it may be taken to be $\int \mu ds$, where μ is the refractive index and the integration is along the course of the *ray* which connects the two points. If the path be varied, the integral is a *minimum* for the actual ray; and from this it readily follows that

$$l = dV/dx, \qquad m = dV/dy, \qquad n = dV/dz, \quad \ldots\ldots\ldots\ldots(1)$$
$$-l' = dV/dx', \quad -m' = dV/dy', \quad -n' = dV/dz', \ldots\ldots\ldots\ldots(2)$$

where l, m, n, l', m', n' are the direction-cosines of the ray at the end and beginning of its course, the terminal points being situated in a part of the system where the refractive index is unity.

In his communication to the British Association (*B. A. Report*, Cambridge 1833, p. 360) Hamilton transforms these equations. As his work is so little known, it may be of interest to quote in full the principal paragraph, with a slight difference of notation :—" When we wish to study the properties of any object-glass, or eye-glass, or other instrument *in vacuo*, symmetric in all respects, about one axis of revolution, we may take this for the axis of z, and we shall have the equations (1), (2), the *characteristic function* V being now a function of the five quantities, $x^2 + y^2$, $xx' + yy'$, $x'^2 + y'^2$, z, z', involving also, in general, the colour, and having its form determined by the properties of the instrument of revolution. Reciprocally, these properties of the instrument are included in the form of the characteristic function V, or in the form of this other connected function,

$$T = lx + my + nz - l'x' - m'y' - n'z' - V, \quad \ldots\ldots\ldots\ldots(3)$$

which may be considered as depending on only three independent variables besides the colour; namely, on the inclinations of the final and initial portions of a luminous path to each other and to the axis of the instrument. Algebraically, T is in general a function of the colour and of the three quantities, $l^2 + m^2$, $ll' + mm'$, $l'^2 + m'^2$; and it may *usually* (though not in every case) be developed according to ascending powers, positive and integer, of these three latter quantities, which in most applications are small, of the order of the squares of the inclinations. We may therefore in most cases confine ourselves to an approximate expression of the form

$$T = T^{(0)} + T^{(2)} + T^{(4)}, \quad \ldots\ldots\ldots\ldots\ldots\ldots\ldots(4)$$

in which $T^{(0)}$ is independent of the inclinations; $T^{(2)}$ is small of the second order, if those inclinations be small, and is of the form

$$T^{(2)} = P(l^2 + m^2) + P_1(ll' + mm') + P'(l'^2 + m'^2); \quad \ldots\ldots(5)$$

and $T^{(4)}$ is small of the fourth order, and of the form

$$\begin{aligned} T^{(4)} = {} & Q(l^2 + m^2)^2 + Q_1(l^2 + m^2)(ll' + mm') \\ & + Q'(l^2 + m^2)(l'^2 + m'^2) + Q_{11}(ll' + mm')^2 \\ & + Q_1'(ll' + mm')(l'^2 + m'^2) + Q''(l'^2 + m'^2)^2; \quad \ldots\ldots(6) \end{aligned}$$

the nine coefficients, $P\ P_1\ P'\ Q\ Q_1\ Q'\ Q_{11}\ Q_1'\ Q''$, being either constant, or at least only functions of the colour. The optical properties of the instrument, to a great degree of approximation, depend usually on these nine coefficients and on their chromatic variations, because the function T may in most cases be very approximately expressed by them, and because the fundamental equations (1), (2) may rigorously be thus transformed;

$$\left.\begin{aligned} x - \frac{l}{n}z = \frac{dT}{dl}, && y - \frac{m}{n}z = \frac{dT}{dm}; \\ x' - \frac{l'}{n'}z' = -\frac{dT}{dl'}, && y' - \frac{m'}{n'}z' = -\frac{dT}{dm'}; \end{aligned}\right\} \quad \ldots\ldots(7)$$

The first three coefficients, $P\ P_1\ P'$, which enter by (5) into the expression of the term $T^{(2)}$, are those on which the focal lengths, the magnifying powers, and the chromatic aberrations depend: the spherical aberrations, whether for direct or inclined rays, from a near or distant object, at either side of the instrument (but not too far from the axis), depend on the six other coefficients, $Q\ Q_1\ Q'\ Q_{11}\ Q_1'\ Q''$, in the expression of the term $T^{(4)}$. Here, then, we have already a new and remarkable property of object-glasses, and eye-glasses, and other optical instruments of revolution; namely, that all the circumstances of their *spherical aberrations*, however varied by distance and inclination, depend (usually) on the values of SIX RADICAL CONSTANTS OF ABERRATION, and may be deduced from these six numbers by uniform and general processes. And as, by employing general symbols to denote the constant coefficients or elements of an elliptic orbit, it is possible to deduce results extending to all such orbits, which can afterwards be particularized for each; so, by employing general symbols for the six constants of aberration, suggested by the fore-going theory, it is possible to deduce general results respecting the aber-rational properties of optical instruments of revolution, and to combine these results afterwards with the peculiarities of each particular instrument by substituting the numerical values of its own particular constants."

Equations (7) are easily deduced. So far as it depends upon the unaccented letters, the total variation of T is

$$dT = l\,dx + m\,dy + n\,dz + x\,dl + y\,dm + z\,dn - \frac{dV}{dx}dx - \frac{dV}{dy}dy - \frac{dV}{dz}dz,$$

or regard being paid to (1),

$$dT = x\,dl + y\,dm + z\,dn,$$

in which

$$l\,dl + m\,dm + n\,dn = 0,$$

so that

$$\frac{dT}{dl} = x - \frac{lz}{n}, \qquad \frac{dT}{dm} = y - \frac{mz}{n};$$

and in like manner by varying the accented letters the second pair of equations (7) follows.

If we agree to neglect the cubes of the inclinations, we may identify n, n' with unity, and (7) becomes

$$x = (z + 2P)\,l + P_1 l', \qquad y = (z + 2P)\,m + P_1 m',$$

$$x' = -P_1 l + (z' - 2P')\,l', \qquad y' = -P_1 m + (z' - 2P')\,m',$$

determining x, x' in terms of z, z', l, l' supposed known, or conversely l, l' in terms of z, z', x, x' supposed known. The case of special interest is that in which x, y, z and x', y', z' are conjugate points, *i.e.* images of one another in the optical system. The ratio $x : x'$ must then be independent of the special values ascribed to l, l'. In order that this may be possible, *i.e.* in order that z, z' may be conjugate planes, the condition is

$$(z + 2P)(z' - 2P') + P_1^2 = 0, \quad\ldots\ldots\ldots\ldots\ldots\ldots(8)$$

and then

$$\frac{x}{x'} = \frac{y}{y'} = -\frac{z + 2P}{P_1} = \frac{P_1}{z' - 2P'}, \quad\ldots\ldots\ldots\ldots\ldots(9)$$

giving the magnification.

Equations (8), (9) express the theory of a symmetrical instrument to a first approximation. In order to proceed further we should have not only to include the terms in (7) arising from $T^{(4)}$, but also to introduce a closer approximation for n. Thus even though $T^{(4)} = 0$, we should have additional terms in the expressions for x, x' equal respectively to

$$\tfrac{1}{2} lz\,(l^2 + m^2) \quad \text{and} \quad \tfrac{1}{2} l'z'\,(l'^2 + m'^2).$$

If the object is merely to express the aberrations for a single pair of conjugate planes, we may attain it more simply by a modification of Hamilton's process.

Supposing that the conjugate planes are $z = 0$, $z' = 0$, we have V a function of the coordinates of the initial point x', y', and of the final point x, y. And if as before l, m, n, l', m', n' are the direction-cosines of the terminal portions of the ray, we still have

$$l = dV/dx, \qquad m = dV/dy, \quad\ldots\ldots\ldots\ldots\ldots(10)$$

$$l' = -dV/dx', \qquad m' = -dV/dy'. \quad\ldots\ldots\ldots\ldots(11)$$

But now instead of transforming to a function of l, m, l', m', from which x', y', x, y are eliminated, we retain x', y' as independent variables, eliminating only x, y, the coordinates of the final or image point*. For this purpose we assume

$$U = lx + my - V. \qquad\qquad (12)$$

The total variation of U is given by

$$dU = x\,dl + l\,dx + y\,dm + m\,dy - \frac{dV}{dx}\,dx - \frac{dV}{dy}\,dy - \frac{dV}{dx'}\,dx' - \frac{dV}{dy'}\,dy',$$

or with regard to (10), (11)

$$dU = x\,dl + y\,dm + l'\,dx' + m'\,dy', \qquad\qquad (13)$$

from which it appears that U is in reality a function of x', y', l, m. As equivalent to (13), we have

$$x = dU/dl, \qquad y = dU/dm, \qquad\qquad (14)$$
$$l' = dU/dx', \qquad m' = dU/dy'. \qquad\qquad (15)$$

So far U appears as a function of the four variables x', y', l, m; but from its nature, as dependent upon $lx + my$ and V, and from the axial symmetry, it must be in fact a function of the *three* variables

$$x'^2 + y'^2, \quad l^2 + m^2, \quad \text{and} \quad lx' + my',$$

the latter determining the angle between the directions of x', y' and l, m. When these quantities are small, we may take

$$U = U^{(0)} + U^{(2)} + U^{(4)} + \dots, \qquad\qquad (16)$$

where $U^{(0)}$ is constant and

$$U^{(2)} = \tfrac{1}{2}L(l^2 + m^2) + M(x'l + y'm) + \tfrac{1}{2}N(x'^2 + y'^2), \qquad (17)$$

L, M, N being constants. If we stop at $U^{(2)}$, equations (14) give

$$x = Ll + Mx', \qquad y = Lm + My', \qquad\qquad (18)$$

determining x, y as functions of x', y', l, m. We have next to introduce the supposition that x, y is conjugate to x', y'. Hence $L = 0$, for to this approximation x, y must be determined by x', y' independently of l, m. Accordingly,

$$x = Mx', \qquad y = My'. \qquad\qquad (19)$$

We are now prepared to proceed to the next approximation. In order to correspond, as far as may be, with the notation of Seidel† we will write

$$U^{(4)} = \tfrac{1}{4}A(l^2 + m^2)^2 + B(l^2 + m^2)(lx' + my')$$
$$+ \tfrac{1}{2}(C - D)(lx' + my')^2 + \tfrac{1}{2}D(l^2 + m^2)(x'^2 + y'^2)$$
$$+ E(lx' + my')(x'^2 + y'^2) + F(x'^2 + y'^2)^2, \qquad\qquad (20)$$

which is the most general admissible function of the fourth degree.

---

* Compare Routh's *Elementary Rigid Dynamics*, § 418.

† Finsterwalder, *München. Sitz. Ber.* Vol. xxvii. p. 408 (1897).

From (20) we obtain by use of (14) the additional terms in x and y dependent on $U^{(4)}$. No generality is lost if at this stage we suppose, for the sake of brevity, $y' = 0$. Accordingly,

$$x = Al\,(l^2 + m^2) + Bx'\,(3l^2 + m^2) + Cx'^2 l + Ex'^3, \dots\dots\dots(21)$$

$$y = Am\,(l^2 + m^2) + 2Bx'\,lm + Dx'^2 m. \dots\dots\dots\dots\dots(22)$$

In order to complete the value of x we must add the expressions in (19) and (21).

Since F disappears from the values of x and y, we see that there are *five* effective constants of aberration of this order, as specified by Seidel. The evanescence of A is the Eulerian condition for the absence of spherical aberration in the narrower sense, *i.e.* as affecting the definition of points lying upon the axis ($x' = 0$). If the Eulerian condition be satisfied, $B = 0$ is identical with what Seidel calls the Fraunhofer condition*. The theoretical investigation of this kind of aberration was one of Seidel's most important contributions to the subject, inasmuch as neither Airy nor Coddington appears to have contemplated it. The conditions $A = 0$, $B = 0$ are those which it is most important to satisfy in the case of the astronomical telescope.

To this order of approximation $B = 0$ is identical with the more general *sine* condition of Abbe, which prescribes that, in order to the good definition of points just off the axis, a certain relation must be satisfied between the terminal inclinations of the rays forming the image of a point situated on the axis. The connexion follows very simply from the equations already found. By (15), (16), (17), (20), with $m = 0$,

$$l' = Ml + Bl^3 + \text{terms vanishing with } x', \ y' \ ;$$

so that for the conjugate points situated upon the axis

$$l' = Ml + Bl^3. \dots\dots\dots\dots\dots\dots\dots(23)$$

The condition $B = 0$ is thus equivalent to a constant value of the ratio l'/l, that is the ratio of the sines of the terminal inclinations of a ray with the axis. And this is altogether independent of the value of A.

On the supposition that the two first conditions $A = 0$, $B = 0$ are satisfied, we have next to consider the significance of the terms multiplied by C and D. Since

$$dx/dl = Cx'^2, \qquad dy/dm = Dx'^2,$$

---

* If A be not equal to zero, it can be shown that the best focusing of points just off the axis requires that

$$Al_0 + Bx' = 0,$$

where l_0 is the value of l for the principal ray. For example, if the optical system reduces to a combination of thin lenses close together, $l_0 = x/f$, where f is the distance of the lenses from the image plane. Since by (19), $x = Mx'$, the condition may be written

$$AM + Bf = 0.$$

we see that C and D represent departures of the primary and secondary foci from the proper plane. In fact if $1/\rho_1$, $1/\rho_2$ be the curvatures of the images, as formed by rays in the two planes,

$$1/\rho_1 = 2C, \qquad 1/\rho_2 = 2D. \quad\quad\quad\quad\quad\quad\quad\quad (24)$$

The condition of astigmatism is then

$$C = D; \quad\quad\quad\quad\quad\quad\quad\quad\quad\quad\quad (25)$$

but unless both constants vanish the image is curved.

Finally the term containing E represents distortion.

If we impose no restriction upon the values of the constants of aberration, we have in general from (21), (22)

$$dx/dl = A\,(3l^2 + m^2) + 6Bx'l + Cx'^2,$$
$$dy/dm = A\,(l^2 + 3m^2) + 2Bx'l + Dx'^2.$$

These equations may be applied to find the curvatures of the image as formed by rays infinitely close to given rays, as for example when the aperture is limited by a narrow stop placed centrally on the axis, but otherwise arbitrarily. The principal ray is then characterized by the condition $m = 0$, and we have

$$dx/dl = 3Al^2 + 6Bx'l + Cx'^2 = 3H + K, \quad\quad\quad\quad (26)$$
$$dy/dm = Al^2 + 2Bx'l + Dx'^2 \ = H + K, \quad\quad\quad\quad (27)$$

equations which determine the curvatures of the images as formed by rays in the neighbourhood of the given one, and deviating from it in the primary and secondary planes respectively.

According to (26), (27),

$$2H = 2Al^2 + 4Bx'l + (C - D)\,x'^2, \quad\quad\quad\quad\quad (28)$$
$$2K = (3D - C)\,x'^2. \quad\quad\quad\quad\quad\quad\quad\quad\quad (29)$$

The requirement of flatness in both images is thus satisfied if $H = 0$, $K = 0$. The former is the condition of astigmatism, and it involves the ratio of $x' : l$, which is dependent upon the position of the stop; but the latter does not depend on this ratio. It corresponds to the condition formulated by Coddington and later by Petzval. From (28), (29) we may of course fall back upon the conditions already laid down for the case where $A = 0$, $B = 0$.

The further pursuit of this subject requires a more particular examination of what occurs when light is refracted at spherical surfaces. Reference may be made to Schwarzschild*, who uses Hamilton's methods as applied to a special form of the characteristic function designated as Seidel's Eikonal. A concise derivation of the Coddington-Petzval condition by elementary methods will be found in Whittaker's tract†.

---

* Göttingen *Abh.* Vol. IV. 1905.

† *Theory of Optical Instruments*, Cambridge, 1907. The optical invariants, introduced by Abbe, are there employed.

Before leaving systems symmetrical about an axis to which all the rays are inclined at small angles, we may remark that, as $U^{(4)}$ contains 6 constants, in like manner $U^{(6)}$ contains 10 constants* and $U^{(8)}$ 15 constants, of which in each case one is ineffective.

The angle embraced by some modern photographic lenses is so extensive that a theory which treats the inclinations as small can be but a rough guide. It remains true, of course, that an absolutely flat field requires the fulfilment of the Coddington-Petzval condition; but in practice some compromise has to be allowed, and this involves a sacrifice of complete flatness at the centre of the image. It will be best to fulfil the conditions $dx/dl = 0$, $dy/dm = 0$, or, what are equivalent,

$$d^2U/dl^2 = 0, \qquad d^2U/dm^2 = 0,$$

not when l is very small but when it attains some finite specified value. If we suppose $y' = 0$, U is a function of x'^2, $l^2 + m^2$, and lx', or say of u, v, w. Hence

$$\frac{d^2U}{dl^2} = 4l^2 \frac{d^2U}{dv^2} + 2 \frac{dU}{dv} + x'^2 \frac{d^2U}{dw^2} + 4x'l \frac{d^2U}{dw\,dv},$$

$$\frac{d^2U}{dm^2} = 4m^2 \frac{d^2U}{dv^2} + 2 \frac{dU}{dv}.$$

After the differentiations are performed, we are to make $m = 0$; so that the two conditions of astigmatism and focus upon the plane, analogous to (28), (29), are

$$\frac{dU}{dv} = 0, \qquad 4l^2 \frac{d^2U}{dv^2} + x'^2 \frac{d^2U}{dw^2} + 4x'l \frac{d^2U}{dv\,dw} = 0,$$

in which v is to be made equal to l^2. But it is doubtful whether such equations could be of service.

Let us now suppose that the system is indeed symmetrical with respect to the two perpendicular planes of x and y, but not necessarily so round the axis of z. In the expression for U no terms can occur which would be altered by a simultaneous reversal of x' and l, or of y' and m. For $U^{(2)}$ we have

$$U^{(2)} = \alpha l^2 + \beta m^2 + \gamma x'l + \delta y'm + \text{terms independent of } l \text{ and } m.$$

Hence, by (14),

$$x = 2\alpha l + \gamma x', \quad y = 2\beta m + \delta y'.$$

If x, y is conjugate to x', y', we must have

$$\alpha = 0, \qquad \beta = 0;$$

so that

$$x = \gamma x', \qquad y = \delta y'. \quad\dots\dots\dots\dots\dots\dots(30)$$

These are the equations of the first approximation, and they indicate that the magnification need not be the same in the two directions.

* Schwarzschild, *loc. cit.*

There are no terms in $U^{(3)}$. As regards $U^{(4)}$, we have

$$U^{(4)} = Al^4 + Bl^2m^2 + Cm^4$$
$$+ Dx'l^3 + Ex'lm^2 + Fy'm^3 + Gy'l^2m$$
$$+ Hx'^2l^2 + Ix'^2m^2 + Jx'y'lm + Ky'^2l^2 + Ly'^2m^2$$
$$+ Mx'^3l + Nx'^2y'm + Ox'y'^2l + Py'^3m$$
$$+ \text{terms independent of } l \text{ and } m. \dots\dots\dots\dots\dots(31)$$

In (31) there are 16 effective constants as compared with 5 in the case where the symmetry round the axis is complete; so that such symmetry implies 11 relations among the constants of (31). For example, in the terms of the first line representing Eulerian aberration, axial symmetry requires that

$$C = \tfrac{1}{2}B = A. \dots\dots\dots\dots\dots\dots\dots\dots(32)$$

We will next suppose that the only symmetry to be imposed is that with respect to the primary plane $y = 0$; so that U is unchanged if the signs of y' and m are both reversed. $U^{(2)}$ is of the same form as in the case of double symmetry, and

$$x = 2al + \gamma x', \quad y = 2\beta m + \delta y'.$$

If x, y is the image of x', y', formed by rays in both planes, $\alpha = 0$, $\beta = 0$, as before. But it may happen, *e.g.* in the spectroscope, that there is astigmatism even in the first approximation. If the points are images of one another as constituted by rays in the *primary* plane, $\alpha = 0$, but β is left arbitrary.

The next term in U may be denoted by $U^{(3)}$. If no conditions of symmetry were imposed, $U^{(3)}$ would include 16 effective terms, *i.e.* terms contributing to x, y; but the symmetry with respect to $y = 0$ excludes 8 of these. We may write

$$U^{(3)} = al^3 + blm^2 + cx'l^2 + dx'm^2 + ey'lm + fx'^2l + gx'y'm + hy'^2l, \dots(33)$$
and, by (14), $\quad x = 3al^2 + bm^2 + 2cx'l + ey'm + fx'^2 + hy'^2. \quad\dots\dots\dots(34)$

If the rays all proceed from the point $x' = 0$, $y' = 0$, the conditions for a well-formed primary focal line are

$$a = 0, \quad b = 0, \dots\dots\dots\dots\dots\dots\dots\dots(35)$$

of which the first expresses that there is no aberration of this order for rays in the primary plane, *i.e.*, that the focal line is *thin*, while the second is the condition that the focal line is *straight**.

But if, while $x' = 0$, y' be left arbitrary, so that the source of light is linear, the evanescence of (34) requires, in addition to (35), that

$$e = 0, \quad h = 0. \quad \dots\dots\dots\dots\dots\dots\dots(36)$$

---

* Compare *Phil. Mag.* Vol. VIII. p. 481 (1879); *Scientific Papers*, Vol. I. p. 440.

330.

VORTICES IN OSCILLATING LIQUID.

[*Proceedings of the Royal Society*, A, Vol. LXXXI. pp. 259—271, 1908.]

IN a paper "On the Circulation of Air observed in Kundt's Tubes, and on some Allied Acoustical Problems*," I applied the equations of viscous incompressible fluid to show that the effect of the bottom of the containing vessel was to generate permanent vortices in the vibrating fluid. It was remarkable that the intensity of the vortical motion, when fully established, proved to be independent of the magnitude of the viscosity, so that the effects could not be eliminated by merely supposing the viscosity to become extremely small. The expression found for the vortices was simple. The horizontal component u of the primary motion near the bottom being $u = u_0 \cos kx \cos nt$, the component velocities of the vortical motion are

$$u' = \tfrac{3}{8} \frac{u_0^2 \sin 2kx}{V} e^{-2ky} (1 - 2ky),$$

$$- v' = \tfrac{3}{8} \frac{u_0^2 \cos 2kx}{V} e^{-2ky} 2ky,$$

y being measured upwards from the bottom, and $V (= n/k)$ the velocity of propagation of waves of the length in question. According to these expressions, the vortical motion is downwards over the places where u has its greatest alternating values. In the case of water contained in a tank and vibrating in its simplest mode, the theoretical motion is downwards in the middle and upwards at the ends. To guard against misinterpretation, it may be well to add that quite close to the bottom the motion, as calculated, is of a quite different character.

In a recent paper†, Mrs Ayrton has examined, with much experimental skill, the vortices arising when water oscillates in a narrow tank, and has

---

* *Phil. Trans.* Vol. CLXXV. p. 1 (1883); *Scientific Papers*, Vol. II. p. 239.
† "On the Non-Periodic or Residual Motion of Water moving in Stationary Waves," *Roy. Soc. Proc.* A, Vol. LXXX. p. 252 (1908).

obtained results which differ somewhat widely from what are indicated in the above formulæ. Near the bottom, and especially when the depth is small, there are indeed vortices of this character; but, in general, the most conspicuous feature consists of vortices revolving in the opposite direction, the water rising in the middle of the tank and falling at the ends. The first thought that occurred to me was that Mrs Ayrton's vortices might be due to defect of freedom in the surface, such as might be supposed to arise from a greasy film opposing extensions and contractions; but in some experiments that I tried, the vortical motion did not seem to be much influenced by cleansing the surface, and the question was suggested as to whether the free surface itself might not originate vortices in somewhat the same way as the bottom does, and more potently on account of the greater velocities of the primary motion there prevailing. I do not remember whether I had any clear view on this question when I wrote the former paper. Vortices originating other-wise than at the bottom were ignored, but I may not have intended to exclude their possibility.

The present paper consists mainly of an attempt to answer the question thus suggested. The investigation is limited to the case of deep water, and even then is rather complicated. If the calculations are correct, we are to conclude that a free surface does *not* generate vortices of this kind, at least if we suppose the viscosity small and include only the square of the motion. How then are the conspicuous vortices, observed by Mrs Ayrton, to be explained? One might attribute them to the ends of the tank, acting much as the bottom does in my former investigation. In the latter case the seat of the forces is near the parts lying midway between the middle of the tank and the ends ($kx = \pm \frac{1}{4}\pi$), and their effect is to push the neighbouring fluid in the direction away from the place of greatest motion ($kx = 0$). A like action at the ends of the tank would push the neighbouring fluid downwards, and thus generate vortices revolving in the observed direction. An objection to this view lies in an observation by Mrs Ayrton on water oscillating in a long tank with several subdivisions (p. 255). Since no solid walls are situated at the intermediate nodes, it may be thought that the action at the ends would be insufficient to establish the whole system of vortices. Probably this would be so. But another influence acting in the same direction may arise from the *faces* of the somewhat narrow tanks employed. It would seem that at the nodes, $kx = \pm \frac{1}{2}(2m + 1)\pi$, the friction in moving up and down along the faces might have the same general effect as the up-and-down motion along the solid walls which constitute the ends.

But I must confess that some observations made with the help of Mrs Sidgwick were not favourable to this view. The ends of the tank were eliminated by using the annular space included between two coaxal cylinders, A, B (beakers), but the vortical motion did not seem to be diminished. The insertion of a strip of glass C held vertically across the annulus, and thus virtually restoring one end, did not make much, if any, difference. This experiment seems to exclude the explanation depending upon the action of the *ends*, and that which would attribute the effect to the *faces* is difficult to reconcile with the highly localised character of the effect, which at first seems to be limited to the immediate neighbourhood of the ends. Can it be that the true explanation would require the retention of terms of higher order than the *square* of the motion?

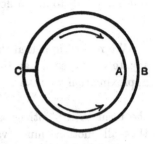

I may mention that on more than one occasion I have witnessed the reversed movement described by Mrs Ayrton on p. 259, "as if a set of water springs had been wound up, and now proceeded to unwind themselves." I presume that the spring depends upon gravity, acting upon unequal densities in the fluid, due either to temperature or to variations in the amount of powder held in suspension. Another possible explanation would lie in the effect of surface contamination.

In the usual notation the equations of motion in two dimensions are

$$\frac{1}{\rho}\frac{dp}{dx} = -\frac{du}{dt} + \nu\nabla^2 u - \frac{1}{2}\frac{d(u^2+v^2)}{dx} - v\left(\frac{du}{dy}-\frac{dv}{dx}\right), \quad \ldots\ldots(1)$$

$$\frac{1}{\rho}\frac{dp}{dy} = -\frac{dv}{dt} + \nu\nabla^2 v - \frac{1}{2}\frac{d(u^2+v^2)}{dy} - u\left(\frac{du}{dy}-\frac{dv}{dx}\right) - g, \quad \ldots\ldots(2)$$

where y is measured vertically upwards, and $\nu\,(=\mu/\rho)$ is the kinematic viscosity. Since the fluid is supposed to be incompressible,

$$\frac{du}{dx} + \frac{dv}{dy} = 0; \quad\ldots\ldots\ldots\ldots\ldots\ldots\ldots\ldots\ldots\ldots(3)$$

or what is equivalent,

$$u = d\psi/dy, \qquad v = -d\psi/dx, \quad \ldots\ldots\ldots\ldots\ldots(4)$$

ψ being the stream-function.

In virtue of (4), we have in (1), (2)

$$\frac{du}{dy} - \frac{dv}{dx} = \nabla^2\psi. \quad \ldots\ldots\ldots\ldots\ldots\ldots\ldots\ldots(5)$$

It may be well to commence with the comparatively simple question of stationary waves, carried to the second order of approximation, when viscosity is neglected. If we eliminate p from (1) and (2), putting at the same time $\nu = 0$, we find

$$-\frac{d}{dt}\nabla^2\psi = u\frac{d\nabla^2\psi}{dx} + v\frac{d\nabla^2\psi}{dy}. \quad \ldots\ldots\ldots\ldots\ldots(6)$$

As was to be expected from the general theory of a frictionless fluid, the solution of (6) to any order of approximation is

$$\nabla^2 \psi = 0. \qquad \dots \dots \dots \dots \dots \dots \dots \dots (7)$$

We now assume that the motion is periodic with respect to x—proportional, say, to $\cos kx$, so far as the first approximation is concerned. Thus for this approximation we take

$$\psi = A e^{ky} \cos kx \cos nt, \qquad \dots \dots \dots \dots \dots \dots (8)$$

the term in e^{-ky}, otherwise admissible, being excluded by the consideration that all motion must vanish when $y = -\infty$, inasmuch as the fluid is supposed infinitely deep. Corresponding to (8),

$$u = d\psi/dy = kAe^{ky} \cos kx \cos nt, \qquad \dots \dots \dots \dots (9)$$

$$v = -d\psi/dx = kAe^{ky} \sin kx \cos nt, \qquad \dots \dots \dots (10)$$

so that

$$u^2 + v^2 = k^2 A^2 e^{2ky} \cos^2 nt. \qquad \dots \dots \dots \dots \dots (11)$$

In virtue of (5), (7), the equations of pressure become

$$\frac{1}{\rho} \frac{dp}{dx} = -\frac{du}{dt} - \frac{1}{2} \frac{d(u^2 + v^2)}{dx}, \qquad \frac{1}{\rho} \frac{dp}{dy} = -\frac{dv}{dt} - \frac{1}{2} \frac{d(u^2 + v^2)}{dy} - g;$$

and thus

$$p/\rho = -gy + nAe^{ky} \sin kx \sin nt - \frac{1}{2} k^2 A^2 e^{2ky} \cos^2 nt + f(t), \qquad \dots \dots (12)$$

where $f(t)$ denotes a function of t which is arbitrary so far as the differential equations of pressure are concerned. The pressure at the surface is to be found from (12) by putting $y = \eta$, where η is the elevation of the surface at the point in question. The relation between η and u, v is thus required accurately to the second order of small quantities.

The differential relation* is

$$\frac{d\eta}{dt} = v - u \frac{d\eta}{dx}. \qquad \dots \dots \dots \dots \dots \dots \dots (13)$$

To the first order we have $d\eta/dt$ equal simply to the value of v at the surface, so that by (10)

$$\eta = kn^{-1} A \sin kx \sin nt, \qquad \dots \dots \dots \dots \dots (14)$$

the origin of y being in the undisturbed surface. This value of η may be used in the small terms of (13), and thus to a second approximation

$$\eta = \frac{kA}{n} \sin kx \sin nt + \frac{k^3 A^2}{4n^2} \cos 2kx \cos 2nt + F(x), \qquad \dots \dots (15)$$

where $F(x)$ is an arbitrary function of x of the second order of small quantities.

We are now prepared to substitute for y its value η in (12). In the principal term we must use the complete value of η from (15). In the second

---

* Cf. Lamb's *Hydrodynamics*, § 10.

term, already containing A as a factor, the first approximation for η suffices, while in the third term we may put $\eta = 0$. The third term thus becomes a function of t only, and may be regarded as cancelled by $f(t)$. We find

$$\frac{p}{\rho} = A\left(n - \frac{gk}{n}\right) \sin kx \sin nt - \frac{k^2 A^2}{4} \cos 2kx$$

$$+ \frac{k^2 A^2}{4}\left(1 - \frac{gk}{n^2}\right) \cos 2kx \cos 2nt - gF(x). \quad \ldots(16)$$

In free vibrations the condition to be satisfied at the surface is $p = 0$. The annulment of the term in nt requires that

$$n^2 = gk, \quad \ldots\ldots\ldots\ldots\ldots\ldots\ldots\ldots\ldots\ldots(17)$$

and the same well-known relation suffices to annul the term in $2nt$. The surface condition is satisfied without any addition to ψ, if besides satisfying (17) we identify $gF(x)$ with $-\frac{1}{4}k^2 A^2 \cos 2kx$. Thus, writing A' for kA/n, we obtain, as the complete value of η,

$$\eta = A' \sin kx \sin nt + \frac{1}{4}kA'^2 \cos 2kx \cos 2nt - \frac{1}{4}kA'^2 \cos 2kx. \quad \ldots(18)$$

We now proceed with the consideration of the problem when viscosity is retained. Eliminating p from (1) and (2), we get, with use of (3) and (5),

$$\nabla^4 \psi - \frac{1}{\nu}\frac{d}{dt}\nabla^2 \psi = \frac{u}{\nu}\frac{d\nabla^2 \psi}{dx} + \frac{v}{\nu}\frac{d\nabla^2 \psi}{dy}. \quad \ldots\ldots\ldots\ldots(19)$$

The terms on the right hand of (19) are of the second order in the amplitude of vibration, and thus for the first approximation we have simply

$$\nabla^4 \psi - \frac{1}{\nu}\frac{d}{dt}\nabla^2 \psi = 0. \quad \ldots\ldots\ldots\ldots\ldots\ldots(20)$$

The solution of (20) may be written

$$\psi = \psi_1 + \psi_2, \quad \ldots\ldots\ldots\ldots\ldots\ldots\ldots(21)$$

where $\qquad \nabla^2 \psi_1 = 0, \qquad \left(\nabla^2 - \frac{1}{\nu}\frac{d}{dt}\right)\psi_2 = 0. \quad \ldots\ldots\ldots\ldots(22)$

We now introduce the suppositions that in the first approximation ψ_1, ψ_2 are proportional to $\cos kx$, and also to e^{int}. The wave-length along x is $2\pi/k$, and the period τ is $2\pi/n$. The equations (22) now become

$$\left(\frac{d^2}{dy^2} - k^2\right)\psi_1 = 0, \qquad \left(\frac{d^2}{dy^2} - k^2 - \frac{in}{\nu}\right)\psi_2 = 0, \quad \ldots\ldots\ldots(23)$$

by which ψ_1 and ψ_2 are to be determined as functions of y. If we write

$$k'^2 = k^2 + in/\nu, \quad \ldots\ldots\ldots\ldots\ldots\ldots(24)$$

we have, as the most general solutions of (23),

$$\psi_1 = A'e^{ky} + B'e^{-ky}, \qquad \psi_2 = C'e^{k'y} + D'e^{-k'y}; \quad \ldots\ldots(25)$$

but if the real part of k' is taken positive, the terms in B' and D' are excluded when the fluid is treated as infinitely deep. Thus for our purpose

$$\psi_1 = A e^{ky} \cos kx \, e^{int}, \qquad \psi_2 = C e^{k'y} \cos kx \, e^{int}, \qquad \dots\dots(26)$$

where A and C are now absolute constants, real or complex. We shall presently find it convenient to suppose C real. From (4) we now find

$$u = (kA \, e^{ky} + k'C e^{k'y}) \cos kx \, e^{int}, \quad v = (kA \, e^{ky} + kC e^{k'y}) \sin kx e^{int}; \ \dots(27, 28)$$

and since to this approximation $d\eta/dt = v$,

$$\eta = (k/in)(A + C) \sin kx \, e^{int}. \quad \dots\dots\dots\dots(29)$$

If we omit the terms of the second order in (1), we find

$$\frac{1}{\rho} \frac{dp}{dx} = - \frac{d}{dy} \left(\frac{d\psi_1}{dt} + \frac{d\psi_2}{dt} \right) + \nu \frac{d}{dy} (\nabla^2 \psi_1 + \nabla^2 \psi_2) = - \frac{d}{dy} \frac{d\psi_1}{dt},$$

in virtue of (22); and in like manner

$$\frac{1}{\rho} \frac{dp}{dy} = \frac{d}{dx} \frac{d\psi_1}{dt} - g.$$

Hence

$$\frac{p}{\rho} = - gy + \int \left(\frac{d^2\psi_1}{dx\,dt} dy - \frac{d^2\psi_1}{dy\,dt} dx \right), \dots\dots\dots\dots(30)$$

the expression to be integrated being a perfect differential by (22).

Applying (30) to the present case, we find

$$p/\rho = - gy - in A e^{ky} \sin kx \, e^{int}. \ \dots \ \dots\dots\dots(31)$$

At the surface we are to suppose $y = \eta$, and η may be neglected in the term already multiplied by A. Thus at the surface

$$\frac{p}{\rho} = i \left\{ \frac{gk}{n} (A + C) - nA \right\} \sin kx \, e^{int}. \dots\dots\dots\dots(32)$$

Now that we have to reckon with viscosity, the stress conditions at the surface can no longer be expressed merely by p. In the usual notation, applicable also in the theory of elastic solids, p_{xx}, p_{yy} denote normal tractions across faces perpendicular to x and y respectively, while p_{xy} denotes the tangential traction which acts parallel to y across the face perpendicular to x, or the equal traction parallel to x across the face perpendicular to y. The expressions for these tractions are*

$$p_{xx} = - p + 2\mu \frac{du}{dx}, \qquad p_{yy} = - p + 2\mu \frac{dv}{dy}, \ \dots\dots\dots(33)$$

$$p_{xy} = \mu \left(\frac{dv}{dx} + \frac{du}{dy} \right). \ \dots\dots\dots\dots(34)$$

When viscosity was neglected, we were able to suppose that the surface of the fluid was entirely free from imposed force. Under such circumstances

* Cf. Lamb's *Hydrodynamics*, § 314.

the vibrations of a viscous fluid could not be maintained. If n is to be real, some maintaining forces are necessary. We will suppose that these forces are exclusively normal in their character, and accordingly make $p_{xy} = 0$; for it is to be observed that in the present approximation a direction parallel to the surface may be identified with the horizontal. By (27), (28), (34) we find, making $y = 0$,

$$2k^2 A + (k^2 + k'^2)\, C = 0, \quad \dots\dots\dots\dots\dots\dots(35)$$

as the condition of no tangential force at the surface. Or, if we substitute for k'^2 its value from (24), (35) becomes

$$A + \left(1 + \frac{in}{2k^2 \nu}\right) C = 0. \quad \dots\dots\dots\dots\dots(36)$$

For the normal traction at the surface we have from (33),

$$n p_{yy}/\rho = A\, \{i\, (n^2 - kg) + 2\nu k^2 n\} \sin kx\, e^{int}$$
$$+\, C\, \{-ikg + 2\nu kk'n\} \sin kx\, e^{int}, \quad \dots\dots\dots(37)$$

and by (36) the expression for η in (29) may be written

$$\eta = -\frac{C}{2k\nu} \sin kx\, e^{int}. \quad \dots\dots\dots\dots\dots(38)$$

These equations constitute the complete symbolical solution of the problem of infinitely small stationary waves maintained by purely normal surface pressures in a fluid of any degree of viscosity*.

In preparing to pass to real quantities, it is simplest to suppose $C = -2k\nu$, so that

$$\eta = \sin kx\, e^{int}. \quad \dots\dots\dots\dots\dots\dots(39)$$

A is then given explicitly by (36), and on substitution in (37) we get

$$p_{yy}/\rho = \{g - n^2/k + 4i\nu nk + 4\nu^2 k^2\, (k - k')\} \sin kx\, e^{int}. \quad \dots\dots(40)$$

The passage to real quantities is now only complicated by the term in k'. If the viscosity be small, this term may be omitted, and we may take

$$p_{yy}/\rho = \{g - n^2/k + 4i\nu nk\} \sin kx\, e^{int} = (g - n^2/k)\, \eta + 4\nu k \cdot d\eta/dt, \dots(41)$$

giving the normal traction necessary to maintain the waves represented by $\eta = \sin kx \cos nt$. If $n^2 = gk$, the part of p_{yy} in the same phase as η disappears, and

$$p_{yy} = 4\mu k \cdot d\eta/dt \dots\dots\dots\dots\dots(42)$$

simply. It is to be remembered that p_{yy} is a *traction*. If, as is usual in hydrodynamics, we use pressure (p'), we see that the pressure has its maximum value when the surface is *falling* fastest, as was to be expected.

---

* Cf. Basset's *Hydrodynamics*, § 520; Lamb, *loc. cit.* § 332.

The accurate expression for the real part of (40) may, of course, be formed. If in (24) we put

$$k^2 = P^2 \cos 2\alpha, \qquad n/\nu = P^2 \sin 2\alpha, \quad \dots\dots\dots\dots(43)$$

then
$$k' = P \cos \alpha + iP \sin \alpha. \quad \dots\dots\dots\dots\dots(44)$$

It is unnecessary to write down the actual form of the real part of (40). In most applications an approximate value of k' suffices. On account of the smallness of ν, n/ν is very large in comparison with k^2, that is to say, the thickness of the stratum through which the tangential motion can be propagated in time τ is very small relatively to the wave-length λ. We may, therefore, usually neglect k^4 in the equation

$$P^4 = k^4 + n^2/\nu^2, \quad \dots\dots\dots\dots\dots\dots(45)$$

and take simply
$$P^2 = n/\nu. \quad \dots\dots\dots\dots\dots\dots(46)$$

Again,
$$(\sin \alpha - \cos \alpha)^2 = 1 - \sin 2\alpha = \tfrac{1}{2}k^4\nu^2/n^2, \quad \dots\dots\dots\dots(47)$$

so that the difference between $\cos \alpha$ and $\sin \alpha$ may often be neglected.

It appears that the terms neglected in (40) when (41) is substituted are of the order $\nu^{3/2}$.

In proceeding to a second approximation we have first to calculate the terms of the second order forming the right-hand member of (19), using the values found in the first approximation. In these

$$\nabla^2\psi = \nabla^2\psi_2 = \frac{1}{\nu}\frac{d\psi_2}{dt} = \frac{inC}{\nu}\, e^{k'y} \cos kx\, e^{int},$$

whence, the imaginary part being rejected,

$$\frac{d\nabla^2\psi}{dx} = \frac{knC}{\nu} \sin kx\, e^{P\cos\alpha.\,y} \sin (nt + P\sin\alpha.\,y), \quad \dots\dots\dots(48)$$

$$\frac{d\nabla^2\psi}{dy} = -\frac{nPC}{\nu} \cos kx\, e^{P\cos\alpha.\,y} \sin (nt + P\sin\alpha.\,y + \alpha). \quad \dots\dots(49)$$

Also, in real quantities by (27), (28), (36),

$$u = C \cos kx \left\{ k e^{ky}\left(-\cos nt + \frac{n}{2k^2\nu}\sin nt \right) \right.$$

$$\left. + P e^{P\cos\alpha\; y} \cos (nt + P\sin\alpha.\,y + \alpha) \right\}, \quad \dots\dots\dots(50)$$

$$v = kC \sin kx \left\{ e^{ky}\left(-\cos nt + \frac{n}{2k^2\nu}\sin nt \right) \right.$$

$$\left. + e^{P\cos\alpha.\,y} \cos (nt + P\sin\alpha.\,y) \right\}. \quad \dots\dots\dots(51)$$

Hence $\quad \dfrac{2\nu\,(u\,d\nabla^2\psi/dx + v\,d\nabla^2\psi/dy)}{nkC^2\sin kx\,\cos kx\,e^{P\cos\alpha\,.\,y}}$

$$= -\,ke^{ky}\{\sin(2nt + P\sin\alpha\,.\,y) + \sin(P\sin\alpha\,.\,y)\}$$

$$+\,\frac{ne^{ky}}{2k\nu}\{\cos(P\sin\alpha\,.\,y) - \cos(2nt + P\sin\alpha\,.\,y)\}$$

$$+\,Pe^{ky}\{\sin(2nt + P\sin\alpha\,.\,y + \alpha) + \sin(P\sin\alpha\,.\,y + \alpha)\}$$

$$-\,\frac{nPe^{ky}}{2k^2\nu}\{\cos(P\sin\alpha\,.\,y + \alpha) - \cos(2nt + P\sin\alpha\,.\,y + \alpha)\}$$

$$-\,2Pe^{P\cos\alpha\,.\,y}\sin\alpha. \dots\dots\dots\dots\dots\dots\dots\dots\dots\dots(52)$$

By (19) the equation with which we have to deal is

$$\nabla^4\psi - \frac{1}{\nu}\frac{d}{dt}\nabla^2\psi = \frac{nkC^2\sin 2kx}{4\nu^2} \times e^{P\cos\alpha\,.\,y} \times \text{right-hand member of (52)}\dots(53)$$

It will be observed that in (52) or (53) the terms on the right, regarded as functions of t, are either independent of t or circular functions of $2nt$. Corresponding terms, proportional also to $\sin 2kx$, will appear in the direct integral of (53), and in addition we must include a "complementary function" representing, so far as required, the complete integral of (53) when the second member is made equal to zero. This part contains the terms of the first approximation (26), which now represent themselves; and we must also be prepared to admit terms of the second order proportional to $\sin 2kx$, and either independent of time or involving $2nt$. In the former case the differential equation reduces to

$$\nabla^4\psi = \left(\frac{d^2}{dy^2} - 4k^2\right)^2\psi = 0, \dots\dots\dots\dots\dots\dots(54)$$

giving as the solution applicable to deep water,

$$\psi = \sin 2kx\,(H + Ky)\,e^{2ky}, \dots\dots\dots\dots\dots\dots(55)$$

where H and K are constants. A term in K would represent vortices of the kind found in the former paper to arise from the action of the bottom (when the liquid is not too deep); and one of the principal objects of the present investigation is to ascertain whether these terms occur as the result of the conditions operative at the free surface. It may be recalled that though such vortices could not arise in an ideal frictionless fluid, their magnitude, when fully established, may be independent of the amount of the friction. In view of the complication of the problem it must suffice to limit the investigation to the case of *small* viscosity, the question being whether vortices can be maintained when the viscosity is reduced without limit.

On the right of (52) there are nine terms, of which five are independent of t. But they are not of equal importance. Since P is of the order $\nu^{-\frac{1}{2}}$, the leading term is the seventh in order. So far as this term is concerned,

$$\left(\frac{d^2}{dy^2} - 4k^2\right)^2 \psi = -\frac{n^2 PC^2 \sin 2kx}{8k\nu^3} \cdot e^{ky} e^{P\cos a \cdot y} \cos(P \sin a \cdot y + a).$$

It is now convenient to revert to complex quantities, regarding $e^{P\cos a \cdot y} \cos(P \sin a \cdot y + a)$ as the real part of $e^{k'y + ia}$. Thus

$$\left(\frac{d^2}{dy^2} - 4k^2\right)^2 \psi = -\frac{k' n^2 C^2 \sin 2kx}{8k\nu^3} \cdot e^{(k'+k)y},$$

giving

$$\psi = -\frac{k' n^2 C^2 \sin 2kx}{8k\nu^3 \{(k'+k)^2 - 4k^2\}^2} \cdot e^{(k'+k)y}. \qquad \ldots\ldots\ldots\ldots(56)$$

To form the surface condition, representing the evanescence of tangential force, we shall require the expression of

$$\frac{du}{dy} + \frac{dv}{dx} = \frac{d^2\psi}{dy^2} - \frac{d^2\psi}{dx^2} = \{(k'+k)^2 + 4k^2\} \psi,$$

in which, since C^2 is already involved as a factor, we may put $y = 0$. Thus

$$\frac{du}{dy} + \frac{dv}{dx} = -\frac{k' n^2 C^2 \sin 2kx \{(k'+k)^2 + 4k^2\}}{8k\nu^3 \{(k'+k)^2 - 4k^2\}^2}, \qquad \ldots\ldots\ldots\ldots(57)$$

from which the imaginary part is to be rejected.

In tracing the value of (57) as ν diminishes, we get ultimately

$$-\frac{n^2 C^2 \sin 2kx}{8kk'\nu^3}. \qquad \ldots\ldots\ldots\ldots\ldots\ldots\ldots\ldots\ldots\ldots(58)$$

In this C^2/ν^2 is of the order A^2, and k' is of order $\nu^{-\frac{1}{2}}$, so that (58) is of order $\nu^{-\frac{1}{2}} A^2$, becoming infinite in comparison with A^2, as ν diminishes without limit. We shall find, however, that this infinite term is compensated by another, to be brought forward later. For our purpose we must retain all terms which do not vanish with ν in comparison with A^2. Hence, with sufficient approximation,

$$\frac{k'\{(k'+k)^2 + 4k^2\}}{\nu\{(k'+k)^2 - 4k^2\}^2} = \frac{k'}{\nu(k'+k)^2} = \frac{1}{\nu k'} - \frac{2k}{\nu k'^2} = \frac{1}{\nu k'} - \frac{2k}{in + \nu k^2} = \frac{1}{\nu k'} + \frac{2ik}{n},$$

of which the second term, being purely imaginary, is to be rejected. Thus

$$\frac{du}{dy} + \frac{dv}{dx} = \text{real part of } -\frac{n^2 C^2 \sin 2kx}{8kk'\nu^3} = -\frac{n^2 C^2 \cos a \sin 2kx}{8k\nu^3 P} \quad \ldots.(59)$$

Referring back to (52) and having regard to the orders of the various terms in respect of ν, we see that the only other term independent of t which needs be retained is the third in order of arrangement. From this term we obtain in like manner,

$$\psi = \frac{n^2 C^2 \sin 2kx}{8\nu^3} \cdot \frac{e^{(k'+k)y}}{\{(k'+k)^2 - 4k^2\}^2}, \qquad \ldots\ldots\ldots\ldots(60)$$

and when $y = 0$,

$$\frac{du}{dy} + \frac{dv}{dx} = \frac{n^2 C^2 \sin 2kx}{8\nu^3} \frac{(k'+k)^2 + 4k^2}{\{(k'+k)^2 - 4k^2\}^2} = \frac{n^2 C^2 \sin 2kx}{8\nu^3 k'^2} \quad \ldots\ldots(61)$$

with sufficient approximation.

In (61), $\nu k'^2$, so far as it need be retained, is a pure imaginary, and accordingly there is no contribution from this source. We are left therefore with (59) as the complete contribution of the direct integral of (53) when $\nu = 0$, so far as the terms independent of t are concerned.

In a similar manner we may treat the terms in $2nt$. We find for the leading term

$$\nabla^4 \psi - \frac{1}{\nu} \frac{d}{dt} \nabla^2 \psi = \text{real part of } \frac{k' \cdot n^2 C^2 \sin 2kx}{8k\nu^3} \cdot e^{(k'+k)y} e^{2int};$$

so that, if we finally reject the imaginary part,

$$\frac{du}{dy} + \frac{dv}{dx} = \frac{n^2 C^2 \sin 2kx \, e^{2int}}{8k\nu^2} \left(-\frac{1}{\nu k'} + \frac{2ik}{n} \right), \quad \ldots\ldots\ldots(62)$$

when $y = 0$. And for the only other term which it is necessary to retain,

$$\frac{du}{dy} + \frac{dv}{dx} = \frac{n C^2}{8\nu^2} \sin 2kx \sin 2nt. \quad \ldots\ldots\ldots\ldots\ldots(63)$$

We have now sufficiently complete expressions for $du/dy + dv/dx$ so far as it results from the direct integral of (53). But to these we have to add terms arising from the complementary function. In this there must, at any rate, be included the terms in nt found in the first approximation. From (27), (28) we get, when y is small,

$$\frac{du}{dy} + \frac{dv}{dx} = \cos kx \cdot e^{int} \{2k^2 A + (k^2 + k'^2)C + 2k^2 A \cdot ky + (k^2 + k'^2)C \cdot k'y\}, \quad (64)$$

and in this we are to substitute for y the value η appropriate to the surface. In the first approximation the terms containing y were neglected, and the surface condition gave

$$2n^2 A + (k^2 + k'^2) C = 0. \quad \ldots\ldots\ldots\ldots\ldots\ldots(65)$$

This relation must still hold, approximately at any rate. Using it in the small terms of (64), we get

$$(k'^2 + k^2) C (k' - k) y \cos kx \, e^{int},$$

in which we may neglect k in comparison with k'. Passing to real quantities, we find

$$-\frac{nPC}{\nu} y \cos kx \sin (nt + \alpha).$$

The real value of y, or η from (38), is

$$\eta = -\frac{C}{2k\nu} \sin kx \cos nt, \quad \ldots\ldots\ldots\ldots\ldots\ldots(66)$$

so that this part of (64) becomes

$$\frac{du}{dy} + \frac{dv}{dx} = \frac{nPC^2}{8k\nu^2} \sin 2kx \{\sin(2nt+\alpha) + \sin\alpha\}. \quad\ldots\ldots\ldots(67)$$

Thus for the terms independent of t we get, from (59), (67),

$$\frac{du}{dy} + \frac{dv}{dx} = \frac{nC^2 \sin 2kx}{8k\nu^2} \left\{ -\frac{n\cos\alpha}{P\nu} + P\sin\alpha \right\}. \quad\ldots\ldots\ldots(68)$$

The factor within braces vanishes with ν, as may be proved from (43), the two infinite terms cancelling without finite residue. Thus altogether $du/dy + dv/dx$ vanishes with ν so far as the terms independent of t are concerned.

The terms in $2nt$ are of less interest. We find, in the same way, when ν is diminished without limit for the complete value of $du/dy + dv/dx$ at the surface,

$$-\frac{nC^2 \sin 2kx \sin 2nt}{8\nu^2}. \quad\ldots\ldots\ldots\ldots(69)$$

But we are not yet in a position to apply the condition which must be satisfied at the surface, viz., that the tangential force shall there vanish; for we must remember that the surface can no longer be treated as parallel to $y=0$. If θ be the angle which the surface at the point under consideration makes with $y=0$, the formulae of transformation are

$$p_{x'x'} = \cos^2\theta\, p_{xx} + \sin^2\theta\, p_{yy} + \sin 2\theta\, p_{xy},$$
$$p_{y'y'} = \sin^2\theta\, p_{xx} + \cos^2\theta\, p_{yy} - \sin 2\theta\, p_{xy},$$
$$p_{x'y'} = (\cos^2\theta - \sin^2\theta)p_{xy} + \sin\theta\cos\theta\,(p_{yy} - p_{xx}).$$

For the present purpose θ may be regarded as a small quantity, equal to $\pm\, d\eta/dx$, whose square may be neglected, and we may take

$$p_{x'y'} = p_{xy} + \theta\,(p_{yy} - p_{xx}). \quad\ldots\ldots\ldots\ldots(70)$$

It is $p_{x'y'}$, and not p_{xy}, which is to be made to vanish to the second order of the vibration.

The expressions for p_{xx}, etc., have already been given in (33), (34). Substituting the values of u, v of the first approximation, we find, when ν is small,

$$-2\left(\frac{du}{dx} - \frac{dv}{dy}\right) = \frac{2nC}{\nu} \sin kx \sin nt, \quad\ldots\ldots\ldots(71)$$

and

$$\frac{d\eta}{dx} = -\frac{C}{2\nu} \cos kx \cos nt. \quad\ldots\ldots\ldots(72)$$

It appears, then, that so far as the terms independent of t are concerned, there is no difference between $p_{x'y'}$ and p_{xy}, and since we have already seen that p_{xy} vanishes, it follows that the surface condition of no tangential force is satisfied to a second approximation, without the addition of any further terms (55), such as would represent permanent vortices. Accordingly, no such vortices exist.

As regards terms in $2nt$, we find in addition to (69) another term of the same form derived from (71) and (72). In a solution complete to the second order these terms would need to be compensated by the introduction of new second-order terms in ψ of the form

$$\psi = (Le^{2ky} + Me^{k''y}) \sin 2kx \, e^{2int}, \quad \ldots\ldots\ldots\ldots\ldots\ldots(73)$$

where $$k''^2 = 4k^2 + 2in/\nu ; \quad \ldots\ldots\ldots\ldots\ldots\ldots\ldots(74)$$

but it is scarcely necessary for our purpose to define them further. Neither does it seem worth while to express at length the equation of pressure when the second-order terms are included. The particular case of no viscosity already considered illustrates the procedure. Terms in the expression of the pressure which are independent of t are balanced by corresponding terms in η not affecting the velocities.

331.

ACOUSTICAL NOTES.—VIII.

[*Philosophical Magazine*, Vol. XVI. pp. 235—246, 1908.]

Partial Tones of Stretched Strings of Variable Density—Maintenance of Vibrations by Impact of Water Drops—Discrimination between Sounds from directly in front and directly behind the Observer—The Acousticon—Pitch of Sibilants—Telephones.

Partial Tones of Stretched Strings of Variable Density.

IF ρ_0 be the longitudinal density of a string of length l stretched with tension T_1, the periodic time (τ_r) of the rth component vibration is given by

$$\tau_r{}^2 = \frac{4l^2\rho_0}{r^2 T_1}. \quad\dots\dots\dots\dots\dots\dots\dots\dots(0)$$

In *Theory of Sound*, 2nd ed. § 140, an approximate theory is given applicable when the density, in place of being strictly uniform, has the value $\rho_0 + \delta\rho$, where $\delta\rho$ is relatively small. We have

$$\tau_r{}^2 = \frac{4l^2\rho_0}{r^2 T_1}\left\{1 + \frac{2}{l}\int_0^l \frac{\delta\rho}{\rho_0}\sin^2\frac{r\pi x}{l}\,dx\right\}. \quad\dots\dots\dots\dots(1)^*$$

"If the irregularity take the form of a small load of mass m at the point $x = b$, the formula may be written

$$\tau_r{}^2 = \frac{4l^2\rho_0}{r^2 T_1}\left\{1 + \frac{2m}{l\rho_0}\sin^2\frac{r\pi b}{l}\right\}. \quad\dots\dots\dots\dots\dots\dots(2)$$

These values of τ_r are correct so far as the first power of the small quantities $\delta\rho$ and m, and give the means of calculating a correction for such slight departures from uniformity as must always occur in practice.

---

* The r^2 in the denominator was inadvertently omitted. It may be worthy of note that the method is not limited to the case of density originally uniform. In general, if Y be the normal function corresponding to a variable density ρ, $\tau_r{}^2$ is altered in the ratio

$$1 + \frac{\int \delta\rho\, Y^2\, dx}{\int \rho\, Y^2\, dx}.$$

As might be expected, the effect of a small load vanishes at nodes, and rises to a maximum at the points midway between consecutive nodes. When it is desired merely to make a rough estimate of the effective density of a nearly uniform string, the formula indicates that attention is to be given to the neighbourhood of loops rather than to that of nodes.

The effect of a small variation of density upon the period is the same whether it occur at a distance x from one end of the string, or at an equal distance from the other end. The *mean* variation at points equidistant from the centre is all that we need regard, and thus no generality will be lost if we suppose that the density remains symmetrically distributed with respect to the centre. Thus we may write

$$\tau_r^2 = \frac{4l^2 \rho_0}{r^2 T_1}(1 + \alpha_r), \quad\quad\quad\quad\quad\quad (3)$$

where

$$\alpha_r = \frac{2}{l}\int_0^{\frac{1}{2}l} \frac{\delta\rho}{\rho_0}\left(1 - \cos\frac{2\pi r x}{l}\right)dx. \quad\quad (4)$$

In this equation $\delta\rho$ may be expanded from 0 to $\frac{1}{2}l$ in the series

$$\frac{\delta\rho}{\rho_0} = A_0 + A_1 \cos\frac{2\pi x}{l} + \ldots + A_r \cos\frac{2\pi r x}{l} + \ldots, \quad\quad (5)$$

where

$$A_0 = \frac{2}{l}\int_0^{\frac{1}{2}l}\frac{\delta\rho}{\rho_0}dx, \quad\quad A_r = \frac{4}{l}\int_0^{\frac{1}{2}l}\frac{\delta\rho}{\rho_0}\cos\frac{2\pi r x}{l}dx. \quad\quad (6, 7)$$

Accordingly,

$$\alpha_r = A_0 - \tfrac{1}{2}A_r. \quad\quad\quad\quad\quad\quad (8)$$

This equation, as it stands, gives the changes in period in terms of the changes of density supposed to be known. And it shows conversely that a variation of density may always be found which will give prescribed arbitrary displacements to all the periods. This is a point of some interest.

In order to secure a reasonable continuity in the density, it is necessary to suppose that $\alpha_1, \alpha_2, \ldots$ are so prescribed that α_r assumes ultimately a constant value when r is increased indefinitely. If this condition be satisfied, we may take $A_0 = \alpha_\infty$, and then A_r tends to zero as r increases.

As a simple example, suppose that it be required so to vary the density of a string that, while the pitch of the fundamental tone is displaced, all other tones shall remain unaltered. The conditions give

$$\alpha_2 = \alpha_3 = \alpha_4 = \ldots = \alpha_\infty = 0.$$

Accordingly,

$$A_0 = A_2 = A_3 = \ldots = 0, \quad\quad A_1 = -2\alpha.$$

Thus by (5)

$$\delta\rho/\rho_0 = -2\alpha_1 \cos(2\pi x/l). \quad\quad\quad\quad\quad\quad (9)\text{''}$$

I have recently made a few observations on the vibrations of loaded wires in illustration of these formulæ. A simple case is afforded by (2). A lump of wax is attached at the centre of a stretched wire, so that $\sin^2(r\pi b/l) = 0$ when r is even, and when r is odd takes the value unity, independently of the particular value of r. It is easily verified on trial that the partial tones of even order are undisturbed, and that those of odd order, though displaced, still constitute a *harmonic series*. The observation is best made with the aid of resonators by comparison with the partial tones of a note on the harmonium. The load may conveniently be chosen so as to depress the pitch by one or more complete semitones. If also the tension (due to a weight) be suitably adjusted, both series of partial tones may be brought into tune with the harmonium, either exactly, or (what is better in practice) approximately.

A more elaborate experiment was next attempted in illustration of the result expressed in (9). But some modification is necessary, inasmuch as in practice we are limited to *positive* loads. But a uniform loading has little significance from our present point of view, lowering all the partial tones by the same musical interval and being capable of compensation by varying the tension (T_1). It suffices therefore to take $\delta\rho/\rho_0$ proportional to $1 - \cos(2\pi x/l)$; and the experiment consists in verifying that the first partial is depressed relatively to all the other tones, and that these latter retain their harmonic relations.

The length of the wire was adjusted to 720 mm., and loads proportional to $1 - \cos(2\pi x/l)$ were attached at intervals of 30 mm., beginning at $x = 0$ and corresponding to angles 0°, 15°, 30°, &c. The unit load (about $\frac{1}{4}$ gm. of wax) was at the points distant from the ends by one quarter of the length (90°, 270°), and the maximum load (2) was at the centre (180°). At the ends (0°, 360°) the loads were zero. By varying the tension the whole was tuned conveniently to suit the harmonium.

The results agreed fairly well with theoretical anticipations. The second and third partials retained very accurately their harmonic relation (fifth). The fourth and fifth partials were too sharp relatively to the second and third by nearly half a semitone. The depression of the first partial relatively to the second and third was 3 semitones or a minor third.

These estimations of pitch were made separately by Mr Enock and myself, and in the case of the higher partials demanded some care. Useful indications are afforded by but partially depressing the key of the harmonium note, whereby the pitch is lowered relatively to the normal. On the other hand it appeared, rather to my surprise, that a lowering of the bellows pressure (sometimes convenient in order to diminish the intensity of sound) slightly *raises* the pitch.

When the loads of wax were all removed, the first partial tone rose nearly an octave, indicating that the loading had been very severe.

In a second experiment the loads were reduced to less than half, the unit being taken equal to 0·1 gm., the distribution of the loads and the length of the wire being as before. Comparisons, either directly with the harmonium or with forks as intermediaries, gave the following results. The second partial of the wire was 5 beats per second flat on (the octave overtone of) harmonium B. The third partial was 4 beats per second flat on the corresponding partial of B. The fourth partial was 2 beats flat on B, and the fifth partial was 2 beats sharp on B. If we regard the two last partials as in tune with B, the second partial is out by 5 vibrations in about 256, or one in 55, viz. between a third and a quarter of a semitone. The first partial of the loaded wire was almost exactly two semitones below harmonium B.

I had supposed that the small deviations from harmonic relations exhibited by the second and higher partials might be due to the discontinuity of the loading or to the fact that the loads were not sufficiently concentrated upon the line of the wire. But subsequent observations upon an unloaded wire showed very similar deviations. The length was the same but the tension was diminished so as to bring the pitch back to B. The first partial of the wire was now one per second flat, the second $1\frac{1}{2}$ per second flat, the third 2 per second flat, the fourth 1 per second *sharp*, the fifth $3\frac{1}{2}$ per second sharp, and the sixth about 8 per second sharp, all referred to the corresponding partials of harmonium B. Since the partials of the harmonium note are necessarily in tune, harmonic relations among the partials of the wire would require beats always upon the same side and numerically as the natural numbers 1, 2, 3, 4, 5, 6. It may be noted that the wooden bridges, by which the vibrating portion of the wire was limited, were provided above and below with sharp edges, and were free to turn. In this way the vibrating portion was well defined, and the tension (due to weights hung over a pulley) was freely transmitted.

Maintenance of Vibrations by Impact of Water Drops.

It is known that jets of liquid tend to resolve themselves under capillary force into more or less uniform processions of drops, the distance between the drops, or the number of drops passing a given point in unit time, depending upon the diameter and velocity of the jet. When a jet, otherwise undisturbed, is under the influence of a regular vibrator of suitable pitch, the resolution becomes absolutely uniform and takes its frequency from the vibrator[*]. If the procession of drops thus regularized is caused to strike a second vibrator of like pitch, the latter will usually be excited. The arrangement is simplified and any question of insufficient agreement of pitch is eliminated, if both parts

[*] The theory of these effects, on lines roughly sketched by Savart and Plateau, is given in *Proc. Roy. Soc.* Vol. xxix. p. 71 (1879); *Scientific Papers*, Vol. i. p. 377; or *Theory of Sound*, 2nd edition, Vol. ii. p. 362.

be assigned to the *same* vibrator which at once regularizes the resolution of the jet and is itself maintained in vibration by the impact of the drops.

The experiment has been tried with tuning-forks of pitch 256 and 320. The reaction between the fork A and the jet issuing at B (fig. 1) is effected through a branch tube E terminating in a metal box D. The box is provided with a wooden socket C cemented on a flexible face, to which the fork is screwed. The vibrations of the fork, transmitted through its stalk, cause the flow at B to be slightly variable, and if the adjustments are suitable determine a stream of drops of the same frequency. The orifice at B, perforated in a thin metal plate, is of about 2 mm. diameter. The tubes are of lead, allowing slight adjustments by bending; the supply of water may be either directly from a tap or preferably from an aspirator bottle. The head of water, about 30 cm., must be adjusted; and it is to be remarked that the question is not merely one of accommodating the natural pitch of the jet to that of the fork. There

Fig. 1.

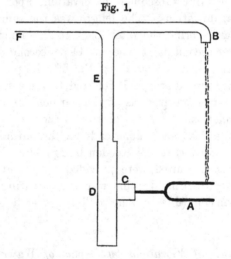

is also the phase to be considered, for the impact of the drops may check, as easily as encourage, an existing vibration. A slight alteration in the distance between A and B may here be useful. In practice attention should be given to the place of resolution of the jet, easily discerned in a suitable light. When this is brought up as near to the orifice as possible, it will be known that the vibration is vigorous and that the phase relation is suitable.

The experiment was quite successful. Both the forks referred to spoke well and steadily when suitable resonators were held near their prongs. But the arrangement is hardly to be recommended for general purposes. The use of water is messy, and unless care be taken is likely to end in rusting the forks. Moreover, the vibrations are not especially vigorous—in comparison, for example, with those which may be obtained electromagnetically. Another objection is to be found in the circumstance that drops of water remaining attached to the fork must render the precise pitch uncertain.

Discrimination between Sounds from directly in front and directly behind the Observer.

As already* mentioned, I am now unable to make the discrimination myself, even in the case of the voice used naturally; so that all that I can report relates to the observations of others. It would seem, however, that even youthful listeners are not always able to pronounce with certainty. In experiments made with some young people in a long corridor, they were able to discriminate among themselves whether a voice came from in front or behind, but when I spoke they made mistakes. The speaker facing towards the listener gave, for example, the numerals one—two—three &c.; and there were enough assistants moving backwards and forwards to eliminate information which might otherwise be given by footsteps. Why my voice afforded less foundation for a judgment was not clear—possibly in consequence of its graver pitch, or because its quality was less familiar. The corridor was so long that the observations were not appreciably disturbed by echos.

As mentioned before, Mr Enock is able in many cases to discriminate front and back when the voice is used naturally. But I find that both indoors and outside he could be deceived. Thus when standing on the lawn only a short distance in front of him, but facing *from* him, I gave the numerals, he judged that I was behind him, and this erroneous judgment was not disturbed even when I conversed freely with him. It would appear that there is not much to go upon, and that when an erroneous impression has once been made it is not easily disturbed by the slight indications available. Probably the turning away of the speaker softens the sibilants and other high elements in the sound, somewhat in the same way as is done by the external ears of the listener when he faces away from the sound. It must be understood that in these experiments the ears were used in a natural manner, without the aid or hindrance of special reflectors.

The repetition and extension of these observations would be of interest; it would be best carried out under the supervision of a physicist young enough to be able himself to form judgments as to the front or back situation of the easier sounds, *e.g.* of the voice. The precautions necessary are indicated in former papers.

The Acousticon.

This instrument, intended to aid the hearing of the partially deaf, is composed of a simple battery, microphone, and telephone circuit. There is

* *Phil. Mag.* Vol. XIII. p. 231 (1907). [This Collection, Vol. v. p. 362.]

nothing special about the battery (one or two dry cells) or the telephone. But the microphone is unusually efficient. The disk which receives the sound is of carbon, about $\frac{1}{2}$ mm. thick, and is clamped at the circumference. Bearing against it are six groups of small ($\frac{1}{2}$ mm.) carbon spheres, having the appearance of shot, held in hemispherical cups cut out in a thick plate of carbon (figs. 2, 3). In use the microphone may be worn like a medal upon the breast, or it may stand upon the table, e.g. at dinner. In a large model there are two microphones, as described, connected in parallel. The instruments upon which I experimented were kindly lent me by a friend who had found them efficient, though fatiguing to the ear.

For my purpose the microphone and telephone were placed in separate rooms, so that nothing could be heard except through the instrument. The reproduction of speech, given at about one foot away from the microphone, was better than anything I had ever heard before. The first impression was

Fig. 2.   Fig. 3.

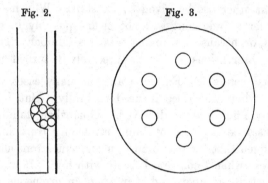

that all the consonantal sounds were completely rendered, but this turned out to be an illusion. In listening to the numerals, given in order, the observer would feel confident that he heard the *f* in *five* and the *s* in *six*. But if the initial sound was prolonged—*fff ive*, *s s s ix*, the observer could not tell until he heard the sequel which it was going to be. Further, if the sounds were given as *s s ive*, *ff ix*, they were heard normally as five and six. It was plain that there was no difference in the rendering of *f* and *s*. I am informed that this is a well-known difficulty in ordinary telephoning, and that in spelling a name containing *f* or *s* it is usual to say "*f* for Friday" or "*s* for Saturday." But the articulation of the acousticon is so superior that it was surprising to find the failure complete. The characterisation of *sh* was not much better, though after a little practice I could distinguish it from *s* or *f*, but probably only by a greater loudness.

These failures might have been ascribed to my rather defective hearing, but other observers with normal hearing did no better. When, however, the iron plate of the telephone was replaced by one three times as thick, a difference between *f* and *s* could be detected, though both were rather weak. The above observations were made with the small model and usually with but one battery cell. In the case of the large model microphone and with two cells in action, it was just possible to hear some difference between f and s, the usual iron plate of the telephone being employed.

The question arises as to *how* the acousticon aids defective hearing. A failure to distinguish f and s seems a bad beginning. I could not find that the general loudness of speech was increased by the instrument. When the speaker stood some 20 feet away from the microphone, I could hear better directly from the situation of the microphone than when in the further room I listened through the instrument. Possibly a clue may be found in the behaviour of the acousticon in presence of low notes. These are very feebly transmitted. Pure tones from forks making 128, 256, 384 vibrations per second, held in front of Helmholtz resonators and close to the microphone, were poorly heard, and of the sound received the greater part seemed to be overtones. Of course low notes are not necessary for speech; otherwise women would be at a disadvantage, contrary to all tradition. But can we suppose that low notes are actually deleterious?

At this point one recalls the observations of A. M. Mayer[*] upon the obliteration of higher sounds by graver ones. These observations have not attracted the attention they deserve. The author himself sufficiently emphasises their importance but he does not appear to have followed them up, as he announced the intention of doing. It was proved that while higher and feebler sounds could be entirely obliterated by louder and graver ones, on the other hand a feeble graver sound remained audible in the presence of a more powerful acuter one. "Indeed in this case as in all others where one sound remains unaffected by intense higher notes, the observer feels as though he had a special sense for the perception of a graver sound—an organ entirely distinct from that which receives the impress of the higher tones.

"That one sonorous sensation cannot interfere with another which is lower in pitch is a remarkable physiological discovery...."

If we suppose, as I think we may, that one type at any rate of deafness involves obtuseness to the higher elements of sound upon which the intelligibility of speech largely depends, while the hearing of graver sounds is unimpaired, Mayer's principle suggests that advantage may ensue from an instrumental suppression of the graver components. A scientific friend has informed me that a relative of his was insensitive to grave sounds and that when addressed in a railway carriage expostulated against being shouted to, as if his hearing was less interfered with by the noises of the train than that of normal persons. There is said indeed to be one type of deafness which finds advantage in such noises, but perhaps only because people speak louder.

[*] *Phil. Mag.* Vol. II. p. 500 (1876).

It would be interesting to experiment upon such a case in detail. For my own part I can hear scarcely anything of what is said to me in the train.

Another indication of the importance of the higher elements in speech is afforded by the advantage experienced by many deaf people from placing the hands behind the ears, palms forward and curved. The tick of a clock, for example, is much enhanced. Perhaps the most striking effect is upon the sound of the wind whistling through trees [or the sound of falling rain]. Artificial reflectors may of course replace the hands, but for the best results they need to be rather nicely shaped.

Pitch of Sibilants.

In connexion with the audibility of sibilants it is desirable to have some idea of their character in respect of pitch. Doubtless this may vary over a considerable range. In my experiments the method was that of nodes and loops*, executed with a sensitive flame and sliding reflector. A hiss given by Mr Enock, which to me seemed very high and not over audible, gave a wave-length (λ) equal to 25 mm. with good agreement on repetition. A hiss which I gave was graver and less definite, corresponding to $\lambda = 32$ mm. The frequency of vibration would be of the order 10,000 per second, more than 5 octaves above middle c†.

Telephones.

Some miscellaneous observations upon telephones may here be recorded. From the fact that an improved articulation of the sibilants and some other consonants accompanies a thickening of the telephone plate, although at considerable cost in other respects, it was thought that a similar advantage might be attained by the introduction of a condenser into the electrical circuit. This entails no further complication if the transmitter, as well as the receiver, be a Bell instrument; but if a microphone be employed as transmitter, a small transformer must also be introduced, whose ratio of transformation may, if desired, be 1 : 1. The line, receiver, and condenser then constitute electrically a secondary circuit, and by choice of a suitable capacity the proper tone of this circuit may be tuned to any desired pitch. The earliest consideration of resonance in an electrical circuit in response to a periodic force was probably by Maxwell‡ in connexion with Grove's

---

* Phil. Mag. Vol. VII. p. 149 (1879); Scientific Papers, Vol. I. p. 406.

† Compare Wiersch, Drude Ann. Vol. XVII. p. 1001 (1905). "Sieht man von ihren tiefsten Partialtönen ab, welche lediglich infolge Resonanz der Mundhöhle beigemischt sind, so entstehen die eigentlichen Reibelaute durch eine Schwingungszahl, welche minimal derjenigen des Grundtones der Luftsäule einer einseitig gedeckten Pfeife von ca. 14 mm. Länge entspricht, maximal aber im Bereiche der Unhörbarkeit liegt." This paper contains some interesting observations upon the influence of the proper tone of a telephone plate upon the articulation.

‡ Phil. Mag. May 1868; Maxwell's Scientific Papers, Vol. II. p. 121.

Experiment in Magneto-Electric Induction. If L, R be respectively the self-induction and resistance of a circuit, C the capacity of the interposed condenser, the current (dx/dt), elicited by the imposed electromotive force $E e^{int}$, is given by

$$\left(-Ln^2 + iRn + \frac{1}{C}\right) x = E;$$

so that the maximum effect occurs when

$$CLn^2 = 1,$$

i.e., when the natural frequency of the circuit coincides with that of the imposed force. *For this pitch* the self-induction is compensated. The effect of short-circuiting the condenser may be represented by taking $C = \infty$. If, apart from phase, the current is unaltered by short-circuiting the condenser, the capacity in action must be equal to one-half of the most favourable capacity, or else must be so great as not to be distinguishable in experiment from infinity. It is to be noted that the accurate compensation of self-induction can be effected for only one pitch, and that in practice the advantage will be limited to a range of pitch not exceeding an octave.

Such experiments as I have been able to make did not exhibit any improvement in articulation as the result of including a condenser in the circuit. It is possible that the simple theory above stated is too much interfered with by complications such as eddy-currents in the iron or hysteresis in the action of the condenser. A distinct resonance could be attained in the region of the higher notes of my harmonium (about 2000 vibrations per second) when an *additional* self-induction was included in the circuit. Such resonance would be practically limited to a region including 3 or 4 semi-tones, and when at the best pitch the condenser was put out of action by a short-circuiting key there was a very marked falling off of intensity. It is probable that this subject has already received attention from those engaged in the endeavour to improve the telephone as a practical instrument; if not, I think it would be worthy of investigation. There should be no great difficulty in securing several electrical resonances at pitches suitably distributed, in addition to the mechanical resonance of the plate.

If my memory serves me, it was observed in the early days of telephony that the sounds remained audible even though the usual ferrotype plate were replaced by one of other material. The experiment is easily tried. When a telephone provided with a copper or aluminium plate was included in a circuit with a battery (of one or two cells) and a simple carbon microphone (after Hughes), the ticking of a watch placed near the microphone was easily audible, and (it happened) with different quality in the two cases. On the other hand a plate of mica was silent under similar conditions, although when the excitation is very violent, as with a make-and-break arrangement, sounds may be heard without a plate at all. The telephone employed was bipolar and of modern manufacture.

In another experiment the permanent steel magnets were removed from the telephone and replaced by a soft-iron **U** which could be magnetized at will by an independent electric current, the coils and pole-pieces of the telephone remaining undisturbed. Without a magnetizing current but little of the ticks of the watch could be heard from the copper plate, but when the soft iron was magnetized the sounds became distinct as before.

I had supposed at first that this experiment might discriminate between the two possible explanations of the sound, the one depending upon traces of iron as an impurity in the copper or aluminium, the other invoking the action of induced currents circulating in the metallic plates. But it appears that in either case the efficiency would be promoted by a high constant magnetization of the pole-pieces.

The copper disk, weighing about one gram, was of a kind unlikely to contain appreciable iron and its action was not affected by washing with hydrochloric acid. When a mica disk, which of itself gave no sound, was dusted over with 1 mg. of fine iron filings attached with varnish, only a very faint sound could be made out. A similar mica disk coated with $\frac{3}{4}$ gm. of copper filings from the same material as that of the disk, yielded no sound. A similar telephone with copper disk was included in the circuit for the sake of rapid comparison throughout the experiments.

From these results it appeared unlikely that the effects were to be attributed to traces of iron in the other metals, and this conclusion was confirmed in a varied form of the experiment tried later. Mr Enock prepared two flat coils of fine covered copper wire weighing together about 2 gms. These were mounted separately on pieces of mica afterwards cemented with wax to the main disk so as to encircle the magnetic poles. When the ends of the flat coils were disconnected, nothing was heard ; but on completing the circuit of one or both coils the effects were very distinct, both with the carbon microphone and watch as before used or with another arrangement giving a more powerful action. We may conclude, I think, that the sounds of a copper disk are due to alternating currents induced in it and reacting upon the nearly constant magnetism of the pole-pieces.

332.

ON REFLEXION FROM GLASS AT THE POLARIZING ANGLE.

[*Philosophical Magazine*, Vol. XVI. pp. 444—449, 1908.]

ACCORDING to Fresnel's theory the polarization is complete when light is reflected at the Brewsterian angle ($\tan^{-1}\mu$) or, as we may put it, light vibrating in the plane of incidence is not reflected at all at the angle in question. It has long been known that this conclusion is but approximately correct. If we attempt to extinguish with a nicol sunlight reflected from ordinary glass, we find that at no angle of incidence and reflexion can we succeed. It is difficult even to fix upon an angle of minimum reflexion with any precision.

The interpretation of these deviations from Fresnel's laws is complicated by uncertainties as to the nature of surfaces of transition from one medium to another. It is certain that many, if not all, surfaces attract to themselves films of moisture and grease from the surrounding atmosphere, and the opinion has been widely held that even in the absence of moisture and grease solid bodies are still coated with films of condensed air. Other complications depend upon possible or probable residues of the polishing material used in the preparation of optical surfaces. It was mainly for these reasons that I gave much attention some years ago* to the case of reflexion from *water*, where at any rate there was no question of a polishing powder and atmospheric moisture could introduce no complication. It was found that Jamin's results, up to that time considered standard, were entirely vitiated by films of grease. Special operations are necessary to remove these films. When proper precautions are taken, the intensity of reflexion at the polarizing angle may be less than $\frac{1}{1000}$ of what Jamin observed. It appeared, however, that the cleanest surfaces were not those which gave the least reflexion. At the highest degrees of purity the light again began to undergo reflexion, though to a very limited amount. The effect had changed *sign*.

---

* *Phil. Mag.* Vol. XXXIII. p. 1 (1892); *Scientific Papers*, Vol. III. p. 496.

The contamination which produces the effects observed by Jamin is but slight, regarded from any other than the optical or capillary point of view. The thickness of the film of olive oil which suffices to stop the movements of camphor fragments deposited upon water, is 2×10^{-7} cm., or about $\frac{1}{300}$ of λ_D. But such a film, or even a much thinner one, entirely disturbs the delicate balance upon which depends the absence of reflexion at the polarizing angle.

For a long time I have intended to make an examination of the corresponding phenomena when light is reflected from a surface of *glass*. I was prepared for complications depending upon moisture and grease, but thought that perhaps I could deal with them. As to the thickness of the films there is little definite information. Theory* indicates that they are likely to be persistent. A long while ago Magnus established the conclusion "that all substances, however different they may be, are raised in temperature when air comes in contact with them which is moister than that surrounding them, and that they are depressed in temperature when they are exposed to air which is drier than that by which they are surrounded†." His experiments included glass, quartz, mica, caoutchouc, metals, and many other substances. In the case of glass, or rather cotton silicate, definite estimates have been given by Parks‡, deduced from actual increases of weight. He finds thicknesses of the order 1.0×10^{-5} cm., about 50 times that of the greasy films which stop the camphor movements upon water and profoundly modify the reflexion of light at the polarizing angle. Even if we allow a good deal for the fact that these films were formed from a saturated atmosphere, enough will remain to explain much optical disturbance.

The deviation from Fresnel's formulæ is best expressed in terms of Jamin's k, representing the ratio of reflected amplitudes for the two principal planes when light, incident at the angle $\tan^{-1}\mu$, is polarized at $45°$ to these planes. According to Fresnel $k = 0$, but Jamin showed that it may assume small finite values, positive or negative. The experimental method employed for the present purpose was substantially the same as in the former observations upon water, and can only be sketched briefly here. Sunlight reflected horizontally from a heliostat was caused to traverse the polarizing nicol mounted in a circle which allowed the rotation to be read to a minute of angle. After reflexion from the plate under examination the light traversed in succession a quarter-wave-plate of mica and the analysing nicol and was then received into the eye, either directly, or with the intervention of a small telescope magnifying about twice. A green glass was also often introduced in order to mitigate chromatic effects. Both the mica and the analysing

---

* *Phil. Mag.* Vol. xxxiii. p. 220 (1892); *Scientific Papers*, Vol. iii. p. 523.

† *Phil. Mag.* Vol. xxvii. p. 245 (1864).

‡ *Phil. Mag.* Vol. v. p. 518 (1903).

nicol were mounted so as to be capable of rotation about the direction of the reflected ray.

The theory of the method is as follows. Fresnel's expressions S and T (*sine*-formula and *tangent*-formula) give the ratios of the reflected to the incident vibrations, for the two principal planes; and their *reality* indicates that there is no change of phase in reflexion (other than 180°). The *ellipticity* is represented by the addition to T of iM, where M is small and $i = \sqrt{(-1)}$. Thus if the incident light be polarized in the plane making an angle α with the principal plane, the reflected vibrations may be represented by

$$(T + iM)\cos\alpha, \quad S\sin\alpha.$$

By the action of the mica, suitably adjusted, a relative change of phase $\frac{1}{2}\pi$ is introduced. This is represented by writing for $S\sin\alpha$, $iS\sin\alpha$. The vibration transmitted by the analyser, set at angle β, is then

$$\cos\alpha\cos\beta\,(T + iM) + iS\sin\alpha\sin\beta;$$

and the intensity of this is

$$T^2\cos^2\alpha\cos^2\beta + (M\cos\alpha\cos\beta + S\sin\alpha\sin\beta)^2.$$

In order that the light may vanish, we must have both $T = 0$ and

$$M + S\tan\alpha\tan\beta = 0,$$

the first of which shows that the dark spot occurs at the Brewsterian angle, while $\tan\alpha\tan\beta$ gives the value of M/S, viz. the k of Jamin. Accordingly if β be set to any convenient angle (such as 45°) and α be then adjusted so as to bring the dark spot to the central position, the product of the tangents of α and β, each measured from the proper zeros, gives k.

In practice it is not necessary to use the zeros. Set β, *e.g.* to $+45°$, and find α; then reset β to $-45°$. The new value of α would coincide with the old one were there no ellipticity; and the difference of values measures α upon a doubled scale. If α' be the second value, so that the difference is $\alpha' - \alpha$, then

$$k = \tan\tfrac{1}{2}(\alpha' - \alpha),$$

or with sufficient approximation in most cases

$$k = \tfrac{1}{2}(\alpha' - \alpha).$$

The sign of $\alpha' - \alpha$ is reversed when the mica is rotated through a right angle, and the absolute sign of k must be found independently.

The subjects of observation have been principally two plates, of which the first is of black glass, *i.e.* glass containing sufficient absorbing material to be opaque. When first examined on May 24 (1907) it had been lying in a box for many years. Carefully cleaned by washing and wiping, it gave $\alpha' - \alpha = +5°$, β as throughout being $\pm 45°$. On polishing rather protractedly with rouge, it gave $\alpha' - \alpha = +35'$, a large reduction. Blowing at the surface while under observation with a stream of chemically dried air had very little effect.

Want of sunshine prevented further observations until Aug. 3, when $\alpha' - \alpha$ was found to be about $+1°$. Treatment with rouge reduced this to $+40'$, and so far as appeared neither heating (with the idea of removing grease) nor treatment with specially moist or specially dry air made much difference. But on further repolishing with rouge the ellipticity practically disappeared. There was no certain change in the position of the dark spot when β was altered between $\pm 45°$.

It seemed that the earlier treatments with rouge had been inadequate. In the last application the polisher was of paper cemented to glass and impregnated with the rouge. After polishing, the glass was breathed upon and carefully wiped with a cloth. In subsequent operations this procedure was always followed.

Further observations showed that even on Aug. 3 the polishing had not been carried far enough. On Aug. 7 $\alpha' - \alpha$ was made to change sign, being reduced (algebraically) to $-10'$ or $-22'$. This was about the limit. Even in this condition the surface did not seem sensitive to moisture, much to my surprise. And it was matter for further surprise when it appeared that 24 hours' exposure to the air of the room—no chemical operations were in progress—sufficed to carry the surface back to the positive side with $\alpha' - \alpha = +10'$, increased in a few days to $+40'$. These changes of sign were observed not only with the black glass, which on the evidence of its polarizing angle is a flint, but also with a piece of patent plate prepared by roughing the hind surface and varnishing it with a cement of nearly the same index. It may be mentioned that another piece of patent plate became good on treatment with hydrofluoric acid and polishing with rouge upon a *soft* tool. After deterioration by exposure the negative values of $\alpha' - \alpha$ could not be recovered by merely cleaning the plate with moisture and wiping; actual repolishing was necessary. The natural inference is that even within 24 hours the substance of the glass is actually attacked by the gases of the atmosphere.

As I was leaving home for some time I arranged an experiment to see whether careful protection would save a glass surface from the above described deterioration, which might probably be attributed to moisture and carbonic acid. Accordingly on Aug. 22 the black glass, giving $\alpha' - \alpha = -14'$, was put away in a tube containing potash and closed, probably airtight, with a rubber cork. On Oct. 15, nearly two months later, the glass was taken out and (with surface untouched) gave $\alpha' - \alpha = -6'$, still on the negative side. The difference of values, though doubtless real, is perhaps no more than may be attributed to moisture and carbonic acid imprisoned with the glass. Four days later the readings gave $+35'$. The plate glass put away at the same time with potash in another tube with $\alpha' - \alpha = -19'$ was examined after two months, on Oct. 19, and gave $\alpha' - \alpha = +4'$, so that in this case the protection seems to have been less efficient.

A large number of further observations were made upon both glasses with the object of ascertaining, if possible, how much of the change which ensues after repolishing is due to a film of foreign (greasy) matter deposited from the atmosphere and how much to an alteration of the glass itself. The negative condition, lost in a day or two after repolishing, is in part recovered under the operation of a careful wiping with moisture, but only to a limited extent. Full recovery requires actual repolishing. It is of course possible that even the mild treatment by wiping may attack the very thin film of altered glass which is all that we can suppose to have been formed in so short a time; but on the whole the evidence pointed to two kinds of contamination, one removable by wiping and the other requiring the more drastic treatment with rouge. In spite of some easily understood irregularities, it appeared that the full effect of wiping was easily produced and that repetition could carry the process no further. The same conclusion is favoured by the results of heating the plate pretty strongly. The same kind of recovery in the direction of the negative condition could thus be attained, but never to the full extent. In some experiments the plate was purposely contaminated. Thus on Nov. 1 a repolished plate at $-13'$ was exposed to the smoke of burning greasy waste, after which $\alpha' - \alpha$ was $+46'$. A very thorough treatment by wiping took it back to $-6'$, but only after repolishing could the original condition $(-10')$ be nearly recovered. In another experiment a stream of air which had passed over petroleum was directed against a repolished surface, but the effect was only momentary. As regards surfaces which have stood a week or two, I think there can be no doubt but that the glass itself has been seriously attacked.

The results here recorded are in many respects very different from what I had anticipated—especially the comparative insensitiveness to grease and moisture. It must be remembered, however, that a surface finished by wiping and in contact with air is certainly contaminated with water and probably with grease. In spite of this it is possible to have the reflexion *free from ellipticity*. As regards grease we may perhaps argue from the manner in which the breath is deposited. A freshly split surface of mica receives the moisture of the breath as an almost invisible film, showing the colours of thin plates as it evaporates, but nothing of the appearance ordinarily associated with dew and dependent upon an irregular deposition. I am not sure whether glass has ever been observed in this condition*, but experience from

* Possibly the path of an electric discharge over a glass surface may be a case in point. [1911. See further, *Nature*, Vol. LXXXVI. p. 416 (1911).]

the days of wet collodion photography convinces me that a wiped glass does not so behave. The best that can be attained is a uniform dull grey appearance, such as under a magnifier would exhibit lenticular drops.

The conclusion which suggests itself is that even a recently repolished surface, which may exhibit but small ellipticity, is in a highly complicated condition. Grease itself may be comparatively inoperative optically on account of its index approximating to that of the glass. But why varying degrees of moisture should make so little difference is not apparent. Surface phenomena generally offer a wide field for investigation, which might lead to results throwing much needed light upon the constitution of matter.

333.

NOTE ON TIDAL BORES.

[*Proceedings of the Royal Society*, A, Vol. LXXXI. pp. 448, 449, 1908.]

IT was shown long ago by Airy that when waves advance over shallow water of depth originally uniform, the crests tend to gain upon the hollows *, so that the anterior slopes become steeper and steeper. Ultimately, if the conditions are favourable, there is formed what may be called a *bore*. Ordinary breakers upon a shelving beach are of this character, but the name is usually reserved for tidal bores advancing up rivers or estuaries. Interesting descriptions of some of these are given in Sir G. Darwin's *Tides* (Murray, 1898).

Although the real bore advances up the channel, we may for theoretical purposes "reduce it to rest" by superposing an equal and opposite motion upon the whole water system. We have then merely to investigate the transition from a relatively rapid and shallow stream of depth l and velocity u to a deeper and slower stream of depth l' and velocity u' (fig. 1). The

Fig. 1.

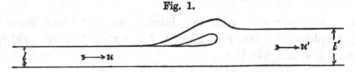

places where these velocities and depths are reckoned are supposed to be situated on the two sides of the bore and at such distances from it that the motions are there sensibly uniform. The problem being taken as in two dimensions, two relations may at once be formulated connecting the depths and velocities. By conservation of matter (" continuity ") we have

$$lu = l'u'. \dots\dots\dots\dots\dots\dots\dots\dots\dots\dots(1)$$

And since the mean pressures at the two sections are $\frac{1}{2}gl$, $\frac{1}{2}gl'$, the equation of momentum is

$$lu(u - u') = \frac{1}{2}g(l'^2 - l^2); \dots\dots\dots\dots\dots\dots(2)$$

whence $\qquad u^2 = \frac{1}{2}g(l + l') \cdot l'/l, \qquad u'^2 = \frac{1}{2}g(l + l') \cdot l/l'. \dots\dots(3)$

* See also *Scientific Papers*, Vol. I. p. 253 (1899).

The loss of energy per unit time at the bore is thus

$$lu\left(\tfrac{1}{2}u^2 + \tfrac{1}{2}gl\right) - lu\left(\tfrac{1}{2}u'^2 + \tfrac{1}{2}gl'\right) = lu \cdot g\left(l' - l\right)\frac{l^2 + l'^2}{4ll'} \cdot \quad\ldots\ldots\ldots(4)$$

That there should be a loss of energy constitutes no difficulty, at least in the presence of viscosity; but the impossibility of a gain of energy shows that the motions here contemplated cannot be reversed.

In order to recur to the natural condition of things where the shallow water is at rest, we have to superpose the velocity u taken negatively upon the above motion. The velocity of the bore is then u and that of the stream above the bore $u - u'$. If l is relatively small, u is much greater than u'.

The reasoning just used is very similar to that applied by Stokes* and by Riemann† to sound waves of expansion moving in one dimension. The matter is discussed in *Theory of Sound*, § 253, where it is shown that the discontinuous solution, obtained from the principles of conservation of mass and momentum, violates the condition of energy. When this was pointed out to Stokes by Kelvin and later by myself‡, he abandoned his solution, which is, however, maintained by a competent German authority§. It is clear, at least, that when the motion is such as to involve a gain of energy, the solution cannot be admitted. The opposite case stands upon a different footing, and we may, perhaps, imagine the redundant mechanical energy to be got rid of somehow at the surface of discontinuity. Even then we should have to face the complication entailed by the development of heat. In the present case of liquid, the heat is of little consequence, and since the motion is not entirely in one dimension, we escape the necessity of dealing with a single plane of discontinuity.

[1911. The reader may refer to a later paper on "Aerial Plane Waves of Finite Amplitude," *Proc. Roy. Soc.*, A, Vol. LXXXIV. p. 247 (1910); this Collection, Vol. v. Art. 346.]

* *Phil. Mag.* Vol. XXXIII. p. 349 (1848).
† *Göttingen Abh.* Vol. VIII. 1860.
‡ Stokes, *Math. and Phys. Papers*, Vol. II. p. 55.
§ Private correspondence.

334.

NOTES CONCERNING TIDAL OSCILLATIONS UPON A ROTATING GLOBE.

[*Proceedings of the Royal Society*, A, Vol. LXXXII. pp. 448—464, 1909.]

SPECULATIONS on tidal questions are much hampered by our ignorance of the peculiar influence of the earth's rotation in any but the simplest cases. The importance of this element was first appreciated by Laplace, and he succeeded in obtaining solutions of various problems relating to a globe completely covered with water to a depth either uniform throughout, or at any rate variable only with latitude. His work has been extended by Kelvin, G. Darwin, and Hough. For an excellent summary, reference may be made to Lamb's *Hydrodynamics*, which includes also important original additions to the theory.

But it must not be overlooked that a theory which supposes the globe to be completely covered with water has very little relation to our actual tides. Indeed, in practice, tidal prediction borrows nothing from Laplace's theory, unless it be to look for tidal periods corresponding with those of the generating forces. And this correspondence, although perhaps first brought into prominence in connection with Laplace's theory, is a general mechanical principle, not limited to hydrodynamics. If the theory of terrestrial tides is to advance, it can only be by discarding the imaginary globe completely covered with water and considering examples more nearly related to the facts, as was done in some degree by Young and Airy in their treatment of tides in canals. It is true that we are unlikely to obtain in this way more than very rough indications, but even such are at present lacking. I am told that opinions differ on so fundamental a question as whether the Atlantic tides are generated in the Atlantic or are derived from the Southern Ocean. Probably both sources contribute; but a better judgment, based on some sort of discussion on *a priori* principles, does not appear hopeless. In this connection, it is interesting to observe that a comparison of spring and neap tides shows that the moon is more effective relatively to the sun than would be expected from the ratio of the generating forces. This indicates some approach to synchronism with a natural free oscillation. That the approach is closest in the case of the moon indicates that the free period·is longer than those of the actual lunar and solar tides.

Were it not for the complication due to the earth's rotation rendering all tidal problems *vortex* problems, as Kelvin put it, questions such as this could be treated without great difficulty, and perhaps illustrated by models. There is nothing improbable in an oscillation backwards and forwards across the Atlantic having a period somewhat exceeding 12 lunar hours, but a treatment at all precise demands the inclusion of the rotation. This suggests the problem of the oscillations of a rotating ocean bounded by two meridians.

The present paper does not profess to make more than a modest contribution to the subject. It commences by developing further the theory of the free vibrations of a plane rectangular sheet of liquid, initiated in a former paper*, but only under the restriction that the angular velocity is relatively small†. Subsequently, the corresponding problem for an ocean on a rotating globe, bounded by two meridians, is attempted, but with limited success. Probably a better command of modern mathematical resources would lead to further results.

Plane Rectangular Sheet.

If ζ be the elevation, u, v the component velocities at any point, the equations of free vibration, when these quantities are proportional to $e^{i\sigma t}$, are‡

$$i\sigma u - 2\omega v = -g\,d\zeta/dx, \\ i\sigma v + 2\omega u = -g\,d\zeta/dy, \Bigg\} \qquad (1)$$

and

$$\frac{d^2\zeta}{dx^2} + \frac{d^2\zeta}{dy^2} + \frac{\sigma^2 - 4\omega^2}{gh}\,\zeta = 0, \qquad (2)$$

in which ω denotes the angular velocity of rotation, h the uniform depth of the water, and g the acceleration of gravity. The boundary walls will be supposed to be situated at $x = \pm\frac{1}{2}\pi$, $y = \pm y_1$.

When ω is evanescent, one of the principal vibrations is represented by

$$u_0 = \cos x, \qquad v_0 = 0; \qquad (3)$$

and ζ_0 is proportional to $\sin x$, so that

$$\sigma_0^2 = gh. \qquad (4)$$

* *Phil. Mag.* Vol. v. p. 297 (1903). [This Collection, Vol. v. p. 93.]

† The condition is satisfied in the case of terrestrial lakes of moderate dimensions especially if they are situated near the equator.

‡ Kelvin, *Phil. Mag.* August, 1880; Lamb, *Hydrodynamics* (3rd ed.), § 206.

This determines the frequency when $\omega = 0$. And since by symmetry a positive and a negative ω must influence the frequency alike, we conclude that (4) still holds in general so long as ω^2 can be neglected. The equation for ζ is at the same time reduced to

$$d^2\zeta/dx^2 + d^2\zeta/dy^2 + \zeta = 0. \quad \ldots\ldots\ldots\ldots\ldots\ldots(5)$$

Taking u_0 and v_0 as given in (3) and the corresponding ζ_0 as the first approximation, we add terms u_1, v_1, ζ_1, proportional to ω, whose forms are to be determined from the equations

$$i\sigma_0 u_1 = - gd\zeta_1/dx, \quad \ldots\ldots\ldots\ldots\ldots\ldots\ldots\ldots(6)$$

$$i\sigma_0 v_1 = - gd\zeta_1/dy - 2\omega\cos x, \quad \ldots\ldots\ldots\ldots\ldots(7)$$

$$(d^2/dx^2 + d^2/dy^2 + 1)(\zeta_1, u_1, v_1) = 0, \quad \ldots\ldots\ldots\ldots(8)$$

v_1, as well as ζ_1 and u_1, satisfying (8), since $\cos x$ does so. They represent in fact a motion that would be possible in the absence of rotation under forces* parallel to v and proportional to $\cos x$. This consideration shows that u_1 is an odd function of both x and y, and v_1 an even function.

The former investigation proceeded from the assumption for u_1 of the form

$$u_1 = A_2 \sin 2x + A_4 \sin 4x + \ldots, \quad \ldots\ldots\ldots\ldots\ldots(9)$$

which provides for the boundary condition to be satisfied at $x = \pm \frac{1}{2}\pi$, whatever functions of y the coefficients A_2, etc., may be. The value of v_1 thus obtained was

$$v_1 = \frac{2\omega i}{\sigma}\left\{\cos x - \frac{2}{\pi}\frac{\cos y}{\cos y_1} - \frac{4}{3\pi}\frac{\cosh(\sqrt{3}.y)}{\cosh(\sqrt{3}.y_1)}\cos 2x\right.$$

$$\left. + \frac{4(-1)^m}{(4m^2-1)\pi}\frac{\cosh\{y\sqrt{(4m^2-1)}\}}{\cosh\{y_1\sqrt{(4m^2-1)}\}}\cos 2mx + \ldots\right\}, \quad \ldots\ldots(10)$$

where $m = 1, 2, 3$, etc.

This value of v_1 may be employed to obtain a correction to σ_0. If we introduce terms u_2, v_2, ζ_2, proportional to ω^2, our equations (1) become, with retention of ω^2,

$$i\sigma(u_0 + u_1 + u_2) - 2\omega v_1 = - g\frac{d}{dx}(\zeta_0 + \zeta_1 + \zeta_2),$$

$$i\sigma(v_1 + v_2) + 2\omega(u_0 + u_1) = - g\frac{d}{dy}(\zeta_1 + \zeta_2);$$

or with regard to the equations satisfied by the terms with zero and unit suffixes,

$$i\sigma u_2 + i(\sigma - \sigma_0)u_0 - 2\omega v_1 = - gd\zeta_2/dx,$$

$$i\sigma v_2 + 2\omega u_1 \qquad\qquad = - gd\zeta_2/dy.$$

These are the equations that would apply to a rotationless sheet under the action of forces parallel to x and y proportional to $i(\sigma - \sigma_0)u_0 - 2\omega v_1$

---

* It will be observed that these forces are not derivable from a potential.

and $2\omega u_1$ respectively, and of speed σ. In order that the motion thus determined should be, as has been supposed, of the second order in ω, it is necessary that these forces should include no component capable of stimulating the principal motion. For this purpose the force parallel to y may be omitted from consideration as operating only upon v_1. Accordingly the condition to be satisfied is

$$\iint \{i\,(\sigma - \sigma_0)\,u_0 - 2\omega v_1\}\,u_0\,dx\,dy = 0, \quad \ldots\ldots\ldots\ldots(11)$$

an equation which may be regarded as giving a correction to σ_0.

In the present case the integration, between the limits $\pm \tfrac{1}{2}\pi$ for x and $\pm y_1$ for y, is straightforward and we get

$$\sigma - \sigma_0 = \frac{64\omega^2}{\pi^2\sigma_0}\left\{\frac{\pi^2}{16} - \frac{\tan y_1}{2y_1} - \frac{1}{1^2\cdot 3^2}\,\frac{\tanh\{y_1\sqrt{(1.3)}\}}{y_1\sqrt{(1.3)}}\right.$$
$$\left. - \frac{1}{3^2\cdot 5^2}\,\frac{\tanh\{y_1\sqrt{(3.5)}\}}{y_1\sqrt{(3.5)}} - \ldots\right\}\ldots\ldots(12)$$

The limiting values of x have been supposed for the sake of brevity to be $\pm\tfrac{1}{2}\pi$. If we denote them by $\pm x_1$, we are to replace x, y, y_1 in (10), (12) by $\tfrac{1}{2}\pi x/x_1$, $\tfrac{1}{2}\pi y/x_1$, $\tfrac{1}{2}\pi y_1/x_1$ respectively. At the same time (4) becomes

$$\sigma_0{}^2 = \frac{\pi^2 g h}{4 x_1{}^2}\,.\quad\ldots\ldots\ldots\ldots\ldots\ldots\ldots\ldots\ldots\ldots(13)$$

The method fails if y_1 is equal to an odd multiple of x_1. It would then become necessary to modify the initial assumption, as formerly explained in treating the case of the square, and $\sigma - \sigma_0$ would rise in magnitude so as to be of the first order in ω.

Equation (12) is not convenient in its application to the case where y_1 is very small. If we expand the tangent and hyperbolic tangents in powers of y_1, we obtain convergent series whose sums are zero for the terms independent of y_1 and proportional to $y_1{}^2$, but for higher powers of y_1 the series are divergent and no satisfactory conclusion can be drawn.

I have applied (12) to calculate the value of $\sigma - \sigma_0$ for the case where $y_1 = \tfrac{1}{2}/\sqrt{3}$. For the various terms of the series within braces involving hyperbolic tangents, we get (with reversed signs) 0·102692, 0·003200, 0·000448, 0·000108, 0·000035, 0·000014, etc., giving in all about 0·106510· Also $\tan 2y_1/2y_1 = 0·514368$, $\pi^2/16 = 0·616850$. Hence

$$\sigma - \sigma_0 = -\frac{64\omega^2}{\pi^2\sigma_0} \times 0·00403.$$

The inadequacy of (12) to deal satisfactorily with the case where y_1 is small led me to seek another solution. Here we assume in the first instance a form for v_1 which satisfies the conditions at $y = \pm y_1$, viz.,

$$v_1 = A_1 \cos \frac{\pi y}{2y_1} + \ldots + A_{2m+1} \cos \frac{(2m+1)\pi y}{2y_1}, \quad \ldots\ldots\ldots(14)$$

where $m = 0, 1, 2$, etc., making $v_1 = 0$ when $y = \pm y_1$. Hence, by (7),

$$-\frac{g}{i\sigma}\zeta_1 = \frac{2\omega y}{i\sigma}\cos x + \frac{2y_1 A_1}{\pi}\sin\frac{\pi y}{2y_1} + \ldots + \frac{2y_1 A_{2m+1}}{(2m+1)\pi}\sin\frac{(2m+1)\pi y}{2y_1},$$

no arbitrary function of x being added, since ζ_1 is odd in y.

Further, by (6),

$$u_1 = \frac{2\omega i}{\sigma}y\sin x + \frac{2y_1}{\pi}\sin\frac{\pi y}{2y_1}\frac{dA_1}{dx} + \ldots + \frac{2y_1}{(2m+1)\pi}\sin\frac{(2m+1)\pi y}{2y_1}\frac{dA_{2m+1}}{dx},$$
$$\ldots(15)$$

which is to be made to vanish when $x = \pm\frac{1}{2}\pi$ for all values of y between $\pm y_1$. Now, between these limits,

$$y = \frac{8y_1}{\pi^2}\left\{\sin\frac{\pi y}{2y_1} - \frac{1}{3^2}\sin\frac{3\pi y}{2y_1} + \ldots + \frac{(-1)^m}{(2m+1)^2}\sin\frac{(2m+1)\pi y}{2y_1}\right\}. \ldots(16)$$

Hence, when $x = \frac{1}{2}\pi$,

$$\frac{dA_1}{dx} = \frac{2\omega}{i\sigma}\frac{4}{\pi}, \qquad \frac{dA_{2m+1}}{dx} = \frac{2\omega}{i\sigma}\frac{4(-1)^m}{(2m+1)\pi}. \quad \ldots\ldots\ldots(17)$$

Now, since v_1 satisfies (8),

$$\frac{d^2 A_{2m+1}}{dx^2} + \left(1 - \frac{\pi^2(2m+1)^2}{4y_1^2}\right)A_{2m+1} = 0*, \ldots\ldots\ldots\ldots(18)$$

whence, v_1 being an even function of x, if $y_1 < \frac{1}{2}\pi$,

$$A_{2m+1} = B_{2m+1}\cosh px, \quad \ldots\ldots\ldots\ldots\ldots\ldots(19)$$

where

$$p_1^2 = \frac{\pi^2}{4y_1^2} - 1, \qquad p_{2m+1}^2 = \frac{(2m+1)^2\pi^2}{4y_1^2} - 1. \ldots\ldots\ldots(20)$$

If $y > \frac{1}{2}\pi$, (19) changes its form for one or more of the values of m. In (19) B_{2m+1} is a constant whose value is to be found from (17). We get, when $x = \frac{1}{2}\pi$,

$$\frac{dA_{2m+1}}{dx} = pB_{2m+1}\sinh\left(\tfrac{1}{2}p\pi\right) = \frac{2\omega}{i\sigma}\frac{4(-1)^m}{(2m+1)\pi};$$

so that finally

$$v_1 = \frac{2\omega}{i\sigma}\frac{4}{\pi}\frac{\cosh p_1 x \cdot \cos(\pi y/2y_1)}{p_1\sinh\left(\tfrac{1}{2}p_1\pi\right)} + \ldots$$

$$+ \frac{2\omega}{i\sigma}\frac{4(-1)^m}{(2m+1)\pi}\frac{\cosh px \cdot \cos\{(2m+1)\pi y/2y_1\}}{p\sinh\left(\tfrac{1}{2}p\pi\right)}. \quad \ldots\ldots(21)$$

---

* The circumstances are such as to justify the differentiation under the sign of summation. (Stokes' *Collected Papers*, Vol. I. p. 281.)

Also from (15)

$$u_1 = \frac{2\omega i}{\sigma}\left[y \sin x - \frac{8y_1}{\pi^2}\frac{\sinh p_1 x}{\sinh\left(\frac{1}{2}p_1\pi\right)}\sin\frac{\pi y}{2y_1} + \ldots \right.$$
$$\left. - \frac{(-1)^m 8y_1}{(2m+1)^2\pi^2}\frac{\sinh px}{\sinh\left(\frac{1}{2}p\pi\right)}\sin\frac{(2m+1)\pi y}{2y_1}\right]. \quad \ldots(22)$$

The introduction of (21) into (11) gives the correction to σ_0 in another form.   We find

$$\sigma - \sigma_0 = -\frac{512\omega^2 y_1^2}{\sigma_0\pi^5}\Sigma\frac{\coth\left(\frac{1}{2}\pi p_{2m+1}\right)}{(2m+1)^4 p_{2m+1}}, \quad \ldots(23)$$

where $m = 0,\ 1,\ 2,\ 3$, etc.   Whatever be the value of y_1, p_{2m+1} becomes larger as m increases, and ultimately $\coth\left(\frac{1}{2}\pi p_{2m+1}\right) = 1$.   If y_1 be small enough, this occurs even for $m = 0$, and we may then omit the coth in (23). If, further, 1 can be neglected in comparison with p_1^2, we may take

$$p_{2m+1} = (2m+1)\,\pi/2y_1,$$

and (23) becomes

$$\sigma - \sigma_0 = -\frac{1024\omega^2 y_1^3}{\sigma_0\pi^6}\Sigma\frac{1}{(2m+1)^5} = -\frac{1024\omega^2 y_1^3}{\sigma_0\pi^6} \times 1\cdot00452, \quad \ldots(24)$$

so that the correction is of the *third* order in y_1, or in y_1/x_1, if we replace y_1 by its general value, viz., $\frac{1}{2}\pi y_1/x_1$.

Comparing (12) and (23), we see that

$$\frac{\tan y}{2y_1} - \frac{1}{1^2.3^2}\frac{\tanh\{y_1\sqrt{(1.3)}\}}{y_1\sqrt{(1.3)}} - \ldots - \frac{\pi^2}{16} = \frac{8y_1^2}{\pi^3}\Sigma\frac{\coth\left(\frac{1}{2}\pi p_{2m+1}\right)}{(2m+1)^4 p_{2m+1}}. \quad \ldots(25)$$

If we take $y_1 = \frac{1}{2}/\sqrt{3}$, we find from (20) $p_1 = 5\cdot3487$, so that all the coths on the right of (25) are nearly equal to unity.   The first term $(m = 0)$ gives $0\cdot0040199$ and the two following are $0\cdot0000165$ and $0\cdot0000013$, so that the right-hand member of (25) is $0\cdot00404$, in sufficient agreement with the number previously calculated from the series on the left.

So far we have supposed that the type of vibration is founded upon $u_0 = \cos\left(\frac{1}{2}\pi x/x_1\right)$.   There is no difficulty in generalising the solution so far as to apply to the type

$$u_0 = \cos\frac{(2l+1)\,\pi x}{2x_1}, \quad \ldots(26)$$

where l is an integer.   We find

$$v_1 = \frac{2\omega}{i\sigma}\Sigma\frac{4\,(2l+1)\,(-1)^{l+m}}{2x_1\,(2m+1)}\frac{\cosh px \cdot \cos\{(2m+1)\,\pi y/2y_1\}}{p\cdot\sinh px_1}, \quad \ldots(27)$$

where the summation relates to m, taking in succession the values $0, 1, 2$, etc., and

$$p^2 = \frac{(2m+1)^2\pi^2}{4y_1^2} - \frac{(2l+1)^2\pi^2}{4x_1^2}. \quad \ldots(28)$$

It is assumed that

$$y_1 < \frac{x_1}{2l+1} \; ; \quad \dotfill \text{(29)}$$

otherwise one or more terms corresponding to the lower values of m will change their form.

A process similar to that already employed when $l = 0$ gives for the corrected value of σ,

$$\sigma - \sigma_0 = -\frac{64\omega^2}{\sigma_0} \frac{(2l+1)^2 y_1^2}{\pi^2 x_1^3} \Sigma \frac{\coth px_1}{(2m+1)^4 p} \; , \quad \dotfill \text{(30)}$$

which agrees with (23) when we put $l = 0$, $x_1 = \tfrac{1}{2}\pi$. It is to be observed that the general value of σ_0^2 is now given by

$$\sigma_0^2 = \frac{(2l+1)^2 \pi^2 gh}{4x_1^2} \; , \quad \dotfill \text{(31)}$$

so that (30) may also be written

$$\frac{\sigma - \sigma_0}{\sigma_0} = -\frac{256 \, \omega^2 y_1^2}{\pi^4 gh} \Sigma \frac{\coth px_1}{(2m+1)^4 px_1} \; , \quad \dotfill \text{(32)}$$

in which l does not appear directly.

There is also another class of primary vibrations in which the motion is parallel to x and is expressed by

$$u_0 = \sin (l\pi x / x_1), \qquad v_0 = 0, \quad \dotfill \text{(33)}$$

l being an integer. For this case we find, in the same way,

$$u_1 = \frac{2\omega}{i\sigma_0} \frac{l\pi y}{x_1} \cos \frac{l\pi x}{x_1} - \frac{2\omega}{i\sigma_0} \frac{8ly_1}{\pi x_1} \Sigma \frac{(-1)^{l+m} \cosh px}{(2m+1)^2 \cosh px_1} \sin \frac{(2m+1)\pi y}{2y_1} \; , \quad \dots \text{(34)}$$

$$v_1 = \frac{2\omega i}{\sigma_0} \Sigma \frac{4l(-1)^{l+m}}{(2m+1) x_1} \frac{\sinh px}{p \cosh px_1} \cos \frac{(2m+1)\pi y}{2y_1} \; , \quad \dotfill \text{(35)}$$

where m takes the values 0, 1, 2, etc., and

$$p^2 = \frac{(2m+1)^2 \pi^2}{4y_1^2} - \frac{l^2 \pi^2}{x_1^2} \; . \quad \dotfill \text{(36)}$$

It is here assumed that $y < 2l/x_1$, so that p^2 is positive even when $m = 0$. In this case u_1 is an even function of x and an odd function of y, while v_1 is odd in x and even in y.

The value of σ_0^2 is now given by

$$\sigma_0^2 = \frac{ghl^2\pi^2}{x_1^2} \; ; \quad \dotfill \text{(37)}$$

and for the correction to σ_0 we have

$$\frac{\sigma - \sigma_0}{\sigma_0} = -\frac{256\omega^2 y_1^2}{\pi^4 gh} \Sigma \frac{\tanh px_1}{(2m+1)^4 px_1} \; . \quad \dotfill \text{(38)}$$

It will not be forgotten that in this formula, as well as in (32), ω^4 is neglected.

In the examples hitherto given the primary motion ($\omega = 0$) is parallel to one of the sides of the rectangle. I will now take an example from the square, where the primary motion is symmetrical with respect to x and y and defined by

$$\zeta_0 \propto \cos x \cos y, \dots\dots\dots\dots\dots(39)$$

the sides of the square being the lines $x = \pm \pi$, $y = \pm \pi$. In harmony with (39) we get

$$u_0 = \sin x \cos y, \qquad v_0 = \cos x \sin y; \dots\dots\dots(40)$$

and since ζ_0 satisfies

$$\frac{d^2\zeta_0}{dx^2} + \frac{d^2\zeta_0}{dy^2} + \frac{\sigma_0^2}{gh} \zeta_0 = 0,$$

we see that

$$\sigma_0^2 = 2gh. \dots\dots\dots\dots\dots(41)$$

Also

$$(d^2/dx^2 + d^2/dy^2 + 2)(\zeta_1, u_1, v_1) = 0. \dots\dots\dots(42)$$

The equations of the next approximation, analogous to (6), (7), are

$$i\sigma_0 u_1 - 2\omega \cos x \sin y = -g d\zeta_1/dx, \dots\dots\dots\dots(43)$$

$$i\sigma_0 v_1 + 2\omega \sin x \cos y = -g d\zeta_1/dy; \dots\dots\dots\dots(44)$$

and they are the same as if impressed forces $2\omega \cos x \sin y$, $-2\omega \sin x \cos y$ acted parallel to u and v respectively and there were no rotation. From this we may infer that u_1 is even in x and odd in y, while v_1 is odd in x and even in y.

The procedure is much the same as before. We assume

$$v_1 = \Sigma V_{2m+1} \cos \tfrac{1}{2}(2m+1)y, \dots\dots\dots\dots(45)$$

where $m = 0, 1, 2$, etc., making $v_1 = 0$ when $y = \pm \pi$. From (43), (44) we deduce

$$u_1 = \frac{4\omega}{i\sigma} \cos x \sin y + \Sigma \frac{2}{2m+1} \frac{dV_{2m+1}}{dx} \sin \frac{2m+1}{2} y; \dots\dots(46)$$

so that

$$0 = \Sigma \frac{2}{2m+1} \frac{dV_{2m+1}}{dx_\pi} \sin \frac{2m+1}{2} y - \frac{4\omega}{i\sigma} \sin y, \dots\dots\dots(47)$$

since $u_1 = 0$ when $x = \pm \pi$.

Now, between the limits $\pm \pi$ for y

$$\sin y = \frac{8}{\pi}\left\{\frac{1}{3}\sin\frac{y}{2} + \frac{1}{1.5}\sin\frac{3y}{2} - \frac{(-1)^m}{(2m-1)(2m+3)}\sin\frac{(2m+1)y}{2}\right\}. \dots(48)$$

And thus

$$\frac{2}{2m+1}\frac{dV_{2m+1}}{dx_\pi} + \frac{4\omega}{i\sigma}\frac{8(-1)^m}{\pi(2m-1)(2m+3)} = 0. \dots\dots\dots(49)$$

But V_{2m+1} satisfies

$$\frac{d^2V_{2m+1}}{dx^2} - \frac{(2m+1)^2}{4}V_{2m+1} + 2V_{2m+1} = 0, \dots\dots\dots(50)$$

and hence, being odd in x, takes the form

$$V_{2m+1} = B_{2m+1} \sinh px, \quad \dots\dots\dots\dots\dots \dots\dots\dots(51)$$

where

$$p^2 = \tfrac{1}{4}(2m+1)^2 - 2. \quad \dots\dots\dots\dots\dots\dots(52)$$

This form obtains when $m > 0$. When $m = 0$, $p = \tfrac{1}{2}i\sqrt{7}$, and

$$V_1 = B_1 \sin(\tfrac{1}{2}\sqrt{7} \cdot x). \quad \dots\dots\dots\dots\dots\dots(53)$$

Using (49), (51) to determine B, we get

$$v_1 = \frac{4\omega}{i\sigma} \left\{ \frac{4\sin(\tfrac{1}{2}\sqrt{7} \cdot x) \cdot \cos\tfrac{1}{2}y}{\pi \cdot 3 \cdot \tfrac{1}{2}\sqrt{7} \cdot \cos(\tfrac{1}{2}\sqrt{7} \cdot \pi)} + \dots \right.$$
$$\left. - \frac{4(-1)^m (2m+1)\sinh px}{\pi(2m+3)(2m-1)p\cosh p\pi} \cos\frac{(2m+1)y}{2} \right\}, \quad \dots\dots\dots(54)$$

in which the first term may be deduced from the general term by putting $m = 0$ if we remember that

$$\sinh(ix) = i\sin x, \qquad \cosh(ix) = \cos x.$$

Also from (46)

$$u_1 = \frac{4\omega}{i\sigma} \left\{ \cos x \sin y - \frac{8(-1)^m \cosh px}{\pi(2m-1)(2m+3)\cosh p\pi} \sin\frac{(2m+1)y}{2} \right\}. \quad \dots(55)$$

It is evident, however, that there must be another expression for u_1 analogous to that given for v_1, and such as would be obtained by starting from

$$u_1 = \Sigma U_{2m+1} \cos\tfrac{1}{2}(2m+1)x \quad \dots\dots\dots\dots\dots(56)$$

instead of from (45). We may, in fact, interchange x and y if we reverse the sign of ω. Thus

$$u_1 = \frac{4\omega}{i\sigma} \left\{ \dots + \frac{4(-1)^m(2m+1)\sinh py}{\pi(2m+3)(2m-1)p\cosh p\pi} \cos\frac{(2m+1)x}{2} + \dots \right\} \quad \dots(57)$$

In applying these results to find a correction to σ_0, we have, much as before,

$$i\sigma_0 u_2 + i(\sigma - \sigma_0) u_0 - 2\omega v_1 = -g\,d\zeta_2/dx,$$
$$i\sigma_0 v_2 + i(\sigma - \sigma_0) v_0 + 2\omega u_1 = -g\,d\zeta_2/dy;$$

and thus

$$\iint \{(\sigma - \sigma_0) u_0 + 2i\omega v_1\} u_0\,dx\,dy + \iint \{(\sigma - \sigma_0) v_0 - 2i\omega u_1\} v_0\,dx\,dy = 0. \quad \dots(58)$$

In accordance with what has been said, if one of these integrals vanishes, so does the other, and we may confine our attention to the former. In the first place

$$\iint u_0^2\,dx\,dy = \pi^2. \quad \dots\dots\dots\dots\dots\dots(59)$$

In integrating $v_1 u_0$ we have

$$\int_{-\pi}^{+\pi} \cos \tfrac{1}{2} (2m+1) y . \cos y \, dy = -\frac{4 (2m+1)(-1)^m}{(2m-1)(2m+3)} \quad \dots\dots\dots (60)$$

and

$$\int_{-\pi}^{+\pi} \sinh px \sin x \, dx = \frac{2 \sinh p\pi}{p^2+1} . \quad \dots\dots\dots\dots (61)$$

Thus

$$\sigma - \sigma_0 = \frac{256\omega^2}{\pi^3 \sigma_0} \Sigma \frac{(2m+1)^2 \tanh p\pi}{(2m-1)^2 (2m+3)^2 \, p\pi \, (p^2+1)} , \quad \dots\dots\dots (62)$$

p being given by (52).

For calculation of the first term under the sign of summation ($m = 0$) the form must be modified.   We find

$$\frac{-8 \tan (\tfrac{1}{2}\pi\sqrt{7})}{27\pi\sqrt{7}} = -0.057309. \quad \dots\dots\dots\dots (63)$$

The most important term is the next for which $m = 1$, $p = \tfrac{1}{2}$.   Under the sign of summation we have

$$\frac{8 . 3^2 \tanh (\tfrac{1}{2}\pi)}{5^3 . \pi} = 0.16816. \quad \dots\dots\dots\dots (64)$$

The following terms are 0.00166, 0.00021, and 0.00005, so that altogether we may take as the sum of the terms under the sign of summation $+ 0.1128$. Accordingly

$$\sigma - \sigma_0 = -\frac{\omega^2}{\sigma_0} \times 2.925 ; \quad \dots\dots\dots\dots (65)$$

and this result, being already of the right dimensions, applies whatever may be the size of the square.   It may be remarked that the sign of the correction is the opposite of that applicable to the *circle*, for which approximately[*]

$$\sigma - \sigma_0 = + 2\omega^2/\sigma_0. \quad \dots\dots\dots\dots (66)$$

These results are, of course, applicable only under the restriction that ω is small compared with σ, the latter quantity depending on the size and depth of the sheet of liquid.   In the case of lakes and seas upon the rotating earth, we have also to remember that ω depends upon the latitude.   At the equator ω vanishes.

Spherical Sheet of Liquid.

An attempt will now be made to apply similar methods to the free vibrations of an ocean on a rotating globe, the water being of uniform depth h, and bounded by vertical walls coincident with two meridians,

---

[*] Lamb, *loc. cit.* p. 306.

$\phi = 0$ and $\phi = 2\phi_1$. Using a similar notation, we have as the general equations *

$$i\sigma u - 2\omega v \cos\theta = -\frac{g}{a}\frac{d\zeta}{d\theta}, \quad \dots\dots\dots\dots\dots(67)$$

$$i\sigma v + 2\omega u \cos\theta = -\frac{g}{a\sin\theta}\frac{d\zeta}{d\phi} \quad \dots\dots\dots\dots(68)$$

with the equation of continuity

$$i\sigma\zeta = -\frac{h}{a\sin\theta}\left\{\frac{d(u\sin\theta)}{d\theta} + \frac{dv}{d\phi}\right\}. \quad \dots\dots\dots(69)$$

Here θ denotes the colatitude, a is the radius of the globe; ω its angular velocity of rotation; u, v the velocities along and perpendicular to the meridian. As is usual, we shall write μ for $\cos\theta$ when convenient.

By (67), (68), u and v may be expressed in terms of ζ, and substitution in (69) will then give an equation in ζ only. When $\omega = 0$, this equation is the well-known one,

$$\frac{d}{d\mu}(1-\mu^2)\frac{d\zeta}{d\mu} + \frac{1}{1-\mu^2}\frac{d^2\zeta}{d\phi^2} + \frac{\sigma^2 a^2}{gh}\zeta = 0. \quad \dots\dots\dots(70)$$

We will suppose that the primary motion—that which would obtain if $\omega = 0$—is represented by ζ_0, u_0, v_0, and that $v_0 = 0$, so that the primary motion is wholly in latitude. And we will begin with the further supposition that

$$\zeta_0 \propto \mu, \qquad u_0 = \sin\theta. \quad \dots\dots\dots\dots(71)$$

Substitution in (70) shows that

$$\sigma_0^2 a^2 = 2gh. \quad \dots\dots\dots\dots\dots(72)$$

The motion is that which might obtain equally over the complete sphere, the liquid heaping itself alternately at the two poles.

It is to be observed that, under the circumstances here contemplated, (70) holds good so long as ω^2 can be neglected, since all that is required in its formation is the omission of ωv and of $\omega du/d\phi$. We write it in the form

$$\frac{d}{d\mu}(1-\mu^2)\frac{d\zeta_1}{d\mu} + \frac{1}{1-\mu^2}\frac{d^2\zeta_1}{d\phi^2} + 2\zeta_1 = 0\dots\dots\dots\dots(73)$$

and we observe that (73) is satisfied also by u_1 and by

$$i\sigma v_1 \sin\theta + 2\omega \sin^2\theta \cos\theta \quad \dots\dots\dots\dots(73\text{A})$$

if ζ_1, u_1, v_1 are the correctional terms proportional to ω. If we substitute $\sin^2\theta \cos\theta$, or $\mu(1-\mu^2)$, in the left-hand member of (73), we get $-6\mu + 10\mu^3$; so that

$$\left[\frac{d}{d\mu}(1-\mu^2)\frac{d}{d\mu} + \frac{1}{1-\mu^2}\frac{d^2}{d\phi^2} + 2\right](v_1\sin\theta) = \frac{2\omega}{i\sigma}(6\mu - 10\mu^3). \quad \dots(74)$$

* Lamb, *loc. cit.* p. 314.

Here $v_1 \sin \theta$ vanishes at the limits of ϕ. If we assume

$$v_1 \sin \theta = \Sigma V_m \sin (m\pi\phi/2\phi_1), \quad \dots\dots\dots(75)$$

we may deduce for the left-hand member of (74)

$$\Sigma \left[\frac{d}{d\mu} (1-\mu^2) \frac{dV_m}{d\mu} - \frac{1}{1-\mu^2} \frac{m^2\pi^2}{4\phi_1^2} V_m + 2V_m \right] \sin \frac{m\pi\phi}{2\phi_1}. \quad \dots\dots(76)$$

For the expansion of the right-hand member we have, between 0 and $2\phi_1$,

$$1 = \frac{4}{\pi} \left\{ \sin \frac{\pi\phi}{2\phi_1} + \frac{1}{3} \sin \frac{3\pi\phi}{2\phi_1} + \frac{1}{5} \sin \frac{5\pi\phi}{2\phi_1} + \dots \right\}; \quad \dots\dots(77)$$

so that for the general term

$$\frac{d}{d\mu} (1-\mu^2) \frac{dV_m}{d\mu} - \frac{1}{1-\mu^2} \frac{m^2\pi^2}{4\phi_1^2} V_m + 2V_m = \frac{2\omega}{i\sigma} \frac{8(3\mu-5\mu^3)}{m\pi}, \quad \dots(78)$$

m being an odd integer. For even values of m, V_m vanishes.

The complete integral of (78) comprises, as complementary function, two functions of μ, one odd and one even, each multiplied by an arbitrary constant. In the present case we have to do only with the odd function, and its coefficient is to be determined by the consideration that V_m remains finite at the poles ($\mu = \pm 1$). A complete treatment presents considerable difficulties. Reference may be made to *Theory of Sound*, § 338. In the present case the n of spherical harmonics is unity, but $s (= m\pi/2\phi_1)$ is not necessarily integral, still less an integer not exceeding 1.

When ϕ_1 is small, the calculation simplifies, for then the second term on the left of (78) predominates, and ultimately we have

$$V_m = -\frac{2\omega}{i\sigma} \frac{32\phi_1^2}{m^3\pi^3} (3\mu - 8\mu^3 + 5\mu^5), \quad \dots\dots\dots(79)$$

which with (75) determines $v_1 \sin \theta$ for the extreme case. We may pursue the approximation with respect to ϕ_1 by substituting from (79) in the first and third terms of (78). For this purpose we may use

$$\left[(1-\mu^2) \frac{d}{d\mu} (1-\mu^2) \frac{d}{d\mu} + 2(1-\mu^2) \right] \mu^n$$
$$= n(n-1)\mu^{n-2} - 2(n^2-1)\mu^n + (n^2+n-2)\mu^{n+2}. \quad \dots(80)$$

As was to be expected, the term in $\mu (n=1)$ contributes nothing.

The result of substituting $3\mu - 8\mu^3 + 5\mu^5$ is thus

$$-48\mu + 228\mu^3 - 320\mu^5 + 140\mu^7;$$

and accordingly the second approximation to V_m is

$$V_m = (79) - \frac{2\omega}{i\sigma} \frac{512\phi_1^4}{m^5\pi^5} (-12\mu + 57\mu^3 - 80\mu^5 + 35\mu^7). \quad \dots\dots(81)$$

When $v_1 \sin \theta$ is known, the corresponding terms in ζ_1 and u_1 may be found from (67), (68).

We will now apply (79) to calculate a correction to σ_0 in the manner already employed for plane sheets. We have to make

$$\iint \{i\,(\sigma - \sigma_0)\,u_0 - 2\omega v_1 \cos\theta\}\,u_0 \sin\theta\,d\theta\,d\phi = 0, \quad\ldots\ldots\ldots(82)$$

where $u_0 = \sin\theta$, and the integrations extend from -1 to $+1$ for μ, and from 0 to $2\phi_1$ for ϕ. The calculation is straightforward, and we find

$$\sigma - \sigma_0 = \frac{16\omega^2\phi_1^2}{35\sigma_0}, \quad\ldots\ldots\ldots\ldots\ldots\ldots(83)$$

where Σm^{-4} (m odd) has been replaced by its equivalent, $\pi^4/96$. A continuance of the approximation from (81) gives another term in (83) involving (as well as ω^2) ϕ_1^4 and Σm^{-6}. It appears that

$$\sigma - \sigma_0 = \frac{4\omega^2}{\sigma_0}\frac{3 \cdot 128\,\phi_1^2}{35\pi^4}\left\{\Sigma m^{-4} - \frac{16 \cdot 11 \cdot \phi_1^2}{9 \cdot \pi^2}\Sigma m^{-6}\right\}$$

$$= \frac{16\omega^2\phi_1^2}{35\sigma_0}\left\{1 - \frac{88\phi_1^2}{45}\right\}, \quad\ldots\ldots\ldots\ldots\ldots\ldots(84)$$

when we substitute for Σm^{-4}, Σm^{-6}, their values, viz., $\pi^4/96$ and $\pi^6/960$. If it were important, the approximation with respect to ϕ_1 could be carried further without much difficulty.

We will now pass on to consider the most important primary mode in which ζ_0 is an *even* function of μ, proportional to the zonal harmonic of order 2, or to $\mu^2 - \tfrac{1}{3}$. In agreement with this we take

$$u_0 = \cos\theta\sin\theta, \qquad v_0 = 0. \quad\ldots\ldots\ldots\ldots(85)$$

Substituting the value of ζ_0 in (70), we see that

$$\sigma_0^2 a^2 = 6gh; \quad\ldots\ldots\ldots\ldots\ldots\ldots(86)$$

and ζ_1 satisfies

$$\frac{d}{d\mu}(1 - \mu^2)\frac{d\zeta_1}{d\mu} + \frac{1}{1 - \mu^2}\frac{d^2\zeta_1}{d\phi^2} + 6\zeta_1 = 0, \quad\ldots\ldots\ldots(87)$$

an equation satisfied also by u_1 and by

$$i\sigma v_1 \sin\theta + 2\omega \cos^2\theta \sin^2\theta. \quad\ldots\ldots\ldots\ldots(88)$$

Substituting (88) in (87), we get

$$\left[\frac{d}{d\mu}(1 - \mu^2)\frac{d}{d\mu} + \frac{1}{1 - \mu^2}\frac{d^2}{d\phi^2} + 6\right](v_1 \sin\theta) = -\frac{4\omega}{i\sigma_0}(1 - 6\mu^2 + 7\mu^4), \ldots(89)$$

the analogue of (74). Retaining (75) and understanding, as before, that m is an odd integer, we get, with use of (77) for the general term,

$$(1 - \mu^2)\frac{d}{d\mu}(1 - \mu^2)\frac{dV_m}{d\mu} - s^2 V_m + 6(1 - \mu^2)V_m$$

$$= -\frac{2\omega}{i\sigma}\frac{8}{m\pi}(1 - 7\mu^2 + 13\mu^4 - 7\mu^6), \ldots(90)$$

where $s = m\pi/2\phi_1$.

This equation may be treated in the same way as was (78); but it may be well to introduce a modification which would be convenient in pursuing the approximation further. We will divide (90) by the factor which multiplies the parenthesis on the right, taking

$$V_m' = V_m \div -\frac{2\omega}{i\sigma}\frac{8}{m\pi} \; ;$$

and assume *

$$V_m' = K_0 + K_2\mu^2 + \ldots + K_n\mu^n, \quad\ldots\ldots\ldots\ldots\ldots(91)$$

where n is even. Substituting in the right-hand member of (90), we find as the coefficient of μ^n

$$(n+1)(n+2)K_{n+2} - (2n^2 - 6 + s^2)K_n + \{(n-1)(n-2) - 6\}K_{n-2}. \;\ldots(92)$$

Hence

$$
\begin{aligned}
n=0, &\quad 1 \cdot 2K_2 - (s^2 - 6)K_0 &&= 1, \\
n=2, &\quad 3 \cdot 4K_4 - (s^2 + 2)K_2 - 6K_0 &&= -7, \\
n=4, &\quad 5 \cdot 6K_6 - (s^2 + 26)K_4 &&= 13, \\
n=6, &\quad 7 \cdot 8K_8 - (s^2 + 66)K_6 + 14K_4 &&= -7, \\
n=8, &\quad 9 \cdot 10K_{10} - (s^2 + 122)K_8 + 36K_6 &&= 0 \; ;
\end{aligned}
$$

the right-hand members being zero for 8 and all higher values of n. It will be seen that one of the coefficients is arbitrary, providing the necessary undetermined element. The problem would be so to choose it as to satisfy the condition at the pole.

When s^2 may be treated as large, we may divide the system of equations by it, obtaining as the first approximation

$$K_0 = -s^{-2}, \qquad K_2 = 7s^{-2}, \qquad K_4 = -13s^{-2}, \qquad K_6 = 7s^{-2}, \;\ldots(93)$$

after which K_8, etc., vanish. To obtain a second approximation we substitute the result of the first approximation in the smaller terms. Thus

$$K_0 = -\frac{1}{s^2 - 6} + \frac{2K_2}{s^2 - 6} = -s^{-2} + 8s^{-4}.$$

In like manner,

$$K_2 = 7s^{-2} - 148s^{-4}, \qquad K_4 = -13s^{-2} + 548s^{-4},$$
$$K_6 = 7s^{-2} - 644s^{-4}, \qquad K_8 = 252s^{-4} \; ;$$

after which the K's are zero to this order. The approximation may be pursued, and at each step another K enters. In this process the difficulty of satisfying the general condition at the pole is evaded.

If we stop at the first approximation (93), we have

$$v_1 \sin\theta = \ldots + \frac{2\omega}{i\sigma}\frac{8\sin s\phi}{m\pi s^2}(1 - 7\mu^2 + 13\mu^4 - 7\mu^6) + \ldots \;\ldots(94)$$

* Thomson and Tait's *Natural Philosophy*, 2nd ed. Part I. p. 210.

Using this value of $v_1 \sin \theta$ in (82), we find

$$\sigma - \sigma_0 = -\frac{16\omega^2\phi_1^2}{63\sigma_0} \quad \dots\dots\dots\dots\dots\dots\dots(95)$$

as the correction applicable for this mode of vibration of a narrow lune.

From some points of view, there is advantage in the use of $v\,(=\sin\theta)$, rather than μ, as independent variable. In place of (90) we have*

$$v^2(1-v^2)\frac{d^2V_m{}'}{dv^2} + v(1-2v^2)\frac{dV_m{}'}{dv} + 6v^2V_m{}' - s^2V_m{}' = 2v^2 - 8v^4 + 7v^6; \quad \dots(96)$$

but for the moment we will take on the right the more general form $av^2 + bv^4 + cv^6$. Assuming

$$V_m{}' = H_0 + H_2v^2 + H_4v^4 + \dots, \quad \dots\dots\dots\dots\dots(97)$$

we find on substitution in the left-hand member of (96) as the coefficient of v^r

$$(r^2 - s^2)\,H_r - \{(r-1)(r-2) - 6\}\,H_{r-2}. \quad \dots\dots\dots\dots(98)$$

Thus for the term depending on a,

$$H_0 = 0, \qquad H_2 = a/(4 - s^2), \qquad H_4 = 0, \text{ etc.};$$

so that

$$V_m{}' = \frac{av^2}{4 - s^2}, \text{ simply.} \quad \dots\dots\dots\dots\dots\dots(99)$$

For the term in b,

$$H_0 = 0, \qquad H_2 = 0, \qquad H_4 = \frac{b}{4^2 - s^2}, \qquad H_6 = \frac{5.4 - 6}{6^2 - s^2}\,H_4, \text{ etc.};$$

so that

$$V_m{}' = \frac{bv^4}{4^2 - s^2} + \frac{b(5.4 - 6)v^6}{(4^2 - s^2)(6^2 - s^2)} + \frac{b(5.4 - 6)(7.6 - 6)v^8}{(4^2 - s^2)(6^2 - s^2)(8^2 - s^2)} + \dots. \text{ (100)}$$

In like manner, for the term in c,

$$V_m{}' = \frac{cv^6}{6^2 - s^2} + \frac{c(7.6 - 6)v^8}{(6^2 - s^2)(8^2 - s^2)} + \dots. \quad \dots\dots\dots\dots(101)$$

Introducing the numerical values of a, b, c, we find for the sum of the three contributions,

$$V_m{}' = \frac{2v^2}{2^2 - s^2} - \frac{8v^4}{4^2 - s^2} - \frac{7s^2v^6}{(4^2 - s^2)(6^2 - s^2)}\left\{1 + \frac{4.9}{8^2 - s^2}v^2 + \frac{4.9}{8^2 - s^2}\frac{6.11}{10^2 - s^2}v^4\right.$$

$$\left. + \frac{4.9}{8^2 - s^2}\frac{6.11}{10^2 - s^2}\frac{8.13}{12^2 - s^2}v^6 + \dots\right\} \quad \dots(102)$$

and there is also to be added the complementary function,

$$V_m{}' = Av^s\left\{1 + \frac{(s-2)(s+3)}{2(2s+2)}v^2 + \frac{(s-2)(s-1)(s+3)(s+4)}{2.4.(2s+2)(2s+4)}v^4 + \dots\right\}, \text{ (103)}$$

in which A is a constant.

_____
* See *Theory of Sound*, § 338.

The expressions in (102), (103) vanish when $\nu = 0$. It is further necessary —and this is the condition determining A—that $dV_m'/d\theta$ should vanish at the equator ($\nu = 1$). Now

$$dV_m'/d\theta = \cos\theta \,.\, dV_m'/d\nu,$$

in which $\cos\theta$ vanishes. So far as regards (99), $dV_m'/d\nu$ is finite when $\nu = 1$, so that no further question arises here. But for (100), (101), (103), $dV_m'/d\nu$ is infinite when $\nu = 1$, and a further scrutiny is called for.

As a first step we may examine (103), taking it in the more general form*

$$\psi_s = \nu^s \left\{ 1 + \frac{(\tfrac{1}{2}s - \tfrac{1}{2}n)(\tfrac{1}{2}s + \tfrac{1}{2}n + \tfrac{1}{2})}{1 \,.\, (s+1)} \nu^2 \right.$$
$$\left. + \frac{(\tfrac{1}{2}s - \tfrac{1}{2}n)(\tfrac{1}{2}s - \tfrac{1}{2}n + 1)(\tfrac{1}{2}s + \tfrac{1}{2}n + \tfrac{1}{2})(\tfrac{1}{2}s + \tfrac{1}{2}n + \tfrac{3}{2})}{1 \,.\, 2 (s+1)(s+2)} \nu^4 + \ldots \right\} . \quad \ldots(104)$$

In Gauss' notation for hypergeometric series,

$$\psi_s = \nu^s F(\alpha, \beta, \gamma, \nu^2), \quad \ldots\ldots\ldots\ldots\ldots(105)$$

where $\quad \alpha = \tfrac{1}{2}s - \tfrac{1}{2}n, \quad \beta = \tfrac{1}{2}s + \tfrac{1}{2}n + \tfrac{1}{2}, \quad \gamma = s + 1.$

Since $\gamma - \alpha - \beta = \tfrac{1}{2} > 0$, F is finite when $\nu = 1$, and accordingly so also is ψ_s. But for $d\psi_s/d\nu$ we have

$$d\psi_s/d\nu = s\nu^{s-1}F + \nu^s dF/d\nu,$$

of which the first part, being finite, need not be regarded. Thus when $\nu = 1$,

$$d\psi_s/d\theta = (1 - \nu^2)^{\frac{1}{2}} dF/d\nu. \quad \ldots\ldots\ldots\ldots\ldots(106)$$

Now $\qquad \dfrac{dF}{d\nu} = \dfrac{\alpha\beta \,.\, 2\nu}{1 \,.\, \gamma} F(\alpha + 1, \beta + 1, \gamma + 1, \nu^2),$

in which†, when $\nu = 1$ nearly,

$$F(\alpha+1, \beta+1, \gamma+1, \nu^2) = \frac{\Gamma(\gamma+1)}{\Gamma(\alpha+1)\,.\,\Gamma(\beta+1)} (1-\nu^2)^{\gamma-\alpha-\beta-1}. \ldots(107)$$

In the present case

$$\gamma - \alpha - \beta - 1 = -\tfrac{1}{2},$$

and thus when $\nu = 1$

$$\frac{d\psi_s}{d\theta} = \frac{2\Gamma(\gamma)}{\Gamma(\alpha)\,.\,\Gamma(\beta)} = \frac{2\Gamma(s+1)}{\Gamma(\tfrac{1}{2}s - \tfrac{1}{2}n)\,.\,\Gamma(\tfrac{1}{2}s + \tfrac{1}{2}n + \tfrac{1}{2})}. \quad \ldots\ldots(108)$$

If $n = 2$, as at present,

$$\frac{d\psi_s}{d\theta} = \frac{2\Gamma(s+1)}{\Gamma(\tfrac{1}{2}s - 1)\,.\,\Gamma(\tfrac{1}{2}s + \tfrac{3}{2})}, \quad \ldots\ldots\ldots\ldots\ldots(109)$$

* *Theory of Sound, loc. cit.* Here $n=2$.
† *Infinite Series*, Bromwich, p. 171. [1911. Mr Bromwich informs me that a factor is missing from the formula referred to, so that on the right of (107) the numerator should include the additional factor $\Gamma(\alpha+\beta-\gamma+1)$, or in the present case $\Gamma(\tfrac{1}{2})$. Corresponding corrections must be made in (108), (109).]

when $\theta = \frac{1}{2}\pi$. Thus $d\psi_s/d\theta$ has a finite value at the equator, as was to be expected.

It may be proved without difficulty that (102) converges when $\nu = 1$ and s is not an even integer. Any finite number of terms which may have negative denominators being excluded, the remainder may be expressed as a hypergeometrical series. But the form is more complicated than before, and the evaluation of $dV_m'/d\theta$ would be rather tedious, even if practicable.

A question obtrudes itself as to what happens when s is an even integer. When $s = 2$, there is synchronism between the primary and a derived vibration, and the occurrence of the infinite denominator $4 - s^2$ is what might have been expected. But in the case of other even integers no synchronism is apparent, and it would seem that the complication is of an analytical character only. The solution compounded of (102) and (103) changes its form. It would be of interest to follow out the process, say for the case $s = 6$, which might roughly represent the circumstances of the Atlantic Ocean, but I am not prepared to undertake the task.

ON THE INSTANTANEOUS PROPAGATION OF DISTURBANCE IN A DISPERSIVE MEDIUM, EXEMPLIFIED BY WAVES ON WATER DEEP AND SHALLOW.

[*Philosophical Magazine*, Vol. XVIII. pp. 1—6, 1909.]

THE solution, first obtained by Cauchy and Poisson, for the propagation in one dimension over deep water of the waves which issue from a concentrated initial disturbance presents peculiar features. One form of it may be written

$$\eta = \frac{1}{\pi x} \left\{ \frac{gt^2}{2x} - \frac{1}{3 \cdot 5} \left(\frac{gt^2}{2x} \right)^3 + \frac{1}{3 \cdot 5 \cdot 7 \cdot 9} \left(\frac{gt^2}{2x} \right)^5 - \dots \right\}, \quad \dots\dots\dots(1)$$

where η denotes the elevation of the surface at time t and place x, $d\eta/dt$ being initially zero throughout and η being initially zero except at $x = 0$. g denotes, as usual, the acceleration of gravity. So far as general mechanical theory is concerned, it might be expected that under these initial conditions the elevation would at first vary as t^2; but that the effect should commence without delay at all points seems at first to conflict with ideas derived from the more familiar theories of sonorous and luminous undulations, according to which a finite time must elapse before any effect reaches places finitely distant from the source.

A plausible explanation of the discrepancy may be found in the reflexion that trains of simple waves move in deep water with velocities proportional to the square roots of the wave-lengths and thus capable of assuming infinite values, and that in accordance with Fourier's theorem the initial disturbance in question must actually be regarded as including trains of infinite wave-length and velocity. But in the sequel we shall find reason for questioning the sufficiency of this explanation.

In the deep water solution the speed σ is proportional to $k^{\frac{1}{2}}$, where $k = 2\pi/\lambda$ and λ is the wave-length. We will now suppose more generally that in a dispersive medium $\sigma = k^m$. In the general Fourier solution

$$\eta = \frac{1}{\pi} \int_0^\infty \cos \sigma t \cdot \cos kx \cdot dk \quad \dots\dots\dots\dots\dots\dots(2)$$

σ is to be given the above value, and for the sake of convergency we must introduce under the integral sign the factor e^{ky}, where y is negative and ultimately is made to vanish. In Prof. Lamb's exposition* of the deep water problem the velocity-potential is employed as an intermediary, and into this the exponential factor enters essentially. All ambiguity is thus avoided; but on the other hand it would seem that it should not be necessary to go behind (2), at any rate if this expression for the elevation is correctly understood. And in the extended problem, where nothing more is known than that $\sigma \propto k^m$, we have no choice.

Expanding the cosine in (2), we get

$$\eta = \frac{1}{\pi}\int_0^\infty \left\{ 1 - \frac{k^{2m}t^2}{1.2} + \frac{k^{4m}t^4}{1.2.3.4} - \dots \right\} \cos kx\, dk, \dots\dots\dots(3)$$

in which on the above understanding

$$\int_0^\infty k^{2mn} \cos kx\, dk = \frac{\Gamma(2mn+1)\cos(mn+\frac{1}{2})\pi}{x^{2mn+1}}. \quad\dots\dots\dots(4)$$

When $n = 0$, (4) vanishes; and in general we have a term in (3) proportional to t^2, predominating when t is small.

We fall back at once upon the simple case of waves in deep water by making $m = \frac{1}{2}$. Then (4) vanishes, if n is even; and if n is odd,

$$\int_0^\infty k^n \cos kx\, dk = \frac{n!}{x^{n+1}}(-1)^{\frac{1}{2}n+\frac{1}{2}}, \quad\dots\dots\dots\dots\dots(5)$$

from which (1) follows ($g = 1$). In this case the series is convergent for all finite values of x and t.

It may be of interest to examine the application to aerial waves for which the velocity of propagation (σ/k) is constant. Here $m = 1$, and (4) vanishes for all (integral) values of n. Hence all the coefficients in the expansion (3) of η in powers of t vanish, and the general conclusion that the elevation commences everywhere without delay fails, as we know it ought to do. At the same time the special character of the failing case becomes apparent. But we have not yet examined the convergency of the series.

Apart from signs and from the cosine factor in (4), the ratio of the term in t^{2n+2} to that in t^{2n} is

$$\frac{t^2}{(2n+1)(2n+2)x^{2m}}\frac{\Gamma(2mn+2m+1)}{\Gamma(2mn+1)}. \quad\dots\dots\dots(6)$$

To find the limit of (6) as n becomes infinite, we may use the formula

$$\Gamma(n+1) = e^{-n}n^{n+\frac{1}{2}}\sqrt{(2\pi)}.$$

---

* *Proc. Lond. Math. Soc.* Ser. 2, Vol. II. p. 373 (1904); *Hydrodynamics*, § 236.

The second fraction in (6) has thus the limiting value

$$(2m)^{2m} (n+1)^{2m},$$

so that (6) becomes ultimately

$$\frac{t^2}{4x^{2m}} (2m)^{2m} (n+1)^{2m-2}. \quad \dots\dots\dots\dots\dots\dots\dots(7)$$

The series is thus certainly convergent if m be less than unity. If $m = 1$, (7) reduces to t^2/x^2, and convergence is not assured unless $x > t$. This marks the moment at which the wave propagated with velocity 1 reaches the point x.

If we were to make $m = 2$, corresponding to the propagation of flexural disturbances along a bar, the cosines in (4) would all vanish, but the lack of convergency indicated by (7) prohibits the conclusion that the disturbance undergoes a finite delay in reaching the point x. From Fourier's special solution* we may see that there is in fact no such delay.

It would be of great interest to examine the influence on water waves due to a limitation of depth (h), but a complete solution for this case analogous to (1) seems hardly practicable. But without much difficulty we may obtain the first term in the series, viz. the term proportional to t^2.

In general from (2)

$$\frac{d\eta}{dt} = -\frac{1}{\pi} \int_0^\infty \sigma \sin \sigma t \cos kx\, dk,$$

which vanishes when $t = 0$, and

$$\frac{d^2\eta}{dt^2} = -\frac{1}{\pi} \int_0^\infty \sigma^2 \cos \sigma t \cos kx\, dk,$$

so that for $t = 0$ $\dfrac{d^2\eta}{dt^2} = -\dfrac{1}{\pi} \displaystyle\int_0^\infty \sigma^2 \cos kx\, dk. \quad \dots\dots\dots\dots\dots\dots(8)$

For waves on water of depth h

$$\sigma^2 = gk \tanh kh, \quad \dots\dots\dots\dots\dots\dots\dots\dots(9)$$

passing when k is great (λ small) into the simple deep water formula $\sigma^2 = gk$, and when k is small (λ great) into the shallow water formula $\sigma^2 = ghk^2$, indicating a uniform velocity of propagation†.

When $a < \pi$, we have‡

$$\int_0^\infty \frac{e^{ax} - e^{-ax}}{e^{\pi x} + e^{-\pi x}} \sin cx\, dx = \frac{(e^{\frac{1}{2}c} - e^{-\frac{1}{2}c}) \sin \frac{1}{2}a}{e^c + e^{-c} + 2\cos a}.$$

---

* *Theory of Sound*, § 192.

† It may be remarked that a modified formula, viz.

$$\sigma^2 = gk\,(1 - e^{-kh}),$$

agreeing with (9) in the extreme cases, would be more amenable to calculation.

‡ De Morgan's *Differential Calculus*, p. 669.

If we differentiate this with respect to c and then make $a = \pi$, to which there seems to be no objection on the understanding postulated,

$$\int_0^\infty \frac{e^{\pi x} - e^{-\pi x}}{e^{\pi x} + e^{-\pi x}}\, x \cos cx\, dx = -\frac{\frac{1}{2}(e^{\frac{1}{2}c} + e^{-\frac{1}{2}c})}{(e^{\frac{1}{2}c} - e^{-\frac{1}{2}c})^2}. \qquad \ldots\ldots\ldots\ldots(10)$$

Adapting (10) to our present purpose, we have

$$\int_0^\infty \sigma^2 \cos kx\, dk = -\frac{g\pi^2}{2h^2}\frac{e^{\pi x/2h} + e^{-\pi x/2h}}{(e^{\pi x/2h} - e^{-\pi x/2h})^2}, \qquad \ldots\ldots\ldots\ldots(11)$$

whence $d^2\eta/dt_0^2$ is given by (8). The first term in the expression for η is accordingly

$$\eta = \frac{g\pi t^2}{8h^2}\frac{\cosh(\pi x/2h)}{\sinh^2(\pi x/2h)}. \qquad \ldots\ldots\ldots\ldots\ldots(12)$$

We learn from (12) that, no more than in the case of deep water, is there any delay in the commencement of disturbance at a finite distance x from the source, and the extension is not without importance, seeing that in truth water cannot be very deep in relation to *all* the wave-lengths concerned, since among these infinite wave-lengths are included. In the present case all the wave-velocities of simple trains are finite, and thus the sudden propagation to a distance is independent of an interpretation earlier suggested.

If in (12) h is small compared with x, we may write

$$\eta = \frac{g\pi t^2}{4h^2}\, e^{-\pi x/2h}, \qquad \ldots\ldots\ldots\ldots\ldots\ldots(13)$$

showing that when x is relatively great the elevation in its earlier stages, though finite, is on a greatly diminished scale.

As regards the general question, I do not think that the instantaneous propagation of disturbance should be considered paradoxical. It is to be remembered that in (9) the water is treated as incompressible, *i.e.* that the velocity of propagation of waves of *expansion* is regarded as infinite.

A more complete treatment of the case of finite depth on the basis of (2) and (9) would be instructive, even if limited to selected values of the ratio $x : h$. So far we have considered only the first stage. When h is small compared with x, an important part of the disturbance arrives under the law applicable to shallow water, viz.

$$\sigma = \sqrt{(gh)}\,.\,k\,(1 - \tfrac{1}{6}k^2h^2). \qquad \ldots\ldots\ldots\ldots\ldots(14)$$

Writing (2) in the form

$$\eta = \frac{1}{2\pi}\int_0^\infty \cos(\sigma t - kx)\, dk + \frac{1}{2\pi}\int_0^\infty \cos(\sigma t + kx)\, dk, \qquad \ldots\ldots\ldots(15)$$

we see that the first integral acquires a specially enhanced value when x

and t are so related that $\sigma t - kx$ is approximately zero for small values of k. The condition is of course

$$t = x/\sqrt{(gh)}, \quad \dots\dots\dots\dots\dots\dots\dots\dots(16)$$

and the important part of the integral will be

$$\eta = \frac{1}{2\pi} \int_0^\infty \cos\left(\tfrac{1}{6} xh^2 k^3\right) dk = \frac{1}{6\pi} \cos\frac{\pi}{6}.\,\Gamma\left(\tfrac{1}{3}\right).\left(\frac{xh^2}{6}\right)^{-\frac{1}{3}}, \quad \dots\dots(17)$$

large when h is relatively small, but still finite.

For larger values of t the most important part of the first integral of (15) occurs in the neighbourhood of such values of k as make $\sigma t - kx$ stationary. This happens when k has a value k_0 such that

$$x - td\sigma/dk_0 = 0. \quad \dots\dots\dots\dots\dots\dots(18)$$

As Kelvin has shown*, the first integral of (15) takes approximately the form

$$\eta = \frac{\cos\left(\sigma_0 t - k_0 x - \tfrac{1}{4}\pi\right)}{\sqrt{\{-2\pi t\, d^2\sigma/dk_0^2\}}}, \quad \dots\dots\dots\dots\dots\dots(19)$$

while the second integral is relatively negligible. For example in the case of deep water waves, where $\sigma = \sqrt{(gk)}$, (18) gives

$$k_0 = gt^2/4x^2, \qquad \sigma_0 = gt/2x,$$

so that

$$\sigma_0 t - k_0 x - \tfrac{1}{4}\pi = gt^2/4x - \tfrac{1}{4}\pi;$$

and finally

$$\eta = \frac{g^{\frac{1}{2}}t}{2\pi^{\frac{1}{2}}x^{\frac{3}{2}}}\cos\left\{\frac{gt^2}{4x} - \tfrac{1}{4}\pi\right\}. \quad \dots\dots\dots\dots(20)$$

If we attempt to fill up the gaps in our solution by applying quadratures to (2), we have to face the difficulty that, as written, the integral is not convergent. Some analytical transformation is called for. One way out of the difficulty might be to calculate the *difference* between the solutions for finite and infinite depth, for it would appear that the non-convergent part of the integral, corresponding to infinitely small wave-lengths, must be the same for both†.

<hr>

* *Proc. Roy. Soc.* Vol. XLII. p. 80 (1887).

† [1911. Reference may be made to Pidduck "On the Propagation of a Disturbance in a Fluid under Gravity," *Proc. Roy. Soc.* A, Vol. LXXXIII. p. 347 (1910).]

336.

ON THE RESISTANCE DUE TO OBLIQUELY MOVING WAVES AND ITS DEPENDENCE UPON THE PARTICULAR FORM OF THE FORE-PART OF A SHIP.

[*Philosophical Magazine*, Vol. XVIII. pp. 414—416, 1909.]

I SUPPOSE that everyone is familiar with the system of oblique waves advancing in *echelon* from the bow of a ship which travels through smooth water. What is not so easily observed from on board is the corresponding wave-profile, *i.e.* the deviation of the water-surface at the side of the ship from the position which it would occupy in a state of rest. Sketches, both of the whole system of waves and of various wave-profiles, have been given by W. and R. E. Froude*, and the influence of the various components of the wave-system in contributing to the aggregate *wave-resistance* has been discussed. Attention has perhaps tended to concentrate upon the directly advancing waves—those whose crests are perpendicular to the ship's motion—and upon the remarkable interaction between the systems originating at the bow and stern. But, apart from its interesting geometrical features, the oblique part of the wave-system also impresses an observer with its mechanical importance as probably contributing in no mean degree to the total wave-making resistance.

From the time of my first acquaintance with drawings of wave-profiles I have been struck with their significance as indicating that the usual form of bow (and perhaps of stern) is not well adapted to minimise the forces of resistance. At the stem and immediately behind, the water is raised above the normal level, and this elevation is undoubtedly the principal feature. The additional pressure thence arising operates normally upon the skin of the ship, and the component in the fore-and-aft direction acting sternwards holds the ship back. It is, I suppose, in order to diminish the force here operative that the bows of many ships are made *hollow*, *i.e.* with a curvature

---

* *Naval Arch. Trans.* 1881; or see Kelvin, *Proc. Mech. Eng.* Aug. 1887; also Lamb's *Hydrodynamics*, 3rd ed. p. 414.

concave outwards. But the question does not stop here. If we follow the wave-profile, we find that somewhat further back the water falls to the normal level and further back still to a level below the normal, that in fact there is a succession of waves of elevation and *depression* of gradually diminishing importance as we recede from the bow. As regards the force acting upon the ship in the fore-and-aft direction, we know that at any point it is the product of two factors, one the pressure acting perpendicularly upon the skin of the ship and secondly the sine of the angle (θ) between the skin and the direction of travel. If, as usual in the fore part of a ship, the skin faces everywhere forwards so that θ is positive, we recognize that if we estimate the pressures from the normal condition of rest, there is a force of retardation from the water-surface downwards in the region of an elevation, but on the other hand a force of acceleration in the region of a depression. And the question at once arises whether we cannot, at least in some degree, accommodate the angle to the pressure so as to diminish the sum total of the fore-and-aft components. In order to attain this end it is evident that θ should be diminished (perhaps even to the extent of becoming negative) in the region of an elevation, and should be increased in the region of a depression.

Several years ago (I think it was in 1902) with the late Mr Gordon, I attempted some experiments in illustration of this theory. The water, to a depth of 4 or 5 inches, was contained in a large flat sponge-bath mounted upon a turn-table and maintained in uniform rotation. Near the circumference the revolving water flowed past a fixed model which represented the ship. The model was of wood with vertical sides and flat bottom, and was curved along its length in conformity with the circular motion of the water. No attempt was made to measure the forces of resistance, but observations were taken of the wave-pattern as affected by changes in the form of the bow. Starting from a somewhat blunt figure, the wood was gradually cut away from the rear of the region of the waves of elevation, thus for one thing sharpening the bow, care being taken not to continue this operation into the rear of the region of the waves of depression. In this way with frequent trials verifying the actual positions of the waves at the various stages, an undulating form with vertical sides, and (it must be admitted) of unprepossessing appearance, was gradually developed, which seemed to have the property of originating a much less pronounced wave-system, with promise of a diminished resistance.

So far as these experiments went, they tended to confirm me in the idea that advantage might arise from a similar figuring of an actual ship, but it was evident that much better trials upon models in a suitable tank was the next step, and my object in writing the present note is to suggest the execution of such trials. The natural course would be to start from a

paraffin model of an actual ship, and after observation of the wave-system
and resistance to add and remove material at and under the places indicated
(see figure), and then after each moderate change to redetermine the form
of the waves and the force of resistance. It is possible of course that
experiments in this direction with discouraging results have been made
already, but I have seen no suggestion of them; and it may be that the
prepossessions of naval architects would be too unfavourable to allow such an
idea to be entertained.

One or two questions are touched upon in conclusion. It may be asked
how far above and below the normal water level should the undulating form
be carried? Observations of the actual pressures at various points with
suitable gauges would be the best guide, but in general the variation of

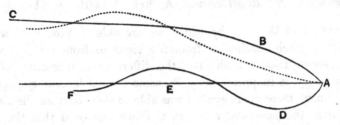

ABC is the original form of one side of the ship; ADEF the wave-profile, D and F
being elevations and E a depression. The dotted line then indicates the
direction of the proposed alterations.

pressure due to waves does not extend far down in comparison with the
wave-length. But here it is to be remembered that most actual ships are
exposed to variations of loading and to encounters with long waves originating
in wind. Another thing that might have to be borne in mind is the danger
of eddy making if the undulations of form were made too pronounced, but
this consideration can hardly be prohibitive of a moderate application of the
principle. Again, it may be noted that strictly speaking the wave-system is
a function of the speed, so that the adjustment could be complete only for
a single speed. But this objection does not seem very serious. To discuss
these questions further at the present stage would be premature.

[1911. It may be remarked that when ships are in narrow waters the
diminution of the oblique waves would often be advantageous in itself,
independently of the power saved.]

337.

ON THE PERCEPTION OF THE DIRECTION OF SOUND.

[*Proceedings of the Royal Society*, A, Vol. LXXXIII. pp. 61—64, 1909.]

THE nature of the clue by which we are able to pronounce whether a sound of low pitch reaches us from the right or from the left was long a mystery, seeing that in such cases the difference of intensities at the two ears, used singly, is inappreciable. By some special laboratory experiments conducted about three years ago*, I was able to show that the discrimination depends upon the *phase*-difference at the two ears, and that the sound is judged to be on that side where the phase is in *advance*. When the pitch is higher (much above *g'*) no distinct lateral effect accompanies a phase-difference, and the discrimination of right and left in ordinary hearing undoubtedly depends upon intensities. Commenting on these results, I remarked [p. 356]: "The conclusion, no longer to be resisted, that when a sound of low pitch reaches the two ears with approximately equal intensities, but with a phase-difference of a quarter of a period, we are able so easily to distinguish at which ear the phase is in advance, must have far-reaching consequences in the theory of audition. It seems no longer possible to hold that the vibratory character of sound terminates at the outer ends of the nerves along which the communication with the brain is established. On the contrary, the processes in the nerve must themselves be vibratory, not, of course, in the gross mechanical sense, but with preservation of the period and retaining the characteristic of phase—a view advocated by Rutherford, in opposition to Helmholtz, as long ago as 1886."

In the *Proceedings*† for January, 1908, Profs. Myers and Wilson detail some interesting experiments, made by another method, which confirm my general conclusion that phase-differences at the outer ears are recognised and give rise to "lateral sensations," but they put a different interpretation upon the fact from that which I suggested in the paragraph above quoted.

---

* *Phil. Mag.* Vol. XIII. p. 214 (1907). [This Collection, Vol. v. p. 347.]

† Vol. LXXX. A, p. 260 (1908); see also *Journal of Psychology*, Vol. II. October, 1908.

In their view the ultimate cause of the lateral effect is still a difference of intensities at the *inner* ears which may occur, in spite of equality at the *outer* ears, in virtue of conduction of sound from one side to the other through the bones of the head. I had already [p. 358] considered the somewhat similar question of the consequences of conduction through the air round the head, and had shown that the observed effects could not be explained upon the intensity theory as the result of such conduction. At first sight there would seem to be but little difference between the two suppositions, and the opposite character of the conclusions depends, in fact, upon the introduction by Profs. Myers and Wilson of a phase-reversal. It may be well to quote their own statement* of the assumptions upon which they base their explanation. These are: "(*a*) that the sound entering one ear is transmitted through the bones of the head to the opposite inner ear, (*b*) that the retardation of phase due to such interaural conduction is small, and (*c*) that the two sets of waves, thus received at each ear directly and by bone conduction, meet coming from opposite directions." It is this arrival from opposite directions which is considered to entail a phase-reversal, and the required conclusion then readily follows. It will be convenient to generalise somewhat the algebraical statement.

Let $y_1 = a \sin (\omega t + \alpha)$ denote the vibration entering the right ear, and $y_2 = b \sin \omega t$ denote the vibration entering the left ear, so that α is the angle (less than π) by which the vibration at the right outer ear is in advance. The resulting effect at the right internal ear may be taken to be

$$y_1' = fa \sin (\omega t + \alpha) - gb \sin (\omega t - \beta).$$

Here f and g are proper fractions, of which f is much greater than g, and β is the retardation in phase due to the passage through the head. The negative sign of the second term expresses the phase-reversal, so that the whole retardation is really $\pi + \beta$. In the same way the effect at the left internal ear will be

$$y_2' = fb \sin \omega t - ga \sin (\omega t + \alpha - \beta).$$

Let I_1 denote the sound intensity at the right internal ear, and I_2 that at the left internal ear. Then

$$I_1 = f^2 a^2 + g^2 b^2 - 2fgab \cos (\alpha + \beta),$$
$$I_2 = f^2 b^2 + g^2 a^2 - 2fgab \cos (\alpha - \beta);$$

and
$$I_1 - I_2 = (f^2 - g^2)(a^2 - b^2) + 4fgab \sin \alpha \sin \beta. \quad \ldots\ldots\ldots\ldots(1)$$

If $b = a$, this reduces to the value given by Myers and Wilson, viz.

$$I_1 - I_2 = 4fga^2 \sin \alpha \sin \beta. \quad \ldots\ldots\ldots\ldots\ldots\ldots(2)$$

In these expressions $\sin \alpha$ is supposed to be positive, that is the vibration at the right outer ear *leads*; and we see that $I_1 > I_2$, if β be positive and less

* *Journ. Psych. loc. cit.* p. 380.

than π. This conclusion depends, of course, upon the assumed reversal of phase, and would itself be reversed if β were the whole retardation incurred.

The question thus raised is of considerable interest. It must be admitted at once that simplicity is on the side of a theory that would attribute the lateral sensation in all cases to a difference of intensity at the two inner ears. Myers and Wilson are evidently much influenced, as I also was, by an *a priori* unwillingness to admit the possibility of a phase-difference being appreciated directly. But I was under the impression that this difficulty would be less felt by physiologists now than in the days of Helmholtz.

When we come to scrutinise closely the intensity theory, as put forward by Myers and Wilson, we meet a good many difficulties, though perhaps none of them are insuperable. It would seem that the positive effect of the second term of (1), where g is very small and β is somewhat small, would be easily overpowered by the first term if b slightly exceeded a, especially if $\sin \alpha$ is also somewhat small. In my experiments, where two tones nearly in unison were led separately to the two ears, the lateral sensation changed over somewhat suddenly from left to right, or *vice versâ*, as the phase-difference passed through zero or 180°. And although the best effects required some regulation of the intensities, it did not appear that the adjustment needed to be at all precise. Myers and Wilson narrate a curious observation in which the intensities at the two outer ears were made unequal. The influence of the change seemed not to be permanent. "We see, then, how rapidly we become adapted to binaural differences of intensity, how susceptible we still are to phase-differences after such adaptation has been established, and how such adaptation, once acquired, persists for a short time after the sound intensity at the two ears has been equalised."

I had already (*loc. cit.* [p. 350]) tried an experiment in order to see whether a small disturbance of relative intensities at the two ears would upset the judgment of an observer listening in the open. Unknown to him, a piece of board could be introduced near his ear on the same side as the sound, by which it was supposed that the intensity on that side would be reduced, perhaps even below the intensity upon the farther side. It was found that no misjudgment of direction could thus be induced. Although not actually conclusive, this result is unfavourable to a theory that would make the lateral sensation dependent upon small differences of intensities. The repetition of the experiment by other observers, and in varied forms, seems to be desirable.

We come now to the third assumption (c). The theory under discussion assumes that two vibrations arriving at the inner ear from opposite directions co-operate if the phases are opposite, and tend to neutralise one another if the phases are the same. But this depends entirely upon the precise mode of action. The assumption is justified if the effect depends upon the velocity of vibration, as when aerial waves impinge from opposite directions upon a sensitive flame. But if the question were the effect upon a resonator, of the kind employed by Helmholtz, the direction of travel is indifferent, and co-operation in all cases requires that the phases be the same*. So far as I see, we have nothing to guide us in choosing between the two alternatives.

And there is a further question. If we are to suppose that the framework of the skull is thrown into vibration by sounds which enter at either ear, the matter is not exhausted when we have included one travel across the head. If a finite system, free from viscosity, is excited by an incident force following the harmonic law, the phases of the resulting vibration are the same (or the opposite) at all points of the system. On this basis the phase-difference β, on which everything depends, disappears. This conclusion may be evaded by appealing to viscous effects, but then it becomes a question whether the propagation across the head represented by g would suffice. It is difficult to form a confident opinion.

Further evidence, if it can be obtained, is very desirable. For the moment, the choice between the competing views is likely to depend upon preconceptions as to the manner in which the nerves act. And it should be remembered that the question is not whether phase-differences at the two outer ears are competent to arouse the lateral sensation, but only as to the manner in which this effect is produced.

* *Phil. Mag.* Vol. VII. p. 153 (1879) [Vol. I. p. 402]; *Theory of Sound*, 2nd edit. Vol. II. p. 403.

338.

THE THEORY OF CROOKES'S RADIOMETER.

[*Nature*, Vol. LXXXI. pp. 69, 70, 1909.]

I HAVE noticed that the theory of this instrument is usually shirked in elementary books, even the best of them confining themselves to an account, and not attempting an explanation*. Indeed, if it were necessary to follow Maxwell's and O. Reynolds's calculations, such restraint could easily be understood. In their mathematical work the authors named start from the case of ordinary gas in complete temperature equilibrium, and endeavour to determine the first effects of a small departure from that condition. So far as regards the internal condition of the gas, their efforts may be considered to be, in the main, successful, although (I believe) discrepancies are still outstanding. When they come to include the influence of solid bodies which communicate heat to the gas and the reaction of the gas upon the solids, the difficulties thicken. A critical examination of these memoirs, and a rediscussion of the whole question, would be a useful piece of work, and one that may be commended to our younger mathematical physicists.

Another way of approaching the problem is to select the case at the opposite extreme, regarding the gas as so attenuated as to lie entirely outside the field of the ordinary gaseous laws. Some suggestions tending in this direction are to be found in O. Reynolds's memoir, but the idea does not appear to have been consistently followed out. It is true that in making this supposition we may be transcending the conditions of experiment, but the object is to propose the problem in its simplest form, and thus to obtain an easy and unambiguous solution—such as may suffice for the purposes of elementary exposition, although the physicist will naturally wish to go further. We suppose, then, that the gas is so rare that the mutual encounters of the molecules in their passage from the vanes to the envelope, or from one part of the envelope to another part, may be neglected, and, further, that the

---

* See for example Poynting and Thomson's *Heat*, p. 150.

vanes are so small that a molecule, after impact with a vane, will strike the envelope a large number of times before hitting the vane again.

Under ordinary conditions, if the vanes and the envelope be all at one temperature, the included gas will tend to assume the same temperature, and when equilibrium is attained the forces of bombardment on the front and back faces of a vane balance one another. If, as we suppose, the gas is very rare, the idea of temperature does not fully apply, but at any rate the gas tends to a definite condition which includes the balance of the forces of bombardment. If the temperature be raised throughout, the velocities of the molecules are increased, but the balance, of course, persists. The question we have to consider is what happens when one vane only, or, rather, one face of one vane, acquires a raised temperature.

The molecules arriving at the heated face have, at any rate in the first instance, the frequencies and the velocities appropriate to the original temperature. As the result of the collision, the velocities are increased. We cannot say that they are increased to the values appropriate to the raised temperature of the surface from which they rebound. To effect this fully would probably require numerous collisions. Any general increase in the velocity of rebound is sufficient to cause an unbalanced force tending to drive the heated surface back, as O. Reynolds first indicated. If we follow the course of the molecules after collision with the heated surface, we see that, in accordance with our suppositions, they will return by repeated collisions with the envelope to the original lower scale of velocities before there is any question of another collision with the heated face. On the whole, then, the heated face tends to retreat with a force proportional both to the density of the gas and to the area of the surface.

A calculation of the absolute value of the excess of pressure cannot be made without further hypothesis. If we were to suppose that the molecules, after collision with the heated face, rebound with the same velocities $(v + dv)$ as they would have were the temperature raised throughout, the pressure would be increased in the ratio $v + (v + dv) : 2v$ or $1 + dv/2v : 1$. On the other hand, if the temperature were actually raised throughout, the pressure, according to the usual gaseous laws, would be increased in the ratio $(v + dv)^2 : v^2$ or $1 + 2dv/v : 1$. On this hypothesis, therefore, the unbalanced increment of pressure on the heated face is one-quarter of the increment that would be caused by a general rise of temperature to the same amount. This estimate is necessarily in excess of the truth, but it is probably of the right order of magnitude.

The supposition upon which our reasoning has been based, viz. that the mean free path of a molecule is large in comparison with the linear dimension of the vessel, has been made for the sake of simplicity, and is certainly a very extreme one. It is not difficult to recognise that in the extreme form it may be dispensed with. All that is really necessary to justify our conclusions is that the mean free path should be very large in comparison with the *vane*. The magnitude and distribution of the velocities with which the molecules impinge will then be independent of the fact that the face of the vane is heated, and this is all that the argument requires. The repulsion by heat of a silk fibre suspended in a moderately rare gas was, it will be remembered, verified by O. Reynolds.

339.

TO DETERMINE THE REFRACTIVITY OF GASES AVAILABLE ONLY IN MINUTE QUANTITIES.

[*Nature*, Vol. LXXXI. p. 519, 1909.]

ON a former occasion* I described a refractometer capable of dealing with rather small quantities (12 c.c.) of gas. The optical tubes, one of which would contain the material under investigation and the other air, were of brass, 20 cm. in length and 6 mm. in bore, and were traversed by two pencils of light from the same origin, subsequently brought to interference in the observing telescope. For this purpose the object-glass of the telescope was provided with two parallel slits opposite the axes of the tubes. The image of the original slit, formed in the focal plane, was examined through a high-power cylindrical lens, constituting the eye-piece of the telescope, and exhibited the familiar pattern of interference bands, the position of which shifts with changes in the densities of the gases occupying the tubes. With this apparatus, and using pressures not exceeding one atmosphere, it was possible to compare refractivities $(\mu - 1)$ with a relative accuracy of about one-thousandth part.

In recent conversation my son, the Hon. R. J. Strutt, raised the question as to the minimum quantity of gas upon which a determination of refractivity could be made, having in mind such rare gases as the radium emanation. Towards answering it I have made a few experiments dealing merely with the optical side of the question.

A reduction of volume in the gas tube implies a reduction of length below the 20 cm. of the apparatus just referred to, and this carries with it a loss of accuracy. A reduction to 2 cm. should leave possible an accuracy of at least 1 per cent., and this was the length chosen. As the inquiry was limited to the optical conditions, it was unnecessary to close the ends, and thus the tubes reduced themselves to two parallel tunnels through a block of paraffin

---

* *Proc. Roy. Soc.* Vol. LXIV. p. 95 (1898); *Scientific Papers*, Vol. IV. p. 364.

2 cm. thick.   They were prepared by casting wax (from a candle) round two similar sewing needles of suitable diameter previously secured in a parallel position.   The rest of the apparatus was merely an ordinary spectroscope arrangement (without prism).   Sunlight admitted through a slit, and rendered parallel by the collimating lens, traversed the double tunnel, and was received by the observing telescope focused, as usual, upon the slit.   It is necessary, of course, that the length of the slit be perpendicular to the plane containing the axes of the tunnels.

The appearance of the bands as seen with a given telescope depends upon the size of the apertures and upon their distance apart.   The width of the bands is inversely as the distance between the centres of the apertures (tunnels), and the horizontal diameter of the luminous field upon which the bands are seen is inversely as the diameter of the apertures themselves. Since a large number of bands is not required, small and rather close apertures are indicated.   The only question is as to the amount of light. If we suppose the apertures and their distance apart to be proportional, we may inquire as to the effect of linear scale L.   Here a good deal may depend upon the relative values of length of slit, focal length of collimator, length as well as diameter of tunnels.   In my apparatus the slit was short, and the height, as well as the width of the field of view, was determined mainly by diffraction.   If we suppose the slit very short, the calculation is simplified, though this cannot be the most favourable arrangement.   With a given width of slit the whole light in the field of view is then proportional to L^2.   Since the angular area of the field practically varies as L^{-2}, it would seem that the brightness varies as L^4.   This would impose an early limit upon the reduction of L; but there are other factors to be regarded.   In order to secure an angular field of given size, we must use an eye-piece the magnifying power of which is proportional to L.   This consideration changes L^4 back to L^2.   Nor is this all.   With a given eye-piece the admissible width of primary slit varies inversely as L, and thus, finally, the brightness of a field of given angular width, and containing a given number of bands, varies as L simply.

In the earlier experiments the tunnels were of $\frac{3}{4}$ mm. bore, and were too widely separated.   In order to see the bands well, a very powerful eye-piece was needed.   An attempt to gain light by substituting a cylindrical lens (very successful in the former apparatus, where the beams are limited by *slits*) for the spherical lenses of the eye-piece showed little advantage. Subsequently smaller tunnels were prepared $\frac{1}{2}$ mm. in bore, and so close that the distance of the nearest parts was rather less than the diameter of either. These gave splendid bands with the ordinary eye-piece of the spectroscope, and I estimated that there should be no difficulty in setting a web correctly to one-twentieth of a band.

The capacity of one of these passages is about 4 cubic millimetres, and I have no doubt a further reduction might be effected, so far as the optics is concerned; but the further such reduction is carried the greater, probably, would become the difficulties of manipulation. The mere closing of the ends of such small tubes with plates of glass would not be an easy matter. In order to prevent encroachment upon the course of the light, it might be necessary to enlarge the ends so as to allow a little more room for overflow of cement. For the present I content myself with showing that it is possible to obtain well-formed black bands on a sufficient angular scale with light which has traversed tubes 2 cm. long and $\frac{1}{2}$ mm. in bore.

340.

NOTE AS TO THE APPLICATION OF THE PRINCIPLE OF DYNAMICAL SIMILARITY.

[*Report of the Advisory Committee for Aeronautics*, 1909–10, p. 38.]

MR LANCHESTER has discussed the application of the principle of dynamical similarity to the problem of the resistances experienced by a plane plate immersed in a stream of fluid. A year or two ago I communicated to Dr Stanton a somewhat more general statement which may be found to possess advantages. We will commence by supposing the plane of the plate perpendicular to the stream and inquire as to the dependence of the forces upon the linear dimension (l) of the plate and upon the density (ρ), velocity (v), and kinematic viscosity (ν) of the fluid. Geometrical similarity is presupposed, and until the necessity is disproved it must be assumed to extend to the thickness of the plate as well as to the irregularities of surface which constitute roughness.

If the above-mentioned quantities suffice to determine the effects, the expression for the mean force per unit area normal to the plate (P), analogous to a pressure, is

$$P = \rho v^2 . f(\nu/vl), \dots\dots\dots\dots\dots\dots\dots\dots\dots\dots\dots(A)$$

where f is an arbitrary function of the *one* variable ν/vl.

It is for experiment to determine the form of this function, or in the alternative to show that the facts cannot be represented at all by an equation of form (A). It is known that somewhat approximately P is proportional to v^2, and again that it is independent of l. If either of these approximations is supposed to hold good absolutely, it follows that f is constant, in which case P is independent of ν, or conversely if P be independent of ν, f must be constant.

The form of f may be determined by experiments in which v is varied while l and ν are constant, or again by varying l while v and ν are constant. A third method, not, it would seem, hitherto applied, is to vary ν keeping

l and v constant, and it would have certain advantages, especially in small scale experiments. The viscosity of water may be diminished (to about $\frac{1}{4}$) by heating it, or increased by admixture of alcohol.

The results of observations are best exhibited in the form of a curve, where the abscissa represents v/lv and the ordinate $P/\rho v^2$. An example of this method will be found in *Phil Mag.* Vol. XLVIII. p. 321, 1899, or *Scientific Papers*, Vol. IV. p. 415, the subject being the size of drops formed under various conditions.

Similar principles apply when the direction of the stream is no longer perpendicular to the plate. Here of course we must have regard to the manner of presentation. Thus, if a rectangular plate is held at obliquity θ, we may suppose that the longer sides are always perpendicular to the current. Equation (A) still applies; only f must be regarded as a function also of θ. We may write

$$P = \rho v^2 . f(\theta, v/vl), \quad\dots\dots\dots\dots\dots\dots\dots\dots\dots\dots(B)$$

where there are now two independent variables. In the case of symmetrical shapes it is evident that f must be an *even* function of θ.

The expression for the mean *tangential* force, reckoned per unit of area, may be written in the same way. For symmetrical shapes

$$T = \rho v^2 . F(\theta, v/vl), \quad\dots\dots\dots\dots\dots\dots\dots\dots\dots\dots(C)$$

where F is now an odd function of θ, vanishing also with ν. If $\theta = \frac{1}{2}\pi$, the plate is parallel to the stream, and the case is one previously considered (*Phil. Mag.* May 1904, p. 66*).

In (B) and (C) the forces are respectively normal and tangential to the plate. The components in other directions may at once be deduced. Thus the component in the direction of the stream is

$$P \cos\theta + T \sin\theta;$$

and it may be expressed in the form (B)

$$\rho v^2 . f_1(\theta, v/vl), \quad\dots\dots\dots\dots\dots\dots\dots\dots\dots\dots(D)$$

where f_1 is an even function of θ. When $\theta = 0$, f and f_1 coincide.

---

* [This Collection, Vol. v. p. 196.]

341.

THE PRINCIPLE OF DYNAMICAL SIMILARITY IN REFERENCE TO THE RESULTS OF EXPERIMENTS ON THE RESISTANCE OF SQUARE PLATES NORMAL TO A CURRENT OF AIR.

[*Report of Advisory Committee*, 1910–11.]

IN a recent paper under the above title, Messrs Bairstow and Booth point out that the influence of the *compressibility* of the air upon the resistance is a question of the ratio v/V, where v is the velocity of the wind, and V that at which sound is propagated. A pretty good idea of the probable effect of compressibility is afforded by the formula which connects pressure and velocity in a gas expanding or contracting adiabatically, viz.:

$$\left(\frac{p}{p_0}\right)^{\frac{\gamma-1}{\gamma}} = 1 + \frac{\gamma-1}{2}\frac{v^2}{V^2}, \quad \dots\dots\dots\dots\dots\dots(a)$$

in which p_0, v are the pressure and velocity at a distance from the obstacle, γ the ratio of specific heats (1·40), and p the pressure of the gas at the centre of the plate on the windward side where the velocity is zero. If v/V is small, (a) gives

$$p - p_0 = \tfrac{1}{2}\rho_0 v^2 (1 + \tfrac{1}{4}v^2/V^2), \quad \dots\dots\dots\dots\dots\dots(b)$$

where ρ_0 is the density at a distance, connected with V by the relation

$$V^2 = \gamma p_0/\rho_0. \quad \dots\dots\dots\dots\dots\dots\dots\dots(c)$$

Until v approaches $\tfrac{1}{4}V$ in value, $p - p_0$ is sufficiently represented by the first term of (b); and we may probably conclude that up to this point the finiteness of V has little effect upon the course of events.

On the general question (*Report for* 1909–10 [Art. 340]) of the form of f in the equation

$$P = \rho v^2 . f(\nu/vl) \quad \dots\dots\dots\dots\dots\dots\dots(d)$$

connecting the mean pressure P on the plate with the density ρ, the velocity v, the linear dimension l, and the kinematic coefficient of viscosity ν, I will add a few remarks.

The first is to point out that in all ordinary cases the ratio ν/vl, which forms the argument of the function f, is exceedingly small. For air, in C.G.S.

measure, $\nu = \cdot 13$. If we take $l = 30$ cm. (one foot) and $v = 670$ cm./sec. (15 miles per hour), we have

$$\nu/lv = 6 \cdot 10^{-6}. \quad \dots\dots\dots\dots\dots\dots\dots\dots\dots\dots\dots(e)$$

This being so, one would not expect $f(\nu/lv)$ to differ in practice from $f(0)$; and if the identification can be made, we know that P is independent of ν and l, and simply proportional to ρv^2.

It will be seen that to make ν/lv comparable with unity, the linear dimension, or the velocity, or both, must be very small.

The argument from the value of ν/lv is of course not demonstrative, for we know nothing of the numerical factors which may enter into the constitution of f; but it is, I think, not without force.

The experimental evidence so far available can hardly be said to support the general applicability of the form (d) at all. According to Stanton and Eiffel a small linear scale carries with it a diminution of P. If (d) is applicable, it follows that the same effect should accompany an increase in ν— a paradoxical conclusion, as Mr Lanchester has remarked. Moreover Dines' observations point in the opposite direction. He writes (*Proc. Royal Soc.* Vol. XLVIII. p. 252, 1890) :—" Barometrical pressure has the result that might be expected, the pressure on the plate varying directly as the height of the mercury. A rise of temperature does not seem to make much difference, but, if anything, it increases the pressure. Experiments have been made through a range of about 40° F., from about 28° F. to 68° F. The greater viscosity, I suppose, at the higher temperatures more than compensates for the decrease of density, for certainly, other circumstances being the same, the pressure is not less at 60° F. than at 30° F., and the lowest values ever obtained were in a thick fog with a temperature below the freezing point." It seems difficult to evade this argument. The viscosity μ, and *a fortiori* ν (which is equal to μ/ρ), necessarily increases with temperature.

If we accept both Stanton's conclusion that P increases with l, and Dines' that P increases with ν, we may find it difficult to defend (d) at all. It would of course be impossible to do so if the statements are accepted as universally true. But we know nothing *a priori* to exclude the possibility that f may increase with its argument at one part of the range, and decrease at another.

In many of the experiments v is known only indirectly from the indications of the Pitot's tubes, in which the measured difference of pressure is taken to represent the square of velocity. I have not studied the experimental evidence concerning Pitot's tubes, but I see no *a priori* reason why v^2 should be represented thus any more than by the total force on a plate exposed to the wind.

342.

NOTE ON THE REGULARITY OF STRUCTURE OF ACTUAL CRYSTALS.

[*Philosophical Magazine*, Vol. XIX. pp. 96—99, 1910.]

THE question must often have presented itself as to how far the mathematical regularities dealt with by the crystallographer are realized in actual crystals. That the natural faces of crystals tend to be plane is fundamental; on the other hand, it is well known that in practice it is difficult to get any but very small faces to stand the roughest optical test*. Explanations of the discrepancy may readily be suggested. The ideal conditions under which alone the tendency to flatness could fully assert itself may be scarcely attainable in practice.

The case of surfaces obtained by cleavage would seem to offer a better chance. To test this one naturally refers to mica. Mr Boys, I think, has somewhere remarked upon the fact that a piece of mica held in front of the object-glass of a telescope does not disturb the definition in the way that a piece of glass does, unless the latter be carefully worked. Mica thin enough to be convenient for such tests is of course too flexible for an examination of flatness. And it is easy to recognize that flexibility is not the only cause of deviation. There are also local irregularities, due possibly to particles of foreign matter or to strains which have exceeded the elastic limit. But these irregularities do not seem seriously to disturb the thickness of a thin plate; and the inquiry suggests itself as to how far the thickness of well split mica is really uniform.

A very delicate test of such uniformity is afforded by reflexion in the mica of a soda-flame. For a preliminary examination, at any rate, it is best to dispense with all optical accessories, simply holding the mica in the hand

---

* [1911. Dr Tutton prefers for the most accurate goniometry surfaces about 2 mm. in diameter. On such surfaces considerable curvature would fail to manifest itself. A phase-error of $\frac{1}{4}\lambda$ between the centre and circumference, about the least that could be seen in the goniometer, corresponds to a radius of curvature of 3 metres only.]

and observing the reflexion of the flame at various parts of the surface and at various moderate angles of incidence. The interposition of a piece of card to shield the eye from the direct light of the flame is convenient. In this way I have examined a number of sheets of superior mica, obtained many years ago from a photographic warehouse. Most of these exhibit serious irregularities in the splitting. The surface is divided into patches where the reflexion varies, the patches themselves appearing uniform and the boundaries sharp to an eye focused upon the plate. In some cases the boundary-lines cross one another—a feature difficult to understand until it is remembered that the irregularities may be upon *both sides* of the plate. The absence of irregularities must not be concluded too hastily. It will happen occasionally that at a particular angle of incidence there is little difference to be recognized upon the two sides of a line of division which is conspicuous enough at a somewhat different incidence of the light. The reason will be apparent to every one familiar with interference phenomena.

In some sheets a considerable area appears uniform; and there were two or three on which no abrupt changes of brightness could be detected however the incidence was varied. One of these was submitted to further observation by the more elaborate method described in *Scientific Papers*, IV. p. 56, in which provision is made for maintaining constant a small angle of incidence. No differences of brightness could be perceived; but I am not sure that in dealing with a flexible and not perfectly flat sheet the method is really more searching. The conclusion that I felt justified in drawing was that there is no abrupt change of thickness capable of producing a shift of $\frac{1}{8}$ or $\frac{1}{10}$ of an interference-band, and no gradual change giving a shift of say $\frac{1}{4}$ of a band. The interpretation is discussed at the close of this note.

In order to submit the equality of thickness to a further test, I divided the sheet of mica into approximately equal parts, cutting it with a pen-knife guided by a straight-edge. A comparison of the relative areas and relative weights of the two parts would then give material for a comparison of relative thicknesses, or at least of relative densities reckoned superficially.

The weights (w, w') of the two parts distinguished as "plain" and "marked" respectively were easily found, the *difference* being determined with special care by weighing them against one another. The weight of plain was ·2466 gm., and the difference was ·00828 gm., so that

$$\frac{w'-w}{w} = \frac{·00828}{·2466} = ·0336.$$

The comparison of areas was a more difficult matter, if only because the edges were not everywhere well defined. Both pieces were very nearly rectangular in shape, the shorter side (a) of plain being 49·1 mm., and the longer (b) being 58·6 mm. The comparison was effected with the aid of a reading microscope, which was used to measure, not the whole width, but always the *difference*. For this purpose the two pieces were approximately superposed in such a manner that the edges were nearly parallel but sufficiently separated to avoid confusion. Thus along the shorter side measurements of the overlap were made at the ends L, N and at the middle M, and on

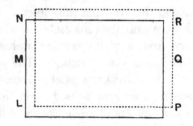

the parallel side at P, Q, R. The differences of these overlaps, suitably averaged by the formulae of the calculus of finite differences, gives $b' - b$. It is hardly necessary to explain the process further. The results were

$$\frac{b' - b}{b} = - \cdot0032, \qquad \frac{a' - a}{a} = + \cdot0379;$$

and thence

$$\frac{a'b' - ab}{ab} = \frac{a' - a}{a} + \frac{b' - b}{b} + \frac{(a' - a)(b' - b)}{ab}$$

$$= + \cdot0379 - \cdot0032 - \cdot0001 = \cdot0346.$$

This is the fraction by which the area of marked exceeds that of plain. The difference between this fraction and that found above for $(w' - w)/w$ indicates that marked is thinner than plain by the fraction ·0010.

The total thickness is most easily found from the weight and area with the aid of an assumed specific gravity (2·8). For plain we get

$$\text{thickness} = \cdot0306 \text{ mm.} = \cdot00120 \text{ inch.}$$

The difference of thickness in the two pieces resulting from the observations on weight and area is thus little over a millionth part of an inch, and might perhaps be considered as devoid of significance. But it is difficult to admit that either the weights or the areas could be in error to the extent of a thousandth part. I am disposed to think that the discrepancy is real in the sense that it would not be eliminated by repetition. Possibly it may be attributed to differences in the condition of the surfaces which had been handled more than was good for them. There are also scratches and other minor irregularities to be considered.

As regards grease it may be worth while to mention the result of a few observations on greasing one side of a " lantern " glass plate of 68 sq. cm. area ($3\frac{1}{4}$ in. × $3\frac{1}{4}$ in.). Two such were opposed in the balance, one face of each being greased alternately over its whole area with heavy marks from a finger which had touched the hair. The difference between the weights of a plate with one face thus greased and clean came out about $\frac{1}{10}$ mg.* The greased area here would exceed that of both faces of one of the mica plates.

* The *mean* thickness of the layer of grease would thus be about $\frac{1}{4}\lambda$.

Not feeling quite satisfied with the edges formed by the knife, which under the microscope compared unfavourably with the original edges, I determined on another set of measurements in which the new edges were made by scissors. The objection in this case is that it is less easy to cut straight, but I thought that the prejudicial effect might be obviated by cutting the two pieces together superposed. The measurements were conducted as before except that now *four* measurements of the overlap were taken along each edge. But the discrepancy resulting from a comparison of the relative weights and areas was not removed, being indeed a little greater than had been found before from what were practically the same pieces.

The total thickness above reckoned is about 50 wave-lengths (λ) of soda light, so that a difference of a thousandth part corresponds to $\frac{1}{20}\lambda$. But as regards the formation of interference-bands in the optical examination, this difference must be multiplied by 3, being doubled in virtue of the double passage of the light reflected by the hinder surface, and increased again in the ratio 3 : 2, or thereabouts, in virtue of the refractive index of the material. The discrepancy suggested by the measurements of weight and area thus amounts to about $\frac{1}{7}$ of a band, and is accordingly not much below the limit marked out on the ground of the merely optical examination.

The conclusion is that some plates of mica are uniform in an extraordinarily high degree and that there is perhaps no reason for doubting that the thickness over finite areas may be as uniform as is consistent with a molecular structure. The stringency of the optical test might probably be increased a few times by silvering the surfaces after the manner of Fabry and Perot (compare *Phil. Mag.* Vol. XII. p. 489, 1906)*.

* [This Collection, Vol. v. p. 341.]

343.

COLOURS OF SEA AND SKY.

[*Royal Institution Proceedings*, Feb. 25, 1910; *Nature*, LXXXIII. p. 48, 1910.]

A RECENT voyage round Africa recalled my attention to interesting problems connected with the colour of the sea. They are not always easy of solution in consequence of the circumstance that there are several possible sources of colour whose action would be much in the same direction. We must bear in mind that the absorption, or proper, colour of water cannot manifest itself unless the light traverse a sufficient thickness before reaching the eye. In the ocean the depth is of course adequate to develop the colour, but if the water is clear there is often nothing to send the light back to the observer. Under these circumstances the proper colour cannot be seen. The much admired dark blue of the deep sea has nothing to do with the colour of water, but is simply the blue of the sky seen by reflection. When the heavens are overcast the water looks grey and leaden; and even when the clouding is partial, the sea appears grey under the clouds, though elsewhere it may show colour. It is remarkable that a fact so easy of observation is unknown to many even of those who have written from a scientific point of view. One circumstance which may raise doubts is that the blue of the deep sea often looks purer and fuller than that of the sky. I think the explanation is that we are apt to make comparison with that part of the sky which lies near the horizon, whereas the best blue comes from near the zenith. In fact, when the water is smooth and the angle of observation such as to reflect the low sky, the apparent blue of the water is much deteriorated. Under these circumstances a rippling due to wind greatly enhances the colour by reflecting light from higher up. Seen from the deck of a steamer, those parts of the waves which slope towards the observer show the best colour for a like reason.

The real colour of ocean water may often be seen when there are breakers. Light, perhaps directly from the sun, may then traverse the crest of the waves and afterwards reach the observer. In my experience such light shows decidedly green. Again, over the screw of the ship a good deal of air is entangled and carried down, thus providing the necessary reflection from under the surface. Here also the colour is green.

The only places where I have seen the sea look blue in a manner not explicable by reflection of the sky were Aden and Suez. Although the sky was not absolutely overcast, it seemed that part at any rate of the copious, if not very deep, blue was to be attributed to the water. This requires not only that the proper colour of the water should here be blue, but also the presence of suspended matter capable of returning the light, unless indeed the sea bottom itself could serve the purpose.

The famous grotto at Capri gives an unusually good opportunity of seeing the true colour of the water. Doubtless a great part of the effect is due to the eye being shielded from external glare and so better capable of appreciating the comparatively feeble light which has traversed considerable thicknesses of water. The question was successfully discussed many years ago by Melloni, who remarks that the beauty of the colour varies a good deal with the weather. The light which can penetrate comes from the sky and not directly from the sun. When the day is clear, the blueness of the sky co-operates with the blueness of the water.

That light reflected from the surface of a liquid does not exhibit the absorption colour is exemplified by brown peaty water such as is often met with in Scotland. The sky seen by reflection is as blue as if the water were pure. But an attempt to illustrate this fact by experiment upon quite a small scale was not at first successful. A large white photographic dish containing dark brown oxidized " pyro " was exposed upon the lawn during a fine day. Although the reflected light certainly came from the clear sky, the colour did not appear pronounced, partly in consequence of the glare of the sunshine from the edges of the dish. The substitution of a dish of glass effected an improvement. But it was only when the eye was protected from extraneous light by the hands, or more perfectly by the interposition of a pasteboard tube held close up, that the blue of the reflected light manifested its proper purity. It would seem that the explanation is to be sought in diffusion of light within the lens of the eye, in consequence of which, especially in elderly persons, the whole field is liable to be suffused with any strong light finding access.

As regards the proper colour of pure water, an early opinion is that of Davy, who, in his *Salmonia*, pronounces in favour of blue, basing his conclusion upon observations of snow and glacier streams. The latter, indeed, are often turbid, but deposit the ground-up rock which they contain when opportunity offers, as in the lake of Geneva. A like conclusion was later put forward by Bunsen on the basis of laboratory observations. The most elaborate experiments are those of Spring, who, in a series of papers published during many years, discusses the difficult questions involved. He tried columns of great length—up to 26 metres; but even when the distance traversed was only 4 or 5 metres, he finds the colour a fine blue only to be compared with the purest sky-blue as seen from a great elevation. But when the tubes contain ordinary water, even ordinary distilled water, the colour is green, or yellow-green, and not blue.

The conversion of the original blue into green is, of course, explicable if there be the slightest contamination with colouring matter of a yellow character—*i.e.* strongly absorbent of blue light. Spring shows that this is the effect of minute traces—down to one ten-millionth part—of iron in the ferric state, or of humus. The greenness of many natural waters is thus easily understood. Another question examined by Spring is not without bearing upon our present subject—viz. the presence of suspended matter. I am the better able to appreciate the work of Spring, that many years ago I tried a variety of methods, including distillation *in vacuo*, in order to obtain water in the condition which Tyndall described as "optically empty," but I met with no success. Spring has shown that the desired result may be obtained by the formation within the body of the liquid of a gelatinous precipitate of alumina or oxide of iron, by which the fine particles of suspended matter are ultimately carried down.

Perhaps the most telling observations upon the colour of water are those of Count Aufsess, who measured the actual transmission of light belonging to various parts of the spectrum. The principal absorption is in the red and yellow. In the case of the purest water, there was practically no absorption above the line F, and a high degree of transparency in this region was attained even by some natural waters. That these waters should show blue, *when in sufficient thickness*, is a necessary consequence.

In my own experiments, made before I was acquainted with the work of Aufsess, the light traversed two glass tubes of an aggregate length of about 4 metres (12 feet). On occasion the light was reflected back so as to traverse this length twice over. I must confess that I have never seen a blue answering to Spring's description, when the original light was white. For final tests I was always careful to employ the light of a completely overcast day, which was reflected into the tubes by a small mirror. The colour, after transmission, showed itself very sensitive to the character of the original source. The palest clear sky of an English winter's day gave a greatly enhanced blue, while, on the other hand, isolated clouds are usually yellowish, and influence the result in the opposite direction. I should myself describe the best colour of the transmitted light on standard days as a greenish blue, but there is some variation in the use of words, and, perhaps, in vision. Some of my friends, but not the majority, spoke of blue simply, but all were agreed that the blueness of a good sky was not approached. The waters tried have been very various. Sea-water from outside the grotto of Capri, from Suez, and from near the Seven Stones Lightship off the Cornish coast, I owe to the kindness of friends. Of these the two former showed a greenish-blue, the latter a full, or perhaps rather yellowish, green, and these colours were not appreciably modified after the water had stood in the tubes for weeks. It is important to remember that the hue may, to some extent, depend upon thickness. It is quite probable that in a greatly increased thickness the Capri and Suez waters would assume a more decided blue colour. But I do not think the Seven Stones water could so behave, the colour, with 12 feet, seeming to involve the absorption of blue light.

Further observations on greater depths of sea-water would be desirable. A naval son informs me that off the coast of Greece a plate lying in 6 fathoms of water looked decidedly blue, although the sky was a dirty grey. I have doubts whether this would be generally the case in the Mediterranean; the green due to moderate thicknesses seems too decided.

Of natural fresh waters that I have tried, none were better than that from a spring in my own garden. This water is hard, but bright and clear, and it shows a greenish-blue, barely distinguishable from that of the Capri and Suez water. Distillation does not improve the blue. Neither did other treatments do any good, such, for example, as partial precipitation of the lime with alkali, or passage of ozone with the idea of oxidising humus. Wishing to try water of high chemical purity, I obtained—through the kind offices of Sir J. Dewar—water twice distilled from alkaline permanganate, and condensed in contact with silver, but the colour was no bluer. In the light of this evidence, I can hardly avoid the conclusion that the blueness of water in lengths of 4 metres has been exaggerated, especially by Spring, although I have no reason to doubt that a fully developed blue may be obtained at much greater thicknesses. I should suppose that sufficient care has not been taken to start with white light. It may be recalled that overcast days are not so common in some parts of the world as in England.

A third possible cause of apparent blueness of the sea must also be mentioned. If a liquid is not absolutely clear, but contains in suspension very minute particles, it will disperse light of a blue character. Although, undoubtedly, this cause must operate to some extent, I have seen no reason to think that it is important. But the existence of three possible causes of blueness complicates the interpretation of the phenomena. Hitherto observers have not been sufficiently upon their guard to distinguish blueness having its origin in the sky from blueness fairly attributable to the water itself.

As regards the light from the sky, the theory which attributes it to dispersal from small particles, many of which are smaller than the wave-length of light, is now pretty generally accepted. To a first approximation at any rate, both the polarization and the colour of the light are easily explained. According to the simplest theory, the polarization should be absolute and a maximum at 90° from the sun, and the colour should be modified from that of the sun according to the factor λ^{-4}. But it is easy to see that there must be complications, even if all the particles are small and spherical. The light illuminating them is not merely the direct light of the sun, but also light diffused from the sky and from the earth's surface. On these grounds alone the polarization must be expected to be incomplete even at 90°, and the certain presence of particles not small in comparison with the wave-length is another cause operating in the same direction. It is rather remarkable that, as I noticed in 1871, the two polarised components show much the same colour*. The observation is best made with a double-image prism mounted near one end of a pasteboard tube, through which a suitable rectangular aperture at the other end is seen double, but with the two images in close juxtaposition. When this is directed to a part of the sky 90° from the sun, and the tube turned until one image is at its darkest, the two polarised components are exhibited side by side in a manner favourable for comparison of colours. The addition at the eye end of a nicol, capable of rotation independently of the tube, gives the means of equalising the brightnesses without altering the colours. This observation, made independently by Spring, is regarded by him as an objection to the theory, and as showing that the causes of the blueness and of the polarization are not the same. The argument would have more weight if the colours of the two components were exactly the same and under all circumstances, but I do not think that this is the case. Observations on the purer sky, to be seen from great elevations, would be of interest†. The question is to what causes the second component is principally due. So far as it depends upon sky illumination, it would be bluer than the first component. Any "residual blue" of the kind described by Tyndall, and due to particles somewhat too big for the simple theory, would make a contribution in the same direction. On the other hand, large particles under the direct light of the sun, and perhaps small ones, so far as illuminated by light from the earth, would contribute a whiter light. In this way an approximate compensation may occur, but the matter is certainly worthy of further attention.

In this connection it should be noticed that, according to the now generally received electro-magnetic theory, complete polarization at 90°

* [*Scientific Papers*, Vol. I. p. 109.]

† [1911. I have had no opportunity of experimenting under specially favourable circumstances; but observations in Essex on the clearest days sometimes show no difference and sometimes that the residual light is the bluer.]

requires that the dispersing particles should behave as if spherical, even although infinitely small. If the shape be elongated, there would be incomplete polarization combined with similarity of colour even under the simplest conditions.

When the particles are no longer very small in comparison with the wave-length, the direction of maximum polarization was found by Tyndall to become oblique, and the deviation is in the opposite direction to that which would have been anticipated from the Brewsterian law for the reflection of light from surfaces of finite area. As I showed in 1881, the gradual precipitation of sulphur from a very weak and acid solution of "hypo" exhibits the phenomena remarkably well*. At a certain stage, depending on the colour of the light, the direction of maximum polarization becomes oblique. Even when the obliquity is well established for blue light, red light still continues to follow the simpler law, and the comparison gives curious information concerning the rate of growth of the particles.

The preferential scattering of light of short wave-length involves of course a gradual yellowing and ultimate reddening of the light transmitted. The formation in this way of sunset colours is well illustrated by the acid hypo.

That Spring rejects this theory in favour of one which would attribute sky-blue to absorption by oxygen or ozone, has been already alluded to. Although one must not conclude too hastily from the behaviour of these bodies when liquefied, it is, of course, possible that their absorbing qualities may influence atmospheric phenomena in some degree. But to attribute the blue of the sky to them seems out of the question. It is sufficient to remark that the setting sun turns red and not blue.

An interesting question remains behind. To what kind of small particles —dispersing short waves in preference—is the heavenly azure due? That small particles of saline or other solid matter, including organic germs, play a part, cannot be doubted, and to them may be attributed much of the bluish haze by which the moderately distant landscape is often suffused. But it seems certain that the very molecules of air themselves are competent to scatter a blue light not very greatly inferior to that which we actually receive. Theory allows a connection to be established between the transparency of air for light of various wave-lengths, and its known refractivity in combination with Avogadro's constant, expressing the number of molecules per cubic centimetre in gas under standard atmospheric conditions. The first estimate of transparency was founded upon Maxwell's value of this constant, viz. $1 \cdot 9 \times 10^{19}$. Recent researches have shown that this number must be raised to $2 \cdot 76 \times 10^{19}$, and that the result is probably accurate to

* [*Scientific Papers*, Vol. I. p. 531.]

within a few per cent.* It has been pointed out by Dr Schuster that the introduction of the raised number into the formula almost exactly accounts for the degree of atmospheric transparency observed at high elevations in the United States, apparently justifying to the full the inference that the normal blue of the sky is due to molecular scattering. But, although there is no reason to anticipate that this general conclusion will be upset, it should not be overlooked that a molecule, especially a diatomic molecule, can hardly be supposed to behave as if it were the dielectric sphere of theory. Questions are here suggested for whose decision the time is perhaps not yet ripe.

P.S.—The question of the colour of the Mediterranean and other waters was long ago discussed by Mr J. Aitken—an excellent observer—in *Proc. Roy. Soc. Edin.* 1881–82. His principal conclusions are very similar to my own. Mr Aitken rightly insists upon the influence of the colour of the suspended matter to which the return of the light is due. Only when this is white, has the proper colour of the water a full chance of manifesting itself. From the heights of Capri, I noticed that the shallow water near the shore showed decidedly green, an effect attributed to the yellowness of the underlying sand.

* It is a curious instance of divergence in scientific opinion that while some still deny the existence of molecules, others have successfully counted them.

344.

THE INCIDENCE OF LIGHT UPON A TRANSPARENT SPHERE OF DIMENSIONS COMPARABLE WITH THE WAVE-LENGTH.

[*Proceedings of the Royal Society*, A, Vol. LXXXIV. pp. 25—46, 1910.]

IN a paper, now nearly thirty years old*, I applied Maxwell's equations of the electro-magnetic field to investigate the disturbance produced by an obstacle upon plane waves of light which travel through a medium otherwise uniform, giving particular attention to the case where the properties of the obstacle differ but little from those of its surroundings. The difference may consist in a variation of $K-$ the specific inductive capacity, or of $\mu-$ the magnetic capacity, or of both; but it was shown that the last supposition leads to results inconsistent with observation, and that the evidence favours the view that μ is to be treated as invariable. Denoting electric displacements by f, g, h, the primary wave was taken to be

$$h_0 = e^{int}e^{ikx}, \dots\dots\dots\dots\dots\dots\dots\dots\dots(23)$$

so that the direction of propagation is along x (negatively), and that of vibration parallel to z. $\Delta\mu$ being omitted, the electric displacements (f_1, g_1, h_1) in the scattered wave, so far as they depend upon the *first power* of ΔK, have at a great distance the values

$$f_1, g_1, h_1 = \frac{k^2 KP}{4\pi r}\left(\frac{\alpha\gamma}{r^2}, \frac{\beta\gamma}{r^2}, -\frac{\alpha^2 + \beta^2}{r^2}\right), \dots\dots(35, 37, 38)$$

in which

$$P = \iiint h_0 \Delta K^{-1} e^{-ikr}\,dx\,dy\,dz. \dots\dots\dots\dots(36)$$

In these equations r denotes the distance between the point (α, β, γ), where the disturbance is required to be estimated, and the element of volume of the obstacle $dx\,dy\,dz$.

It is evidently implied that in a direction ($\alpha = 0$, $\beta = 0$) parallel to that of primary vibration, the scattered vibration vanishes, whatever may be the

---

* *Phil. Mag.* Vol. XII. p. 81 (1881); *Scientific Papers*, Vol. I. p. 518. In the second term of Equation (32) and following, ΔK^{-1} should read $\Delta\mu^{-1}$.

size and shape of the obstacle, so that any light which may appear in this direction must depend upon powers of ΔK higher than the first.

If ΔK be uniform throughout the obstacle, it may be removed from under the integral sign in (36), and if the value of K for the external medium be taken as unity, we may write

$$K . \Delta K^{-1} = - (K - 1).$$

If the obstacle is very small, h_0 and r may be treated as constants in the integration, and

$$P = - (K - 1) h_0 e^{-ikr} . T, \quad \dots\dots\dots\dots(40)$$

if T be the whole volume of the small obstacle. Using this in (35), etc., we get

$$f_1, g_1, h_1 = \frac{\pi T (K-1)}{\lambda^2 r} e^{i (nt-kr)} \left[-\frac{\alpha\gamma}{r^2}, -\frac{\beta\gamma}{r^2}, \frac{\alpha^2+\beta^2}{r^2} \right], \quad \dots(41, 42, 43)$$

λ being the wave-length $(2\pi/k)$. These equations apply, whatever may be the shape of the small obstacle; and, of course, they undergo no change if the shape be spherical. In that case it was shown that the restriction upon the value of $K - 1$ may be removed, provided we replace $K - 1$ by $3 (K - 1)/(K + 2)$.

We have seen that the scattered light vanishes to a first approximation in $(K - 1)$ for the direction parallel to z. In the case of the sphere it was shown that for this direction we have to a second approximation for a small obstacle

$$f_2 = \frac{\pi T (K - 1)^2}{\lambda^2 r} e^{i (nt-kr)} . \frac{k^2 R^2}{25}, \quad \dots\dots\dots\dots(60)$$

with $g_2 = 0$, $h_2 = 0$. In (60) R denotes the radius of the sphere, and r is measured from the centre. "Comparing (60) and (41) we see that the amplitude of the light scattered along z is not only of higher order in ΔK, but is also of the order $k^2 R^2$ in comparison with that scattered in other directions. The incident light being white, the intensity of the component colours scattered along z varies as the inverse eighth power of the wave-length, so that the resultant light is a rich blue." It is obvious from (41) that in general the intensity of the scattered light varies as λ^{-4}.

"There is another point of importance to be noticed. Although when the terms of the second order are included, the scattered light does not vanish along the axis of z, the peculiarity is not lost but merely transferred to another direction. Putting together the terms of the first and second orders we see that the scattered light will vanish in a direction in the plane of xz, inclined to z (towards $+ x$) at a small angle θ, such that

$$\theta = \frac{\Delta K}{K} \frac{k^2 R^2}{25} = (K - 1) \frac{k^2 R^2}{25}." \quad \dots\dots\dots\dots(61)$$

Some experiments in illustration were then described.

Returning to the case where $(K-1)^2$ is neglected, we consider a spherical obstacle of any radius R, whose centre is at the origin of co-ordinates. From (23), (36)

$$P = -(K-1)\, e^{int} \iiint e^{ik\,(x-r)}\, dx\, dy\, dz,$$

r being the distance between $dx\,dy\,dz$, and the (distant) point at which $f_1,\ g_1,\ h_1$ are to be estimated. It is evident that, so far as the secondary ray is concerned, P depends only upon the angle (χ) which this ray makes with the primary ray. We will suppose that $\chi = 0$ in the direction backwards along the primary ray, and that $\chi = \pi$ along the primary ray continued. The integral may then be found in the form

$$\frac{2\pi R^2 e^{-ikr}}{k\cos\frac12\chi} \int_0^{\frac12\pi} J_1(2kR\cos\tfrac12\chi\cos\phi)\cos^2\phi\, d\phi;\ \ldots\ldots\ldots(48)$$

r_0 being the distance of the point of observation from the centre of the sphere. If we expand the Bessel's function and drop the suffix in r_0 as no longer required, we get

$$P = -\frac{4\pi R^3 (K-1)\, e^{i\,(nt-kr)}}{3} \left\{ 1 - \frac{m^2}{2\cdot5} + \frac{m^4}{2\cdot4\cdot5\cdot7} \right.$$
$$\left. - \frac{m^6}{2\cdot4\cdot6\cdot5\cdot7\cdot9} + \frac{m^8}{2\cdot4\cdot6\cdot8\cdot5\cdot7\cdot9\cdot11} - \ldots \right\},\ \ \ldots\ldots(49)$$

in which m is written for $2kR\cos\frac12\chi$. From this $f_1,\ g_1,\ h_1$ follow at once by (35), etc. Along the continuation of the primary ray $\cos\frac12\chi = 0$, and P reduces to

$$-\frac{4\pi}{3}\, R^3(K-1)\, e^{i\,(nt-kr)},$$

as was to be expected. It is to be observed that in this solution there is no limitation upon the value of R if $(K-1)^2$ is neglected absolutely.

Prof. Love, in a valuable paper on the "Scattering of Electric Waves by a Dielectric Sphere*," has treated this problem by a different method, limited on the one hand to the sphere, but on the other applicable whatever may be the value of $(K-1)$, and (so far as the general analytical expressions are concerned) whatever may be the size of the sphere. In the 1881 paper I had treated in this way the problem of a dielectric *cylinder*. From these expressions Love deduces first and second approximations applicable when the radius of the sphere may be treated as small in comparison with the wave-length. It will be of interest to compare these approximations with those already referred to.

---

* *Lond. Math. Soc. Proc.* Vol. xxx. p. 308 (1899).

In Prof. Love's notation the electric forces are represented by X, Y, Z; and, so far as relates to the medium outside the sphere, they may be identified with f, g, h used above. The co-ordinates of the point of observation are denoted by x, y, z instead of by α, β, γ. Again, Love supposes the direction of propagation to be along $-z$ (instead of $-x$), and the direction of vibration to be parallel to y (instead of z), changes represented by taking one step backwards in the cycle x, y, z, x. Thus (with omission of a constant multiplier which runs through) the primary wave is represented by

$$Y = e^{ik(ct+z)}. \qquad \ldots\ldots\ldots\ldots\ldots\ldots\ldots\ldots\ldots\ldots(X)$$

Corresponding to this, the complete expressions, so far as terms of order $k^4 R^5$, for the forces in the scattered wave at a great distance are found to be

$$(X, Y, Z) = \frac{K-1}{K+2} \frac{k^2 R^3 e^{ik(ct-r+R)}}{r} \left(-\frac{xy}{r^2}, \frac{x^2+z^2}{r^2}, -\frac{yz}{r^2} \right)$$

$$\times \left\{ 1 - ikR - k^2 R^2 \left(\frac{19}{18} - \frac{6}{5} \frac{K}{K+2} \right) \right\}$$

$$+ (K-1) \frac{k^4 R^5 e^{ik(ct-r+R)}}{30r} \left(0, -\frac{z}{r}, \frac{y}{r} \right)$$

$$+ \frac{K-1}{2K+3} \frac{k^4 R^5 e^{ik(ct-r+R)}}{6r} \left(0, -\frac{z}{r}, \frac{y}{r} \right)$$

$$+ \frac{K-1}{2K+3} \frac{k^4 R^5 e^{ik(ct-r+R)}}{3r} \frac{yz}{r^2} \left(\frac{x}{r}, \frac{y}{r}, \frac{z}{r} \right). \quad \ldots\ldots\ldots(XLII)$$

So far as the first approximation involving $k^2 R^3$, this agrees *mutatis mutandis* with (41, 42, 43), since $T = \frac{4}{3}\pi R^3$ and $k = 2\pi/\lambda$.

We may next consider the application of (XLII) to the secondary ray in the direction of primary vibration ($x = 0$, $z = 0$, $y = r$), for which the first approximation vanishes. We have

$$(X, Y, Z) = \frac{(K-1) k^4 R^5 e^{ik(ct-r)}}{30r} \left[(0, 0, 1) + \frac{5}{2K+3} (0, 0, -1) \right]$$

$$= \frac{(K-1)^2 k^4 R^5 e^{ik(ct-r)}}{15(2K+3)r} (0, 0, 1);$$

and this agrees with (60) when we introduce the supposition that $(K-1)$ is small, so that $2K + 3$ in the denomination may be identified with 5.

I have already commented upon this agreement[*]; but recently in pursuing the comparison I came upon a discrepancy. It is evident from the theory already given that a great general simplification should attend the

---

[*] *Scientific Papers*, Vol. I. p. 536 (1899).

supposition that $(K-1)$ is very small. Introducing this into (XLII) and omitting for the moment the factor

$$(K-1)\,k^2R^3\frac{e^{ik\,(ct-r+R)}}{r},$$

we find for the three last sets of terms

$$-\frac{k^2R^2}{15}\frac{z}{r}\left(-\frac{xy}{r^2},\ \frac{x^2+z^2}{r^2},\ -\frac{yz}{r^2}\right),$$

showing within the bracket the same dependence on x, y, z as does the first set. Hence altogether, expanding the exponential e^{ikR}, we may write in this case

$$(X,\,Y,\,Z)=\frac{(K-1)\,k^2R^3\,e^{ik\,(ct-r)}}{3r}\left(-\frac{xy}{r^2},\ \frac{x^2+z^2}{r^2},\ -\frac{yz}{r^2}\right)$$
$$\times\left\{1+k^2R^2\left(\tfrac{1}{2}-\tfrac{19}{18}+\tfrac{2}{5}\right)-\frac{k^2R^2}{5}\frac{z}{r}\right\}.$$

This should correspond with (35), (49) of the former investigation if in the latter we stop at the term in m^2 where

$$m^2=4k^2R^2\cos^2\tfrac{1}{2}\chi=2k^2R^2\left(1+\frac{z}{r}\right),$$

so that

$$P=-\frac{4\pi R^3\,(K-1)\,e^{i\,(nt-kr)}}{3}\left\{1-\frac{k^2R^2}{5}\left(1+\frac{z}{r}\right)\right\}.$$

It will be seen that in all respects except one the agreement is complete. For example, i has disappeared otherwise than as contained in $e^{ik\,(ct-r)}$. But there remains a numerical discrepancy, since $(\tfrac{1}{2}-\tfrac{19}{18}+\tfrac{2}{5})$ is not equal to $-\tfrac{1}{5}$.

For some time I was at a loss to account for this discrepancy, until I noticed that the omission of a term $\tfrac{2}{45}k^2R^2$ in Love's equation (XXXIV) would produce harmony, and this led me to scrutinise equation (XVIII), where (I believe) the error lies. It appears to me that these equations should be

$$\phi_n=\frac{ik\,e^{ikct}}{n\,(n+1)\,\psi_n\,(kR)}\left[\frac{R^2}{2n+3}\,\psi_{n+1}\,(kR)\frac{dV_{n+1}}{dx}\right.$$
$$\left.-\frac{r^{2n+1}}{2n-1}\,\psi_{n-1}\,(kR)\frac{d}{dx}\left(\frac{V_{n-1}}{r^{2n-1}}\right)\right],$$
$$\chi_n=-\frac{ik\,e^{ikct}}{n\,(n+1)\,\psi_n\,(kR)}\left[\frac{R^2}{2n+3}\,\psi_{n+1}\,(kR)\frac{dV_{n+1}}{dy}\right.$$
$$\left.-\frac{r^{2n+1}}{2n-1}\,\psi_{n-1}\,(kR)\frac{d}{dy}\left(\frac{V_{n-1}}{r^{2n-1}}\right)\right],$$

$\left.\vphantom{\begin{array}{c}a\\b\\c\\d\end{array}}\right\}$ ...(XVIII*)

the change consisting in the substitution of $(2n+3)$ for $(2n+1)$ in the denominators of the first terms of ϕ_n, χ_n. We should then obtain

$$\chi_1=-\tfrac{1}{2}ik\,e^{ikct}.y\quad\text{simply,}$$

instead of (XXXIV), which involves also the factor $(1 + \frac{2}{45} k^2 R^2)$. This is a matter to which I shall presently return.

In view of numerical calculations, it may be well to complete the statement of Prof. Love's *results*, and to show that, especially when corrected as above, they admit of considerable simplification.

For the functions which occur in (XVIII) we have $V_0 = 1$, and for n greater than unity

$$V_n = \frac{(ikr)^n}{1 . 3 . 5 \ldots (2n-1)} P_n \left(\frac{z}{r}\right), \quad \ldots\ldots\ldots\ldots(XIV)$$

P_n denoting as usual Legendre's function, so that V_n is a solid harmonic of degree n. For ψ_n we have

$$\psi_n(\eta) = (-)^n 1 . 3 . 5 \ldots (2n+1) \left(\frac{1}{\eta} \frac{d}{d\eta}\right)^n \frac{\sin \eta}{\eta}$$

$$= 1 - \frac{\eta^2}{2 . 2n+3} + \frac{\eta^4}{2 . 4 . 2n+3 . 2n+5} - \frac{\eta^6}{2 . 4 . 6 . (2n+3)(2n+5)(2n+7)}$$
$$+ \ldots, \quad \ldots(IV)^*$$

and the functions $\psi_n(\eta)$ satisfy certain sequence equations which may be written

$$\frac{1}{\eta} \frac{d}{d\eta} \psi_n(\eta) = -\frac{\eta^2}{2n+3} \psi_{n+1}(\eta) = (2n+1)\{\psi_{n-1}(\eta) - \psi_n(\eta)\}. \quad \ldots(VI)$$

We have also to deal with another function, $E_n(\eta)$, such that

$$E_n(\eta) = (-)^n 1 . 3 . 5 \ldots (2n+1) \left(\frac{1}{\eta} \frac{d}{d\eta}\right)^n \frac{e^{-i\eta}}{\eta}, \quad \ldots\ldots(VII)$$

and the functions $E_n(\eta)$ satisfy the same sequence equations as the functions $\psi_n(\eta)$. It will be seen that $-i\psi_n(\eta)$ is the imaginary part of $E_n(\eta)$. As included in (VII),

$$E_0(\eta) = \frac{e^{-i\eta}}{\eta}, \qquad E_1(\eta) = 3 \frac{1+i\eta}{\eta^3} e^{-i\eta}. \quad \ldots\ldots(XXXVI)$$

This being understood, the disturbance in the secondary wave, so far as it depends upon ϕ_n, χ_n, may be written

$$X_n = M_n E_n(kr) \left(y \frac{d\phi_n}{dz} - z \frac{d\phi_n}{dy}\right)$$

$$+ \frac{N_n}{ik}\left[-(n+1) E_{n-1}(kr) \frac{d\chi_n}{dx} + \frac{nk^2 r^{2n+3}}{(2n+1)(2n+3)} E_{n+1}(kr) \frac{d}{dx} \frac{\chi_n}{r^{2n+1}}\right], \quad \ldots(XV)$$

with two similar equations for Y_n, Z_n. The complete expressions for X, Y, Z involve a summation with respect to n which takes the values 1, 2, 3, etc.

---

* This series is the same as that which occurs in the expression for $J_{n+\frac{1}{2}}(\eta)$.

All is now defined except the constants M_n, N_n, which depend upon the boundary conditions to be satisfied at the surface of the sphere. For the present case of a dielectric sphere of refractive index $\sqrt{K}$, or k'/k, Prof. Love finds the equivalent of

$$M_n = \frac{\psi_{n-1}(kR) - \dfrac{\psi_{n-1}(k'R)}{\psi_n(k'R)}\,\psi_n(kR)}{-E_{n-1}(kR) + \dfrac{\psi_{n-1}(k'R)}{\psi_n(k'R)}\,E_n(kR)}, \qquad \dots\dots\dots\dots(\text{XXIII})$$

$$N_n = \frac{K\psi_{n-1}(kR) - \left\{(K-1)\dfrac{n}{2n+1} + \dfrac{\psi_{n-1}(k'R)}{\psi_n(k'R)}\right\}\psi_n(kR)}{-KE_{n-1}(kR) + \left\{(K-1)\dfrac{n}{2n+1} + \dfrac{\psi_{n-1}(k'R)}{\psi_n(k'R)}\right\}E_n(kR)},$$

$$\dots(\text{XXIII } bis)$$

which completes the analytical solution.

Returning now to (XVIII*) we see that the first terms dV_{n+1}/dx, dV_{n+1}/dy are solid harmonics of degree n. Also V_{n-1}/r^{2n-1} is a solid harmonic of degree $-n$. When differentiated with respect to x or y, it becomes a solid harmonic of degree $-n-1$, and when further multiplied by r^{2n+1}, a solid harmonic of degree n. Hence both terms in (XVIII*) are solid harmonics of degree n, and it will appear (as might, indeed, be anticipated) that they are of the same form. Writing μ for z/r, we have from (XIV)

$$\frac{dV_{n+1}}{dx} = \frac{(ik)^{n+1}r^{n-1}x}{1.3.5\dots(2n+1)}\left\{(n+1)P_{n+1}(\mu) - \mu\frac{dP_{n+1}}{d\mu}\right\}. \qquad \dots\dots\dots(a)$$

Again

$$r^{2n+1}\frac{d}{dx}\frac{V_{n-1}}{r^{2n-1}} = \frac{(ik)^{n-1}r^{n-1}x}{1.3.5\dots(2n-3)}\left\{-nP_{n-1}(\mu) - \mu\frac{dP_{n-1}}{d\mu}\right\}. \qquad \dots\dots(b)$$

Now (Todhunter's *Laplace's Functions*, pp. 43, 46) the quantities within { } in (a) and (b) are equal to one another and to $-dP_n/d\mu$, so that

$$r^{2n+1}\frac{d}{dx}\frac{V_{n-1}}{r^{2n-1}} = -\frac{(2n-1)(2n+1)}{k^2}\frac{dV_{n+1}}{dx}, \qquad \dots\dots\dots\dots(c)$$

and

$$\frac{R^2}{2n+3}\psi_{n+1}(kR)\frac{dV_{n+1}}{dx} - \frac{r^{2n+1}}{2n-1}\psi_{n-1}(kR)\frac{d}{dx}\frac{V_{n-1}}{r^{2n+1}}$$

$$= \frac{dV_{n+1}}{dx}\frac{2n+1}{k^2}\left\{\frac{k^2R^2}{2n+1.2n+3}\psi_{n+1}(kR) + \psi_{n-1}(kR)\right\}$$

$$= \frac{dV_{n+1}}{dx}\frac{2n+1}{k^2}\psi_n(kR),$$

by (VI). We thus obtain the greatly simplified forms

$$\phi_n,\ \chi_n = \frac{(2n+1)\,ik^{-1}\,e^{ikct}}{n.n+1}\left\{\frac{dV_{n+1}}{dx},\ -\frac{dV_{n+1}}{dy}\right\}, \qquad \dots\dots\dots\dots(d)$$

in which R does not appear at all.

Referring to (XIV) and introducing the value of $P_n(z/r)$, we find

$$V_0 = 1, \qquad V_2 = -\tfrac{1}{6}k^2(2z^2 - x^2 - y^2),$$
$$V_1 = ikz, \qquad V_3 = -\frac{ik^3}{30}\{2z^3 - 3z(x^2 + y^2)\}. \qquad \Bigg\} \quad \ldots\ldots(\text{XXVI})$$

Thus in the case of $n = 1$ we have

$$\chi_1 = -\tfrac{1}{2}iky\, e^{ikct},$$

without any approximation depending on the smallness of R.

For further reductions it will be convenient to introduce a modified form of V_n, viz. U_n, defined by

$$V_n = \frac{(ik)^n\, U_n}{1.3.5\ldots(2n-1)} = \frac{(ikr)^n}{1.3.5\ldots(2n-1)}\, P_n\left(\frac{z}{r}\right). \quad \ldots(\text{XIV}*)$$

We have now to consider the solution expressed in (XV), and we may at once introduce the simplifying condition that r is very great. In this case from (VII)

$$E_n(kr) = i^n . 1 . 3 . 5 \ldots (2n+1)\frac{e^{-ikr}}{(kr)^{n+1}}; \quad \ldots\ldots(\text{XXVIII})$$

and we find from (XV), (d)

$$X_n = \frac{(-)^{n+1}(2n+1)\,e^{ik(ct-r)}}{n.n+1.\,kr^{n+1}}\left[M_n\left(y\frac{d}{dz} - z\frac{d}{dy}\right)\frac{dU_{n+1}}{dx} - N_n\left(r\frac{d}{dx} - \frac{nx}{r}\right)\frac{dU_{n+1}}{dy}\right];$$
$$\ldots(e)$$

and in the cyclical changes by which Y and Z are deduced from X, dU_{n+1}/dx, dU_{n+1}/dy are to stand unaltered.

For systematic calculation it is best to express U_{n+1} in terms of P. As in (a)

$$\frac{dU_{n+1}}{dx} = xr^{n-1}\{(n+1)P_{n+1} - \mu P'_{n+1}\} = -xr^{n-1}P'_n; \quad\ldots\ldots\ldots(f)$$

so that we may write

$$\frac{dU_{n+1}}{dx}, \frac{dU_{n+1}}{dy} = -(x, y)\, r^{n-1}P'_n. \quad \ldots\ldots\ldots\ldots\ldots\ldots(g)$$

The differential coefficients which appear in (e) are now readily formed. We find

$$X_n = \frac{(-)^{n+1}(2n+1)\,e^{ik(ct-r)}}{n.n+1.\,kr}\frac{xy}{r^2}[-M_n P''_n - N_n(P'_n + \mu P''_n)], \quad \ldots\ldots\ldots(h)$$

$$Y_n = \frac{(-)^{n+1}(2n+1)\,e^{ik(ct-r)}}{n.n+1.\,kr}\left[M_n\left(-\mu P'_n + \frac{x^2}{r^2}P''_n\right)\right.$$
$$\left. + N_n\left\{\left(1 - \frac{y^2}{r^2}\right)P'_n - \frac{y^2}{r^2}\mu P''_n\right\}\right], \quad \ldots\ldots(i)$$

$$Z_n = \frac{(-)^{n+1}(2n+1)\,e^{ik(ct-r)}}{n.n+1.\,kr}\frac{y}{r}[M_n P'_n + N_n\{-\mu P'_n + (1 - \mu^2)P''_n\}]. \quad \ldots(j)$$

The { } in (j) may alternatively be written $\mu P'_n - n(n+1)P_n$, in virtue of the differential equation satisfied by P_n.

It should be observed that the above values satisfy $xX_n + yY_n + zZ_n = 0$, as of course they ought to do.

At this stage it may be convenient to write down the values of P'_n for the earlier values of n.

	P'_n
$n=1$	1
$n=2$	3μ
$n=3$	$-\dfrac{3}{2} + \dfrac{3.5}{2}\mu^2$
$n=4$	$-\dfrac{3.5}{2}\mu + \dfrac{5.7}{2}\mu^3$

Thus from (h), (i), (j),

$$X_1,\ Y_1,\ Z_1 = \frac{3\,e^{ik\,(ct-r)}}{2\,.\,kr}\left[-\frac{xy}{r^2}\,N_1,\ -\mu M_1 + N_1\left(1 - \frac{y^2}{r^2}\right),\ \frac{y}{r}\,(M_1 - \mu N_1)\right];$$

$$X_2,\ Y_2,\ Z_2 = -\frac{5\,e^{ik\,(ct-r)}}{2\,.\,kr}\left[-\frac{xy}{r^2}\,(M_2 + 2\mu N_2),\ M_2\left(1 - 2\mu^2 - \frac{y^2}{r^2}\right) + \mu N_2\left(1 - \frac{2y^2}{r^2}\right),\right.$$
$$\left.\frac{y}{r}\{\mu M_2 + N_2(1 - 2\mu^2)\}\right];$$

$$X_3 = \frac{7\,e^{ik\,(ct-r)}}{4\,.\,kr}\,\frac{xy}{r^2}\left\{-5\mu M_3 + N_3\left(\frac{1}{2} - \frac{3.5}{2}\mu^2\right)\right\},$$

$$Y_3 = \frac{7\,e^{ik\,(ct-r)}}{4\,.\,kr}\left[\mu M_3\left\{\frac{1}{2} - \frac{3.5}{2}\mu^2 + 5\left(1 - \frac{y^2}{r^2}\right)\right\}\right.$$
$$\left. + N_3\left\{\left(1 - \frac{y^2}{r^2}\right)\left(\frac{3.5}{2}\mu^2 - \frac{1}{2}\right) - 5\mu^2\right\}\right],$$

$$Z_3 = \frac{7\,e^{ik\,(ct-r)}}{4\,.\,kr}\cdot\frac{y}{r}\left[M_3\left(\frac{5}{2}\mu^2 - \frac{1}{2}\right) + \mu N_3\left(\frac{11}{2} - \frac{15}{2}\mu^2\right)\right];$$

$$X_4 = -\frac{9\,e^{ik\,(ct-r)}}{4\,.\,kr}\,\frac{xy}{r^2}\left[M_4\left(\frac{3}{2} - \frac{3.7}{2}\mu^2\right) + N_4\,(3\mu - 14\mu^3)\right],$$

$$Y_4 = -\frac{9\,e^{ik\,(ct-r)}}{4\,.\,kr}\left[M_4\left\{-\frac{7}{2}\mu^4 + \frac{3}{2}\mu^2 - \frac{x^2}{r^2}\left(\frac{3}{2} - \frac{3.7}{2}\mu^2\right)\right\}\right.$$
$$\left. + N_4\left\{\frac{7}{2}\mu^3 - \frac{3}{2}\mu + \frac{y^2}{r^2}(3\mu - 14\mu^3)\right\}\right],$$

$$Z_4 = -\frac{9\,e^{ik\,(ct-r)}}{4\,.\,kr}\,\frac{y}{r}\left[M_4\left(\frac{7}{2}\mu^3 - \frac{3}{2}\mu\right) + N_4\left(-14\mu^4 + \frac{3.9}{2}\mu^2 - \frac{3}{2}\right)\right].$$

Our expressions naturally simplify when we limit ourselves to secondary rays which lie in the principal planes. No real loss of generality is thereby incurred, since whatever may be the direction of primary vibration, that vibration may be resolved into two, respectively in and perpendicular to the plane which contains the primary and secondary rays. Thus, if $y = 0$, that is for secondary rays whose direction is perpendicular to that of primary vibration, we have generally $X_n = 0$, $Z_n = 0$, and

$$Y_n = \frac{(-)^{n+1}(2n+1)\,e^{ik\,(ct-r)}}{n\,.\,n+1\,.\,kr}\,[M_n\{\mu P'_n - n(n+1)\,P_n\} + N_n P'_n].\quad\ldots(k)$$

Again, if $x = 0$, that is for secondary rays lying in the plane which contains the directions of primary vibration and propagation, $X_n = 0$, and

$$Y_n = \frac{(-)^{n+1}(2n+1)\,e^{ik\,(ct-r)}}{n\,.\,n+1\,.\,kr}\,[-M_n\mu P'_n - N_n\mu\{\mu P'_n - n(n+1)\,P_n\}],$$

while Z_n retains the form (j). The direction of vibration is in the plane $x = 0$, and is, of course, perpendicular to the secondary ray. Its magnitude is given by

$$\frac{yZ_n - zY_n}{r} = \frac{(-)^{n+1}(2n+1)\,e^{ik\,(ct-r)}}{n\,.\,n+1\,.\,kr}\,[M_n P'_n + N_n\{\mu P'_n - n(n+1)\,P_n\}].\quad\ldots(l)$$

It will be seen that (k) and (l) differ only by an interchange of M_n, N_n.

It may be remarked that when $\mu = \pm 1$, (k) and (l) should become identical, except as to sign. If $\mu = -1$, the signs are the same; if $\mu = +1$, they are opposite. Thus when $\mu = \pm 1$, the [ ] in (l) becomes $P'_n(M_n \mp N_n)$ simply. And in this case by the differential equation,

$$\mu P'_n = \tfrac{1}{2}n(n+1)\,P_n.$$

When n is even, $P_n = 1$, and then $P'_n = \pm \tfrac{1}{2}n(n+1)$; when n is odd, $P_n = \pm 1$, and then $P'_n = \tfrac{1}{2}n(n+1)$.

Accordingly, (l) gives in the case of $\mu = \pm 1$,

$$\frac{yZ - zY}{r} = \frac{e^{ik\,(ct-r)}}{kr}\,[\tfrac{3}{2}M_1 + \tfrac{5}{2}N_2 + \tfrac{7}{2}M_3 + \tfrac{9}{2}N_4 + \ldots$$
$$\mp (\tfrac{3}{2}N_1 + \tfrac{5}{2}M_2 + \tfrac{7}{2}N_3 + \tfrac{9}{2}M_4 + \ldots)],\quad\ldots(m),$$

the upper sign in the ambiguities relating to $\mu = +1$.

Arranged in powers of μ the general expression from (l) as far as M_5, N_5 inclusive is

$$\frac{yZ - zY}{r} = \frac{e^{ik\,(ct-r)}}{kr}\left[\frac{3}{2}M_1 - \frac{5}{2}N_2 - \frac{7}{8}M_3 + \frac{27}{8}N_4 + \frac{11}{16}M_5\right.$$
$$\left.+ \mu\left\{-\frac{3}{2}N_1 - \frac{5}{2}M_2 + \frac{77}{8}N_3 + \frac{27}{8}M_4 - \frac{11\,.\,29}{16}N_5\right\}\right.$$

$$+ \mu^2 \left\{ 5N_2 + \frac{35}{8} M_3 - \frac{9 \cdot 27}{8} N_4 - \frac{77}{8} M_5 \right\}$$

$$+ \mu^3 \left\{ - \frac{7 \cdot 15}{8} N_3 - \frac{9 \cdot 7}{8} M_4 + \frac{33 \cdot 21}{8} N_5 \right\}$$

$$+ \mu^4 \left\{ \frac{9 \cdot 7}{2} N_4 + \frac{11 \cdot 21}{16} M_5 \right\} - \mu^5 \frac{11 \cdot 105}{16} N_5 \Big]. \quad \ldots\ldots\ldots\ldots(n)$$

The corresponding expression for Y from (k) is derivable by interchange of M and N.

When $\mu = 0$, we have in (l) to consider

$$M_n P'_n - n(n+1) N_n P_n.$$

In this when n is odd, $P_n(0) = 0$ and $P'_n(0)$ takes in succession the values

$$1, \quad -\frac{3}{2}, \quad \frac{3 \cdot 5}{2 \cdot 4}, \quad -\frac{3 \cdot 5 \cdot 7}{2 \cdot 4 \cdot 6}, \quad \frac{3 \cdot 5 \cdot 7 \cdot 9}{2 \cdot 4 \cdot 6 \cdot 8}, \text{ etc.,}$$

while when n is even, $P'_n(0) = 0$, and $P_n(0)$ takes the values

$$[1], \quad -\frac{1}{2}, \quad \frac{1 \cdot 3}{2 \cdot 4}, \quad -\frac{1 \cdot 3 \cdot 5}{2 \cdot 4 \cdot 6}, \quad \frac{1 \cdot 3 \cdot 5 \cdot 7}{2 \cdot 4 \cdot 6 \cdot 8}, \text{ etc.}$$

Using these in (l), we find for $\mu = 0$

$$\frac{yZ - zY}{r} = \frac{e^{ik(ct-r)}}{kr} \Big[\frac{3}{2} M_1 - \frac{5}{2} N_2 - \frac{7}{2 \cdot 4} M_3 + \frac{9 \cdot 3}{2 \cdot 4} N_4 + \frac{11 \cdot 3}{2 \cdot 4 \cdot 6} M_5$$

$$- \frac{13 \cdot 3 \cdot 5}{2 \cdot 4 \cdot 6} N_6 - \frac{15 \cdot 3 \cdot 5}{2 \cdot 4 \cdot 6 \cdot 8} M_7 + \frac{17 \cdot 3 \cdot 5 \cdot 7}{2 \cdot 4 \cdot 6 \cdot 8} N_8 + \ldots \Big], \quad \ldots(o)$$

and the corresponding value of Y in (k) is obtained by interchange of M and N.

We now pass on to consider the values of M_n, N_n as given in (XXIII). By (VII), when η is small, $E_n(\eta)$ is of the order $\eta^{-(2n+1)}$, so that N_n is of the order η^{2n+1}. M_n is two orders higher, viz., η^{2n+3}, since the numerator is itself a small quantity of the order η^2. When η is small, the most important terms in X, Y, Z depend upon N_1 and are of order η^3 or R^3. If we are satisfied with an approximation as far as R^5 inclusive, we have only to consider N_1, M_1, and N_2. For R^7 we need N_1, M_1, N_2, M_2, N_3, and so on.

Let us begin by supposing that $(K - 1)$ is small, and retain only the first power of this quantity. For example, in $\psi_n(k'R)$, we may write

$$\psi_n(k'R) = \psi_n(kR) + \psi'_n(kR) \cdot (k' - k) R.$$

We find, omitting for brevity the argument in $\psi_n(kR)$, etc.,

$$M_n = \frac{(K - 1) kR}{2} \frac{\psi_{n-1} \psi'_n - \psi_n \psi'_{n-1}}{\psi_{n-1} E_n - \psi_n E_{n-1}}, \quad \ldots\ldots\ldots\ldots\ldots\ldots\ldots\ldots\ldots(p)$$

$$N_n = (K - 1) \frac{\left(\psi_{n-1} - \frac{n}{2n+1} \psi_n \right) \psi_n + \frac{1}{2} kR (\psi_{n-1} \psi'_n - \psi_n \psi'_{n-1})}{\psi_{n-1} E_n - \psi_n E_{n-1}}. \quad \ldots(q)$$

The denominator in (p), (q) appears at first sight complex, since E is so. If, however, we separate real and imaginary parts, writing in accordance with the definitions (IV), (VII)

$$E_n(\eta) = \Psi_n(\eta) - i\psi_n(\eta), \quad \dots\dots\dots\dots\dots\dots(r)$$

we see that

$$\psi_{n-1}E_n - \psi_n E_{n-1} = \psi_{n-1}\Psi_n - \psi_n\Psi_{n-1}, \quad \dots\dots\dots\dots(s)$$

from which the imaginary part has disappeared. In this case accordingly M_n, N_n are wholly real.

The denominator (s) admits of simple evaluation. By (VI)

$$\psi_{n-1}E_n - \psi_n E_{n-1} = \frac{\eta}{2n+1}(\psi'_n E_n - \psi_n E'_n).$$

Now, if V_n be a solid harmonic, the same in both cases, $V_n\psi_n$ and $V_n E_n$ both satisfy the equation $(\nabla^2 + k^2) = 0$, and hence by Green's theorem (*Theory of Sound*, Vol. II. p. 252) as applied to the space included between two concentric spheres, we find that

$$\psi'_n E_n - \psi_n E'_n = C_n \eta^{-2n-2},$$

where C_n is independent of η. To determine C_n we may suppose η very small, when

$$\psi_n = 1, \quad \psi'_n = -\frac{\eta}{2n+3}, \quad E_n = (2n+1)\frac{1^2 . 3^2 \dots (2n-1)^2}{\eta^{2n+1}}.$$

Thus

$$C_n = 1^2 . 3^2 . 5^2 \dots (2n+1)^2,$$

and

$$\psi_{n-1}E_n - \psi_n E_{n-1} = 1^2 . 3^2 . 5^2 \dots (2n-1)^2 \times (2n+1)\,\eta^{-2n-1}. \quad \dots(t)$$

It will be understood that (t) is true without approximation as regards η, if $(K-1)^2$ is neglected.

It is now a simple matter to calculate M_n, N_n to a moderately high power of η. So far as η^7 inclusive, we have

$$N_1 = \frac{2(K-1)\eta^3}{9}\left(1 - \frac{\eta^2}{5} + \frac{7\eta^4}{10 . 5 . 7}\right), \quad M_1 = \frac{(K-1)\eta^5}{45}\left(1 - \frac{\eta^2}{7}\right),$$

$$N_2 = \frac{(K-1)\eta^5}{5 . 15}\left(1 - \frac{\eta^2}{7}\right), \quad M_2 = \frac{(K-1)\eta^7}{35 . 45},$$

$$N_3 = \frac{4(K-1)\eta^7}{7 . 35 . 45}.$$

Thus

$$Z = Z_1 + Z_2 + Z_3 = \frac{e^{ik(ct-r)}}{kr} . \frac{y}{r}\left[\frac{3}{2}M_1 - \frac{5}{2}N_2\right.$$

$$\left. + \mu\left\{-\frac{3}{2}N_1 - \frac{5}{2}M_2 + \frac{77}{8}N_3\right\} + 5\mu^2 N_2 - \frac{7 . 15}{8}\mu^3 N_3\right],$$

in which $\frac{3}{2}M_1 - \frac{5}{2}N_2 = 0$, so that μ becomes a factor of the whole expression as we know it ought to do.   Also

$$X = \frac{e^{ik\,(ct-r)}}{kr}\,\frac{xy}{r}\left[-\frac{3}{2}N_1 + \frac{5}{2}M_2 + \frac{7}{8}N_3 + \mu\left(5N_2 - \frac{7\cdot5}{4}M_3\right) - \frac{7\cdot3\cdot5}{2\cdot4}N_3\mu^2\right].$$

In X the multiplier of xy should be the same as the multiplier of yz in Z. In order that this may be so, we must have $M_2 = \frac{7}{4}N_3$, a relation satisfied by the tabular values.   In the expression for Y, the part which does not contain the factor $1 - y^2/r^2$, or $(x^2 + z^2)/r^2$, vanishes in virtue of the relations already mentioned.   Finally, arranged in powers of η, or kR,

$$X,\,Y,\,Z = \frac{e^{ik\,(ct-r)}\,k^2R^3\,(K-1)}{3r}\left(-\frac{xy}{r^2},\,\frac{x^2+z^2}{r^2},\,-\frac{yz}{r^2}\right)$$

$$\times\left\{1 - \tfrac{1}{5}k^2R^2\,(1+\mu) + \frac{k^4R^4}{2\cdot5\cdot7}\,(1+\mu)^2\right\},$$

in agreement with (35), (49) so far as the present solution extends.   The former method is incomparably the more appropriate in this particular case, giving the result for the sphere to any degree of approximation in kR, as well as admitting of application to obstacles of other forms.   The general conclusion that (o) vanishes when $(K-1)^2$ is neglected would probably be difficult to establish by the present method.

Abandoning the restriction as to the smallness of $(K-1)$, we find from (XXIII) as far as R^5 inclusive

$$N_1 = \frac{2\,(K-1)\,k^3R^3}{3\,(K+2)}\left\{1 - \tfrac{3}{5}k^2R^2 + \frac{6Kk^2R^2}{5\,(K+2)}\right\},$$

$$M_1 = \frac{(K-1)\,k^5R^5}{45},\qquad N_2 = \frac{(K-1)\,k^5R^5}{15\,(2K+3)};\quad\ldots\ldots\ldots\ldots(u)$$

where it will be seen that to this order M, N are still real.   The corresponding values of X, Y, Z are given by

$$X,\,Y,\,Z = \frac{3\,e^{ik\,(ct-r)}}{2\cdot kr}\,N_1\left(-\frac{xy}{r^2},\,\frac{x^2+z^2}{r^2},\,-\frac{yz}{r^2}\right) + \frac{3\,e^{ik\,(ct-r)}}{2\cdot kr}\,M_1\left(0,\,-\frac{z}{r},\,\frac{y}{r}\right)$$

$$-\frac{5\,e^{ik\,(ct-r)}}{2\cdot kr}\,N_2\left\{-\frac{2xyz}{r^3},\,\frac{z}{r}\left(1-\frac{2y^2}{r^2}\right),\,\frac{y}{r}\left(1-\frac{2z^2}{r^2}\right)\right\}.\quad\ldots(v)$$

This agrees with Prof. Love's solution, except in the first set of terms, as already mentioned.

The special case where the secondary ray lies in one or other principal plane is given by (n) and its derivative, when the approximate values of M, N are introduced.

When kR, or η, can no longer be treated as small, we must calculate M and N from the general formulae (XXIII) and (XXIII bis).   Separating E

into its real and imaginary parts, we see that the denominator of M_n is equal to

$$-\Psi_{n-1}(\eta) + \frac{\psi_{n-1}(\eta')}{\psi_n(\eta')}\Psi_n(\eta) + i \times \text{numerator}.$$

Also that the numerator for N_n exceeds the numerator for M_n by

$$(K-1)\psi_{n-1}(\eta) - (K-1)\frac{n}{2n+1}\psi_n(\eta);$$

and that the real part of the denominator of N_n exceeds the real part of the denominator of M_n by

$$(K-1)\frac{n}{2n+1}\Psi_n(\eta) - (K-1)\Psi_{n-1}(\eta),$$

while the imaginary part of the denominator of N_n is $i \times$ the numerator of N_n. When the numerators and denominators have been computed, the values of M_n, N_n must be reduced to the standard complex form.

In selecting special cases for calculation we notice that η and η' (kR and $k'R$) are both arbitrary. With the view of reaching something distinctive, it seemed desirable to choose a refractive index (k'/k) not too small. For this purpose $k'/k = 1\cdot5$ seemed suitable, as large enough, and yet not outside the experimental range. This corresponds to $K = 2\cdot25$. By further choosing $\eta = 1$, $\eta = 3/2$, $\eta = 9/4$ we make the tabulation of ψ serve a double purpose.

The most obvious procedure for calculating $E_n(\eta)$, equivalent to $\Psi_n(\eta) - i\psi_n(\eta)$, is to use the sequence formula (VI),

$$E_{n+1} = \frac{2n+1 \cdot 2n+3}{\eta^2}\{E_n - E_{n-1}\}, \quad \ldots\ldots\ldots\ldots(VI)$$

starting from E_0 and E_1 as given by (XXXVI). So far as regards the real part, represented by Ψ, this course is satisfactory; but the same cannot be said of the application to ψ. As n increases, the successive ψ's tend to equality, and any errors are multiplied at the next step by the large factor $(2n+1)(2n+3)$. So much is this the case that although we use seven-figure logarithms throughout, even the first figure in ψ_6 or ψ_7 becomes doubtful.

It is true that these higher orders in ψ do not need to be known very accurately, for they contribute but little to the final result; nevertheless, an improvement in the calculation is called for.

An easy escape from the difficulty is provided by the series for ψ_n given in (IV). When η is moderate, as here, and n somewhat large, the convergence is good, and we may calculate ψ_6 and ψ_5 by a perfectly straightforward process to a high degree of accuracy. Having obtained ψ_6 and ψ_5, we may then use the sequence formula (VI) in the reverse direction to deduce in turn ψ_4, ψ_3, etc., without loss of accuracy. The values of Ψ_n and ψ_n, calculated for the various cases, are here tabulated. It is possible that they may prove useful in kindred enquiries. From them are deduced the complex values of M and N as already explained. In this case the tables give the *logarithms* of the real and imaginary parts.

Tables of Ψ and ψ.

n	Ψ	ψ	Ψ	ψ
	$\eta=1$		$\eta=1\cdot5$	
0	0·54030	0·84147	0·04716	0·66499
1	4·1453	0·90351	1·3929	0·79235
2	54·075	0·93053	8·9713	0·849C0
3	1747·5	0·94569	117·89	0·88121
4	—	0·95541	3049·8	0·90204
5	—	—	—	0·91664
6	—	—	—	0·92743
	$\eta=2\cdot25$		$\eta=27/8$	
0	−0·279188	0·345811	−0·288262	−0·068531
1	+0·295636	0·577175	−0·136838	+0·238183
2	+1·70318	0·685527	+0·199406	+0·403904
3	+9·73119	0·749095	+1·03318	+0·509211
4	+99·9041	0·791075	+4·61147	+0·582438
5	+1933·51	0·820928	+31·1003	+0·636441
6	+51793·7	0·843272	+332·545	+0·677969
	$\eta=2$		$\eta=3$	
0	−0·208073	0·454649	−0·329997	0·047040
1	+0·525919	0·653096	−0·062959	0·345677
2	+2·75247	0·744180	+0·445063	0·497729
3	+19·4823	0·796977	+1·97564	0·591312
4	+263·495	0·831564	+10·7141	0·655080
5	+6039·32	0·856018	+96·1227	0·701448
6	+206485	0·874244	+1357·05	0·736733
	$\eta=7/4$		$\eta=21/8$	
0	−0·101855	0·562278	−0·331241	0·188160
1	+0·864129	0·725411	+0·070827	0·460481
2	+4·73135	0·799021	+0·875249	0·592807
3	+44·1968	0·841250	+4·08595	0·672133
4	+811·861	0·868721	+29·3550	0·725271
5	+24815·9	0·888047	+363·049	0·763446
6	—	0·902393	—	0·792237

Tables of M and N.

n	M_n	N_n
	$\eta=1$	
1	$[\overline{2}\cdot4483]-i[\overline{4}\cdot900]$	$[\overline{1}\cdot2635]-i[\overline{2}\cdot5424]$
2	$[\overline{4}\cdot879]$	$[\overline{2}\cdot0109]-i[\overline{4}\cdot0218]$
3	$[\overline{5}\cdot08]$	$[\overline{4}\cdot43]$
	$\eta=3/2$	
1	$[\overline{1}\cdot35853]-i[\overline{2}\cdot74171]$	$[\overline{1}\cdot6160]-i[\overline{1}\cdot3389]$
2	$[\overline{2}\cdot100]\;\cdot-i[\overline{4}\cdot200]$	$[\overline{2}\cdot8566]-i[\overline{3}\cdot715]$
3	$[\overline{4}\cdot633]$	$[\overline{3}\cdot622]\;-i[\overline{5}\cdot243]$
4	$[\overline{6}\cdot99]$	$[\overline{4}\cdot16]$
	$\eta=9/4$	
1	$[\overline{1}\cdot52056]-i[\;\overline{1}\cdot94164]$	$[\overline{1}\cdot69832]-i[\;\overline{1}\cdot72202]$
2	$[\overline{1}\cdot36224]-i[\;\overline{2}\cdot74960]$	$[\overline{1}\cdot58138]-i[\;\overline{1}\cdot24719]$
3	$[\overline{2}\cdot18132]-i[\;\overline{4}\cdot36275]$	$[\overline{2}\cdot77736]-i[\;\overline{3}\cdot55628]$
4	$[\overline{4}\cdot87043]-i[\;\overline{7}\cdot7391]$	$[\overline{3}\cdot6612]\;-i[\;\overline{5}\cdot322]$
5	$[\overline{5}\cdot377]\;-i[\overline{10}\cdot75]$	$[\overline{4}\cdot3277]\;-i[\;\overline{8}\cdot66]$
6	$[7\cdot80]\;\;-i[\overline{13}\cdot6]$	$[\overline{6}\cdot88]\;\;-i[\overline{11}\cdot77]$
	$\eta=2$	
1	$[\overline{1}\cdot68350]-i[\overline{1}\cdot80011]$	$[\overline{1}\cdot69493]-i[\overline{1}\cdot63560]$
2	$[\overline{2}\cdot98851]-i[\overline{3}\cdot98120]$	$[\overline{1}\cdot40882]-i[\overline{2}\cdot84950]$
3	$[\overline{3}\cdot73029]-i[\overline{5}\cdot46059]$	$[\overline{2}\cdot44577]-i[\overline{4}\cdot89188]$
4	$[\overline{4}\cdot32695]-i[\overline{8}\cdot654]$	$[\overline{3}\cdot23189]-i[\overline{6}\cdot4638]$
5	$[\overline{6}\cdot7764]$	$[\overline{5}\cdot8390]$
6	$[7\cdot0937]$	$[\overline{6}\cdot2901]$
	$\eta=7/4$	
1	$[\overline{1}\cdot63910]-i[\overline{1}\cdot40580]$	$[\overline{1}\cdot67342]-i[\overline{1}\cdot52299]$
2	$[\overline{2}\cdot57067]-i[\overline{3}\cdot14194]$	$[\overline{1}\cdot16474]-i[\overline{2}\cdot33907]$
3	$[\overline{3}\cdot22163]-i[\overline{6}\cdot44]$	$[\overline{2}\cdot06571]-i[\overline{4}\cdot1315]$
4	$[\overline{5}\cdot708]$	$[\overline{4}\cdot7389]$
5	$[\overline{6}\cdot042]$	$[\overline{5}\cdot228]$

For a systematic calculation it appears preferable to employ the forms (k), (l) rather than the expansions in powers of μ. The functions involving P, once calculated for the various values of n, are then available for any value of η. The following table exhibits the algebraic forms:—

n	$\dfrac{(2n+1)\,P'_n}{n\,(n+1)}$	$\dfrac{(2n+1)\,\mu P'_n}{n\,(n+1)}-(2n+1)\,P_n$
1	$\tfrac{3}{2}$	$-\tfrac{3}{2}\mu$
2	$\tfrac{5}{2}\mu$	$\tfrac{5}{2}(1-2\mu^2)$
3	$\tfrac{7}{5}(-1+5\mu^2)$	$\tfrac{7}{5}(11\mu-15\mu^3)$
4	$\tfrac{9}{5}(-3\mu+7\mu^3)$	$\tfrac{9}{5}(-3+27\mu^2-28\mu^4)$
5	$\tfrac{11}{35}(1-14\mu^2+21\mu^4)$	$\tfrac{11}{35}(-29\mu+126\mu^3-105\mu^5)$
6	$\tfrac{13}{35}(5\mu-30\mu^3+33\mu^5)$	$\tfrac{13}{35}(5-100\mu^2+285\mu^4-198\mu^6)$

The particular cases $\mu=0$, $\mu=\pm1$ have already been considered. It will be observed that both functions are either entirely odd or entirely even.

The tables give the logarithmic values for $\mu=0,\ \tfrac14,\ \tfrac12,\ \tfrac34,\ 1$. Thus, in the first table for $\mu=\tfrac14$, $n=4$, we have the number whose logarithm is $\bar{1}\cdot85776$, and this is to be taken negatively when $\mu=+\tfrac14$. Moreover, since this function of μ is odd, the sign is to be reversed, *i.e.*, made positive, when $\mu=-\tfrac14$.

Table for $\dfrac{(2n+1)\,P'_n}{n\,(n+1)}$.

n	$\mu=0$	$\mu=\tfrac14$	$\mu=\tfrac12$	$\mu=\tfrac34$	$\mu=1$
1	$+[0\cdot17609]$	$+[0\cdot17609]$	$+[0\cdot17609]$	$+[0\cdot17609]$	$+[0\cdot17609]$
2	0	$+[\bar{1}\cdot79588]$	$+[0\cdot09691]$	$+[0\cdot27300]$	$+[0\cdot39794]$
3	$-[\bar{1}\cdot94201]$	$-[\bar{1}\cdot77928]$	$+[\bar{1}\cdot33995]$	$+[0\cdot20029]$	$+[0\cdot54407]$
4	0	$-[\bar{1}\cdot85776]$	$-[\bar{1}\cdot84703]$	$+[\bar{1}\cdot89819]$	$+[0\cdot65321]$
5	$+[\bar{1}\cdot83727]$	$+[\bar{1}\cdot15331]$	$-[\bar{1}\cdot91191]$	$-[\bar{1}\cdot19988]$	$+[0\cdot74036]$
6	0	$+[\bar{1}\cdot82016]$	$-[\bar{1}\cdot24977]$	$-[\bar{1}\cdot94131]$	$+[0\cdot81291]$

Table for $\dfrac{\mu P'_n(2n+1)}{n\,(n+1)}-(2n+1)\,P_n$.

n	$\mu=0$	$\mu=\tfrac14$	$\mu=\tfrac12$	$\mu=\tfrac34$	$\mu=1$
1	0	$-[\bar{1}\cdot57403]$	$-[\bar{1}\cdot87506]$	$-[0\cdot05115]$	$-[0\cdot17609]$
2	$+[0\cdot39794]$	$+[0\cdot33995]$	$+[0\cdot09691]$	$-[\bar{1}\cdot49485]$	$-[0\cdot39794]$
3	0	$+[0\cdot34265]$	$+[0\cdot50132]$	$+[0\cdot22573]$	$-[0\cdot54407]$
4	$-[0\cdot52827]$	$-[0\cdot20401]$	$+[0\cdot35218]$	$+[0\cdot57335]$	$-[0\cdot65321]$
5	0	$-[0\cdot56836]$	$-[0\cdot14504]$	$+[0\cdot64946]$	$-[0\cdot74036]$
6	$+[0\cdot60879]$	$-[\bar{1}\cdot17713]$	$-[0\cdot63256]$	$+[0\cdot47638]$	$-[0\cdot81291]$

The way is now clear for the last step, viz., the calculation of $(yZ-zY)/r$ and Y by summation with respect to n of the values given in (l) and (k).

As to this there is little to call for special remark. Of course the odd and even terms are added separately.

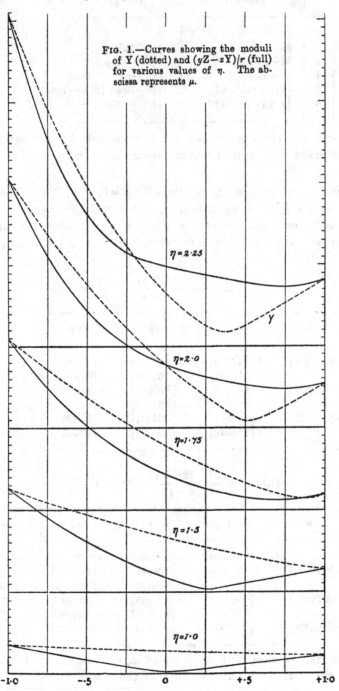

FIG. 1.—Curves showing the moduli of Y (dotted) and $(yZ-zY)/r$ (full) for various values of η. The abscissa represents μ.

$\eta = 2\cdot25$

y

$\eta = 2\cdot0$

$\eta = 1\cdot75$

$\eta = 1\cdot5$

$\eta = 1\cdot0$

-1·0   -·5   0   +·5   +1·0

The tables [on p. 566] exhibit the complex values and the modulus, subject to multiplication by the factor $(kr)^{-1} e^{ik\,(ct-r)}$. Thus for $\eta = 1, \mu = -1$, we have

$$\frac{yZ - zY}{r} = \frac{e^{ik\,(ct-r)}}{kr}(0\cdot3457 - i \times 0\cdot0538) = 0\cdot3499\,\frac{e^{ik\,(ct-r-\rho)}}{kr},$$

where ρ influences the phase only.

As regards the values of η, or kR, or $2\pi R/\lambda$, the numbers 1, 3/2, 9/4 were at first chosen. Subsequently it appeared desirable to fill up the gap at the more interesting stage, and so calculations for $\eta = 1\frac{3}{4}$, $\eta = 2$ were added. The values of the *modulus* are plotted in the curves, fig. 1, as functions of μ for the various values of η. It will be remembered that in all cases the refractive index is supposed to be 1·5.

When $\eta = 1$, *i.e.*, when the circumference of the sphere equals the wavelength in the general medium, there is comparatively little departure from the conditions appropriate to the infinitely small sphere. When the secondary ray is in a direction perpendicular to that of primary vibration, viz., when $y = 0$, the vibration (Y) varies but little with μ. It is otherwise when the secondary ray is in the plane containing the directions of the primary ray and of primary vibration. The second column shows that in that case the secondary vibration varies greatly with μ, and that in the perpendicular direction ($\mu = 0$) the intensity is very small. This is the direction in which there would be no effect, however large the sphere, if the square of the difference of optical quality could be neglected absolutely.

The departure from the law of the infinitely small sphere becomes more apparent when $\eta = 1\cdot5$. The principal feature here is the displacement of the maximum polarisation from $\mu = 0$ to the neighbourhood of $\mu = +\frac{1}{4}$, *i.e.*, to a direction inclining *backwards* along the course of the incident light. When the incident light is unpolarised, the diffracted light is polarised at all angles (other than $\mu = \pm 1$), and in the same sense as from infinitely small particles.

The latter law ceases to hold when η rises to about 1·75. A third neutral point (direction of no polarisation) enters at $\mu = +1$ and moves inwards. When $\eta = 2$, the change has proceeded so far that the neutral point is in the neighbourhood of $\mu = 0$, the polarisation for positive μ being the *reverse* of that found for infinitely small particles. At $\mu = +\frac{1}{2}$ the reversed polarisation is very pronounced. When η attains 2·25, the neutral point is still further on (near $\mu = -\frac{1}{4}$), and the maximum reverse polarisation is near $\mu = \frac{1}{3}$. It will be seen that these changes occur very rapidly in the neighbourhood of $\eta = 2$.

It would be possible to follow these calculations to greater values of η, and such an extension would not be without interest, but the arithmetical work

μ	$\dfrac{yZ-zY}{r}$	Modulus	Y	Modulus
		$\eta=1$		
-1	$0\cdot3457-i\times0\cdot0538$	$0\cdot3499$	$0\cdot3457-i\times0\cdot0538$	$0\cdot3499$
$-\frac{3}{4}$	$0\cdot2525-i\times0\cdot0404$	$0\cdot2557$	$0\cdot3266-i\times0\cdot0534$	$0\cdot3310$
$-\frac{1}{2}$	$0\cdot1669-i\times0\cdot0272$	$0\cdot1691$	$0\cdot3083-i\times0\cdot0530$	$0\cdot3128$
$-\frac{1}{4}$	$0\cdot0884-i\times0\cdot0141$	$0\cdot0895$	$0\cdot2904-i\times0\cdot0527$	$0\cdot2951$
0	$0\cdot0165-i\times0\cdot0009$	$0\cdot0165$	$0\cdot2731-i\times0\cdot0523$	$0\cdot2781$
$+\frac{1}{4}$	$-0\cdot0490+i\times0\cdot0121$	$0\cdot0505$	$0\cdot2566-i\times0\cdot0519$	$0\cdot2618$
$+\frac{1}{2}$	$-0\cdot1083+i\times0\cdot0250$	$0\cdot1111$	$0\cdot2405-i\times0\cdot0516$	$0\cdot2460$
$+\frac{3}{4}$	$-0\cdot1619+i\times0\cdot0380$	$0\cdot1663$	$0\cdot2250-i\times0\cdot0512$	$0\cdot2307$
$+1$	$-0\cdot2103+i\times0\cdot0508$	$0\cdot2164$	$0\cdot2103-i\times0\cdot0508$	$0\cdot2164$
		$\eta=1\cdot5$		
-1	$1\cdot1902-i\times0\cdot4237$	$1\cdot2634$	$1\cdot1902-i\times0\cdot4237$	$1\cdot2634$
$-\frac{3}{4}$	$0\cdot8463-i\times0\cdot3302$	$0\cdot9084$	$1\cdot0212-i\times0\cdot3992$	$1\cdot0965$
$-\frac{1}{2}$	$0\cdot5647-i\times0\cdot2401$	$0\cdot6136$	$0\cdot8643-i\times0\cdot3751$	$0\cdot9422$
$-\frac{1}{4}$	$0\cdot3388-i\times0\cdot1533$	$0\cdot3719$	$0\cdot7190-i\times0\cdot3510$	$0\cdot8001$
0	$0\cdot1629-i\times0\cdot0698$	$0\cdot1772$	$0\cdot5844-i\times0\cdot3270$	$0\cdot6697$
$+\frac{1}{4}$	$0\cdot0316+i\times0\cdot0105$	$0\cdot0333$	$0\cdot4602-i\times0\cdot3032$	$0\cdot5511$
$+\frac{1}{2}$	$-0\cdot0597+i\times0\cdot0875$	$0\cdot1059$	$0\cdot3453-i\times0\cdot2793$	$0\cdot4441$
$+\frac{3}{4}$	$-0\cdot1159+i\times0\cdot1614$	$0\cdot1987$	$0\cdot2390-i\times0\cdot2556$	$0\cdot3499$
$+1$	$-0\cdot1414+i\times0\cdot2321$	$0\cdot2718$	$0\cdot1414-i\times0\cdot2321$	$0\cdot2718$
		$\eta=1\cdot75$		
-1	$1\cdot8683-i\times0\cdot9404$	$2\cdot0917$	$1\cdot8683-i\times0\cdot9404$	$2\cdot0917$
$-\frac{3}{4}$	$1\cdot2802-i\times0\cdot7661$	$1\cdot4919$	$1\cdot4992-i\times0\cdot8280$	$1\cdot7127$
$-\frac{1}{2}$	$0\cdot8330-i\times0\cdot6059$	$1\cdot0301$	$1\cdot1668-i\times0\cdot7165$	$1\cdot3693$
$-\frac{1}{4}$	$0\cdot5081-i\times0\cdot4597$	$0\cdot6852$	$0\cdot8694-i\times0\cdot6060$	$1\cdot0598$
0	$0\cdot2885-i\times0\cdot3273$	$0\cdot4363$	$0\cdot6041-i\times0\cdot4964$	$0\cdot7819$
$+\frac{1}{4}$	$0\cdot1592-i\times0\cdot2085$	$0\cdot2623$	$0\cdot3682-i\times0\cdot3878$	$0\cdot5347$
$+\frac{1}{2}$	$0\cdot1067-i\times0\cdot1032$	$0\cdot1484$	$0\cdot1594-i\times0\cdot2801$	$0\cdot3223$
$+\frac{3}{4}$	$0\cdot1191-i\times0\cdot0112$	$0\cdot1196$	$-0\cdot0251-i\times0\cdot1734$	$0\cdot1752$
$+1$	$0\cdot1858+i\times0\cdot0675$	$0\cdot1977$	$-0\cdot1858-i\times0\cdot0675$	$0\cdot1977$
		$\eta=2$		
-1	$2\cdot4767-i\times1\cdot7984$	$3\cdot0608$	$2\cdot4767-i\times1\cdot7984$	$3\cdot0608$
$-\frac{3}{4}$	$1\cdot4988-i\times1\cdot4716$	$2\cdot1005$	$1\cdot8327-i\times1\cdot4949$	$2\cdot3650$
$-\frac{1}{2}$	$0\cdot8053-i\times1\cdot1919$	$1\cdot4384$	$1\cdot2909-i\times1\cdot1980$	$1\cdot7611$
$-\frac{1}{4}$	$0\cdot3478-i\times0\cdot9583$	$1\cdot0195$	$0\cdot8417-i\times0\cdot9076$	$1\cdot2379$
0	$0\cdot0839-i\times0\cdot7699$	$0\cdot7745$	$0\cdot4759-i\times0\cdot6235$	$0\cdot7844$
$+\frac{1}{4}$	$-0\cdot0228-i\times0\cdot6257$	$0\cdot6261$	$0\cdot1855-i\times0\cdot3459$	$0\cdot3925$
$+\frac{1}{2}$	$-0\cdot0040-i\times0\cdot5247$	$0\cdot5247$	$-0\cdot0372-i\times0\cdot0747$	$0\cdot0834$
$+\frac{3}{4}$	$+0\cdot1132-i\times0\cdot4660$	$0\cdot4795$	$-0\cdot1988+i\times0\cdot1901$	$0\cdot2751$
$+1$	$+0\cdot3056-i\times0\cdot4487$	$0\cdot5429$	$-0\cdot3056+i\times0\cdot4487$	$0\cdot5429$
		$\eta=2\cdot25$		
-1	$3\cdot0635-i\times2\cdot6979$	$4\cdot0822$	$3\cdot0635-i\times2\cdot6979$	$4\cdot0822$
$-\frac{3}{4}$	$1\cdot5157-i\times2\cdot0593$	$2\cdot5569$	$1\cdot9791-i\times2\cdot1286$	$2\cdot9065$
$-\frac{1}{2}$	$0\cdot4857-i\times1\cdot5448$	$1\cdot6193$	$1\cdot1464-i\times1\cdot5972$	$1\cdot9661$
$-\frac{1}{4}$	$-0\cdot1392-i\times1\cdot1497$	$1\cdot1581$	$0\cdot5364-i\times1\cdot1036$	$1\cdot2270$
0	$-0\cdot4540-i\times0\cdot8696$	$0\cdot9810$	$0\cdot1234-i\times0\cdot6473$	$0\cdot6590$
$+\frac{1}{4}$	$-0\cdot5383-i\times0\cdot6999$	$0\cdot8830$	$-0\cdot1157-i\times0\cdot2281$	$0\cdot2558$
$+\frac{1}{2}$	$-0\cdot4585-i\times0\cdot6364$	$0\cdot7844$	$-0\cdot2018+i\times0\cdot1544$	$0\cdot2541$
$+\frac{3}{4}$	$-0\cdot2689-i\times0\cdot6745$	$0\cdot7261$	$-0\cdot1531+i\times0\cdot5003$	$0\cdot5232$
$+1$	$-0\cdot0139-i\times0\cdot8101$	$0\cdot8102$	$+0\cdot0139+i\times0\cdot8101$	$0\cdot8102$

would soon become heavy. Also, without increasing η, the refractive index might be varied. Some data for such calculations are included in the tables already given.

So far the primary light has been supposed to be either unpolarised, or, if polarised, to be plane polarised in one or other principal plane. In the more general case recourse must be had to the complex expressions, which allow the effects of the two principal components of the primary light to be combined, with the necessary allowance for any phase-difference. But it is scarcely necessary to pursue the treatment of this part of the subject.

Experimental.

So far as I know, the only experimental work bearing at all closely upon the preceding calculations is that of Govi and Tyndall (*Proc. Roy. Soc.*, January, 1869), who observed the light scattered from smoke of various kinds. Tyndall established the existence of neutral points at angles dependent upon the condition of the smoke, but his indications are not very precise. For example, he does not distinguish between density due to increased number of particles in a given space and that due to increasing size, of which the effects may be expected to be very different. Nor, again, do I see any suggestion that the position of the neutral point varies with the colour of the light. Nevertheless, Tyndall's description is suggestive, and it formed the starting-point of my own work upon the subject.

In continuation of the experiments already referred to at the commencement of this paper, I have lately repeated the observation on the nascent precipitate formed when a little acid is added to a dilute and well-filtered solution of hyposulphite. The liquid was contained in a small beaker placed in a dark room and exposed to a beam of sunlight. The observation of the light dispersed at various angles may be made with a nicol, but a double-image prism is better. This may be mounted at one end of a tube of which the other carries a small aperture of such dimensions that two oppositely polarised rectangular fields are seen in close juxtaposition. In the early stage the double-image prism, suitably rotated, shows complete polarisation of the light scattered along BC, fig. 2, at right angles to the primary rays AB. Somewhat later the minimum light seen along BC shows Tyndall's "residual blue," indicating that for this colour the polarisation is no longer complete. But if the direction of observation be changed towards BD, the polarisation is improved.

Fig. 2.

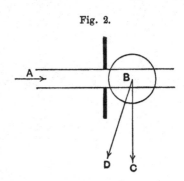

As the particles grow, a stage is reached in which the effects for blue and. red light are well contrasted. If a blue glass be inserted in the course either of the primary or of the secondary light, that polarised component which originally vanished (along *BC*) is found along *BD* to be the brighter of the two, showing that the polarisation is *reversed*. The substitution of a red for the blue glass restores the original character of the polarisation, the increase in the wave-length compensating the growth of the particles. It is not pretended that the observed contrast is as great as the theoretical curves might suggest; nor could this reasonably be expected, seeing that the particles are not all of the same size.

April 30.—I have recently learned from Prof. Love that he had become aware of the error in equation XVIII, and that a correction was published in *Math. Soc. Proc.*, Vol. XXXI. p. 489.

345.

ON COLOUR VISION AT THE ENDS OF THE SPECTRUM.

[*Nature*, Vol. LXXXIV. pp. 204, 205, 1910.]

IT is half a century since Maxwell* investigated the chromatic relations of the spectral colours and exhibited the results on Newton's diagram. The curve "forms two sides of a triangle with doubtful fragments of the third side. Now, if three colours in Newton's diagram lie in a straight line, the middle one is a compound of the two others. Hence all the colours of the spectrum may be compounded of those which lie at the angles of this triangle. These correspond to the following—scarlet, wave-length (in Fraunhofer's measure), 2328; green, wave-length, 1914; blue, wave-length, 1717. All the other colours of the spectrum may be produced by combinations of these; and since all natural colours are compounded of the colours of the spectrum, they may be compounded of these three primary colours. I [Maxwell] have strong reason to believe that these are the three primary colours corresponding to three modes of sensation in the organ of vision, on which the whole system of colour, as seen by the normal eye, depends."

Later observations, such as those of König and Dieterici†, have in the main confirmed Maxwell's conclusions. The green corner is indeed more rounded off than he supposed. It is with regard to the "doubtful fragments of the third side" that I have something to say. According to Maxwell's results with both of his observers the extreme red deviates from the less extreme by a tendency towards blue. Neither my friends‡ nor I can perceive anything of this. When the extreme and the less extreme red are seen in juxtaposition in the colour-box, no difference whatever can be perceived after the brightnesses are adjusted to equality. I have not any precise measurements of wave-length, but the extreme red passed a cobalt glass while the less extreme was stopped. Observations at the ends of the

---

* *Phil. Trans.* 1860.

† Helmholtz, *Phys. Optik*, 2nd edition, p. 340.

‡ Mr Gerald Balfour included.

spectrum are more difficult than elsewhere. Owing to deficiency of illumination at these parts there is more danger of false light finding access. To get satisfactory results I found it desirable to supplement the action of the prisms by placing red glass over the slits. It is probable that Maxwell was misled by some defect of this sort, since the differences he found would appear to lie outside the errors of observation. The German observers, it should be added, also found the colour constant at the red end.

At the other extreme the tendency of the violet towards red is, to my vision, not in the least doubtful. Some remarks made a few years ago by Dr Burch, who speaks of violet in terms which I could not possibly use, were the occasion of a more particular examination. Although, so far as I remembered, I had never made the trial, I was confident that I should be able to match violet approximately with blue *plus* red, and full blue with violet *plus* green. And it seemed further that this must be the general estimation, as there is no widely spread protest against describing the upper extreme of the spectrum as " violet "—a name which would be quite inappropriate in the absence of an approach towards red. The light which the flower of that name sends to the eye undoubtedly includes red rays.

The apparatus employed is on the model of the first described in an early paper*, the only difference worth mentioning being that the side upon which the movable slits are disposed is made oblique, to meet the variation in focal length along the spectrum. By this means any desired mixture of spectrum colours can be exhibited in juxtaposition with any other. For example, the violet can be shown alongside the blue, and any addition can be made to either. A few trials in 1907 confirmed my anticipations, an approximate match being easily attained by addition of *red* to the blue or of *green* to the violet. The slits by which the light entered were protected with suitable coloured glasses, cobalt glass being used for the blue and violet slits. In this way, as already mentioned, the danger of false light is obviated. I do not affirm that the mixture of blue and red looked *exactly* the same as the violet. I thought that I could recognise the violet as being more saturated, but the difference, if real, was very small and certainly a mere fraction of the original difference between blue and violet. Needless to say, the blue chosen was a full blue, showing no approximation to green.

The point of greatest interest lies in the contrast between my observations and those of Mr Gerald Balfour, who was with me at the time. Mr G. Balfour is one of the three brothers whom I found in 1881 (*loc. cit.*) to make anomalous matches of mixtures of red and green with spectrum yellow. To effect the match they use much smaller amounts of red than is required by normal eyes. But their colour vision is as acute as usual, and the abnormality is

* *Nature*, Vol. xxv. p. 64 (1881); *Scientific Papers*, Vol. i. p. 544.

quite distinct from what is called colour-blindness. To Mr Balfour's vision the violet of the spectrum is *not* redder than the blue, and such addition of red to blue as I required to make the match gave, in his estimation, a "reddish purple." Curiously enough, Mr Enock, who was my assistant at that time, bore similar testimony, no addition of red on either side improving the match, which was indeed nearly complete as it stood. It is probably not a coincidence that Mr Enock is also abnormal in his red *plus* green = yellow match, coming perhaps about half-way between myself and the Balfours.

When a few months ago I commenced to write out an account of these observations, it occurred to me that it would sound strange if I described my own judgments as normal and those of two other male observers as abnormal, and I sought to confirm my own judgment by that of others, especially of women. As to this, there was no difficulty. I usually showed first the simple blue and violet with about equal illumination* and asked the observer to describe them. In nearly every case the names blue and violet were correctly given. Can you describe one as redder than the other? was the next question. In most cases the answer came, "the violet is the redder"; but in some others all I could get at this stage was a negative. When, however, the same addition of red light that I require was made to the blue, every female observer that I have tried agreed that now the difference had practically disappeared. I can say with confidence that in this matter my own vision is normal.

Lately I have had another opportunity of repeating the observation with Mr G. Balfour. It is certain that he sees no colour difference at all between the blue and violet. When to the blue an addition of red (less than I require for a match) is made, he describes the mixture as a reddish-purple, strongly distinguished from the violet. Mr A. J. Balfour also could see no difference between the blue and violet, but he seemed rather less sensitive to additions of red. A determination of wave-lengths gave for the (mean) violet 415 (above G), and for the blue 440. The red was rather extreme.

That ordinary normal vision is very approximately trichromic cannot be doubted; but a question may be raised as to the possible existence of a very subordinate fourth element of colour. Thus Dr Burch's descriptions might suggest that in his vision the sensation of violet depended upon such a fourth element. I am speaking here of fundamental sensations, not of such judgments as make yellow appear a distinct sensation to normal eyes, although certainly resolvable into red and green. The only way to get a final answer to such questions is by making matches with superposed colours; but to this

* This adjustment can be made by partially cutting off the light on the side required by means of strips of glass interposed. By varying the number (up to 5 or 6) or inclination, the proportion of light transmitted can be regulated. This procedure was found more convenient than altering the widths of the slits.

method some workers seem singularly averse.   In my own case I am certain that there is no fourth element of colour practically operative.

The character of the three primary sensations in normal vision is another and a much more difficult question.   Perhaps in recent years we have rather lost sight of the argument which weighed with Maxwell in the passage above quoted.   The better to see its significance, let us suppose that the spectrum is *accurately* represented on Newton's diagram by two sides of a triangle, and inquire into the significance of this disposition.   The only explanation which does not involve highly improbable coincidences seems to be that in each spectrum colour only two of the three elements are involved.   If the third is involved at all, how comes it to be involved in such a way as to make the spectrum straight?   And the fact that near the red end variation of wavelength entails no variation of colour, makes in the same direction.   That the green corner is rounded off and that (if it be so) the sides are not quite straight, may diminish, but cannot destroy, the cogency of the argument, while the less precise character of the conclusion is not without advantages.

346.

AERIAL PLANE WAVES OF FINITE AMPLITUDE.

[*Proceedings of the Royal Society*, A, Vol. LXXXIV. pp. 247—284, 1910.]

Waves of Finite Amplitude without Dissipation.

IN the investigations which follow, we are concerned with the motion of an elastic fluid in one dimension, say, parallel to x. It is implied not only that there are no component velocities perpendicular to x, but that the motion is the same in any perpendicular plane, so that it is a function of x and of the time (t) only. If u be the velocity at any point x, p the pressure, ρ the density, X an impressed force, the dynamical equation for an inviscid fluid is

$$\frac{du}{dt} + u\frac{du}{dx} = X - \frac{1}{\rho}\frac{dp}{dx}. \quad\dots\dots\dots\dots\dots\dots\dots(1)$$

At the same time the "equation of continuity" takes the form

$$\frac{d\rho}{dt} + \frac{d(\rho u)}{dx} = 0. \quad\dots\dots\dots\dots\dots\dots(2)$$

The first step, and it was a very important one, in the treatment of waves of finite amplitude is due to Poisson*. Under the assumption of Boyle's law, $p = a^2\rho$, he proved that for waves travelling in one direction (positive) the circumstances of the propagation are expressed by

$$u = f\{x - (a+u)t\}, \quad\dots\dots\dots\dots\dots\dots(3)$$

in which f denotes an arbitrary function. When u can be neglected in comparison with a, this reduces to the familiar law of undisturbed propagation applicable to infinitesimal waves.

Poisson does not discuss the significance of (3) further than to show that the boundaries of a continuous wave, limited to a finite range along x, are propagated with the ordinary velocity a, and that accordingly the length of the wave does not alter as it advances. The meaning of (3) is that in general u advances with a velocity equal, not to a, but to $a + u$, and that this might be expected is very easily seen (Earnshaw). From the ordinary

---

* "Mémoire sur la Théorie du Son," *Journ. de l'École Polytechnique*, 1808, Vol. VII. p. 319.

theory we know that an infinitely small disturbance is propagated with a certain velocity a, which velocity is relative to the parts of the medium undisturbed by the wave. Let us consider now the case of a wave so long that the variations of velocity and density are insensible for a considerable distance along it, and at a place where the velocity (u) is finite let us imagine a small secondary wave to be superposed. The velocity with which the secondary wave is propagated through the surrounding medium is a, but on account of the local motion of the medium itself the whole velocity of advance is $a + u$, and depends upon the part of the long wave at which the small wave is placed. What has been said of the secondary wave applies also to the parts of the long wave itself, and thus we see that after a time t the place where a certain velocity u is to be found is in advance of its original position by a distance equal, not to at, but to $(a + u) t$, or, as we may express it, u is propagated with velocity $(a + u)$.

A closer discussion of the solution represented by Poisson's integral was given by Stokes*, who pointed out the difficulty which ultimately arises from the motion becoming discontinuous. If we draw a curve to represent the distribution of velocity, taking x for abscissa and u for ordinate, we may find the corresponding curve after the lapse of time t by the following construction:—Through any point on the original curve draw a straight line in the positive direction parallel to x, and of length equal to $(a + u) t$, or, as we are concerned with the shape of the curve only, equal to ut. The locus of the ends of these lines is the velocity-curve after a time t.

But this law of derivation cannot hold good indefinitely. The crests of the velocity-curve gain continually on the troughs and must at last overtake them. After this the curve would indicate two values of u for one value of x, ceasing to represent anything that could actually take place. In fact we are not at liberty to push the application of the integral beyond the point at which the velocity becomes discontinuous, or the velocity-curve has a vertical tangent. In order to find when this happens, let us take two neighbouring points on any part of the curve which slopes downwards in the positive direction, and inquire after what time this part of the curve becomes vertical. If the difference of abscissae be dx, the hinder point will overtake the forward point in the time $-dx/du$. Thus the motion, as determined by Poisson's integral, becomes discontinuous after a time equal to the reciprocal, taken positively, of the greatest negative value of du/dx.

For example, let us suppose that

$$u = U \cos \frac{2\pi}{\lambda} \{x - (a + u) t\}, \quad \dots\dots\dots\dots\dots\dots(4)$$

* "On a Difficulty in the Theory of Sound," *Phil. Mag.* November, 1848.

where U is the greatest initial velocity. When $t = 0$, the greatest negative value of du/dx is $-2\pi U/\lambda$, so that discontinuity will commence at the time $t = \lambda/2\pi U$.

The only kind of wave travelling in the positive direction which can escape ultimate discontinuity is one which has no forward slope. This is the case of a wave forming the transition between a larger constant value of u when x exceeds a certain value, and a smaller constant value when x falls short of a certain value. As time passes, the slope everywhere becomes easier. We shall see presently that this wave is a wave of rarefaction, in the sense that during its passage the gas passes from a greater to a less density.

It is worthy of remark that, although we may of course conceive a wave of finite disturbance to exist at any moment, there is in general a limit to the duration of its previous independent existence. By drawing lines in the negative instead of in the positive direction we may trace the history of the velocity-curve; and we see that as we push our inquiry further and further into past time the forward slopes become easier and the backward slopes steeper. At a time equal to the greatest positive value of dx/du, antecedent to that at which the curve is first contemplated, the velocity would be discontinuous. The exception is now a wave of condensation, involving a passage always from a less to a greater density.

When discontinuity sets in, a state of things exists to which the usual differential equations are inapplicable; and the subsequent progress of the motion has not been determined. It is probable, as suggested by Stokes, that some sort of reflection would ensue. In regard to this matter we must be careful to keep purely mathematical questions distinct from physical ones. We shall see later how the tendency to discontinuity may be held in check by forces of a dissipative character. But this has nothing directly to do with the mathematical problem of determining what would happen to waves of finite amplitude in a medium, free from viscosity, whose pressure is under all circumstances proportional to the density*. To suppose that the problem has no solution would seem to be tantamount to admitting an inherent contradiction in the assumption, usually made in hydrodynamics, of a continuous fluid subject to Boyle's law. It would be strange if the necessity of a molecular constitution for gases could be established by such an argument.

With Poisson's integral (3), showing how the velocity is propagated, there is associated another law connecting the velocity and the density in a positive progressive wave. In the case of a fluid obeying Boyle's law this relation is

$$u - a \log \rho = \text{const.} \quad\quad\quad\quad\quad (5)$$

It does not occur explicitly in Poisson's memoir, and Earnshaw considers that Poisson did not discover it. Certainly it is remarkable that he omitted

* *Theory of Sound*, 1878, § 251.

to formulate the law, but at the same time it is difficult to suppose him ignorant of it, seeing that it follows by simple subtraction from two of his equations*. A formula equivalent to (5) was given explicitly, so far as I know, for the first time by Airy†, who attributes it to De Morgan.

The assumption that in a progressive wave there is a definite relation between u and ρ forms the basis of Earnshaw's investigation‡. That such a relation is to be expected may be shown by a line of argument analogous to that already employed in connection with Poisson's integral.

Whatever may be the law of pressure as a function of density, the velocity of propagation of small disturbances is according to the usual theory equal to $\sqrt{(dp/d\rho)}$, and in a positive progressive wave the relation between velocity and condensation (s) is

$$u : s = \sqrt{(dp/d\rho)}, \dots\dots\dots\dots\dots\dots\dots\dots(6)$$

where $s = \delta\rho/\rho$. If this relation be violated at any point, a wave will emerge, travelling in the negative direction. Let us now picture to ourselves the case of a positive progressive wave in which the changes of velocity and density are very gradual but become important by accumulation, and let us inquire what condition must be satisfied in order to prevent the formation of a negative wave. It is clear that the answer to the question whether or not a negative wave will be generated at any point will depend upon the state of things in the immediate neighbourhood of the point, and not upon the state of things at a distance from it, and will therefore be determined by the criterion applicable to small disturbances. In applying this criterion we are to consider the velocities and condensations, not absolutely, but relatively to those prevailing in the neighbouring parts of the medium, so that the form of (6) proper for the present purpose is

$$du = \sqrt{\left(\frac{dp}{d\rho}\right)} \cdot \frac{d\rho}{\rho}, \dots\dots\dots\dots\dots\dots\dots(7)$$

whence

$$u = \int \sqrt{\left(\frac{dp}{d\rho}\right)} \cdot \frac{d\rho}{\rho}, \dots\dots\dots\dots\dots\dots\dots(8)$$

which is the relation between u and ρ generally necessary for a positive progressive wave, as laid down by Earnshaw§.

Earnshaw worked with the so-called Lagrangian form of the equations, in which the motions of particular particles are followed, and he obtained complete solutions for a wave progressive in one direction. In the case of

---

* Equations (1), p. 364, and (b), p. 367.

† *Phil. Mag.* 1849, Vol. xxxiv. p. 402. The corresponding formula for long tidal waves of finite amplitude was also given.

‡ *Roy. Soc. Proc.* January 6, 1859; *Phil. Trans.* 1860, p. 133.

§ *Theory of Sound*, 1878, § 251.

Boyle's law the relation between velocity and density is that already given (5), and in the case of the adiabatic law, where

$$\frac{p}{p_0} = \left(\frac{\rho}{\rho_0}\right)^\gamma, \dots\dots\dots(9)$$

Earnshaw finds
$$\left(\frac{\rho}{\rho_0}\right)^{\frac{1}{2}(\gamma-1)} = 1 + \frac{(\gamma-1)u}{2a}, \dots\dots(10)$$

where a is the velocity of infinitesimal disturbances under the condition represented by p_0, ρ_0, viz. $a^2 = \gamma p_0/\rho_0$. In (9) γ denotes as usual the ratio of the two specific heats; and in (10), applicable to a *positive* progressive wave, the constant of integration has been so chosen that $u = 0$ corresponds to $\rho = \rho_0$.

The generalised form of Poisson's integral, appropriate when p is *any* given function of ρ, does not appear quite explicitly in Earnshaw's memoir. The line of argument already used shows that it must be

$$u = f[x - \{u + \sqrt{(dp/d\rho)}\}t]. \dots\dots(11)$$

In the case of a gas obeying Boyle's law,

$$\sqrt{(dp/d\rho)} = \text{const.}, \dots\dots(12)$$

and (11) reduces to Poisson's form.

In the case of the adiabatic law, we have from (9), (10),

$$\sqrt{\left(\frac{dp}{d\rho}\right)} = a\left(\frac{\rho}{\rho_0}\right)^{\frac{1}{2}(\gamma-1)} = a + \frac{\gamma-1}{2}u, \dots\dots(13)$$

so that (Earnshaw)
$$u + \sqrt{(dp/d\rho)} = a + \tfrac{1}{2}(\gamma+1)u. \dots\dots(14)$$

Thus (11) assumes the form
$$u = f[x - \{a + \tfrac{1}{2}(\gamma+1)u\}t], \dots\dots(15)$$

and this with (10) may be considered to constitute the solution of the problem up to the point where discontinuity sets in. We may fall back upon Boyle's law by putting in (15) $\gamma = 1$.

It appears that whether the relation of pressure to density be isothermal or adiabatic, there is a change of type as the wave advances. There can be no escape from such a change unless $u + \sqrt{(dp/d\rho)}$ be constant. Using (8), we may deduce in this case

$$\sqrt{(dp/d\rho)} = B/\rho,$$

B being a constant, whence
$$p = A - B^2/\rho \dots\dots(16)$$

expresses the only law of pressure under which waves of finite amplitude can be propagated without undergoing a change of type (Earnshaw). A simpler derivation of (16) will be given presently.

Earnshaw further considers the genesis of disturbance in a gas originally at rest by the motion of a piston (supposed to be contained in a tube), but some of his conclusions appear to need revision. All that is required in these problems is virtually contained in (8), (11). If X denote the position of the piston at time T, the velocity of its motion is $U = dX/dT$, and this velocity is shared by the gas in contact with it. On the positive side the velocity of propagation of U (equal to u) is, by (11),

$$U + \sqrt{(dp/d\rho)};$$

so that if at time t (greater than T), U is to be found at x, we must have

$$x = X + \{U + \sqrt{(dp/d\rho)}\}(t - T). \quad\ldots\ldots\ldots\ldots(17)$$

Among the problems which naturally suggest themselves would be to determine what happens when the piston originally at rest at $X = 0$ begins to move at time $T = 0$ with a constant velocity. But if this velocity be positive, the discontinuity, which immediately ensues, causes the failure of our equations. On the other hand, if the constant velocity be negative, say $- V$, the initial discontinuity disappears forthwith, and the subsequent motion may be traced.

Take, for example, the case of Boyle's law. We have

$$X = UT, \qquad U + \sqrt{(dp/d\rho)} = U + a,$$

so that $\qquad x = Ut + a(t - T) \qquad (t > T),$

and we have to consider where the velocities 0 and $- V$ are to be found at time t. Now $U = 0$ corresponds to the range of T from $- \infty$ to 0, so that x ranges from at to ∞. Again, $U = - V$ corresponds to the range of T from 0 to t, so that x ranges from $(a - V)t$ to $- Vt$. The whole range of x on the positive side of the piston is now accounted for, except the interval from $x = (a - V)t$ to $x = at$. This is occupied by the transition of velocity from 0 to $- V$, and we infer, from what was said in the discussion of Poisson's integral, that this transition must take place linearly.

Under Boyle's law the relation between velocity and density (5) is such that, however fast the piston may recede (u negative), a complete vacuum can never be formed behind it. It is otherwise under the adiabatic law (10), where $\rho = 0$ corresponds to

$$u = - 2a/(\gamma - 1) \quad \text{(Earnshaw)}.$$

It may be of interest to consider further a few examples of (17). Still assuming Boyle's law, let us suppose that the piston is at rest ($X = 0$) until $T = 0$, and then moves with uniform acceleration (g), so that ($T +$)

$$U = gT, \qquad X = \tfrac{1}{2}gT^2 = U^2/2g.$$

The use of these in (17) gives

$$x = \frac{a^2}{2g} + (U+a)t - \frac{(U+a)^2}{2g}, \quad\ldots\ldots\ldots\ldots\ldots(18)$$

showing that the relation between x and U is parabolic. In (18) we see that dx/dU vanishes, when $t = (U+a)/g$. Thus, if g be positive, i.e. if the wave be one of condensation, discontinuity sets in at the front after an interval, reckoned from the beginning of the motion, equal to a/g. But if g be negative, there is no discontinuity, and (18) remains valid for an indefinite time.

In general, as in the last example $(g +)$, the discontinuity sets in locally at one point of the velocity-curve, while other parts are temporarily exempt. It is of interest to inquire under what law the piston must advance so as to generate a linear velocity-curve. For then, under the adiabatic law, which includes Boyle's, a velocity-curve, once linear, remains linear, and if discontinuity enters, it must affect the whole curve simultaneously.

It may be worth while to pause here for a moment to inquire what law of pressure is implied in the permanence of the linear character of the velocity-curve. By (7)

$$\rho \frac{du}{d\rho} = \sqrt{\left(\frac{dp}{d\rho}\right)};$$

so that if $u + \sqrt{(dp/d\rho)}$ is a linear function of u, $du/d\log\rho$ must also be a linear function of u. This requires that

$$\rho\, du/d\rho = C\rho^n,$$

where C and n are constants, and the most general relation between p and ρ consistent with the requirements is

$$p = A + B\rho^\gamma, \quad\ldots\ldots\ldots\ldots\ldots\ldots\ldots(19)$$

where A, B, γ are constants. The relation (19) may be regarded as a kind of generalised adiabatic law; it includes the special law (16) under which a velocity-curve is absolutely permanent in type.

Supposing the motion to commence at $T = 0$, we have

$$X = \int_0^T U\, dT = UT - \int_0^U T\, dU,$$

and hence from (17), under the supposition of Boyle's law,

$$x = -\int T\, dU - (U+a)t - aT; \quad\ldots\ldots\ldots\ldots(20)$$

and the question before us is so to determine T as a function of U that (20) may be linear in U. From (20) when t is constant

$$\frac{dx}{dU} = -T + t - a\frac{dT}{dU} = t_0, \text{ say,}$$

where t_0 is a constant, whence

$$T = t - t_0 + He^{-U/a},$$

H being the constant of integration. But, since $U = 0$ when $T = 0$, this assumes the form

$$T = T'(1 - e^{-U/a}), \quad \dots\dots\dots\dots\dots\dots(21)$$

T' being written for $t - t_0$. Or, if we express U in terms of T,

$$U = -a \log \left(1 - \frac{T}{T'}\right). \quad \dots\dots\dots\dots\dots\dots(22)$$

In (21), (22), T' is positive, if U is positive; and U becomes infinite when $T = T'$. We must therefore regard the law as limited to values of T less than T'.

From (22) we find

$$X = \int_0^T U\,dT = a\left\{(T' - T)\log\left(1 - \frac{T}{T'}\right) + T\right\}, \quad \dots\dots\dots(23)$$

which completely expresses the motion of the piston. The corresponding velocity-curve at time t may be verified by means of (20), (21). It is expressed by

$$U = \frac{at - x}{T' - t}, \quad \dots\dots\dots\dots\dots\dots\dots\dots(24)$$

exhibiting the linear character of the slope of velocity. Evidently the slope becomes vertical throughout when $t = T'$.

In the above example it is not necessary to suppose the law of motion of the piston continued up to $T = T'$. On the contrary, we may imagine U to increase up to some prescribed finite value and then to remain constant. In this case the slope expressed by (24) forms the transition between $U = 0$ for values of x greater than at and the finite value at which the acceleration of the piston stops.

If the wave be one of rarefaction we must take T' negative, say $-T''$. In this case the analogue of (22) shows that U constantly increases in numerical value but does not become infinite in any finite time. The analogue of (24) is

$$U = \frac{x - at}{t + T''}, \quad \dots\dots\dots\dots\dots\dots\dots\dots(25)$$

representing a slope which ever grows easier as time passes.

The problem also admits of solution when the gas follows the adiabatic law (9). As in (15), (20),

$$x = X + \{a + \tfrac{1}{2}(\gamma + 1)\,U\}\,(t - T); \quad \dots\dots\dots\dots(26)$$

and (26) is to satisfy the condition of making dx/dU constant. In this

$$dX/dU = U \cdot dT/dU,$$

so that
$$\frac{dT}{dU}\{(\gamma - 1)\,U + 2a\} + (\gamma + 1)\,(T - t) = \text{const.}$$

On integration we obtain, under the condition that U and T vanish together,

$$\left(1 - \frac{T}{T'}\right)\left[1 + \frac{\gamma - 1}{2a}\,U\right]^{\frac{\gamma+1}{\gamma-1}} = 1, \quad\ldots\ldots\ldots\ldots\ldots(27)$$

T' being a constant which, if positive, corresponds to $U = \infty$, thereby determining U as a function of T.

For the value of $X = \int_0^T U\,dT$, we have

$$\frac{\gamma - 1}{2a}\frac{X}{T'} = -\frac{\gamma + 1}{2}\left[\left(1 - \frac{T}{T'}\right)^{\frac{2}{\gamma+1}} - 1\right] - \frac{T}{T'}$$

$$= \frac{\gamma - 1}{2} - \frac{\gamma + 1}{2}\left(1 + \frac{\gamma - 1}{2a}\,U\right)^{-\frac{2}{\gamma-1}} + \left(1 + \frac{\gamma - 1}{2a}\,U\right)^{-\frac{\gamma+1}{\gamma-1}}. \ \ldots(28)$$

It may be observed that, although U is infinite when $T = T'$, X remains finite. Using this in (26), we get finally, on reduction,

$$x = at - \tfrac{1}{2}(\gamma + 1)\,U\,(T' - t), \quad\ldots\ldots\ldots\ldots\ldots(29)$$

or
$$U = \frac{2}{\gamma + 1}\frac{at - x}{T' - t}. \quad\ldots\ldots\ldots\ldots\ldots(30)$$

If $x < at$, U is a linear function of x, and (T' being positive) the slope of the velocity-curve increases until it becomes vertical when $t = T'$.

If T' is negative, the wave is one of rarefaction, and (30) applies however great t may be.

By putting $\gamma = 1$ we fall back from (30) to (24), and less simply from (28) to (23)*.

Riemann's work† is of somewhat later date than Earnshaw's, but in one important respect is more general. It may be convenient briefly to recall the principal result.

Taking p a given function of ρ, say $\phi(\rho)$, and putting $X = 0$, we have from (1) and (2)

$$\frac{du}{dt} + u\frac{du}{dx} = -\phi'(\rho)\frac{d\log\rho}{dx}, \qquad \frac{d\log\rho}{dt} + u\frac{d\log\rho}{dx} = -\frac{du}{dx}.$$

---

* I have since found that this problem was successfully treated by Hugoniot.

† *Göttingen Abhandlungen*, 1860, Vol. VIII.

If the second of these equations be multiplied by $\pm \sqrt{\{\phi'(\rho)\}}$ and be added to the first, we find

$$\frac{dr}{dt} = -\{u + \sqrt{\phi'(\rho)}\}\frac{dr}{dx}, \qquad \frac{ds}{dt} = -\{u - \sqrt{\phi'(\rho)}\}\frac{ds}{dx}, \quad \ldots\ldots(31)$$

where
$$2r = f(\rho) + u, \qquad 2s = f(\rho) - u, \quad \ldots\ldots\ldots\ldots(32)$$

and
$$f(\rho) = \int \sqrt{\phi'(\rho)} \, . \, d \log \rho. \quad \ldots\ldots\ldots\ldots\ldots(33)$$

From these follow

$$dr = \frac{dr}{dx}[dx - \{u + \sqrt{\phi'(\rho)}\}\,dt],$$

$$ds = \frac{ds}{dx}[dx - \{u - \sqrt{\phi'(\rho)}\}\,dt];$$

so that r remains constant when x and t change in such a manner that $dx = \{u + \sqrt{\phi'(\rho)}\}\,dt$, and s remains constant when x and t change so that $dx = \{u - \sqrt{\phi'(\rho)}\}\,dt$. In the case of a positive progressive wave $s = 0$, whence $f(\rho) = u$ and also $r = u$. The velocity with which u travels in such a wave is accordingly $u + \sqrt{(dp/d\rho)}$, of which fact (11) is merely another form of statement. Riemann's equations are more general than anything previously given, as not limited to a single progressive wave.

Since Riemann's equations do not seem to have been applied in any example of continuous motion, I have thought it worth while to inquire whether they can be satisfied when r and s are both linear functions of x, Boyle's law being assumed, so that in (32), (33)

$$\sqrt{\phi'(\rho)} = a, \qquad f(\rho) = a \log \rho.$$

If we suppose
$$r = Ax + B, \qquad s = Cx + D, \quad \ldots\ldots\ldots\ldots\ldots(34)$$

we obtain, on substitution in (31), equations for the determination of A, B, C, D, as functions of the time. In the first instance we find

$$A - C = 1/t, \quad \ldots\ldots\ldots\ldots\ldots\ldots(35)$$

in which to t a constant may be added, and further

$$A = \frac{H+1}{2t}, \qquad C = \frac{H-1}{2t}, \quad \ldots\ldots\ldots\ldots(36)$$

when H is an arbitrary constant. Also

$$B - D = -aH + L/t, \quad \ldots\ldots\ldots\ldots\ldots(37)$$

L being another arbitrary constant, and thence

$$B = \tfrac{1}{2}a(H^2 - 1)\log t + \frac{L(H+1)}{2t} + \tfrac{1}{2}M, \quad \ldots\ldots\ldots(38)$$

$$D = \tfrac{1}{2}a(H^2 - 1)\log t + \frac{L(H-1)}{2t} + \tfrac{1}{2}N, \quad \ldots\ldots\ldots(39)$$

with
$$M - N = -2aH. \quad \ldots\ldots\ldots\ldots\ldots(40)$$

If we allow the origin of x to be arbitrary, as well as that of t, we may write

$$2r = (H+1)\,x/t + (H^2-1)\,a \log t + M, \quad\ldots\ldots\ldots(41)$$

$$2s = (H-1)\,x/t + (H^2-1)\,a \log t + N, \quad\ldots\ldots\ldots(42)$$

$$u = r - s = x/t - aH, \quad\ldots\ldots\ldots\ldots\ldots\ldots(43)$$

$$a \log \rho = r + s = Hx/t + (H^2-1)\,a \log t + \tfrac{1}{2}(M+N). \quad\ldots\ldots(44)$$

If $H = \pm 1$, the logarithmic term disappears, and either r or s is constant. In these cases we fall back upon single progressive waves.

If $H = 0$, $u = x/t$ in (43), and (44) gives $1/\rho$ proportional to t. The density is thus uniform with respect to x, but the volume of a given mass grows proportionally with t. The uniform expansion occurs in such a manner that the gas remains unmoved at the origin of co-ordinates. Since in this case $du/dt + u\,du/dx = 0$, we see that every part of the gas moves with unaccelerated velocity.

Waves of Permanent Regime.

When waves are propagated in one dimension without change of type, the circumstances are dynamically the same as in steady motion, as appears at once by impressing on the system a velocity equal and opposite to that of wave-propagation. The problem may conveniently be considered under this form.

From the general equation of continuity (2), by making $d\rho/dt$ equal to zero, or independently, we have

$$\rho u = \rho_0 u_0 = m, \quad\ldots\ldots\ldots\ldots\ldots\ldots\ldots(45)$$

where m is a constant which Rankine called the mass-velocity. The dynamical equation (1) reduces to

$$u \frac{du}{dx} + \frac{1}{\rho}\frac{dp}{dx} = X. \quad\ldots\ldots\ldots\ldots\ldots(46)$$

If $X = 0$, (46) may be written

$$\int_{p_0}^{p} \frac{dp}{\rho} = \tfrac{1}{2}u_0^2 - \tfrac{1}{2}u^2, \quad\ldots\ldots\ldots\ldots\ldots(47)$$

where u_0 is the velocity corresponding to p_0.

Eliminating u, we get

$$\int_{p_0}^{p} \frac{dp}{\rho} = \tfrac{1}{2}u_0^2\left(1 - \frac{\rho_0^2}{\rho^2}\right), \quad\ldots\ldots\ldots\ldots(48)$$

determining the law of pressure under which alone it is possible for a stationary wave to maintain itself in fluid moving (outside the wave) with velocity u_0. From (48)

$$\frac{dp}{d\rho} = u_0{}^2 \frac{\rho_0{}^2}{\rho^2}, \quad \dots\dots\dots\dots\dots\dots\dots\dots(49)$$

or
$$p + m^2/\rho = p_0 + m^2/\rho_0, \dots\dots\dots\dots\dots\dots(50)$$

the law found by Earnshaw.

Since, under the adiabatic law, the relation between density and pressure differs from (50), we conclude that a self-maintaining stationary aerial wave is an impossibility, unless it be in virtue of impressed forces, or of viscosity, or other dissipative agencies not now regarded.

When the changes of density concerned are *small*, (50) may be satisfied approximately; and we see from (49) that the velocity of the stream (outside the wave) necessary to keep the wave stationary is given by

$$u_0 = \sqrt{(dp/d\rho)},$$

which is the same as the velocity of the wave reckoned relatively to the fluid at a distance.

This way of regarding the subject shows, perhaps more clearly than any other, the nature of the relation between velocity and density. In a stationary wave-form a loss of velocity accompanies an augmented density, according to the principle of energy, and therefore the fluid composing the condensed parts of a wave moves forward more slowly than the undisturbed portions. Relatively to the fluid at a distance, the motion of the condensed parts is in the same direction as that in which the waves travel.

By means of (46), we can find what impressed force is required in order to ensure a stationary wave-form when (50) is not satisfied. For example, if $p = a^2\rho$, we find from (45), (46),

$$X = u \frac{du}{dx} + a^2 \frac{d \log \rho}{dx} = (u^2 - a^2) \frac{d \log u}{dx}, \dots\dots\dots\dots(51)$$

showing that an impressed force is necessary at every place where u is variable and unequal to a. In (51) X is the accelerating force so called. The actual force operative upon the element of mass $\rho\,dx$ is $X\rho\,dx$. Thus, on integration,

$$\int X\rho\,dx = m \int \left(1 - \frac{a^2}{u^2}\right) du = m\,(u_2 - u_1) \left\{1 - \frac{a^2}{u_1 u_2}\right\}, \dots\dots\dots(52)$$

if the range of integration extend from the place where the velocity is u_1 to the place where it becomes equal to u_2. The integral applied force vanishes if the terminal velocities are such that their geometric mean is a. We may apply this to the case of a velocity-curve giving a simple gradual transition from one constant velocity u_1 to another constant velocity u_2. Under the above condition the integral force vanishes, but finite forces are required at all points of the slope, except the particular point where $u = a$.

It is of some importance to notice that although, under the condition $u_1 u_2 = a^2$, the applied forces contribute on the whole no *momentum*, yet they do contribute *energy*, positive or negative. To find the work done in unit of time by the forces we have

$$2/m \cdot \int X\rho \cdot u \cdot dx = u_2^2 - u_1^2 - 2u_1 u_2 \log (u_2/u_1). \quad \ldots\ldots\ldots\ldots (53)$$

The better to interpret this let us suppose that u_1 and u_2 are positive, and in the first instance that $u_2 > u_1$, so that the fluid passes from a less to a greater velocity, or by (45) from a greater to a less density. In the case of a wave of rarefaction we have therefore to consider the sign of

$$y^2 - 1 - 2y \log y, \quad \ldots\ldots\ldots\ldots\ldots\ldots\ldots\ldots\ldots (54)$$

when $y > 1$. It is not difficult to prove that this sign is always positive. When $y - 1$ is small, the approximate value of (54) is $\frac{1}{3} (y - 1)^3$, and is therefore positive when $y > 1$. Again, if we remove the positive factor y from (54) and then differentiate, we obtain $(1 - 1/y)^2$, which is positive. Hence, when $y > 1$, (54) is necessarily positive. The propagation of the wave of rarefaction without change of type requires that the impressed forces, contributing on the whole no momentum, should nevertheless do work upon, *i.e.* communicate energy to, the gas.

In like manner, if the wave be one of condensation, *i.e.*, if the gas passes from a less to a greater density, the operation of the impressed forces is to remove energy from the gas forming the wave. It follows that although dissipative forces, such as those arising from viscosity, may possibly constitute a machinery capable of maintaining the type of a wave of condensation, in no case can they maintain the type of a wave of rarefaction.

It is desirable to extend this argument to waves propagated under the adiabatic law. In general, from (45), (46),

$$X\rho = m \frac{du}{dx} + \frac{dp}{dx};$$

so that $\int X\rho\, dx = m (u_2 - u_1) + p_2 - p_1 = m^2/\rho_2 - m^2/\rho_1 + p_2 - p_1.$

As in (50), the condition that on the whole no momentum is communicated is

$$m^2/\rho_2 - m^2/\rho_1 + p_2 - p_1 = 0. \quad \ldots\ldots\ldots\ldots\ldots\ldots (55)$$

Again,

$$m^{-1} \cdot \int X\rho \cdot u \cdot dx = \tfrac{1}{2}(u_2^2 - u_1^2) + \int_{p_1}^{p_2} \frac{dp}{\rho} = \int_{p_1}^{p_2} \frac{dp}{\rho} - \frac{p_2 - p_1}{2}\left(\frac{1}{\rho_2} + \frac{1}{\rho_1}\right). \ldots (56)$$

The question in which we are interested is the sign of (56). If we regard v (the volume of unit mass), viz., $1/\rho$, as the ordinate, and p as the abscissa of a curve, the first term on the right of (56) represents the area of the curve bounded by two ordinates and the axis of p, while the second term is what the area would be if the ordinate retained throughout the mean of the terminal values.

So far the argument is general. If the relation between p and v be adiabatic, and $p_2 > p_1$, the expression (56) is negative, since v proportional to $p^{-1/\gamma}$ makes d^2v/dp^2 positive*. The final pressure exceeding the initial pressure denotes a wave of condensation, and we conclude, as before, that maintenance of type in such a wave requires removal of energy from the wave, while in the contrary case of a wave of rarefaction additional energy would need to be supplied.

The problem now under discussion is closely related to one which has given rise to a serious difference of opinion. In his paper of 1848 already referred to, Stokes considered the *sudden* transition from one constant velocity to another, and concluded that the necessary conditions for a permanent regime could be satisfied. Results equivalent to his may be deduced from (45) in connection with the condition $(u_1 u_2 = a^2)$ already found from (52) to express that there is no change of momentum on the whole. Thus,

$$u_1 = a \sqrt{(\rho_2/\rho_1)}, \qquad u_2 = a \sqrt{(\rho_1/\rho_2)}. \dots\dots\dots\dots(57)$$

Similar conclusions were put forward by Riemann in 1860 (*loc. cit.*). Commenting on these results in the *Theory of Sound* (1878), I pointed out that although the conditions of *mass* and *momentum* were satisfied, the condition of *energy* was violated, and that therefore the motion was not possible; and in republishing this paper† Stokes admitted the criticism, which had indeed already been made privately by Kelvin. On the other hand, Burton‡ and H. Weber§ maintain, at least to some extent, the original view.

Inasmuch as they ignored the question of energy, it was natural that Stokes and Riemann made no distinction between the cases where energy is gained or lost. As I understand, Weber abandons Riemann's solution for the discontinuous wave (or *bore*, as it is sometimes called for brevity) of rarefaction, but still maintains it for the case of the bore of condensation. No doubt there is an important distinction between the two cases; nevertheless, I fail to understand how a loss of energy can be admitted in a motion

---

* Compare Lamb's *Hydrodynamics*, § 280.
† *Collected Works*, Vol. ii. p. 55.
‡ *Phil. Mag.* 1893, Vol. xxxv. p. 316.
§ *Die Partiellen Differentialgleichungen der Mathematischen Physik*, Braunschweig, 1901, Vol. ii. p. 496.

which is supposed to be subject to the isothermal or adiabatic laws, in which
no dissipative action is contemplated. In the present paper the discussion
proceeds upon the supposition of a *gradual* transition between the two
velocities or densities. It does not appear how a solution which violates
mechanical principles, however rapid the transition, can become valid when
the transition is supposed to become absolutely abrupt. All that I am able
to admit is that under these circumstances dissipative forces (such as viscosity)
that are infinitely small may be competent to produce a finite effect.

If we suppose that under the influence of small dissipative forces the
bore of Stokes and Riemann can be propagated, at least approximately, we
naturally inquire whether it can be regarded as the complete outcome of the
simple progressive wave with a straight velocity slope which, as we have
found, tends after a definite interval of time to assume the character of
a bore. It would seem that the answer must be in the negative. Taking
Boyle's law, we recognise from (5) that in the progressive wave, just before
the formation of the bore, the relation between the velocities and densities is

$$u_2 - u_1 = a \log \rho_2 - a \log \rho_1,$$

while in (57) the relation is

$$u_2 - u_1 = a \sqrt{(\rho_2/\rho_1)} - a \sqrt{(\rho_1/\rho_2)}.$$

The two functions of ρ on the right, which are independent of any common
addition to u_1 and u_2, cannot be identified (unless $\rho_2 = \rho_1$), as we have found
already in discussing (54). This incompatibility may be regarded as a
confirmation of Stokes' opinion that something of the nature of reflection
must ensue.

Permanent Regime under the influence of Dissipative Forces.

The first investigation to be considered under this head is a very
remarkable one by Rankine "On the Thermodynamic Theory of Waves of
Finite Longitudinal Disturbance[*]," which (except a limited part expounded
by Maxwell in his *Theory of Heat*) has been much neglected[†]. Conduction of heat is here for the first time taken into account and although
there are one or two serious deficiencies, not to say errors, presently to be
noticed, the memoir marks a very definite advance.

The first step is the establishment of an equation equivalent (when the
wave is reduced to rest) to (45), and of Earnshaw's relation (50), in which

---

[*] *Phil. Trans.* 1870, Vol. CLX. Part II. p. 277.

[†] I must take my share of the blame. Rankine is referred to by Lamb (*Hydrodynamics*,
1906, p. 466). The body of Rankine's memoir seems to have been composed without acquaintance
with the writings of his predecessors; but in a supplement he notices the work of Poisson, Stokes,
Airy, and Earnshaw.

equations we shall usually substitute v, the volume of unit mass, for $1/\rho$. Rankine remarks that "no substance yet known fulfils the condition expressed by (50) between finite limits of disturbance, at a constant temperature, nor in a state of non-conduction of heat (called the *adiabatic* state). In order, then, that permanency of type may be possible in a wave of longitudinal disturbance, there must be both change of temperature and conduction of heat during the disturbance." However, we shall see later that even under Boyle's law *viscosity* is competent to endow a wave with permanency.

The question is, how can Earnshaw's law be satisfied? Obviously not (in the absence of viscosity) if the expansions are adiabatic; but if at every stage the right quantity of heat is added or subtracted, the gas may be made to follow any prescribed law. This is the idea underlying Rankine's investigation. For the unit mass of a *perfect gas* we have, as usual, $pv = R\theta$, θ denoting absolute temperature. The condition of the gas is defined by any two of the three quantities p, v, θ, and the third may be expressed in terms of them. The relation between simultaneous variations is

$$d\theta/\theta = dp/p + dv/v. \quad\quad\quad\quad (58)$$

In order to effect the changes specified by dp and dv, it is in general necessary to communicate heat to the gas. Calling the necessary quantity of heat dQ, we may write

$$dQ = \left(\frac{dQ}{dv}\right) dv + \left(\frac{dQ}{dp}\right) dp. \quad\quad\quad\quad (59)$$

Suppose now (a) that $dp = 0$. Equations (58), (59), give

$$\frac{dQ}{d\theta} (p \text{ constant}) = \left(\frac{dQ}{dv}\right) \frac{v}{\theta},$$

where $dQ/d\theta$ (p constant) expresses the specific heat of the gas under constant pressure. Denoting this by C, we have

$$C = \left(\frac{dQ}{dv}\right) \frac{v}{\theta}.$$

Again, suppose (b) that $dv = 0$. We find in a similar manner that if c denote the specific heat under constant volume,

$$c = \left(\frac{dQ}{dp}\right) \frac{p}{\theta}.$$

Thus, in general, $$dQ/\theta = Cdv/v + cdp/p. \quad\quad\quad\quad (60)$$

If between (58) and (60) we eliminate dp, there results

$$dQ = (C - c)\frac{pdv}{R} + cd\theta. \quad\quad\quad\quad (61)$$

In (61) $dQ = 0$ corresponds to adiabatic expansion, when according to

Mayer's principle the cooling effect $-cd\theta$ is equal to the external work done by the gas pdv. Hence

$$C - c = R, \quad \dots\dots\dots\dots\dots\dots\dots\dots\dots(62)$$

and therefore

$$\gamma = \frac{C}{c} = \frac{C}{C - R}, \quad \dots\dots\dots\dots\dots\dots\dots(63)$$

a relation discovered by Rankine himself in 1850.

Rankine then applies (60) to find what heat must be communicated in order that the gas may follow Earnshaw's law making $dp = -m^2 dv$. With regard to (62), it appears that

$$dQ = \frac{dp}{m^2(\gamma - 1)} \{p_0 + m^2 v_0 - (\gamma + 1)p\}. \quad \dots\dots\dots\dots(64)$$

It will be understood that under the condition now imposed of Earnshaw's relation, as well as of the ordinary gas law, there remains but one independent variable, and that the state of the gas may be expressed in terms of any *one* of the three quantities p, v, θ.

We have next to consider how far the necessary supply of heat defined by (64) can be effected by *conduction*. If the initial state (distinguished by suffix 1) and final state (with suffix 2) be of uniformity with respect to x, the total quantity of heat received by the gas during its passage must be zero, or $\int dQ = 0$. Hence from (64) Rankine finds

$$p_0 + m^2 v_0 = \tfrac{1}{2}(\gamma + 1)(p_1 + p_2). \quad \dots\dots\dots\dots\dots(65)$$

This is a necessary condition; but of course there is nothing so far to show that it is sufficient.

In (65) p_0, v_0 are any corresponding values of p and v within the wave, and we may identify them with p_1, v_1 *.

Thus

$$m^2 v_1 = \tfrac{1}{2}(\gamma - 1)p_1 + \tfrac{1}{2}(\gamma + 1)p_2. \quad \dots\dots\dots\dots(66)$$

The velocity u_1, equal to mv_1, is that with which the wave advances relatively to the fluid in state (1). And

$$u_1^2 = m^2 v_1^2 = v_1\{\tfrac{1}{2}(\gamma - 1)p_1 + \tfrac{1}{2}(\gamma + 1)p_2\} \quad \dots\dots\dots\dots(67)$$

gives the square of the velocity of wave-propagation relatively to fluid (1). The velocity of propagation of infinitely small disturbances (p_2 nearly equal to p_1) is given by $u_1^2 = \gamma p_1 v_1$; and thus a wave of finite condensation is propagated faster than an infinitesimal wave, and according to (67) a wave of finite rarefaction would be propagated slower than an infinitesimal wave. Moreover, there is no limit to the velocity of a wave of condensation.

* We may, of course, [alternatively] identify p_0, v_0 with p_2, v_2.

Rankine proceeds to express the absolute temperature (θ) at a point where the pressure is p in a wave of permanent type. By Earnshaw's law (50) in combination with (65)

$$\frac{\theta}{\theta_0} = \frac{pv}{p_0 v_0} = \frac{p}{p_0} \cdot \frac{(\gamma+1)(p_1+p_2)-2p}{(\gamma+1)(p_1+p_2)-2p_0}, \quad \dots\dots\dots(68)$$

and for the ratio of terminal temperatures

$$\frac{\theta_2}{\theta_1} = \frac{p_2}{p_1} \cdot \frac{(\gamma+1)p_1+(\gamma-1)p_2}{(\gamma+1)p_2+(\gamma-1)p_1}. \quad \dots\dots\dots\dots(69)$$

The second fraction on the right of (69) obviously represents the ratio of volumes v_2/v_1, or of densities ρ_1/ρ_2.

In order to justify (65), it is not necessary that the terminal states be states of absolute uniformity. It will suffice that the temperature be there stationary ($d\theta/dx=0$), which secures that no conduction of heat takes place there, and a state of stationary temperature usually involves a stationary pressure. To make the most of (65) we must apply it to the smallest ranges, i.e. between consecutive places where dp/dx vanishes.

But here a question arises which Rankine does not seem to have considered. In order to secure the necessary transfers of heat by means of conduction it is an indispensable condition that the heat should pass from the hotter to the colder body. If maintenance of type be possible in a particular wave as the result of conduction, a reversal of the motion will give a wave whose type cannot be so maintained. We have seen reason already for the conclusion that a dissipative agency can serve to maintain the type only when the gas passes from a less to a more condensed state. If this be so, the application which Rankine makes to a periodic wave is evidently prohibited.

According to the *second* law of thermodynamics, the criterion whether the transformation is possible as the result of dissipative action is the sign of $\int dQ/\theta$. If this be negative, the transformation is not possible. From (64) with use of (65)

$$dQ = \frac{(\gamma+1)\,dp}{2m^2(\gamma-1)}(p_1+p_2-2p). \quad \dots\dots\dots\dots(70)$$

In (68) we may give p_0, θ_0 any corresponding values found in the wave. Thus p_0 lies between p_1 and p_2, and ($\gamma>1$)

$$(\gamma+1)(p_1+p_2)-2p_0 \text{ is positive.}$$

Accordingly $\int dQ/\theta$ takes the same sign as

$$\int_{p_1}^{p_2} \frac{dp\,(p_1+p_2-p)}{p\,\{(\gamma+1)(p_1+p_2)-2p\}}. \quad \dots\dots\dots\dots(71)$$

The integral (71) is evaluated without difficulty. Dropping the factor $1/(\gamma+1)$, and writing $\varpi = p_2/p_1$, we get

$$\log \varpi + \gamma \log \frac{\gamma+1+(\gamma-1)\,\varpi}{\gamma-1+(\gamma+1)\,\varpi}\,. \dotfill (72)$$

It is evident that (72) changes sign when we substitute $1/\varpi$ for ϖ.

If we expand (72) in powers of $(\varpi-1)$ we find

$$(72) = \frac{(\gamma^2-1)\,(\varpi-1)^3}{12\gamma^2} + \cdots,$$

the terms in $(\varpi-1)$, $(\varpi-1)^2$, disappearing. Thus, when $\varpi-1$ is positive, (72) begins positive. Differentiating (72) with respect to ϖ, we get

$$\frac{1}{\varpi} + \frac{\gamma\,(\gamma-1)}{\gamma+1+(\gamma-1)\,\varpi} - \frac{\gamma\,(\gamma+1)}{\gamma-1+(\gamma+1)\,\varpi}\,. \dotfill (73)$$

When (73) is reduced to a single fraction, the denominator is positive, and the numerator is

$$(\gamma^2-1)\,(\varpi-1)^2.$$

We infer that if $\varpi>1$, (72) is always positive, and that if $\varpi<1$, (72) is always negative. Hence if $p_2 > p_1$, i.e. if the wave be one of condensation, the communications of heat required are such as may arise from conduction; but if the wave be one of rarefaction, its permanency can in no wise be attained as the result of conduction. A wave of condensation here means a wave such that during its progress the gas passes always from a less dense to a more dense state, and the most important case is when the limits are finite, so that the passage constitutes the transition from one uniform density to a greater uniform density.

Rankine proceeds to examine more particularly under what conditions a wave can be permanent. " In order that a particular type of disturbance may be capable of permanence during its propagation, a relation must exist between the temperatures of the particles and their relative positions, such that the conduction of heat between the particles may effect the transfers of heat required by the thermodynamic conditions of permanence of type."

The equation of conduction is readily found. The heat conducted in unit time across a layer of the gas is represented by $k\,d\theta/dx$, where k is a coefficient of conductivity which may be a function of the condition of the gas, here dependent on one variable. The equation of conduction is $(u = +)$

$$v\frac{d}{dx}\left(k\,\frac{d\theta}{dx}\right) = \frac{D}{Dt}\,Q = u\frac{dQ}{dx} = mv\frac{dQ}{dx},$$

whence, if we reckon Q from the initial condition of constant pressure p_1,

$$k\,\frac{d\theta}{dx} = mQ. \dotfill (74)$$

And from (70)

$$Q = \frac{\gamma+1}{2m^2(\gamma-1)} \int_{p_1}^{p} (p_1 + p_2 - 2p)\, dp = \frac{\gamma+1}{2m^2(\gamma-1)} (p-p_1)(p_2-p). \quad \ldots(75)$$

Also from (50), (65),

$$p + m^2v = \tfrac{1}{2}(\gamma+1)(p_1+p_2), \qquad\qquad\qquad\ldots\ldots(76)$$

whence
$$\theta = \frac{pv}{R} = \frac{p}{2m^2R}\{(\gamma+1)(p_1+p_2)-2p\}, \ldots\ldots\ldots\ldots(77)$$

and
$$\frac{d\theta}{dp} = \frac{(\gamma+1)(p_1+p_2)-4p}{2m^2R}. \qquad\ldots\ldots\ldots\ldots(78)$$

Using these, we find with regard to (62)

$$dx = \frac{k}{mQ}\frac{d\theta}{dp}\, dp = \frac{k\, dp}{mc(\gamma+1)}\frac{(\gamma+1)(p_1+p_2)-4p}{(p-p_1)(p_2-p)}, \qquad\ldots\ldots(79)$$

by which is determined the distribution of pressure (and thence of density and temperature) along the line of propagation.

On the supposition that k is constant, Rankine integrates (79) in terms of logarithms. Writing

$$p - \tfrac{1}{2}(p_1 + p_2) = q, \qquad \tfrac{1}{2}(p_2 - p_1) = q_1,$$

he obtains
$$\frac{dx}{dq} = \frac{k}{mc(\gamma+1)}\frac{(\gamma-1)(p_1+p_2)-q}{q_1^2-q^2},$$

and
$$x = \frac{k}{mc(\gamma+1)}\left\{\frac{(\gamma-1)(p_1+p_2)}{2q_1}\log\frac{q_1+q}{q_1-q} + 2\log\left(1-\frac{q^2}{q_1^2}\right)\right\}, \quad\ldots(80)$$

x being measured from the place where $q=0$. Mathematically the wave is infinitely long; but practically the transition of pressure is effected in a distance comparable with $k/mc(\gamma+1)$, which may be small in terms of ordinary standards. It is to be observed that the general character of the result does not depend upon the constancy of k.

Reverting to (79), we see that the denominator on the right is positive, and that the numerator is also positive for that part of the wave where p is nearly equal to $\tfrac{1}{2}(p_1+p_2)$. Thus, for this part of the wave at any rate, p and x increase together; or, since u is positive, the gas passes to a condition of greater density—the wave must be one of condensation. This consideration, as we have seen, Rankine overlooked. And a further limitation presents itself: since there cannot be two pressures in one place, it is evident that dx/dp must not change sign. The numerator in (79) must be positive over its *whole* range from p_1 to p_2, and this will not be the case if p_2/p_1 exceed $(\gamma+1)/(3-\gamma)$, equal for common gases to 1·61. The conclusion is that the only kind of wave, involving a transition from one uniform pressure to another, which can be maintained with the aid of conduction is a wave of condensation, and then only when the ratio of pressures does not exceed a moderate value.

The next contribution to the subject upon which I have to comment is contained in a long and ably written memoir by Hugoniot[*]. This author, though he covers to a great extent the same ground, makes no reference to Stokes, Earnshaw, Riemann, or Rankine, and but a very slight one to Poisson— a circumstance which increases the difficulty of comparison. Since Hugoniot uses the Lagrangian form of equation, his investigation runs naturally on the same lines as Earnshaw's, whose general solution for a single progressive wave is reproduced. I have already alluded to the solution of special problems relating to the propagation of a wave of variable type.

The most original part of Hugoniot's work has been supposed to be his treatment of discontinuous waves involving a sudden change of pressure, with respect to which he formulated a law often called after his name by French writers. But a little examination reveals that this law is *precisely the same* as that given 15 years earlier by Rankine, a fact which is the more surprising inasmuch as the two authors start from quite different points of view. Rankine's investigation, as we have seen, is expressly based upon conduction of heat in the gas, but Hugoniot supposes his gas to be non-conducting. A question of some delicacy is here involved, which will repay careful examination. It will be convenient to give a paraphrase of Hugoniot's argument[†].

This argument depends upon an application of the principle of energy to a region bounded by two fixed planes, including the place of discontinuity. The work done by the fluid as it emerges with volume v_2 against the pressure p_2 is $p_2 v_2$. On the whole, therefore, the external work done by the passage of the unit of mass is $p_2 v_2 - p_1 v_1$. The increase of kinetic energy of the fluid is

$$\tfrac{1}{2}(u_2{}^2 - u_1{}^2) = \tfrac{1}{2} m^2 (v_2{}^2 - v_1{}^2) = \tfrac{1}{2}(v_2 + v_1)(p_1 - p_2),$$

in virtue of (50), which requires that

$$p_1 - p_2 + m^2 (v_1 - v_2) = 0. \quad\ldots\ldots\ldots\ldots\ldots\ldots(81)$$

The sum of these is

$$p_2 v_2 - p_1 v_1 + \tfrac{1}{2}(p_1 - p_2)(v_2 + v_1) = \tfrac{1}{2}(v_2 - v_1)(p_2 + p_1). \quad\ldots\ldots\ldots(82)$$

We have next to consider the internal energy of unit of mass in the initial and final states. For this purpose we suppose the gas to expand adiabatically from its actual volume v to an infinite volume. In this expansion the work done by the gas is

$$\int_v^\infty p\,dv = \frac{pv}{\gamma - 1}, \quad\ldots\ldots\ldots\ldots\ldots\ldots\ldots(83)$$

* *Journal de l'École Polytechnique*, 1887, 1889.
† Compare Lamb's *Hydrodynamics*, 1906, § 280.

so that the difference of internal energy in the two states is

$$\frac{p_2 v_2}{\gamma - 1} - \frac{p_1 v_1}{\gamma - 1}. \quad \dots\dots\dots\dots\dots\dots\dots(84)$$

The principle of energy requires that the sum of this and (82) be zero, whence

$$\gamma = \frac{(p_2 - p_1)(v_1 + v_2)}{(p_1 + p_2)(v_1 - v_2)} \quad \dots\dots\dots\dots\dots\dots(85)$$

is the relation between the pressures and volumes in the two states. The result thus found by Hugoniot is the same as Rankine's. From Rankine's equation (65)

$$p_1 + m^2 v_1 = p_2 + m^2 v_2 = \tfrac{1}{2}(\gamma + 1)(p_1 + p_2) = \tfrac{1}{2}(p_1 + p_2) + \tfrac{1}{2}m^2(v_1 + v_2),$$

it follows that

$$m^2 = \gamma \frac{p_1 + p_2}{v_1 + v_2} = \frac{p_2 - p_1}{v_1 - v_2}, \quad \dots\dots\dots\dots\dots\dots(86)$$

which is identical with (85).

The first remark that I will make is that, although Hugoniot assumes that the transition between the two states is sudden, there is nothing in his argument which requires this, all that is really necessary being that the *régime* is permanent. The next remark is that, however valid (85) may be, its fulfilment does not secure that the wave so defined is possible. As a matter of fact, a whole class of such waves is certainly impossible, and I would maintain, further, that a wave of the kind is never possible under the conditions, laid down by Hugoniot, of no viscosity or heat-conduction.

A closer examination of the process by which (85) was obtained will show that while the first law of thermodynamics has been observed, the second law has been disregarded. The crux of the matter lies in the comparison of the internal energies of the incoming and outgoing gas expressed in (84). If (p_2, v_2) and (p_1, v_1) lie upon the same adiabatic, the work corresponding to the passage from the one state to the other is given without ambiguity by (84). But in the present case the two states do not lie upon the same adiabatic, and the work required is deduced upon the assumption that nothing is involved in the passage at $v = \infty$ from one adiabatic to the other. What is actually there required is the communication (positive or negative) of an infinitesimal quantity of heat. From the point of view of the first law the infinitesimal quantity of heat may be neglected, but not so from the point of view of the second law, since the transfer is supposed to take place at the zero of temperature. When heat and work are distinguished, infinitesimal heat at zero may have a finite value. The imaginary passage to infinity has the advantage of leading rapidly to the required conclusion, but it rather tends to obscure the real nature of the process. While all the other items of the account are mechanical work, the passage from one adiabatic to the other (which may take place at constant finite volume) is a question of *heat* as distinguished from work. If during a complete cycle work would be lost and corresponding heat gained, the operation is dissipative and there need be no contradiction if viscosity or heat-conduction enter, but the opposite contingency of a gain of work at the expense of heat is excluded in all cases. The conclusion is the same as before. While a wave of condensation may, perhaps, maintain a permanent regime as the result of dissipative agencies, a permanent wave of rarefaction is excluded.

It is remarked by Hugoniot that even when the ratio p_1/p_2 is infinite, v_2/v_1 does not exceed $(\gamma+1)/(\gamma-1)$, which for common gases is equal to about 6. A similar remark is made by Duhem[*], who discusses the whole question with great generality. With regard to perfect gases "lorsqu'une quasi-onde de choc se propage au sein d'un gas parfait, le fluide le plus condensé est toujours en amont de l'onde et le fluide le moins condensé en aval." But, so far as I see, neither of these authors proves that the propagation is possible in any case.

It is a question of great interest to inquire what is the influence of viscosity and especially whether alone, or in co-operation with heat-conduction, it allows a wave of condensation to acquire a permanent regime. We proceed to consider this question on the basis of the usual equations, although it must be admitted that their application to conditions which are somewhat extreme raises points of uncertainty.

Reverting to our original equations, we recognise that (45) is unaffected by the inclusion of viscosity, and that the change required in (46) is represented by writing[†]

$$X = \frac{4}{3\rho} \frac{d}{dx} \left(\mu \frac{du}{dx} \right),$$

so that (46) takes the form

$$m \frac{du}{dx} + \frac{dp}{dx} - \frac{4}{3} \frac{d}{dx} \left(\mu \frac{du}{dx} \right) = 0,$$

whence ($v = 1/\rho$)

$$p + m^2 v - \frac{4}{3} m\mu \frac{dv}{dx} = p_1 + m^2 v_1 = p_2 + m^2 v_2, \quad \dots\dots\dots(87)$$

the terminal states (p_1, v_1), (p_2, v_2), being of uniformity, so that dv/dx there vanishes. From this it appears that (81), relating to the terminal states, holds good equally when viscosity is regarded.

---

[*] *Zeitschrift f. Physikal. Chem.* Vol. LXIX. p. 169 (1909).
[†] Lamb's *Hydrodynamics*, §§ 314, 316.

A simple example under the head of viscosity is to suppose the temperature maintained uniform, as by a powerful radiation, so that the gas follows Boyle's law, making

$$pv = p_1 v_1 = p_2 v_2 = a^2.$$

From this and (87) we get

$$m^2 v_1 v_2 = a^2,$$

and

$$p_1 + m^2 v_1 = a^2/v_1 + a^2/v_2.$$

Using these in (87), we find

$$\frac{3m\,dx}{4\mu} = -\frac{v\,dv}{(v_1 - v)(v - v_2)}, \quad\dots\dots\dots\dots\dots(88)$$

as governing the distribution of v along the line of propagation. In a wave of condensation $v_1 > v > v_2$, so that the denominator on the right of (88) is positive. Thus when m is positive, dv/dx is negative, as should be the case. On integration (μ constant)

$$\frac{3mx}{4\mu} = \frac{1}{v_1 - v_2}\{v_1 \log (v_1 - v) - v_2 \log (v - v_2)\}, \quad\dots\dots\dots(89)$$

the origin of x being chosen suitably.

The transition of volumes from v_1 to v_2 occupies, mathematically speaking, the whole range from $x = -\infty$ to $x = +\infty$, but practically it may be very sudden. Since in (88) dx/dv never changes sign, the condition of permanency for a condensational wave can always be satisfied, whatever may be the value of the ratio v_1/v_2* or p_1/p_2, contrasting in this respect with the limitation found to be necessary on Rankine's conclusion relative to heat-conduction.

As regards the velocity of wave propagation into the rarer medium, we have for its square

$$u_1^2 = m^2 v_1^2 = a^2 v_1/v_2. \quad\dots\dots\dots\dots\dots\dots(90)$$

Returning to the case where heat development and viscosity are both regarded, we see that in virtue of (81) Hugoniot's reasoning is still applicable without change, and it leads to the same final relation (85) as was found by Rankine when heat-conduction is alone considered.

In endeavouring to apply Rankine's method to the more general case where viscosity is retained, we shall find it more convenient to treat v, or $(1/\rho)$, rather than p, as independent variable. If, as before, dQ denotes the total quantity of heat received by unit mass of the gas, we have from (60), (62),

$$(\gamma - 1)\frac{dQ}{dx} = \gamma p \frac{dv}{dx} + v \frac{dp}{dx};$$

---

* But the limitation pointed out by Hugoniot still obtains; otherwise, one of the pressures would be negative.

or, on elimination of p by means of (87),

$$(\gamma-1)\frac{dQ}{dx}=\gamma\left\{p_1+m^2v_1-m^2v+\tfrac{4}{3}m\mu\frac{dv}{dx}\right\}\frac{dv}{dx}+v\left\{-m^2\frac{dv}{dx}+\tfrac{4}{3}m\frac{d}{dx}\left(\mu\frac{dv}{dx}\right)\right\}.$$
$$\text{...(91)}$$

In (91) dQ consists of two parts, the first (dQ_1), with which alone Rankine dealt, the heat received by conduction, and the second (dQ_2) the heat developed internally under viscosity. As regards the latter, the heat developed in volume v and time dt is $\tfrac{4}{3}v\mu\,(du/dx)^2\,dt^*$, in which we are to replace dt by dx/u, and u by mv, so that

$$\frac{dQ_2}{dx}=\tfrac{4}{3}m\mu\left(\frac{dv}{dx}\right)^2.\qquad\qquad\text{...(92)}$$

Multiplying this by $(\gamma-1)$ and subtracting it from (91), we get

$$(\gamma-1)\frac{dQ_1}{dx}=\gamma\,(p_1+m^2v_1)\frac{dv}{dx}-(\gamma+1)\,m^2v\frac{dv}{dx}+\tfrac{4}{3}m\frac{d}{dx}\left(\mu v\frac{dv}{dx}\right).\text{ ...(93)}$$

As in Rankine's investigation, the whole heat received by conduction in passing from one uniform state v_1 to another uniform state v_2 must vanish. Hence, on integrating between these limits, and dividing out the factor (v_2-v_1), we have

$$(\gamma+1)\,m^2\,(v_1+v_2)=2\gamma\,(p_1+m^2v_1)=2\gamma\,(p_2+m^2v_2)$$
$$=\gamma\,(p_1+p_2)+\gamma m^2\,(v_1+v_2),$$

or, as in (86), $\qquad\qquad m^2\,(v_1+v_2)=\gamma\,(p_1+p_2),$

the same relation as was found by Rankine. Introducing it into (93), we get

$$(\gamma-1)\frac{dQ_1}{dx}=\tfrac{1}{2}\,(\gamma+1)\,m^2\left\{(v_1+v_2)\frac{dv}{dx}-\frac{dv^2}{dx}\right\}+\tfrac{4}{3}m\frac{d}{dx}\left(\mu v\frac{dv}{dx}\right).\text{ ...(94)}$$

A particular case arises when we suppose the conductivity to be zero, so that dQ_1/dx vanishes throughout. We have then

$$\tfrac{4}{3}\mu v\frac{dv}{dx}+\tfrac{1}{2}\,(\gamma+1)\,m\,(v_1-v)\,(v-v_2)=0,\qquad\text{...(95)}$$

differing from (88) only by the factor $\tfrac{1}{2}\,(\gamma+1)$. On the supposition that μ is constant, the solution is nearly the same as in (89), and, in fact, reduces to it when $\gamma=1$, which represents Boyle's law. This case of no conduction is thus satisfactorily disposed of. Whatever be the ratio of pressures, a wave of condensation is always possible.

It should be remarked, however, that the supposition of constant μ does not consist with the facts as known for actual gases when γ differs from

---

* Lamb's *Hydrodynamics*, § 341.

unity. For such gases viscosity, though independent of *density*, varies with *temperature*, so that μ will not be constant in (95). But since μ is always positive, this complication merely affects the particular form of the integral and not the general conclusion as to the possibility of a permanent wave.

In general, from (94), if we reckon Q_1 from the terminal state v_1,

$$(\gamma - 1) Q_1 = \tfrac{1}{2} (\gamma + 1) m^2 (v_1 - v)(v - v_2) + \tfrac{4}{3} m \mu v \frac{dv}{dx} \quad \ldots\ldots\ldots (96)$$

The equation of conduction is the same (74) as before. And for θ, from (87),

$$\theta = \frac{pv}{R} = \frac{v}{R} \left\{ \frac{\gamma + 1}{2\gamma} m^2 (v_1 + v_2) - m^2 v + \tfrac{4}{3} m \mu \frac{dv}{dx} \right\};$$

so that with regard to (62) the equation of conduction becomes

$$\frac{k}{mc} \left[m \frac{dv}{dx} \left\{ \frac{\gamma + 1}{2\gamma} (v_1 + v_2) - 2v \right\} + \tfrac{4}{3} \frac{d}{dx} \left(\mu v \frac{dv}{dx} \right) \right]$$
$$= \tfrac{1}{2} (\gamma + 1) m (v_1 - v)(v - v_2) + \tfrac{4}{3} \mu v \frac{dv}{dx}. \quad \ldots (97)$$

By omitting the terms containing μ we may of course fall back on Rankine's problem.

Equation (97), in its general form, is much more complicated than when either viscosity or heat-conduction is alone regarded, in consequence of the occurrence of the differential coefficient of the second order. In general, both k and μ are functions of temperature, and therefore of v; but, according to Maxwell's theory, which assumes a molecular repulsion inversely as the fifth power of the distance, $c\mu/k$ is independent of temperature (as well as of density), and takes the value $\tfrac{2}{3}$. And it would seem that this independence of temperature and density is general, seeing that the ratio is of no dimensions, at least so long as the repulsive force can be represented by an inverse power of the distance*. We shall write h for the above ratio and assume that for a given gas it is an absolute constant. Thus, μ' being written for μ/m, (97) takes the form—

$$\mu' \frac{d}{dx} \left(\mu' \frac{dv^2}{dx} \right) + \mu' \frac{dv^2}{dx} \left\{ \frac{3(\gamma + 1)}{8\gamma} \frac{v_1 + v_2}{v} - \tfrac{3}{2} - h \right\} = \tfrac{3}{4} h (\gamma + 1)(v_1 - v)(v - v_2);$$
$$\ldots (98)$$

in which v^2 may be regarded as the dependent variable. For v^2 we shall write ξ, and if

$$U = \mu' d\xi/dx, \ldots\ldots\ldots\ldots\ldots\ldots\ldots (99)$$

---

* Compare *Roy. Soc. Proc.* Vol. LXVI. p. 68 (1900); *Scientific Papers*, Vol. IV. p. 453.

our equation, since it contains x only through dx, may be reduced to one of the first order in U and ξ, i.e.,

$$U \frac{dU}{d\xi} + U f(\xi) = F(\xi), \quad\ldots\ldots\ldots\ldots\ldots\ldots\ldots(100)$$

where

$$f(\xi) = \frac{3(\gamma+1)}{8\gamma} \frac{\sqrt{\xi_1}+\sqrt{\xi_2}}{\sqrt{\xi}} - \tfrac{3}{2} - h, \quad\ldots\ldots\ldots\ldots(101)$$

and

$$F(\xi) = \tfrac{3}{4} h (\gamma+1)(\sqrt{\xi_1}-\sqrt{\xi})(\sqrt{\xi}-\sqrt{\xi_2}). \quad\ldots\ldots\ldots(102)$$

If U can be found as a function of ξ from (100), x follows by simple integration of (99).

In considering equation (100) we may conveniently regard ξ as the linear co-ordinate of a material particle of unit mass moving in a straight line with velocity U. The first term, $U dU/d\xi$, then represents the acceleration of the particle; the second, $U f(\xi)$, may be regarded as a *resistance*, proportional to the velocity, and at the same time variable with the position (ξ); and the third term on the right hand represents a force, which is also a function of position. If t be the time in this subsidiary problem, $U = d\xi/dt$, and (100) may be written

$$\frac{d^2\xi}{dt^2} + f(\xi)\frac{d\xi}{dt} = F(\xi), \quad\ldots\ldots\ldots\ldots\ldots\ldots(103)$$

while by (99)

$$dx = \mu' dt. \quad\ldots\ldots\ldots\ldots\ldots\ldots\ldots(104)$$

If μ' be constant, the substitution of $\mu't$ for x in (98) is obvious.

If we take $h = 0\cdot4$, $\gamma = 1\cdot41$, (101), (102), become

$$f(\xi) = 0\cdot641 \frac{\sqrt{\xi_1}+\sqrt{\xi_2}}{\sqrt{\xi}} - 1\cdot900, \quad\ldots\ldots\ldots\ldots(105)$$

$$F(\xi) = 0\cdot723 (\sqrt{\xi_1}-\sqrt{\xi})(\sqrt{\xi}-\sqrt{\xi_2}). \quad\ldots\ldots\ldots\ldots(106)$$

It will be observed that, over the range from ξ_1 to ξ_2, $F(\xi)$ is positive, but that the sign of $f(\xi)$ is doubtful. If ξ has the greater terminal value ξ_1, f is negative; but, when it has the smaller terminal value ξ_2, the sign depends upon the ratio ξ_1/ξ_2. If this ratio $< 1\cdot21$, f is negative; otherwise it is positive.

I suppose that a complete analytical solution of our equation is not to be expected, and it is, indeed, hardly necessary for our purpose. What we most wish to know is whether a solution is possible which satisfies the prescribed conditions. Among these is the requirement that U in (100) vanish at both limits; and even then the manner of evanescence must be such as to secure that x, as determined by (99), shall be infinite at these limits. As the problem originally presents itself, we should have the representative particle travelling in the negative direction from ξ_1 to ξ_2, starting with no velocity and arriving with no velocity. It seems simpler to consider it in a modified form, *i.e.* with the motion reversed, so that it takes place in the direction of the force F. There is, then, no question of the particle stopping between the limits and returning upon its course. We may make this change, if in (105) we reverse the sign of f. We consider, then, the motion of the particle to be in the positive direction, from ξ_2 to ξ_1, with zero velocity at both limits, the motion between ξ_2 and ξ_1 being aided by the force F, which itself vanishes at these limits, and being also subject to a force of the nature of resistance, proportional to velocity. When ξ_1/ξ_2 does not exceed 1·21, the force is a resistance in the ordinary sense, *i.e.* it opposes the motion, and, in any case, it has this character near (and beyond) the arrival end ξ_1. But when ξ_1/ξ_2 exceeds 1·21, the force becomes what we may call a counter-resistance, and aids the motion near the initial end ξ_2. As regards F, in the neighbourhood of each limit it becomes a force proportional to distance therefrom, repulsive near ξ_2, and attractive near ξ_1. Thus, when ξ is nearly equal to ξ_2,

$$F(\xi) = 0\cdot723 \frac{\sqrt{\xi_1} - \sqrt{\xi_2}}{2\sqrt{\xi_2}} (\xi - \xi_2); \quad \ldots\ldots\ldots\ldots\ldots(107)$$

and when ξ is nearly equal to ξ_1,

$$F(\xi) = 0\cdot723 \frac{\sqrt{\xi_1} - \sqrt{\xi_2}}{2\sqrt{\xi_1}} (\xi_1 - \xi). \quad \ldots\ldots\ldots\ldots\ldots(108)$$

The particle, starting from ξ_2, is bound to go through to ξ_1. If it arrives at ξ_1 with zero velocity, we shall have, presumably, a solution of our problem. It is possible, however, that on first arrival at ξ_1, it may pass through, and only settle down after a number of oscillations. To this there does not appear to be any objection; but if on the return from ξ_1 it overshoots ξ_2, it can never again return to ξ_1, since on the left of ξ_2 the sign of f is negative; and then our problem has no solution. On the other hand, from the nature of F, it is not possible for the particle passing through ξ_1 in the positive direction to escape returning.

The character of the start from ξ_2 can be investigated with the aid of approximate equations. Thus, in (100) we may treat $f(\xi)$ as constant, say 2α, where α may be either positive or negative, and take, as in (107), $F(\xi) = \beta(\xi - \xi_2)$, where β is positive, so that

$$U\frac{dU}{d\xi} + 2\alpha U - \beta(\xi - \xi_2) = 0. \quad \ldots\ldots\ldots\ldots\ldots(109)$$

If in (109) we assume

$$U = \lambda(\xi - \xi_2), \ldots\ldots\ldots\ldots\ldots\ldots\ldots\ldots(110)$$

we find that the equation is satisfied provided that

$$\lambda = -\alpha \pm \sqrt{(\alpha^2 + \beta)}, \quad \ldots\ldots\ldots\ldots\ldots\ldots(111)$$

one value (λ_1) being positive and one (λ_2) negative. In the present case, where U must be positive when $\xi > \xi_2$, λ_1 is to be chosen.

The differential equation (109) can be made homogeneous, and its general solution*, when λ is real, can be put into the form

$$\frac{\{U - \lambda_1(\xi - \xi_2)\}^{\lambda_1}}{\{U - \lambda_2(\xi - \xi_2)\}^{\lambda_2}} = C, \quad\quad\quad\quad\quad (112)$$

where C is an arbitrary constant. This solution, of course, covers the cases where the particle starts from ξ_2 with a finite velocity (U_0), and it appears that $C = U_0^{\lambda_1 - \lambda_2}$. We might conclude from this that when $U_0 = 0$, then $C = 0$, but the conclusion is not safe. If, however, U and $\xi - \xi_2$ are of the same order of magnitude, we may write $U = r(\xi - \xi_2)$, where r is not infinite. Substituting this in (112), we get

$$(r - \lambda_1)^{\lambda_1} \cdot (r - \lambda_2)^{-\lambda_2} \cdot (\xi - \xi_2)^{\lambda_1 - \lambda_2} = C. \quad\quad\quad (113)$$

When $\xi - \xi_2$ vanishes, the third factor on the left is zero, and, since the first and second factors are not infinite, the conclusion follows that $C = 0$. This takes us back to (110), (111), the second solution (involving λ_2) relating to the case where U is negative, the motion being one of *approach* from the positive side to ξ_2.

These conclusions may be arrived at more easily from (103), of which the general solution in the present case is

$$\xi - \xi_2 = A e^{\lambda_1 t} + B e^{\lambda_2 t}, \quad\quad\quad\quad\quad (114)$$

giving

$$U = \lambda_1 A e^{\lambda_1 t} + \lambda_2 B e^{\lambda_2 t}. \quad\quad\quad\quad\quad (115)$$

From these we may deduce

$$U - \lambda_1(\xi - \xi_2) = (\lambda_2 - \lambda_1) B e^{\lambda_2 t},$$

$$U - \lambda_2(\xi - \xi_2) = (\lambda_1 - \lambda_2) A e^{\lambda_1 t};$$

whence (112) follows by elimination of t. For our present purpose, $\xi - \xi_2$ is to vanish when $t = -\infty$, so that $B = 0$; and (110) follows with $\lambda = \lambda_1$. It will be remarked that (110) makes x, as determined by (99), infinite when $\xi = \xi_2$. The circumstances of the start from ξ_2 are thus definite and suitable. The question is as to the arrival at ξ_1.

In the neighbourhood of ξ_1 the approximate equation is

$$U \frac{dU}{d\xi} + 2\alpha' U - \beta'(\xi - \xi_1) = 0, \quad\quad\quad\quad (116)$$

where α' is positive and, by (108), β' negative. If now

$$U = \lambda'(\xi - \xi_1), \quad\quad\quad\quad\quad (117)$$

the values of λ' are

$$\lambda' = -\alpha' \pm \sqrt{(\alpha'^2 + \beta')}; \quad\quad\quad\quad\quad (118)$$

---

* See, for example, Boole's *Differential Equations*, p. 33.

so that both values, if real, are negative. On the supposition of reality, (112) retains its form (with ξ_1 for ξ_2). If the velocity of arrival (U_0) be finite, $C = U_0{}^{\lambda'_1 - \lambda'_2}$, as before; and it might be supposed that, if U_0 vanishes when $\xi - \xi_1 = 0$, C would have to vanish or become infinite. Such a conclusion would be incorrect. If in (113) we suppose λ'_1 to be numerically the smaller of the two values, the third factor indeed vanishes with ($\xi - \xi_1$) as before; but the conclusion that $C = 0$ is evaded if ultimately $r = \lambda'_1$.

The situation is most easily understood from the solution in terms of t as in (114), (115). Since λ'_1, λ'_2 are *both* negative, the condition that $\xi - \xi_1$ and U shall vanish together when $t = \infty$ is satisfied, whatever may be the values of A and B. There are now an infinite number of possible types of solution, instead of only one as in the former case. And it appears that the two simple types included under (117) are not at all upon an equal footing. Except in the *particular* case where $A = 0$, the solution always tends ultimately to the form $U = \lambda_1 (\xi - \xi_1)$, and of course it may assume this form throughout. All these solutions satisfy the condition as to the infinitude of x when $U = 0$.

Whether the values of λ' be real or not, the particle must ultimately settle down at ξ_1, unless it escape from the region to which the approximate equation applies. For in (118) the *real part* of λ' is always negative.

Returning to (100) in its general form, let us consider the variation of U for a given ξ as dependent upon variations in f and F. We have

$$U \frac{d\delta U}{d\xi} + \delta U \frac{dU}{d\xi} + f . \delta U + U . \delta f - \delta F = 0,$$

or
$$\frac{d\delta U}{d\xi} + P \delta U - Q = 0, \quad \dots\dots\dots\dots\dots(119)$$

where P and Q are supposed to be known functions of ξ, viz.,

$$P = \frac{f + dU/d\xi}{U}, \qquad Q = \frac{\delta F - U . \delta f}{U}. \quad \dots\dots\dots(120)$$

The solution of the linear equation (119) is

$$\delta U = e^{-\int P d\xi} \left(\int e^{\int P d\xi} Q d\xi + c \right). \quad \dots\dots\dots\dots(121)$$

If $Q = 0$, δU has the same sign as c, so that an increment of velocity communicated at any point remains throughout of the same sign. Again, if Q be throughout of one sign, δU, as dependent upon it, has the same sign. For example, if U be positive over the range considered, δf positive, and δF negative, then δU is certainly negative. The increments δf, δF may be local, vanishing over any part of the range.

The application to the present problem is obvious. If the particle passing any point between ξ_2 and ξ_1, with velocity U, arrives at ξ_1 for the first time without velocity, it will still arrive at ξ_1 without velocity (it must in any case arrive, since F is positive), if U be diminished, or if f be increased, or if F be diminished, or if all these changes occur together. And in the limit, when ξ_1 is closely approached, the ratio of U to $(\xi_1 - \xi)$ is in general the same.

By use of this principle we may assure ourselves as to the possibility of a solution in certain cases where ξ_1/ξ_2 does not greatly exceed unity. We imagine a simplified problem which admits of analytical solution and is derived from the actual one by alterations which everywhere (over the range from ξ_2 to ξ_1) increase F and diminish f. If this modified problem admits of the solution required, a fortiori will the original problem do so.

If we consider the curve which according to (106) represents F as a function of ξ, we see that it is concave downwards, and that F will everywhere be increased if we substitute for the curve the two terminal tangents at ξ_2 and ξ_1, whose equations are given in (107), (108). The abscissa of K, the point of intersection, is $\xi = \sqrt{(\xi_2 \xi_1)}$. (Fig. 1.)

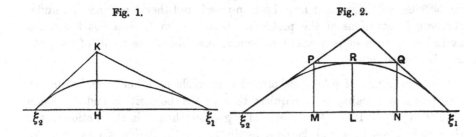

Fig. 1.   Fig. 2.

As regards f, its value is given by (105) with sign reversed, and is diminished when ξ is diminished. The changes will therefore be in the required direction if we represent f over $\xi_2 H$ by its value at ξ_2, viz.,

$$f_2 = 2\alpha = 1\cdot259 - 0\cdot641s, \dots\dots\dots\dots\dots(122)$$

and from H to ξ_1 by its value at H, viz.,

$$f_1 = 2\alpha' = 1\cdot900 - 0\cdot641 (s^{\frac{1}{2}} + s^{-\frac{1}{2}}), \dots\dots\dots(123)$$

if for brevity we write s for $\sqrt{(\xi_1/\xi_2)}$, so that s is the ratio of terminal densities in the original problem.

As regards the first portion of the course, the solution already given (110), (111), determines the value of U on arrival at H. We have

$$U = \lambda_1 \xi_2 (s - 1), \quad \text{where} \quad \lambda_1 = \sqrt{(\alpha^2 + \beta)} - \alpha,$$

in which α is given by (122) while by (107) $\beta = 0\cdot361 (s - 1)$. Using these, we get

$$U = (s - 1) \xi_2 [\sqrt{\{(0\cdot630 - 0\cdot320s)^2 + 0\cdot361 (s - 1)\}} - 0\cdot630 + 0\cdot320s]. \dots(124)$$

If $s = 1 + \sigma$, where σ is small, (124) becomes

$$U = \sigma \xi_2 \times \frac{\beta}{2\alpha} = 0.584 \xi_2 \sigma^2. \quad \therefore \dots \dots \dots (125)$$

As regards the second portion of the course, the appropriate solution is provided by (117), (118), where λ' is restricted to be real. And in accordance with our suppositions α' is given by (123) while from (108) $\beta' = -0.361(1 - s^{-1})$. In choosing between the values of λ' we are at liberty to take that which gives the largest value of U at H consistent with $U = 0$ at ξ_1, viz.,

$$\lambda'_2 = -\alpha' - \surd(\alpha'^2 + \beta').$$

Thus at H $\qquad U = (s^2 - s) \xi_2 \{\alpha' + \surd(\alpha'^2 + \beta')\}, \dots \dots \dots (126)$

or approximately, in terms of σ (supposed small),

$$U = \sigma \xi_2 (0.618 + 0.033\sigma). \quad \dots \dots \dots (127)$$

From (125), (127) we may infer that when σ is small, *i.e.* when s does not much exceed unity, the particle, starting from ξ_2, arrives at H with a velocity small enough to admit of its being stopped on arrival at ξ_1, even under the simplifying conditions that have been imposed, and therefore *a fortiori* under the actual conditions of the problem. Hence when ξ_1 does not too much exceed ξ_2, a wave of permanent regime *is* possible as the result of viscosity and heat-conduction.

But the range of ξ_1/ξ_2 thus proved admissible is rather severely limited. The postulated reality of λ' requires that $\alpha'^2 + \beta'$ be positive, and this again requires that $s < 1.34$. For values of s greater than this the motion in the simplified problem would become oscillatory. Calculation shows that, for $s = 1.34$, (124) gives

$$U = (s - 1) \xi_2 \times 0.20,$$

while (126) gives $\qquad U = (s - 1) \xi_2 \times 0.43 ;$

so that up to this limit the particle starting from ξ_2 in the simplified problem, and therefore also in the actual problem, would arrive at ξ_1 with zero velocity. Up to a ratio of densities equal to 1.34, the wave of permanent regime is certainly possible.

The next step in this line of procedure will be to replace the curve representing F by the broken line $\xi_2 PQ\xi_1$ formed by *three* of its tangents, of which two are the same terminal tangents as before, while the third, PQ, may be taken to be the horizontal tangent parallel to $\xi_1 \xi_2$. (Fig. 2.) By (106) the point L where F is a maximum is determined by

$$\surd \xi = \tfrac{1}{2} (\surd \xi_1 + \surd \xi_2) = \tfrac{1}{2} \surd \xi_2 (s + 1),$$

and the corresponding value of F is

$$0.1875 \xi_2 (s - 1)^2.$$

The abscissae of the points of intersection of the horizontal tangent with the terminal tangents are given by (107), (108). For M

$$\xi = \tfrac{1}{2}\xi_2(s+1),$$

and for N

$$\xi = \tfrac{1}{2}\xi_2(s^2+s).$$

As for f we are to take along $\xi_2 M$ the value at ξ_2; along MN the value at M; and along $N\xi_1$ the value at N, as given by (105), with sign changed.

For the two terminal portions the solutions are of the same form as before, but for the middle portion a new form is required. Making F constant in (100) and writing $2\alpha''$ for f, we find on integration

$$- 4\alpha''^2 \xi = F \log (F - 2\alpha'' U) + 2\alpha'' U + C, \dots\dots\dots\dots(128)$$

where C is an arbitrary constant, to be determined so as to suit the velocity with which the particle arrives at M.

As before, the limiting value of s is determined by the consideration that, for our purpose, the arrival at ξ_1 must not be oscillatory. We get $s = 1\cdot633$ about, and with this value of s it is not difficult to show that the velocity of arrival at N is below the value prescribed by the solution for $N\xi_1$. The necessary conditions are thus fulfilled and we infer that the wave of uniform regime is possible so long as $s < 1\cdot63$. Although this ratio is moderate, it exceeds that found admissible in Rankine's problem, which leaves viscosity out of account. We there found that the greatest admissible ratio of *pressures* was $1\cdot61$, which by (68) corresponds to $1\cdot40$ for the ratio of *densities*. Of course, in Rankine's problem the solution definitely fails at this point, while in the present problem all that we have so far proved is that the limit exceeds $1\cdot63$.

From the low limiting values of s found necessary in these two cases in order to secure the reality of the roots of (118), it might be inferred that the reality would fail for the limiting motion in the neighbourhood of ξ_1, when s had a considerable value, but such is not the case. If we use the value of f from (105) appropriate to the terminal point ξ_1 itself, we have

$$2\alpha' = 1\cdot900 - 0\cdot641 (1 + s^{-1}) = 1\cdot259 - 0\cdot641 s^{-1};$$

and

$$\beta' = - 0\cdot3615 (1 - s^{-1}).$$

Thus, even when $s = \infty$,

$$\alpha'^2 + \beta' = 0\cdot3963 - 0\cdot3615 = + 0\cdot0048;$$

and the roots are always real. Hence, subsidence at ξ_1 is not ultimately oscillatory, but there is nothing in this argument to exclude a finite number of oscillations before subsidence into the region governed by the approximate equation.

The method of approximation already followed might be pushed further, but it seems preferable to use the general method of numerical calculation for the solution of differential equations as formulated by Runge*. The equation is that numbered (100), in which f (whose sign is to be reversed) and F are given by (105), (106). If we write $U' = U/\xi_2$, $\xi' = \xi/\xi_2$, our equation takes the form

$$\frac{dU'}{d\xi'} = \frac{0.723}{U'}(s - \sqrt{\xi'})(\sqrt{\xi'} - 1) - 1.900 + \frac{0.641(1+s)}{\sqrt{\xi'}} \quad \ldots\ldots(129)$$

The value of s being given, it is required to trace the connection between U' and ξ', simultaneous values being denoted respectively by a and b. If a receive the increment h, we have to calculate the corresponding increment k for b. If we call the function on the right of (129) $\phi(\xi', U')$, Runge gives as a first approximation to k

$$k_1 = \phi(a + \tfrac{1}{2}h, \ b + \tfrac{1}{2}\phi_0 . h)h, \quad \ldots\ldots\ldots\ldots\ldots(130)$$

where $\phi_0 = \phi(a, b)$. The next approximation is

$$k = k_1 + \tfrac{1}{3}(k_2 - k_1), \quad \ldots\ldots\ldots\ldots\ldots\ldots(131)$$

where $\quad k_2 = \tfrac{1}{2}(k' + k''')$ and $k' = \phi_0 . h$,

$$k'' = \phi(a + h, \ b + k')h, \quad k''' = \phi(a + h, \ b + k'')h. \ \ldots\ldots\ldots(132)$$

Having determined the new simultaneous values $a + h, b + k$, we make a fresh departure therefrom, and so trace out the function step by step.

In the present application the starting point is $a = 1$ (i.e. $\xi = \xi_2$), $b = 0$, and we have to trace the function until $\xi' = s^2$. The initial value of ϕ is to be found from (129) by putting $\xi' = 1$. Writing $U' = \phi_0(\xi' - 1)$, we get

$$\phi_0^2 + \{1.900 - 0.641(1+s)\}\phi_0 - \tfrac{1}{2} \times 0.723(s - 1) = 0, \ \ldots\ldots(133)$$

of which the positive root is to be chosen.

The extreme admissible value of s in our present problem is 6, and the first and rather elaborate calculation that I have made relates to this case. From (133) $\phi_0 = 3.1591$, so that taking $a = 1, b = 0, h = 1$, we have $k' = 3.1591$. Calculating from (130) we find $k_1 = 2.255$, and from (132)

$$k'' = 1.7076, \qquad k''' = 2.0772,$$

making $k_2 = 2.6182$. Hence the correction to k_1, viz., $\tfrac{1}{3}(k_2 - k_1)$, is equal to 0.121, and $k = 2.376$. Thus, corresponding to $\xi' = 2$, we get $U' = 2.376$. The following are the values of U' obtained successively in this way:—

ξ'	U'	ξ'	U'	ξ'	U'
1	0·000	14	7·502	33	1·2942
2	2·376	18	6·807	34	0·8532
3	3·903	22	5·674	35	0·4160
4	4·986	26	4·241	$35\tfrac{1}{2}$	0·2043
6	6·380	30	2·610	$35\tfrac{3}{4}$	0·1012
10	7·526	32	1·736	36	0·0004

---

* See Forsyth's *Differential Equations*, p. 51.

The correction to k_1 is everywhere subordinate. In the last step from $35\frac{3}{4}$ to 36 no correction to k_1 is applied.

There is a little difficulty in tracing by this method and with full accuracy the final progress to zero when $\xi' = 36$. If any doubt be left, it may be removed by applying the former method to the course from 35 to 36, using the value of f appropriate to 35 and the terminal tangent as the representative of the curve for F. From this it appears that even if U' at $\xi' = 35$ were as great as 0·7468, the moving particle could not pass $\xi' = 36$. The conclusion is that even in this extreme case of $s = 6$ the solution exists, and that a wave of permanent regime is possible. Further, from (99) we see that since U and μ' are both positive, $d\xi/dx$ is positive throughout, and the transition from the one density to the other takes place *without alternation.*

After what has been proved little doubt could remain but that a solution is possible when s has any value lower than 6. I have, however, thought it desirable to add a rough calculation (rough on account of the relative magnitude of the steps) for the case of $s = 3$:—

ξ'	1	2	3	5	7	8	$8\frac{1}{2}$	9
U'	0·00	0·90	1·19	1·26	0·82	0·40	0·19	0·01

It is a question of some importance to consider what is the thickness of the transitional layer in the waves of uniform regime which have been proved to be possible. Mathematically speaking, the transition occupies an infinite space; but if we understand the expression to refer to a transition approximately complete, the thickness involved is finite, and indeed extremely small. Reference to (98) shows that x is of the order μ', or μ/m, or $\mu/\rho u$, where u is the velocity of the wave. For the present purpose we may take u as equal to the usual velocity of sound, *i.e.* 3×10^4 cm. per second. For air under ordinary conditions the value of μ/ρ in C.G.S. measure is 0·13; so that x is of the order $\frac{1}{3} \times 10^{-5}$ cm. That the transitional layer is in fact extremely thin is proved by such photographs as those of Boys, of the aerial wave of approximate discontinuity which advances in front of a modern rifle bullet; but that according to calculation this thickness should be well below the microscopic limit may well occasion surprise.

Resistance to Motion through Air at High Velocities.

According to the adiabatic law the pressures and velocities in a compressible fluid free from external force, see (47), are related by

$$\left(\frac{p_2}{p_1}\right)^{\frac{\gamma-1}{\gamma}} = 1 + \frac{\gamma-1}{2} \frac{u_1{}^2 - u_2{}^2}{a_1{}^2}, \dots\dots\dots\dots\dots(134)$$

in which p_1, ρ_1, u_1 denote the pressure, density, and velocity at one point of the path; p_2, ρ_2, u_2 the corresponding quantities at another point. Also $a_1{}^2 = \gamma p_1/\rho_1$, so that a_1 is the velocity of infinitesimal disturbances in the condition (1). In an early paper[*] I suggested the application of this formula to bodies moving through air at high velocities. Regarding the obstacle as stationary and the fluid in motion with velocity u_1 and pressure p_1, the pressure p_2 corresponding to the loss of this velocity is given by putting $u_2 = 0$ in (134), a_1 being the ordinary velocity of sound. This is the pressure which should obtain at the axial point on the nose of a symmetrical bullet, and although this value in strictness represents the *maximum* pressure, the analogy of an incompressible fluid suggests that the mean pressure on a flat surface would not be greatly inferior. But in a recent discussion[†], Mr Mallock has shown that this formula immensely overestimates the resistance actually experienced by a bullet, and (so far as I am aware) the discrepancy remains unexplained.

If indeed the adiabatic law really prevailed throughout, there could be no escape from the conclusion formulated. A consideration of the photographs by Boys[‡] will suggest the required explanation. At a short constant distance in front of the bullet there is an aerial bore, or place of approximate discontinuity. Along the axis, the fluid moving up to the bullet changes its density, and therefore pressure and temperature, *suddenly*, so that there is here a special opportunity for viscosity and heat-conduction to take effect. The pressures and velocities on the two sides of the bore are related, not according to the adiabatic law, but according to Rankine's law already discussed. The changes which occur may be separated into two stages. The first is the sudden one in which the fluid passes from the atmospheric condition p_0, ρ_0, with velocity u_0 to the condition denoted by p_1, ρ_1, u_1. After passing the bore the fluid changes gradually according to the adiabatic law already stated until at the nose of the bullet the condition is represented by p_2, ρ_2, with $u_2 = 0$.

---

[*] *Phil. Mag.* Vol. II. p. 430 (1876); *Scientific Papers*, Vol. I. p. 289.

[†] *Roy. Soc. Proc.* A, Vol. LXXIX. p. 266 (1907).

[‡] *Nature*, Vol. XLVII. p. 440 (1893). The particular photograph reproduced by Mallock does not exhibit well the feature in question.

We are now in a position to calculate the final pressure p_2. For the first stage we have Rankine's formula (67), making

$$\rho_0 u_0^2 = \tfrac{1}{2}(\gamma - 1) p_0 + \tfrac{1}{2}(\gamma + 1) p_1,$$

or, if $a^2 = \gamma p_0/\rho_0$,
$$\frac{p_1}{p_0} = \frac{2\gamma}{\gamma + 1}\frac{u_0^2}{a^2} - \frac{\gamma - 1}{\gamma + 1}, \quad\ldots\ldots\ldots\ldots\ldots(135)$$

determining the pressure just inside the bore in terms of u_0 (the velocity of the bullet through quiescent air) and a, the ordinary velocity of sound. When $v_0 = a$, $p_1 = p_0$. For values of u_0 less than a, the first stage does not exist and we may suppose $p_1 = p_0$, $u_1 = u_0$.

In the second stage we use (134) with $u_2 = 0$. Thus

$$\left(\frac{p_2}{p_1}\right)^{\frac{\gamma-1}{\gamma}} = 1 + \frac{\gamma - 1}{2}\frac{\rho_1 u_1^2}{\gamma p_1}, \quad\ldots\ldots\ldots\ldots\ldots(136)$$

in which, by a formula analogous to (66),

$$\rho_1 u_1^2 = \tfrac{1}{2}(\gamma + 1) p_0 + \tfrac{1}{2}(\gamma - 1) p_1.$$

Hence
$$\left(\frac{p_2}{p_0}\right)^{\frac{\gamma-1}{\gamma}} = \frac{(\gamma+1)^2}{4\gamma}\left(\frac{p_1}{p_0}\right)^{\frac{\gamma-1}{\gamma}}\left\{1 + \frac{\gamma - 1}{\gamma + 1}\frac{p_0}{p_1}\right\}. \quad\ldots\ldots\ldots(137)$$

When $u_0 > a$, p_1/p_0 is to be calculated from (135). When the resulting value is substituted in (137), p_2/p_0 is determined.

When $u_0 < a$, we have simply

$$\left(\frac{p_2}{p_0}\right)^{\frac{\gamma-1}{\gamma}} = 1 + \frac{\gamma - 1}{2}\frac{u_0^2}{a^2}. \quad\ldots\ldots\ldots\ldots\ldots(138)$$

If u_0/a be *small*, (138) reduces to

$$\frac{p_2}{p_0} = 1 + \frac{\rho_0 u_0^2}{2p_0},$$

or
$$p_2 - p_0 = \tfrac{1}{2}\rho_0 u_0^2, \quad\ldots\ldots\ldots\ldots\ldots\ldots(139)$$

as for an incompressible fluid.

When $u_0 = a$, both systems give

$$\left(\frac{p_2}{p_0}\right)^{\frac{\gamma-1}{\gamma}} = \frac{\gamma + 1}{2}. \quad\ldots\ldots\ldots\ldots\ldots(140)$$

When u_0/a is *large*, the second terms on the right of (135) and (137) may be neglected, and we obtain

$$\frac{p_2}{p_0} = \frac{\gamma + 1}{2}\frac{u_0^2}{a^2}\left\{\frac{(\gamma+1)^2}{4\gamma}\right\}^{\frac{1}{\gamma-1}}; \quad\ldots\ldots\ldots\ldots(141)$$

or, when we put $\gamma = 1\cdot41$,

$$p_2/p_0 = 1\cdot30 u_0^2/a^2. \quad\ldots\ldots\ldots\ldots(142)$$

The following are some corresponding values of p_2/p_0 and u_0/a, calculated from (135), (137), with $\gamma = 1\cdot41$:—

u_0/a ...	1	2	3	4
p_2/p_0...	1·90	4·49	11·7	20·7

From this point onwards the approximate formula (142) may suffice. The values found are in good agreement with Mr Mallock's curve.

The question as to the linear interval between the bore and the nose of the bullet cannot be answered from the results of the present paper. In strictly one-dimensional motion, the bore and the plane wall constituting the obstacle could not move at the same speed.

347.

NOTE ON THE FINITE VIBRATIONS OF A SYSTEM ABOUT A CONFIGURATION OF EQUILIBRIUM.

[*Philosophical Magazine*, Vol. xx. pp. 450—456, 1910.]

THE theory of the infinitesimal free vibrations of a system, depending on any number of independent coordinates, about a position of stable equilibrium has long been familiar. In my book on the *Theory of Sound* (2nd ed. Vol. II. p. 480) I have shown how to continue the approximation when the motion can no longer be regarded as extremely small, and the following conclusions were arrived at :—

(*a*) The solution obtained by this process is periodic, and the frequency is an even function of the amplitude (H_1) of the principal term ($H_1 \cos nt$).

(*b*) The Fourier series expressive of each coordinate contains cosines only, without sines, of the multiples of nt. Thus the whole system comes to rest at the same moment of time, *e.g.* $t = 0$, and then retraces its course.

(*c*) The coefficient of $\cos rnt$ in the series for any coordinate is of the rth order (at least) in the amplitude (H_1) of the principal term. For example, the series of the third approximation, in which higher powers of H_1 than H_1^3 are neglected, stop at $\cos 3nt$.

(*d*) There are as many types of solution as degrees of freedom; but, it need hardly be said, the various solutions are not superposable.

One important reservation (it was added) has yet to be made. It has been assumed that all the factors, such as $(c_2 - 4n^2a_2)^*$, are finite, that is, that no coincidence occurs between an harmonic of the actual frequency and the natural frequency of some other mode of infinitesimal vibration. Otherwise, some of the coefficients, originally assumed to be subordinate, become infinite, and the approximation breaks down.

* See below.

I have lately had occasion to consider more closely what happens in these exceptional cases; and I propose to take as an example a system with two degrees of freedom, so constituted that the frequencies of infinitesimal vibration are exactly as 2 : 1.   In the absence of dissipative and of impressed forces, everything may be expressed by means of the functions T and V, representing the kinetic and potential energies.   In the case of infinitely small motion in the neighbourhood of the configuration of equilibrium, T and V reduce themselves to quadratic functions of the velocities and displacements with constant coefficients, and by a suitable choice of coordinates the terms involving *products* of the several coordinates may be made to disappear.   Even though we intend to include terms of higher order, we may still avail ourselves of this simplification, choosing as coordinates those which have the property of reducing the terms of the second order to sums of squares. We will further suppose that T is completely expressed as a sum of squares of the velocities with *constant* coefficients, a case which will include the vibrations of a particle moving in two dimensions about a place of equilibrium. We may then write

$$T = \tfrac{1}{2} a_1 \dot{\phi}_1{}^2 + \tfrac{1}{2} a_2 \dot{\phi}_2{}^2, \quad \dots\dots\dots\dots\dots\dots\dots\dots\dots(1)$$

$$V = \tfrac{1}{2} c_1 \phi_1{}^2 + \tfrac{1}{2} c_2 \phi_2{}^2 + V_3 + \dots, \quad \dots\dots\dots\dots\dots\dots(2)$$

where

$$V_3 = \gamma_1 \phi_1{}^3 + \gamma_2 \phi_1{}^2 \phi_2 + \gamma_3 \phi_1 \phi_2{}^2 + \gamma_4 \phi_2{}^3, \quad \dots\dots\dots\dots(3)$$

giving as Lagrange's equations

$$a_1 d^2\phi_1/dt^2 + c_1 \phi_1 + 3\gamma_1 \phi_1{}^2 + 2\gamma_2 \phi_1 \phi_2 + \gamma_3 \phi_2{}^2 = 0, \quad \dots\dots\dots(4)$$

$$a_2 d^2\phi_2/dt^2 + c_2 \phi_2 + 3\gamma_4 \phi_2{}^2 + 2\gamma_3 \phi_1 \phi_2 + \gamma_2 \phi_1{}^2 = 0. \quad \dots\dots\dots(5)$$

To satisfy these equations we assume

$$\phi_1 = H_0 + H_1 \cos nt + H_2 \cos 2nt + H_3 \cos 3nt + \dots, \quad \dots\dots\dots(6)$$

$$\phi_2 = K_0 + K_1 \cos nt + K_2 \cos 2nt + K_3 \cos 3nt + \dots. \quad \dots\dots\dots(7)$$

In general we may take as one approximate solution

$$\phi_1 = H_1 \cos nt, \qquad \phi_2 = 0 \quad \dots\dots\dots\dots\dots\dots\dots\dots(8)$$

with

$$n^2 = c_1/a_1; \quad \dots\dots\dots\dots\dots\dots\dots\dots\dots\dots\dots\dots(9)$$

and in proceeding to a second approximation we may regard all the other coefficients as small relatively to H_1. On this supposition the 4th and 5th terms in (4) may be omitted, so that ϕ_1 is separated from ϕ_2.   Substituting from (6) and equating the terms containing the various multiples of nt, we get

$$c_1 H_0 + \tfrac{3}{2} \gamma_1 H_1{}^2 = 0,$$

$$(c_1 - n^2 a_1) H_1 = 0,$$

$$(c_1 - 4n^2 a_1) H_2 + \tfrac{3}{2} \gamma_1 H_1{}^2 = 0;$$

so that

$$\phi_1 = -\frac{3\gamma_1 H_1{}^2}{2c_1} + H_1 \cos nt - \frac{3\gamma_1 H_1{}^2}{c_1 - 4n^2 a_1} \cos 2nt, \quad \dots\dots\dots(10)$$

with

$$c_1 = n^2 a_1,$$

as in the first approximation.   In like manner

$$c_2 K_0 + \tfrac{1}{2}\gamma_2 H_1^2 = 0,$$
$$(c_2 - n^2 a_2) K_1 = 0,$$
$$(c_2 - 4n^2 a_2) K_2 + \tfrac{1}{2}\gamma_2 H_1^2 = 0.$$

Thus, if c_2 differs both from $n^2 a_2$ and from $4n^2 a_2$, we have

$$\phi_2 = -\frac{\gamma_2 H_1^2}{2c_2} - \frac{\gamma_2 H_1^2}{2(c_2 - 4n^2 a_2)} \cos 2nt. \quad \dots\dots\dots(11)$$

But if
$$c_2 - n^2 a_2 = 0,$$
the inference that $K_1 = 0$ does not follow; and if
$$c_2 - 4n^2 a_2 = 0,$$
the terms in $\cos 2nt$ in (10), (11) assume infinite values.   Accordingly these two cases demand further consideration.   We will commence with that where
$$c_2 - n^2 a_2 = 0,$$
that is, where both modes of infinitesimal vibration have the same frequency.

We must now discard the supposition that $\phi_2 = 0$ approximately and be prepared to allow K_1, as well as H_1, to be quantities of the first order of smallness.   The other coefficients in (6), (7) are still of the second order at least.   Substituting in (4), (5) and retaining only terms not above the second order, we get

$$c_1 H_0 + (c_1 - n^2 a_1) H_1 \cos nt + (c_1 - 4n^2 a_1) H_2 \cos 2nt + \dots$$
$$+ 3\gamma_1 H_1^2 \cos^2 nt + 2\gamma_2 H_1 K_1 \cos^2 nt + \gamma_3 K_1^2 \cos^2 nt = 0,$$
$$c_2 K_0 + (c_2 - n^2 a_2) K_1 \cos nt + (c_2 - 4n^2 a_2) K_2 \cos 2nt + \dots$$
$$+ 3\gamma_4 K_1^2 \cos^2 nt + 2\gamma_3 H_1 K_1 \cos^2 nt + \gamma_2 H_1^2 \cos^2 nt = 0;$$

whence

$$c_1 H_0 + \tfrac{1}{2}(3\gamma_1 H_1^2 + 2\gamma_2 H_1 K_1 + \gamma_3 K_1^2) = 0, \quad \dots\dots\dots(12)$$
$$c_2 K_0 + \tfrac{1}{2}(3\gamma_4 K_1^2 + 2\gamma_3 H_1 K_1 + \gamma_2 H_1^2) = 0, \quad \dots\dots\dots(13)$$
$$(c_1 - 4n^2 a_1) H_2 + \tfrac{1}{2}(3\gamma_1 H_1^2 + 2\gamma_2 H_1 K_1 + \gamma_3 K_1^2) = 0, \quad \dots\dots\dots(14)$$
$$(c_2 - 4n^2 a_2) K_2 + \tfrac{1}{2}(3\gamma_4 K_1^2 + 2\gamma_3 H_1 K_1 + \gamma_2 H_1^2) = 0. \quad \dots\dots\dots(15)$$

Also
$$H_3, \&c. = 0, \qquad K_3, \&c. = 0.$$

These equations, arising from the terms independent of t and proportional to $\cos 2nt$, $\cos 3nt$, &c., determine H_0, K_0, H_2, K_2, &c. when H_1, K_1, and n are known.   The term in $\cos nt$ gives further

$$H_1(c_1 - n^2 a_1) = 0, \qquad K_1(c_2 - n^2 a_2) = 0.$$

Thus when
$$n^2 = c_1/a_1 = c_2/a_2, \quad \dots\dots\dots\dots\dots\dots\dots(16)$$

K_1 as well as H_1 is an arbitrary quantity of the first order. And this completes the solution to the second approximation.

When the process is pursued to the next stage, the ratio H_1/K_1 may become determinate. In illustration of this let us suppose that V is an even function of both ϕ_1 and ϕ_2. Thus $V_3 = 0$, and

$$V_4 = \delta_1 \phi_1^4 + \delta_3 \phi_1^2 \phi_2^2 + \delta_5 \phi_2^4. \quad \ldots\ldots\ldots\ldots\ldots(17)$$

Using this as before, we obtain

$$H_0 = 0, \qquad K_0 = 0, \qquad H_2 = 0, \qquad K_2 = 0, \qquad H_4, \&c. = 0, \qquad K_4, \&c. = 0.$$

To determine H_3, K_3 we have

$$(c_1 - 9n^2 a_1) H_3 + \delta_1 H_1^3 + \tfrac{1}{2} \delta_3 H_1 K_1^2 = 0, \quad \ldots\ldots\ldots\ldots(18)$$

$$(c_2 - 9n^2 a_2) K_3 + \delta_5 K_1^3 + \tfrac{1}{2} \delta_3 K_1 H_1^2 = 0. \quad \ldots\ldots\ldots\ldots(19)$$

Also from the terms in $\cos nt$

$$H_1 [c_1 - n^2 a_1 + 3\delta_1 H_1^2 + \tfrac{3}{2} \delta_3 K_1^2] = 0, \quad \ldots\ldots\ldots\ldots(20)$$

$$K_1 [c_2 - n^2 a_2 + 3\delta_5 K_1^2 + \tfrac{3}{2} \delta_3 H_1^2] = 0. \quad \ldots\ldots\ldots\ldots(21)$$

Equations (20), (21) can be satisfied by supposing either H_1 or K_1 to vanish while the other remains finite. Thus if $H_1 = 0$, (20) is satisfied and (21) gives

$$c_2 - n^2 a_2 + \tfrac{3}{2} \delta_3 H_1^2 = 0, \quad \ldots\ldots\ldots\ldots\ldots(22)$$

determining n. From (19) we see that in this case $K_3 = 0$, while H_3 is given by (18) with $K_1 = 0$.

There is also another solution in which both H_1 and K_1 are finite. Since by supposition

$$c_2/a_2 = c_1/a_1,$$

$$\frac{2\delta_1 H_1^2 + \delta_3 K_1^2}{\delta_3 H_1^2 + 2\delta_5 K_1^2} = \frac{c_1 - n^2 a_1}{c_2 - n^2 a_2} = \frac{c_1}{c_2}, \quad \ldots\ldots\ldots\ldots(23)$$

which determines K_1^2/H_1^2, and then either (20) or (21) gives n^2. Equations (18), (19) determine H_3, K_3 with two alternatives according to the sign of K_1/H_1.

In certain cases the ratio K_1/H_1 may remain arbitrary; for example, if

$$c_2 = c_1 \quad \text{and} \quad 2\delta_1 = 2\delta_5 = \delta_3,$$

making V_4 a complete square.

The other class of cases demanding further examination arises when

$$c_2/a_2 = 4c_1/a_1, \quad \ldots\ldots\ldots\ldots\ldots\ldots(24)$$

and it requires that K_2 should be treated as a quantity of the first order as well as H_1, the remaining coefficients being still of the second order. The substitution of (6), (7) in (4), (5) then gives

$$c_1 H_0 + (c_1 - n^2 a_1) H_1 \cos nt + (c_1 - 4n^2 a_1) H_2 \cos 2nt + \dots$$
$$+ 3\gamma_1 H_1^2 \cos^2 nt + 2\gamma_2 H_1 K_2 \cos nt \cos 2nt + \gamma_3 K_2^2 \cos^2 2nt = 0, \quad \dots(25)$$

$$c_2 K_0 + (c_2 - n^2 a_2) K_1 \cos nt + (c_2 - 4n^2 a_2) K_2 \cos 2nt + \dots$$
$$+ 3\gamma_4 K_2^2 \cos^2 2nt + 2\gamma_3 H_1 K_2 \cos nt \cos 2nt + \gamma_2 H_1^2 \cos^2 nt = 0. \quad \dots(26)$$

From the terms independent of t we get

$$2c_1 H_0 + 3\gamma_1 H_1^2 + \gamma_3 K_2^2 = 0, \qquad 2c_2 K_0 + \gamma_2 H_1^2 + 3\gamma_4 K_2^2 = 0; \quad \dots(27)$$

from the terms in $3nt$

$$(c_1 - 9n^2 a_1) H_3 + \gamma_2 H_1 K_2 = 0, \qquad (c_2 - 9n^2 a_2) K_3 + \gamma_3 H_1 K_2 = 0; \quad \dots(28)$$

from the terms in $4nt$

$$(c_1 - 16n^2 a_1) H_4 + \tfrac{1}{2}\gamma_3 K_2^2 = 0, \qquad (c_2 - 16n^2 a_2) K_4 = \tfrac{3}{2}\gamma_4 K_2^2; \quad \dots\dots(29)$$

while coefficients with higher suffixes than 4 vanish. Further, from the terms in nt, $2nt$

$$(c_1 - n^2 a_1) H_1 + \gamma_2 H_1 K_2 = 0, \qquad (c_2 - n^2 a_2) K_1 + \gamma_3 H_1 K_2 = 0, \quad \dots(30)$$

$$(c_1 - 4n^2 a_1) H_2 + \tfrac{3}{2}\gamma_1 H_1^2 = 0, \qquad (c_2 - 4n^2 a_2) K_2 + \tfrac{1}{2}\gamma_2 H_1^2 = 0. \quad \dots(31)$$

These equations determine H_0, K_0, K_1, H_2, H_3, K_3, H_4, K_4 as functions of H_1 and K_2 of the second order, when n is known. To find n and the ratio K_2/H_1 we have the first equation of (30) and the second of (31). With regard to (24) these may be written

$$c_1 - n^2 a_1 + \gamma_2 K_2 = 0, \quad \dots\dots\dots\dots\dots(33)$$

$$\frac{c_2}{c_1} (c_1 - n^2 a_1) K_2 + \tfrac{1}{2}\gamma_2 H_1^2 = 0; \quad \dots\dots\dots\dots(34)$$

of which the first may be considered to determine n. Eliminating $(c_1 - n^2 a_1)$, we get

$$K_2/H_1 = \pm \sqrt{(c_1/2c_2)}. \quad \dots\dots\dots\dots\dots\dots(35)$$

This completes the solution to the second order of small quantities.

If $V_3 = 0$, the above solution reduces itself to that of the first approximation. In this case, especially if V is an even function of ϕ_1 and ϕ_2, see (17), a solution, correct to the third order of small quantities, is readily developed; it is hardly necessary to give the details.

In *Theory of Sound* (*loc. cit.*) I remarked upon the failure of the simple theory to deal with the apparently simple problem of the vibrations in one dimension of a column of gas, obeying Boyle's law, and contained in a cylindrical tube with stopped ends. So far as I am able to see, the present extension does not help the matter. In this case there are an infinite number of coincidences between natural frequencies of infinitesimal vibration and harmonics of the fundamental vibration. From what we know of the behaviour of *progressive* waves of finite amplitude, it is perhaps not surprising if no solutions exist of the character contemplated. Probably after the lapse of a finite time discontinuity will ensue.

348.

THE PROBLEM OF THE WHISPERING GALLERY.

[*Philosophical Magazine*, Vol. xx. pp. 1001—1004, 1910.]

THE phenomena of the whispering gallery, of which there is a good and accessible example in St Paul's cathedral, indicate that sonorous vibrations have a tendency to cling to a concave surface. They may be reproduced upon a moderate scale by the use of sounds of very high pitch (wave-length = 2 cm.), such as are excited by a bird-call, the percipient being a high pressure sensitive flame*. Especially remarkable is the narrowness of the obstacle, held close to the concave surface, which is competent to intercept most of the effect.

The explanation is not difficult to understand in a general way, and in *Theory of Sound*, § 287, I have given a calculation based upon the methods employed in geometrical optics. I have often wished to illustrate the matter further on distinctively wave principles, but only recently have recognized that most of what I sought lay as it were under my nose. The mathematical solution in question is well known and very simple in form, although the reduction to numbers, in the special circumstances, presents certain difficulties.

Consider the expression in plane polar coordinates (r, θ)

$$\psi_n = J_n(kr) \cos(kat - n\theta), \quad \dots\dots\dots\dots\dots\dots(1)$$

applicable to sound in two dimensions, ψ denoting velocity-potential; or again to the transverse vibrations of a stretched membrane, in which case ψ represents the displacement at any point†. Here a denotes the velocity of propagation, $k = 2\pi/\lambda$, where λ is the wave-length of straight waves of the given frequency, n is any integer, and J_n is the Bessel's function usually so denoted. The waves travel circumferentially, everything being reproduced when θ and t receive suitable proportional increments. For the present purpose we suppose that there are a large number of waves round the circumference, so that n is great.

---

* *Proc. Roy. Inst.* Jan. 15, 1904. [*Scientific Papers*, Vol. v. p. 171.]

† *Theory of Sound*, §§ 201, 339. [1911. Electrical vibrations in two dimensions are also included, if the reflecting wall be supposed perfectly conducting. When the electrical vibrations are perpendicular to the plane, the analogy is with the membrane; when they are performed in the plane, they are analogous to the aerial vibrations. (See Vol. iv. p. 295.)]

As a function of r, ψ is proportional to $J_n(kr)$. When z is great enough, $J_n(z)$, as we know, becomes oscillatory and admits of an infinite number of roots. In the case of the membrane held at the boundary any one of these roots might be taken as the value of kR, where R is the radius of the boundary. But for our purpose we suppose that kR is the *first* or lowest root (after zero) which we may call z_1. In this case $J_n(z)$ remains throughout of one sign. For the aerial vibrations, in which we are especially interested, the boundary condition, representing that $r = R$ behaves as a fixed wall, is that $J_n'(kR) = 0$. We will suppose that k and R are so related that kR is equal to the *first* root (z_1') of this equation. The character of the vibrations as a function of r thus depends upon that of $J_n(z)$, where n is very large and z less than z_1 or z_1'. And we know that in general, n being integral,

$$J_n(z) = \frac{1}{\pi} \int_0^\pi \cos(z \sin \omega - n\omega) \, d\omega. \quad \ldots\ldots\ldots\ldots\ldots\ldots(2)$$

Moreover, the well known series in ascending powers of z shows that in the neighbourhood of the origin $J_n(z)$ is very small, the lowest power occurring being z^n.

The tendency, when n is moderately high, may be recognized in Meissel's tables*, from which the following is extracted :—

z	$J_{18}(z)$	$J_{21}(z)$	z	$J_{18}(z)$	$J_{21}(z)$
24	-0.0931	$+0.2264$	16	$+0.0668$	$+0.0079$
23	$+0.0340$	0.2381	15	0.0346	0.0031
22	0.1549	0.2105	14	0.0158	0.0010
21	0.2316	0.1621	13	0.0063	0.0003
20	0.2511	0.1106	12	0.0022	0.0001
19	0.2235	0.0675	11	0.0006	0.0000
18	0.1706	0.0369	10	0.0002	
17	0.1138	0.0180	9	0.0000	

From the second column we see that the first root of $J_{18}(z) = 0$ occurs when $z = 23.3$. The function is a maximum in the neighbourhood of $z = 20$, and sinks to insignificance when z is less than 14, being thus in a physical sense limited to a somewhat narrow range within $z = 23.3$.

The above applies to the membrane problem. In the case of aerial waves the third column shows that $J_{21}(z)$ is a maximum when $z = 23.3$, so that $J_{21}'(23.3) = 0$. This then is the value of kR, or z_1'. It appears that the important part of the range is from 23.3 to about 16.

The course of the function $J_n(z)$ when n and z are both large and nearly equal has recently been discussed by Dr Nicholson†. Under these

---

* Gray and Mathews' *Bessel's Functions*.

† *Phil. Mag.* Vol. xvi. p. 271 (1908); Vol. xviii. p. 6 (1909).

circumstances the important part of (2) evidently corresponds to small values of ω. If $z = n$ absolutely, we may write ultimately

$$J_n(n) = \frac{1}{\pi}\int_0^\pi \cos n\,(\omega - \sin \omega)\,d\omega = \frac{1}{\pi}\int_0^\infty \cos n\,(\omega - \sin \omega)\,d\omega$$

$$= \frac{1}{\pi}\int_0^\infty \cos \frac{n\omega^3}{6}\,d\omega = \frac{1}{\pi}\left(\frac{6}{n}\right)^{\frac{1}{3}}\int_0^\infty \cos \alpha^3\,d\alpha$$

$$= \Gamma(\tfrac{1}{3})\,.\,2^{-\frac{2}{3}}3^{-\frac{1}{6}}\pi^{-1}n^{-\frac{1}{3}}, \quad\dots\dots\dots\dots\dots\dots\dots(3)$$

one of Nicholson's results.

In like manner when $n - z$, though not zero, is relatively small, (1) may be made to depend upon Airy's integral. Thus

$$J_n(z) = \frac{1}{\pi}\int_0^\infty \cos\{(n-z)\,\omega + \tfrac{1}{6}z\omega^3\}\,d\omega. \quad\dots\dots\dots\dots(4)$$

In the second of the papers above cited Nicholson tabulates $z^{\frac{1}{3}}J_n(z)$ against $2\cdot1123\,(n-z)/z^{\frac{1}{3}}$. It thence appears that

$$z_1 = n + \frac{2\cdot4955}{2\cdot1123}n^{\frac{1}{3}} = n + 1\cdot1814\,n^{\frac{1}{3}}. \quad\dots\dots\dots\dots(5)$$

The maximum (about $0\cdot67$) occurs when

$$z = n + \cdot51\,n^{\frac{1}{3}}, \quad\dots\dots\dots\dots\dots\dots\dots(6)$$

and the function sinks to insignificance ($0\cdot01$) when

$$z = n - 1\cdot5\,n^{\frac{1}{3}}. \quad\dots\dots\dots\dots\dots\dots\dots(7)$$

Thus in the membrane problem the practical range is only about $2\cdot7\,n^{\frac{1}{3}}$.

In like manner

$$z_1' = n + \frac{1\cdot0845}{2\cdot1123}n^{\frac{1}{3}} = n + \cdot51342\,n^{\frac{1}{3}}; \quad\dots\dots\dots\dots(8)$$

so that in the aerial problem the practical range given by (7) and (8) is about $2\cdot1\,n^{\frac{1}{3}}$.

To take an example in the latter case, let $n = 1000$, representing approximately the radius of the reflecting circle. The vibrations expressed by (1) are practically limited to an annulus of width 20, or one fiftieth part only of the radius. With greater values of n the concentration in the immediate neighbourhood of the circumference is still further increased.

It will be admitted that this example fully illustrates the observed phenomena, and that the clinging of vibrations to the immediate neighbourhood of a concave reflecting wall may become exceedingly pronounced.

Another example might be taken from the vibrations of air within a *spherical* cavity. In the usual notation for polar coordinates $(r,\ \theta,\ \phi)$ we have as a possible velocity-potential $\psi = (kr)^{-\frac{1}{2}} J_{n+\frac{1}{2}}(kr) \sin^n \theta \cos(kat - n\phi)$, and the discussion proceeds as before.

So far as I have seen, the ultimate form of $J_n(z)$ when n is very great and z a moderate multiple of n has not been considered. Though unrelated to the main subject of this note, I may perhaps briefly indicate it.

The form of (2) suggests the application of the method employed by Kelvin in dealing with the problem of water waves due to a limited initial disturbance. Reference may also be made to a recent paper of my own*.

When n and z are great the only important part of the range of integration in (2) is the neighbourhood of the place or places, where $z \sin \omega - n\omega$ is *stationary* with respect to ω. These are to be found where

$$\cos \omega_1 = n/z, \quad \dots\dots\dots\dots\dots\dots\dots\dots\dots\dots(9)$$

from which we may infer that when z is decidedly less than n, the total value of the integral is small, as we have already seen to be the case. When $z > n$, ω_1 is real, and according to (9) would admit of an infinite series of values. Only one, however, of these comes into consideration, since the actual range of integration is from 0 to π. We suppose that z is so much greater than n that ω_1 has a sensible value.

The application of Kelvin's method gives at once

$$J_n(z) = \sqrt{\left(\frac{2}{\pi z}\right)} \frac{\cos\{z \sin \omega_1 - n\omega_1 - \frac{1}{4}\pi\}}{\sqrt{\{\sin \omega_1\}}}. \quad \dots\dots\dots(10)$$

We may test this by applying it to the familiar case where z is so much greater than n as to make $\omega_1 = \frac{1}{2}\pi$. We find

$$J_n(z) = \sqrt{\left(\frac{2}{\pi z}\right)} . \cos\{z - \frac{1}{2}n\pi - \frac{1}{4}\pi\}, \quad \dots\dots\dots\dots(11)$$

the well known form.

As an example of (10),

$$J_n(2n) = \sqrt{\left(\frac{2}{n\pi\sqrt{3}}\right)} . \cos\{(\sqrt{3} - \frac{1}{3}\pi)n - \frac{1}{4}\pi\}. \quad \dots\dots\dots(12)$$

Although in (2) n is limited to be integral, it is not difficult to recognize that results such as (3), (5), (12), applicable to large values of n, are free from this restriction.

---

* *Phil. Mag.* Vol. xviii. p. 1, immediately preceding Nicholson's paper just quoted. [*Scientific Papers*, Vol. v. p. 514.]

349.

ON THE SENSIBILITY OF THE EYE TO VARIATIONS OF WAVE-LENGTH IN THE YELLOW REGION OF THE SPECTRUM.

[*Proceedings of the Royal Society*, A, Vol. LXXXIV. pp. 464—468, 1910.]

DR EDRIDGE-GREEN* has introduced a method of classifying colour-vision by determining the number of separate parts or divisions in the spectrum within each of which the observer can perceive no colour difference. Movable screens are provided in the focal plane of the spectroscopic telescope, by which the part admitted to the eye is limited and the limits measured in terms of wave-length. Beginning at the extreme visible red, more and more of the spectrum is admitted until a change of colour (not merely of brightness) is just perceptible. This gives the first division. The second division starts from the place just determined, and is limited in the direction of shorter wave-length by the same condition. In this way the whole spectrum is divided into a number of contiguous divisions, or patches, which Dr Edridge-Green terms monochromatic. It will be observed that the delimitation of these patches includes an arbitrary element depending on the point from which the start is made—in this case the extreme red.

"Tested with this instrument a normal individual will, as a rule, name six distinct colours (viz., red, orange, yellow, green, blue, violet), and will mark out by means of the shutters about 18 monochromatic patches. Occasionally we come across individuals with a greater power of differentiating hues, to whom, as to Newton, there is a distinct colour between the blue and violet, which Newton called indigo. Such individuals will mark out a greater number of monochromatic patches, from 22 up to 29. The limited number of monochromatic patches which can be marked out in this way is at first surprising when we consider how insensibly one part of the spectrum seems to shade into the next when the whole of the spectrum is looked at. The number and position of the patches present, however, great uniformity from one case to another."

---

* *Roy. Soc. Proc.* B, Vol. LXXXII. p. 458 (1910), and earlier writings.

Being curious to know into what class my own vision would fall on this system, I was glad to be tested by Dr Edridge-Green last July.   The number of patches proved to be 17, a little short of the number he lays down in the passage above quoted as normal.   The slight deficiency appears to be in the high violet.   I have known for some years that I required more light in the violet to measure interference-rings than did my assistant, Mr Enock, and that the deficiency of sensibility was greatest for my right eye, used with Dr Green's apparatus.   The limits of the actual patches were as follows :—

780—635½—624—612—603—595—586—576—560—541—521—509—
500—489½—477—462—443—426.

Thus in the region of the D lines a patch including wave-lengths between 595 and 586 did not manifest a difference of colour.   The interval between the D lines on the above scale being 0·60, it appears that my "mono-chromatic patch" was 15 times this interval.

While it is undoubtedly true that in this way of working no colour-difference was perceptible as the eye travelled backwards and forwards over the patch, my experience with colour discs and other colour-mixing arrange-ments made me feel certain that under more favourable conditions I could discriminate much smaller differences of wave-length.   Special experiments have since proved that I can in fact discriminate by colour between points in the spectrum as close together as the two D lines.

In order to compare two colours with advantage it is necessary that each should extend with uniformity over a considerable angular area, and that the two areas should be in close juxtaposition.   The requirements of the case are sufficiently met by a colour-box (after Maxwell) such as I described nearly 30 years ago*.   In this form of apparatus a second slit, placed at the focus, allows a narrow width of the spectrum to pass; but instead of regarding the transmitted portion with an eyepiece, the eye is brought close to the slit and focussed upon the prism, which thus appears uniformly lighted with such rays as the second slit allows to pass.   The light thus presented is of course not absolutely homogeneous; it includes a mixture of neighbouring spectrum rays, the degree of purity augmenting as the slits are narrowed.   With the aid of a refracting prism of small angle (set per-pendicularly to the dispersing prisms) the field of view is divided into two parts which correspond to any desired colours according to the situation of the two primary slits.   For the present purpose these primary slits lie nearly in one straight line, inasmuch as the two spectrum colours to be compared are close together.

A detail of some importance in delicate work may here be mentioned. It is known that in many cases, e.g. in lantern projection, the spectrum lines

* Nature, Vol. xxv. pp. 64—66 (1881); Scientific Papers, Vol. i. p. 543. See also Nature, August 18, 1910. [Scientific Papers, Vol. v. p. 569.]

corresponding to a straight primary slit are sensibly curved. Mr Madan proposed many years ago to cure this defect by counter-curving the primary slit. In the kind of instrument under discussion it is desirable to retain straight primary slits, but there is nothing to forbid curvature of the second or eye slit, which is a fixture, and such curvature is necessary for the most effective working. A deficiency in this respect, or in focussing, may entail objectionable changes of colour as the eye moves about behind its slit. The simplest way of making the adjustment is to illuminate a somewhat narrow primary slit with soda light and to fit the jaws of the secondary slit to the image thus obtained and examined with a lens as eyepiece.

In making the observations on sensitiveness, one primary slit, as well as the eye slit, remains fixed, the position being chosen so as to provide yellow light from the neighbourhood of D. The second slit can be moved as a whole while retaining its width. The shutters necessary were cut from thin zinc sheet and were held by sealing or soft wax, in a manner which need not be minutely described. The movements of the shutter which carries the second slit were measured by callipers.

The procedure is quite simple. If the colours seen are strongly contrasted, the movable slit is displaced until the difference is moderate. Marks may then be given; 0, denoting that the difference is uncertain; R_1, that it is just distinct in the direction of making the second patch the redder; G_1, that it is just distinct in the opposite direction. Similarly, R_2, G_2 denote differences in the two directions which are more than distinct, and so on. After each observation worth recording, the position of the movable slit is measured.

One further precaution ought to be mentioned. In making a decision when the difference of colour is slight, care should be taken that the brightnesses are nearly equal. When, as in my experiments, daylight is employed, the passage of clouds may cause a disturbance in this respect, even if the two primary slits are of equal width. The interposition of a piece of ground glass a little behind the primary slits is usually a remedy. But this reduces the illumination, and it is sometimes preferable to adjust the brightness otherwise. It may be done conveniently by cutting off some of the light on the preponderating side by the interposition of one or more strips of glass held at varying angles of obliquity.

In this manner, as the result of sets of observations made on several days, it was found that a movement of the second slit through 0·15 mm. was sufficient to carry the variable colour from being distinctly redder than the standard to distinctly greener. No doubt the result might have been arrived at quicker with a more refined apparatus, in which the movements of the slit were controlled and measured by a micrometer screw, but I do not think it would be any more certain. Probably the distance is something of an over-estimate. In several of the measurements included, the distinctness of the difference was unnecessarily pronounced. We may conclude that the eye is capable of appreciating without fail a difference of situation represented by 0·07 mm.

It remains to interpret the result in terms of wave-lengths. By allowing light to enter at the eye slit, or rather at a narrower slit superposed upon it, a spectrum is formed at the other end whose scale has to be determined. It appeared that the distance from D to E was 7 mm. The difference of wave-length between these lines is 62·3. The perceptible difference is 1/100 of this, corresponding nearly enough to the difference between the D lines. I think I am safe in saying that I could distinguish the colours of the two D lines if favourably presented to the eye.

This degree of sensitiveness, though not higher than I had expected, is a little difficult to reconcile with the monochromatic appearance of a portion of the spectrum 15 times wider. I suppose that the gradual character of the transition is an obstacle to the recognition of differences. The question of angular magnitude may also enter. No doubt a very small apparent magnitude would be unfavourable. It is possible that in Dr Green's apparatus an eyepiece of higher power, with a corresponding augmentation in the intrinsic brilliancy of the source of light, would allow of an increase in the number of distinguishable patches. The experiment would be worth a trial.

It will be seen that the existence of "monochromatic patches" in the spectrum is far from meaning that the eye is incapable of making chromatic distinctions within their range. I do not infer from this that the results of the method are without significance. Undoubtedly it is possible by means of it to classify colour-vision, and such a classification cannot be without interest, even if we fail as yet to understand exactly what it means.

In conclusion, I will remark that those who lay great stress upon the number of principal colours recognized by any particular eye seem to me to overlook too much the colours not represented in the spectrum. To most of us *white* is a sensation quite as distinct from any other as yellow can be. In my estimation purple has a better claim than orange to be reckoned a principal colour. The fact, too, that dark orange reveals its character so little as to be called by another name (brown) seems to indicate that these distinctions are not of fundamental importance.

CAMBRIDGE: PRINTED BY JOHN CLAY, M.A. AT THE UNIVERSITY PRESS.

Printed in the United States
By Bookmasters